I0072196

# A Comprehensive Study of Genetics
## Volume I

# A Comprehensive Study of Genetics
# Volume I

Edited by **Rosanna Mann**

New York

Published by Callisto Reference,
106 Park Avenue, Suite 200,
New York, NY 10016, USA
www.callistoreference.com

**A Comprehensive Study of Genetics: Volume I**
Edited by Rosanna Mann



International Standard Book Number: 978-1-63239-007-3 (Hardback)



Printed in the United States of America.

# Contents

# Preface

Every since the dawn of civilisation, man has tried to observe and understand the fundamentals of heredity, trait inheritance and the various subtle variations in living organisms. This information was used for selective breeding and to improve yields.

With advancements in technology and biological sciences, further inroads were made into the field of genetics, defined as the study of genes. Path-breaking work by scientists like Gregor Mendel not only earned him the title 'Father of Modern Genetics', but also shed light on multiple aspects of genetics that were unknown earlier.

Genetic processes work in sync with the surrounding environment and experiences to influence development and behaviour of the subject, often referred to as 'Nature Versus Nurture'. Studies and researches have revealed through their results that nutrition and maintenance substantially affects growth in all living beings, including humans.

This book is about all that and more. Here, we will look at the fascinating world of genetics and take a closer look at the developments that this field has undergone in the recent years. From students to experts, professionals to amateurs, everyone is invited to read this book and know more about how our genes work. I do believe that the content in this book will be helpful to many researchers in this field around the world. I hope that this textbook will enhance the knowledge of scientists in this area of study.

I wish to thank all the great minds who have contributed their researches in this book. I would also like to thank my publisher for the endless support and cooperation.

**Editor**

# The "Bringing into Cultivation" Phase of the Plant Domestication Process and Its Contributions to *In Situ* Conservation of Genetic Resources in Benin

**R. Vodouhè[1] and A. Dansi[2,3]**

[1] *Bioversity International, Office of West and Central Africa, 08 BP 0931 Cotonou, Benin*
[2] *Laboratory of Agricultural Biodiversity and Tropical Plant Breeding, Department of Genetics, Faculty of Sciences and Technology (FAST), University of Abomey-Calavi (UAC), 071BP28 Cotonou, Benin*
[3] *Department of Crop Science (DCS), Crop, Aromatic and Medicinal Plant Biodiversity Research and Development Institute (IRDCAM), 071BP28 Cotonou, Benin*

Correspondence should be addressed to A. Dansi, adansi2001@gmail.com

Academic Editors: D. W. Archer and V. C. Concibido

All over the world, plant domestication is continually being carried out by local communities to support their needs for food, fibre, medicine, building materials, etc. Using participatory rapid appraisal approach, 150 households were surveyed in 5 villages selected in five ethnic groups of Benin, to investigate the local communities' motivations for plant domestication and the contributions of this process to *in situ* conservation of genetic resources. The results indicated differences in plant domestication between agroecological zones and among ethnic groups. People in the humid zones give priority to herbs mainly for their leaves while those in dry area prefer trees mostly for their fruits. Local communities were motivated to undertake plant domestication for foods (80% of respondents), medicinal use (40% of respondents), income generation (20% of respondents) and cultural reasons (5% of respondents). 45% of the species recorded are still at early stage in domestication and only 2% are fully domesticated. Eleven factors related to the households surveyed and to the head of the household interviewed affect farmers' decision making in domesticating plant species. There is gender influence on the domestication: Women are keen in domesticating herbs while men give priority to trees.

## 1. Introduction

Plant domestication is the evolutionary process whereby a population of plants becomes accustomed to human provision and control [1]. For many authors [2, 3], domestication is generally considered to be the end-point of a continuum that starts with exploitation of wild plants, continues through cultivation of plants selected from the wild but not yet genetically different from wild plants (initial phase of bringing into cultivation), and ends with the adaptation to the agroecology through conscious or unconscious human morphological selection and hence genetic differences distinguishing the domesticated species from its wild progenitor. According to local communities, the collection of plants from the wild for cultivation on farm (fields or home gardens) is a common practice continually being carried out under diverse agroecosystems. Many varieties, landraces, and cultivars of plants have been developed through this process to meet human (and/or animal) demand for food, fibre, medicine, building materials, and so forth [4].

Throughout the world, the process of plant domestication has been either broadly analysed [5–9] or studied for species or group of species including acacias [10], yam [11, 12], tomatoes [13], barley [1], rice [4, 14], baobab [15], leafy vegetables [16], and fonio [17, 18]. These studies revealed the existence of different steps in the domestication process and highlighted that the practices used to highly vary with the species and the sociolinguistic groups across countries. Therefore, it is useful to document the process at country level.

This study aims to investigate plant domestication in different ethnic groups and agroecological zones of the Republic of Benin in order to

(i) document the species diversity, the domestication levels, and the use of the species under domestication;

(ii) understand the motives of the domestication and the factors affecting farmers' decision making in domesticating plant species;

(iii) analyse the gender influence on plant domestication.

## 2. Material and Methods

*2.1. The Study Area.* The Republic of Benin is situated in west Africa, between the latitudes 6°10′ N and 12°25′ N and longitudes 0°45′ E and 3°55′ E [19]. It covers a total land area of 112,622 km$^2$ with a population estimated at about 7 millions [20]. The country is partitioned into 12 departments inhabited by 29 ethnic groups [19]. The south and the centre are relatively humid agroecological zones with two rainy seasons and mean annual rainfall of 1500 mm/year [19]. The north is situated in arid and semiarid agroecological zones characterized by unpredictable and irregular rainfall oscillating between 800 and 950 mm/year with only one rainy season. Mean annual temperatures range from 26 to 28°C and may exceptionally reach 35 to 40°C in the far northern localities [20, 21]. The country has about 2,807 plant species [21]. Vegetation types are semideciduous forest (south), woodland and savannah woodland (centre east and northeast), dry semideciduous forest (centre west and south of northwest), and tree and shrub savannahs (far north).

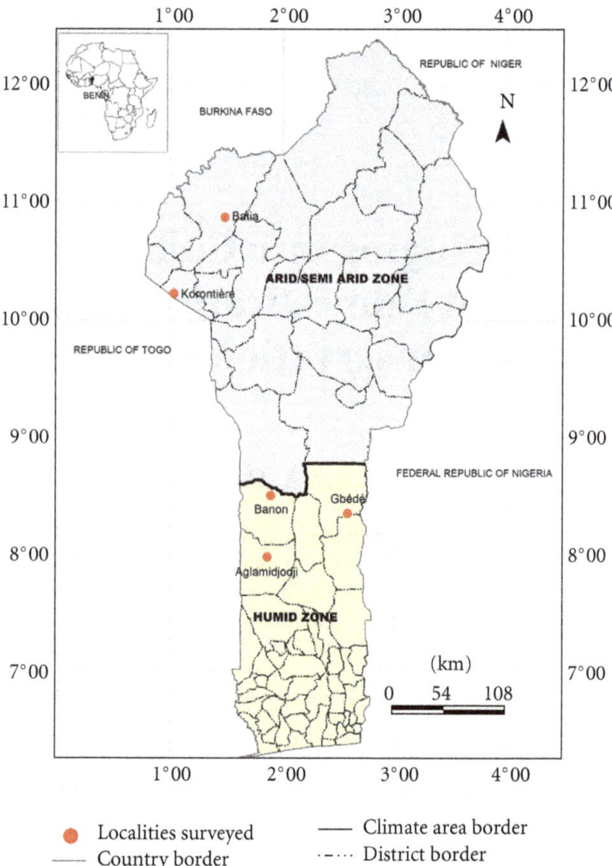


FIGURE 1: Benin map showing the location of the surveyed sites.

*2.2. Site Selection and Survey.* For the study, five villages (Aglamidjodji, Banon, Batia, Gbédé, and Korontière) were selected in the two contrasting agroecological zones of the country (Figure 1). Aglamidjodji, Banon, and Gbédé are located in the central region of Benin (humid zone), while Batia and Korontière are in the north (arid zone). In term of the vegetation type, Aglamidjodji and Korontière are entirely degraded; Banon and Gbédé are forested, while Batia is located in a savannah zone (Pendjari Park; Figure 1). Aglamidjodji, Banon, Batia, and Gbédé are inhabited, respectively, by the ethnic groups Mahi, Nago-Fè, Gourmanché, and Nago-Tchabè. Korontière is shared by two ethnic groups: the Ditamari (local and dominant) and the Lamba (originated from the Republic of Togo and in minority).

Data were collected during expeditions from the different sites through the application of participatory research appraisal tools and techniques such as direct observation, group discussions, individual interviews, and field visits using a questionnaire [16]. Interviews were conducted with the help of translators from each area. In each site, local farmers' organizations were involved in the study to facilitate the organization of group meetings (details of the research objectives were presented to the farmers, and general discussion was held on the steps of the plant domestication process) and assist in the data collection at household level.

In each village, 30 households (total of 150 for the study zone) were randomly selected using the transect method described by Dansi et al. [16]. At household level, interview was conducted only with the head of family and his wife. However, in case of polygamy, all wives were involved in the discussions taking into consideration key roles played by women in plant domestication and biodiversity management and conservation on farm [22–25]. During each interview, sociodemographic data of the surveyed household (size, total area available, total area cultivated, number of crops practiced, area occupied by the major crops, number of food shortages experienced during the last ten years) and of its head (age, number of wives, number of the social groups to which he belongs, education level, age of his wife or first wife when many) were first collected. Then, the household head and his wife were asked to list (vernacular name) the species being domesticated by their household.

Field visits were conducted to see and document the listed species in their natural habitats (bushes, shallows) or where they are being cultivated (home gardens, cultivated fields). On each species inventoried, information recorded through discussions were related to status (wild, cultivated), life form (tree, shrub, and herb), habitat, part of the plant used and season of availability, importance (food, nutrition, medicinal values, etc.), reasons for domestication, and person (husband or his wife; gender issue) responsible for its domestication. Scientific names were determined by the plant taxonomist of the research team using the Analytic Flora of Benin [21], and pictures were taken for report.

Different steps exist in the bringing into cultivation phase of the plant domestication process. For each species, the level reached in this phase was determined and quoted using a seven-step model modified following Dansi et al. [16] and described as follows.

*Step 1.* Species entirely wild and collected only when needed.

*Step 2.* Wild species maintained in the fields when found during land preparation (clearance, burning, and weeding) due to its proved utility and regular need, its scarcity around habitations, and the difficulties for getting it on time, in quality and in quantity. These preserved plants are subject to regular observations for the understanding of their reproductive biology.

*Step 3.* Farmers start paying more attention to the preserved plants (weeding, protection against herbivorous) for their survival and their normal growth. A sort of ownership on the plants start.

*Step 4.* The reproductive biology of the species is known, and multiplication and cultivation of the species in the home gardens or in selected parts of cultivated fields are undertaken by farmers or healers. At this stage, farmers tend to conduct diverse experiments (date of planting, sowing or planting density, pest and diseases management, etc.) in order to master mass production of the species in the future. The ownership on the plant is more rigorous.

*Step 5.* The species is cultivated and harvested using traditional practices.

*Step 6.* To improve the quality of the product, farmers adopt specific criteria to select plants that better satisfied people needs. The best cultivars/plants (good grain/fruit quality, resistant/tolerance to diseases and pests) are known, and technical package is adopted for their development and multiplication. At this stage, access to market is considered and some species benefit from traditional postharvest technologies (method for processing, cooking or conservation, etc.) to meet consumers' needs.

*Step 7.* Selection initiatives continue with cooking qualities, protection against pests, and diseases in cultivation and storage. Income generation is more clearly taken care of: market demands (quantity and quality) are also taken into account, and species varieties that meet consumers' preferences are selected and produced.

*2.3. Data Analysis.* Data were analysed through descriptive statistics (frequencies, percentages, means, etc.) in order to generate summaries and tables at different (villages, ethnic groups, households) levels. To compare the mean numbers of species in domestication recorded per household between ethnic groups or agroecological zones, the nonparametric tests of Wilcoxon and of Kruskal-Wallis were computed using SAS [26]. To analyse the relationships between villages in term of species in domestication, villages surveyed were considered as individuals and the plant species under domestication as variables and scored, for each village, as 1 when present or 0 if not. Using this methodology, 69 variables (corresponding to the species inventoried) were created and a binary matrix was compiled. Pairwise distances between villages were computed by NTSYS-pc 2.2 [27], using Jaccard coefficient of similarity [28]. Similarity matrix was used to design a dendrogram using UPGMA cluster analysis [29, 30]. The same process was used to examine the distribution of the species with regards to their levels of domestication and habitats. Here, the 69 species inventoried were still considered as individuals and the different domestication levels and habitats recorded as variables and also scored as 1 when present or 0 when absent. The binary matrix compiled was used to perform a principal coordinate analysis (PCA) and generate a dendrogram as described above using the same software packages. Spearman coefficient of correlation was calculated using SAS statistical package [26] to test the influence of six variables related to the households surveyed (size, number of crops practiced, total area available, total area cultivated, total area occupied by the major crops, number of food shortages experienced the last ten years) and of five parameters linked to the head of the household interviewed (age, education level, number of wives, age of the first wife, number of the social groups to which he belongs) on the household decision making with regard to the number of species to domesticate.

## 3. Results

*3.1. Sociodemographic Profile of the Households Surveyed.* The size of the households surveyed varied from 1 to 40 with 9 on average. The maximum size (40) was obtained at Banon and the minimum (1) at Aglamidjodji and at Batia. Among the 150 respondents, 25.34% were women and 74.66% were men; 51.66% have never been to school, 30.83% went to primary school, and 17.51% attended secondary school. The average age of the respondents was 40 years (minimum 20 years; maximum 75 years). The majority (79.16%) of the men respondents had one to two wives. Most of the respondents (71%) did not belong to any farmers' association (group), 22% belong to one, two, three, or four groups, and a very few number (7%) are members of 5 to 6 groups.

*3.2. Diversity of the Species under Domestication.* Throughout the five villages surveyed, a great diversity of plant species under domestication was found. A total of 69 species belonging to 62 genera and 40 families (Table 1) were inventoried and documented. Among the 40 families, the five most important were the Leguminosae-Caesalpinioideae (7 species), the Lamiaceae (5 species), the Asteraceae (4 species), the Moraceae (3 species), the Bombacaceae (3 species), and the Asclepiadaceous (3 species). The remaining families (34) have only one to two species. For these 69

Table 1: Diversity, vernacular names, and utilisation of the species under domestication across ethnic groups.

| Number | Scientific names | Family | Vernacular name | Part of the plant used |
|---|---|---|---|---|
| 1 | *Adansonia digitata* | Bombacaceae | Otché (Fè, Nago), Télou (Lamba), Zouzon (Mahi), Boutouobou (Gourmantché) | Gourmanthé, Nago, Fè (Fruits and Leaves); Lamba (fruits) |
| 2 | *Agelanthus dodoneifolius* | Loranthaceae | Ayapou (Lamba) | Lamba (bark) |
| 3 | *Annona senegalensis* | Annonaceae | Alilou (Lamba) | Lamba (Leaves, fruits) |
| 4 | *Anogeissus leiocarpus* | Combretaceae | Kolou (Lamba) | Lamba (bark) |
| 5 | *Balanites aegyptiaca* | Balanitaceae | Boukpanwounkpôhôbou (Gourmantché) | Gourmantché (fruits) |
| 6 | *Bixa orellana* | Bixaceae | Timinti-éssô (Fè) | Fè (fruits) |
| 7 | *Blighia sapinda* | Sapidaceae | N'tchin (Nago) | Nago (fruits) |
| 8 | *Bombax costatum* | Bombacaceae | Kpahoudèhouin (Mahi), Houlou (Lamba) | Mahi, Lamba (Leaves) |
| 9 | *Caesalpinia bonduc* | Fabaceae- caesalpinioideae | Adjikoun (Mahi), Ogrounfè (Nago), Fèo (Fè) | Fè (Leaves, roots, seeds), Tchabè (Roots), Mahi (Root, |
| 10 | *Calotropis procera* | Asclepiadaceae | Touloukou (Lamba) | Lamba (Leaves) |
| 11 | *Ceiba pentandra* | Bombacaceae | Ogoun Fè (Fè) | Fè (Leaves) |
| 12 | *Celosia argentea* | Amaranthaceae | Tchôkôyôkôtô (Nago), Sôman (Mahi) | Nago, Mahi (Leaves) |
| 13 | *Celosia trigyna* | Amaranthaceae | Adjèmanwofô (Nago, Fè), | Nago, Mahi (Leaves) |
| 14 | *Ceratotheca sesamoides* | Pedaliaceae | Agbôssou (Mahi), Koumonkoun (Fè), Idjabô (Nago), Assoworou (Lamba) | Mahi, Fè, Gourmantché, Nago, Lamba (Leaves) |
| 15 | *Corchorus tridens* | Tiliaceae | Ountcho (Nago) | Nago (Leaves) |
| 16 | *Cissus populnea* | Vitaceae | Tchôkougbôlô (Fè), Kpôgôlô (Nago), Anyar (Lamba) | Fè, Nago, Lamba (roots) |
| 17 | *Clausena anisata* | Rutaceae | Oroukôgbo (Fè) | Fè (Leaves and roots) |
| 18 | *Cleome ciliata* | Capparaceae | Aiya (Mahi) | Mahi (Leaves) |
| 19 | *Cleome gynandra* | Capparaceae | Akaya (Nago) | Nago (Leaves) |
| 20 | *Cochlospermum tinctorium* | Cochlospermaceae | Boussôrôbou (Gourmantché) | Gourmanthé (Roots) |
| 21 | *Crassocephalum rubens* | Asteraceae | Akôgbo (Mahi), Gboolo (Nago, Fè) | Fè, Nago, Mahi (Leaves) |
| 22 | *Cymbopogon giganteus* | Poaceae | Kpalman mihou (Lamba) | Lamba (Leaves) |
| 23 | *Detarium microcarpum* | Leguminosae | Kpôr (Lamba), Bounankpôhôbou (Gourmantché) | Gourmantché, Lamba (Roots, fruits) |
| 24 | *Dichrostachys cinerea* | Leguminosae | Nanha sèhô (Lamba) | Lamba (Roots) |
| 25 | *Diospyros mespiliformis* | Ebenaceae | Ankalé (Lamba), Bougaabou (Gourmantché) | Lamba, Gourmantché (fruits) |
| 26 | *Dioscorea abyssinica* | Dioscoreaceae | Koudjabouwoungou (Gourmantché) | Gourmantché (Tuber) |
| 27 | *Dioscorea praehensilis* | Dioscoreaceae | Ichou (Fè) | Fè (Tuber) |
| 28 | *Echinops longifolius* | Asteraceae | Koumantchaintchain (Wama) | Wama (Roots) |
| 29 | *Eriosema pellegrinii* | Leguminosae | Kassimintê (Wama) | Wama (Roots) |
| 30 | *Ficus abutilifolia* | Moraceae | Agbèdè (Fè), Okpoto (Nago) | Fè, Nago (Leaves) |
| 31 | *Ficus ingens* | Moraceae | Boukankanbou (Gourmantché) | Gourmantché (Leaves) |
| 32 | *Ficus sycomorus* | Moraceae | Oukankanmou (Gnindé) | Gnindé (Leaves) |
| 33 | *Gardenia erubescens* | Rubiaceae | Bounansôôbou (Gourmantché), kaou (Lamba) | Gourmantché (Fruits), Lamba (Fruits, stem) |

Table 1: Continued.

| Number | Scientific names | Family | Vernacular name | Part of the plant used |
|---|---|---|---|---|
| 34 | *Haumaniastrum caeruleum* | Lamiaceae | Atingbinnintingbin (Fè) | Fè (Leaves) |
| 35 | *Heteropteris leona* | Malpigluaceae | Nansikôr (Lamba) | Lamba (Leaves and Roots) |
| 36 | *Hibiscus sabdariffa* | Malvaceae | Kpakpala (Nago), Kpakpa (Fè) | Fè, Nago (Leaves) |
| 37 | *Indigofera bracteolata* | leguminosae | Tikouyè ogoutè (Gnindé) | Gnindé (Leaves and roots) |
| 38 | *Justicia tenella* | Acanthaceae | Djagou-djagou (Fè) | Fè (Leaves) |
| 39 | *Lagenaria siceraria* | Cucurbitaceae | kaka (Nago) | Nago (Leaves) |
| 40 | *Lannea microcarpa* | Anacardiaceae | Bougbantchabou (Gourmantché) | Gourmantché (fruits) |
| 41 | *Launeae taraxacifolia* | Asteraceae | Odôdô (Nago, Fè), Gnantotoé (Mahi) | Fè, Nago, Mahi (Leaves) |
| 42 | *Lippia multiflora* | Verbenaceae | Aglaala (Mahi), Tchaga (Fè) | Fè, Mahi (Leaves, flowers) |
| 43 | *Momordica charantia* | Cucurbitaceae | Tchaati (Fè), Gnissikin (Mahi) | Fè, Mahi (Leaves) |
| 44 | *Ocimum americanum* | Lamiaceae | Ofin (Fè) | Fè (Leaves) |
| 45 | *Ocimum basilicum* | Lamiaceae | Ounkpèhoun (Fè), Gbogbotyin (Nago), Hissin-hissin (Mahi) | Nago (Leaves) |
| 46 | *Ocimum gratissimum* | Lamiaceae | Simonba (Fè), Kioyo (Mahi) | Fè, Mahi (Leaves) |
| 47 | *Parkia biglobosa* | Leguminosae | Ayoya (Mahi), Ougba (Nago), Igba (Fè), Boudoubou (Gourmantché), S'lou (Lamba) | Mahi, Fè, Nago, Lamba (fruits); Gourmantché (Fruits, Bark) |
| 48 | *Pergularia daemia* | Asclepiadaceae | Agbonfoun-foun (Fè) | Fè (Leaves) |
| 49 | *Phyllanthus muellenianus* | Euphorbiaceae | Akanmankogou (Mahi) | Mahi (Leaves) |
| 50 | *Piliostigma thonningii* | Leguminosae | Wôkou (Lamba) | Lamba (Leaves, Roots) |
| 51 | *Platostoma africanum* | Lamiaceae | Kouloubi (Fè), Gouloubi (Nago) | Nago, Fè (Leaves) |
| 52 | *Pseudocedrela kotschyi* | Meliaceae | Asntélémr (Lamba) | Lamba (Bark) |
| 53 | *Psorospermum alternifolium* | Clusiaceae | Kpinon-kpinon (Fè) | Fè (Leaves) |
| 54 | *Raphionacme brownii* | Asclepiadaceae | Kousséligou (Gourmantché), Kohounsèhounta (Wama) | Gourmantché, Wama (Tuber) |
| 55 | *Rauvolfia vomitoria* | Apocynaceae | Essô èyèdjè (Fè) | Fè (Leaves) |
| 56 | *Saba comorensis* | Apocynaceae | Louou (Lamba) | Lamba (Fruits) |
| 57 | *Sarcocephalus latifolius* | Rubiaceae | Bounangnibou (Gourmantché), Athithélou (Lamba) | Lamba (Leaves, Roots, fruits); Gourmantché (Fruits) |
| 58 | *Sclerocarya birrea* | Anacardiaceae | Mounannikmon (Otamari), Bounanmag'bou (Gourmantché) | Otamari (Fruits, Leaves); Gourmantché (fruits) |
| 59 | *Sesamum radiatum* | Pedaliaceae | Dossé (Nago), Koumonkoun-adjagbalè (Fè), Ungangoun (Gourmantché), Natawourou (Lamba), Agbô (Mahi) | Mahi, Fè, Gourmantché, Nago, Lamba (Leaves) |
| 60 | *Solanum erianthum* | Solanaceae | Mon (Fè) | Fè (Leaves) |
| 61 | *Sterculia tragacantha* | Sterculiaceae | Akèmonkodjèko (Fè) | Fè (Leaves) |
| 62 | *Strychnos spinosa* | Loganiaceae | Fountoumdrô (Lamba) | Lamba (fruits and Roots) |
| 63 | *Talinum triangulare* | Portulacaceae | Odondon (Nago), Odondon (Fè), Glassoéman (Mahi) | Nago, Fè, Mahi (Leaves) |
| 64 | *Tamarindus indica* | Leguminosae | Boupouguibou/Boupouobou (Gourmantché), Timtélém (Lamba) | Gourmantché (Fruits, Leaves); Lamba (Fruits) |

Table 1: Continued.

| Number | Scientific names | Family | Vernacular name | Part of the plant used |
|---|---|---|---|---|
| 65 | *Vernonia colorata* | Asteraceae | Arikoro (Nago) | Nago (Leaves) |
| 66 | *Vitellaria paradoxa* | Sapotaceae | Kotoblè (Mahi), Emin (Fè, Nago), Boussanbou (Gourmantché), Sèmou (Lamba) | Mahi, Fè, Nago, Lamba (fruits), Gourmantché (fruits, bark) |
| 67 | *Vitex doniana* | Verbenaceae | Bougaanbou (Gourmantché), Akpagnarou (Lamba), Fonman (Mahi), Ewa (Fè), Akoumanlapka (Nago) | Mahi, Fè, Gourmantché, Nago, Lamba (Leaves, fruits) |
| 68 | *Ximenia americana* | Oleracea | Klivovoé (Mahi), Boumirinbou (Gourmantché) | Mahi (fruits); Gourmantché (Fruits, Leaves, Roots) |
| 69 | *Zanthoxylum zanthoxyloides* | Rutaceae | Tchanouwèlè (Fè) | Fè (Leaves, Roots, Bark, Thorns) |

Table 2: Number of plant species under domestication per village and their distribution per type of plant and by habitat.

| Villages | Total | Types of plants | | | Habitat | | | |
|---|---|---|---|---|---|---|---|---|
| | | Trees | Shrubs | Herbs | Forest | Fallow | Cultivated field | Home garden |
| Banon | 33 | 8 | 4 | 21 | 4 | 5 | 7 | 2 |
| Gbédé | 22 | 6 | 2 | 14 | 10 | 12 | 8 | 3 |
| Aglamidjodji | 18 | 5 | 3 | 10 | 8 | 7 | 5 | 1 |
| Korontière | 27 | 14 | 6 | 7 | 8 | 7 | 6 | 3 |
| Batia | 21 | 12 | 3 | 6 | 10 | 8 | 7 | 3 |

species inventoried, 138 vernacular names (Table 1) were recorded. They vary from place to place and sometime within the same ethnic group (Table 1). Per village, the total number of species under domestication inventoried varies from 18 (Aglamindjodji) to 32 (Banon) with 24 species on average per village (Table 2). The species found consisted of 19 trees (27.53%), 11 shrubs (16%) and 39 (56.47%) erect, creeping or climbing herbs (Table 1). A higher proportion of trees was observed in the northern region (Korontière and Batia) in comparison to the southern zone (Table 2).

Geographic distribution of the species inventoried showed high variability (Table 1). Some species such as *Adansonia digitata, Parkia biglobosa, Sesamum radiatum, Vitellaria paradoxa,* and *Vitex doniana* were found under domestication in all the villages surveyed, while many others like *Celosia trigyna, Cleome ciliate,* and *Lippia multiflora* were restricted to only one or two sites (Table 1). The great majority (50 to 71%) of the species was found in forests or fallows (Table 2). Only a few numbers were found in cultivated fields or in the home gardens. The mean number of species found under domestication per household significantly ($P = 0.0002$) varied between agroecological zones and among ethnic groups, but no significant difference was obtained between savannah and forest zones. In the humid zone, the mean number of species per household recorded was 8, while, in the arid zone, it was 5. At 30% of similarity level, the dendrogram constructed to analyse the relationships between surveyed villages in term of species under domestication led to two groups, namely, G1 and G2 (Figure 2): G1 gathers Batia and Korontière, the two villages of the north, while G2 assembles the three villages of the centre (Aglamidjodji, Banon, and Gbédé).

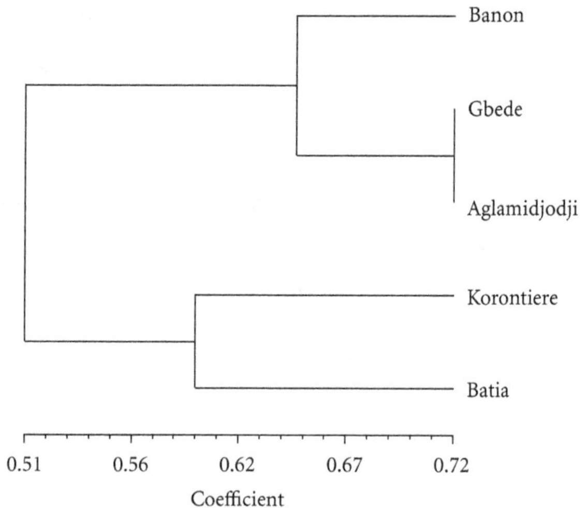


Figure 2: UPGMA dendrogram based on Jaccard coefficient of similarity showing the grouping of the villages.

In all the villages surveyed, most of the species (61.90 to 77.77%) under domestication were well known to the local communities at both taxonomical and biological (growth, ecological requirements, reproduction) levels (Table 3). Among the species inventoried, three were reported as under threat due to over exploitation by people. These were *Caesalpinea bonduc, Launeae taraxacifolia,* and *L. multiflora.*

Table 3: Knowledge of the species and of their biology by the local communities.

| Ethnic groups | Total | Knowledge of the species | | Knowledge of the species' biology | | Period of availability | | |
|---|---|---|---|---|---|---|---|---|
| | | Widely known | Little known | Known | Unknown | AS | RS | DS |
| Ditamari/Lamba | 27 | 17 | 10 | 18 | 9 | 8 | 16 | 3 |
| Gourmantché | 21 | 13 | 8 | 15 | 6 | 5 | 9 | 7 |
| Mahi | 18 | 12 | 6 | 10 | 8 | 5 | 11 | 2 |
| Nago | 36 | 28 | 8 | 27 | 9 | 9 | 25 | 2 |

*AS: all seasons, RS: rainy season, DS: dry season.*

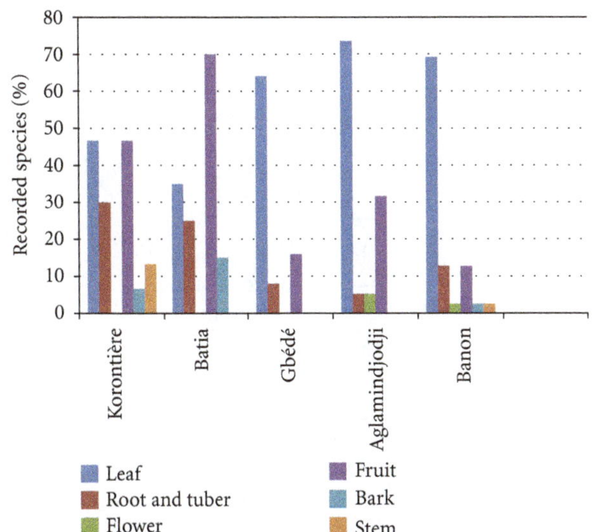


Figure 3: Relative importance of the species under domestication with regard to their organs used across villages.

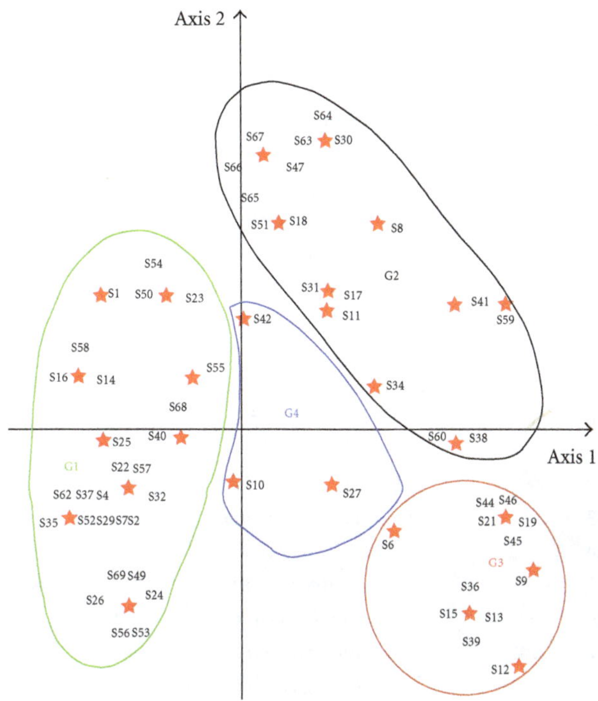


Figure 4: Principal coordinate analysis showing grouping of the species in relation to habitat and domestication levels. Species codes are those used in Table 1.

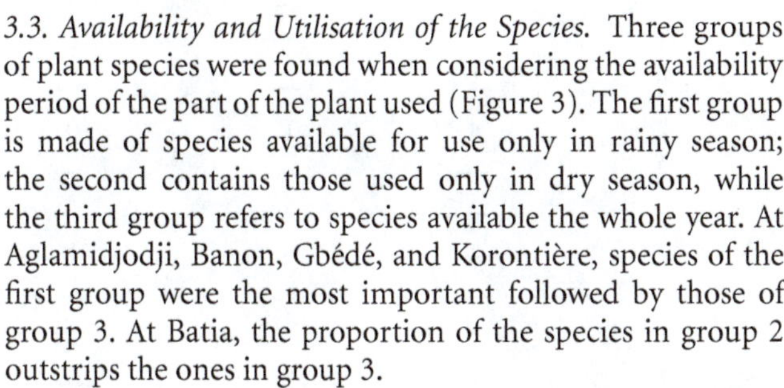

*3.3. Availability and Utilisation of the Species.* Three groups of plant species were found when considering the availability period of the part of the plant used (Figure 3). The first group is made of species available for use only in rainy season; the second contains those used only in dry season, while the third group refers to species available the whole year. At Aglamidjodji, Banon, Gbédé, and Korontière, species of the first group were the most important followed by those of group 3. At Batia, the proportion of the species in group 2 outstrips the ones in group 3.

The organs (leaves, fruits, bark, roots, tuber, and flowers) of the different species inventoried used by the local communities vary considerably with the species and ethnic groups (Table 1). At Batia (Gourmantché zone), the species domesticated for their fruits are the most important followed by those domesticated for their leaves (Figure 3). In the other four villages (Aglamidjodji, Banon, Gbédé, and Korontière), the situation is opposite: species from which leaves are the most useful parts were the most numerous followed by those used for their fruits (Figure 3). Out of the 69 species inventoried, fourteen were domesticated only for medicinal purposes, three (*Cochlospermum tinctorium*, *L. taraxacifolia* and *L. multiflora*) were typically nutraceutical (as they have medicinal properties beside their nutritional value), and the others (52 in total) are used for food or medicine depending on the part of the plant considered (Table 1).

*3.4. Domestication Levels of the Species.* The domestication levels recorded for the species inventoried vary from 0 to 5. The number of species decreased with the domestication level. The majority of these (31 species, 45%) was found at Step 1 in all the villages where they have been signalled, and only one species (*Dioscorea praehensilis*) was found at Step 6 (Table 4). For most of the species (38 in total, Table 4) other than those found at Step 1 in all the villages, the domestication level is not consistent from one village to the other (Table 4). *S. radiatum*, for example, is at Step 1 at Korontière, Step 2 at Gbédé, Step 3 at Aglamidjodji, and Banon and Step 5 at Batia (Table 4).

The principal coordinate analysis carried out to analyse the relationships among species in terms of habitat and domestication levels led to four groups, namely, G1, G2, G3 and G4 (Figure 4).

Table 4: Domestication levels of the species and their variations across villages (species found only at Step 1 are not included).

| Number | Scientific name | Domestication levels | | | | |
|---|---|---|---|---|---|---|
| | | Aglamidjodji | Banon | Batia | Gbédé | Korontière |
| 1 | *Adansonia digitata* | 0 | 0 | 2 | 0 | 1 |
| 2 | *Bixa orellana* | — | 3 | — | — | — |
| 3 | *Bombax costatum* | 2 | — | — | — | 0 |
| 4 | *Caesalpinea bonduc* | 2 | 4 | — | 3 | — |
| 5 | *Calotropis procera* | — | — | — | — | 4 |
| 6 | *Ceiba pentandra* | — | 2 | — | — | — |
| 7 | *Celosia argentea* | 4 | — | — | 3 | — |
| 8 | *Celosia trigyna* | — | 3 | — | 3 | — |
| 9 | *Ceratotheca sesamoides* | 0 | 1 | — | 1 | 1 |
| 10 | *Corchorus tridens* | — | — | — | 3 | — |
| 11 | *Clausena anisata* | — | 1 | — | — | — |
| 12 | *Cleome ciliata* | 1 | — | — | — | — |
| 13 | *Cleome gynandra* | — | — | — | 2 | — |
| 14 | *Crassocephalum rubens* | 3 | 2 | — | 3 | — |
| 15 | *Detarium microcarpum* | — | — | 1 | — | 0 |
| 16 | *Dioscorea praehensilis* | — | 5 | — | — | — |
| 17 | *Ficus abutilifolia* | — | 2 | — | 1 | — |
| 18 | *Ficus ingens* | — | — | 1 | — | — |
| 19 | *Haumaniastrum caeruleum* | — | 2 | — | — | — |
| 20 | *Hibiscus sabdariffa* | — | 3 | — | 3 | — |
| 21 | *Justicia tenella* | — | 2 | — | — | — |
| 22 | *Lagenaria siceraria* | — | — | — | 3 | — |
| 23 | *Launeae taraxacifolia* | 2 | 2 | — | 2 | — |
| 24 | *Lippia multiflora* | 4 | 1 | — | — | — |
| 25 | *Ocimum americanum* | — | 3 | — | — | — |
| 26 | *Ocimum basilicum* | 0 | 1 | — | 2 | — |
| 27 | *Ocimum gratissimum* | 3 | 2 | — | — | — |
| 28 | *Parkia biglobosa* | 1 | 1 | 1 | 2 | 2 |
| 29 | *Piliostigma thonningii* | — | — | — | — | 1 |
| 30 | *Platostoma africanum* | — | 1 | — | 1 | — |
| 31 | *Rauvolfia vomitoria* | — | 2 | — | — | — |
| 32 | *Sesamum radiatum* | 2 | 2 | 4 | 1 | 0 |
| 33 | *Solanum erianthum* | — | 2 | — | — | — |
| 34 | *Talinum triangulare* | 1 | 2 | — | 2 | — |
| 35 | *Tamarindus indica* | — | — | 2 | — | 1 |
| 36 | *Vernonia colorata* | — | — | — | 1 | — |
| 37 | *Vitellaria paradoxa* | 1 | 1 | 2 | 1 | 2 |
| 38 | *Vitex doniana* | 1 | 1 | 1 | 1 | 0 |

(i) G1 gathers the wild species which naturally occur in the forests, savannahs and fallows and which are at Step 1.

(ii) G2 is the group of the species spared in the fields when found during land preparation and which received no or very little management attention from farmers for their survival (species found at Step 2 or 3).

(iii) G3 assembles all the species found at Step 4 of the overall domestication process. It is the group of the species under cultivation in home gardens or in specific parts of cultivated fields.

(iv) G4 pulls together the cultivated species found at Step 5 (*Calotropis procera*/S10; *L. multiflora*/S42) and at Step 6 (D. praehensilis/S27).

At 60% of similarity, the dendrogram (Figure 5) of the UPGMA cluster analysis performed on the same data revealed tree classes (C1, C2, C3) of which two (C1 and C2) correspond, respectively, to G1 and G2, while the third one (C3) is G3 and G4 pulled together.

TABLE 5: Contribution of some species under domestication to household income generation.

| Species | Minimum (US$) | Maximum (US$) |
|---|---|---|
| *Caesalpinea bonduc* | 7 | 8 |
| *Celosia argentea* | 100 | 140 |
| *Celosia trigyna* | 2 | 5 |
| *Cochlospermum tinctorium* | 20 | 144 |
| *Ceratotheca sesamoides* | 10 | 90 |
| *Crassocephalum rubens* | 3 | 10 |
| *Dioscorea praehensilis* | 9 | 30 |
| *Haumaniastrum caeruleum* | 4 | 8 |
| *Launeae taraxacifolia* | 120 | 192 |
| *Lippia multiflora* | 2 | 10 |
| *Parkia biglobosa* | 400 | 600 |
| *Sesamum radiatum* | 50 | 96 |

*3.5. Motivations behind the Plant Domestication.* According to farmers, the domestication of a plant starts, when its usefulness is proved, its demand is confirmed and regular, its availability around dwellings is seriously decreasing and when getting the desired quantity on time for use becomes problematic. They reported that plant domestication is generally done by simple curiosity or for dietary, medicinal, economic, or cultural reasons. Among these reasons, the most important is food security (50.85% of respondents) followed by medicinal use (30.5% of respondents), economic reasons (14.41% of respondents), and cultural reasons (4.24% of respondents).

In fact, many of the species recorded are sold in the markets and their annual contribution to household income generation and poverty reduction is appreciable (Table 5). A comparison between economic values and domestication levels of twelve species (Table 5) revealed that species such as *Ceratotheca sesamoides*, *C. tinctorium*, *L. taraxacifolia*, and *P. biglobosa* although having a relatively high economic value (in the rural areas surveyed), are still at very low domestication levels. *C. tinctorium*, for example, is still at Step 1 of the domestication process, while its root (dried and grinded to a powder) is highly valued as nutraceutical vegetable (treatment of malaria, diabetes) in the northern regions of Benin. One species (*Agelanthus dodoneifolius*) was domesticated only for cultural reasons. In Lamba ethnic zone, one believes that it protects houses against evil spirits. Several factors affect farmers' decision making in domesticating plants. A correlation analysis revealed that among eleven (11) parameters related to the households surveyed and to the head of the household interviewed, eight are significantly correlated ($P < 0.0001$) with the number of species domesticated per household either positively (size of the household, age of the head of the household, age of the household wife, total area available, total area cultivated, area occupied by the major crops) or negatively (education level of the head of the household, number of food shortages experienced during the last ten years) while three (Number of wives, number of the social groups, number of crops practiced) showed no significant correlation.

*3.6. Gender and Plant Domestication.* The number of species found under domestication varied according to the gender (Table 6). Out of the 69 species recorded throughout the five villages surveyed, 31 (44.92%) were found under domestication with only women, 18 (26.08%) with only men, and 20 (28.98%) with both men and women. Some differences were observed between ethnic zones (Table 6). Hence, in the cultural areas Nago and Mahi (central Benin), the number of species being domesticated by women (50 to 55.55% of the total) is higher than the ones under the control of men. Contrary to Nago and Mahi ethnic groups, in the Gourmantché, Ditamari, and Lamba ethnic groups in northern Benin, men domesticated more species (42.85 to 59.25% of the total) than women. The classification of the species recorded according to both gender and use revealed that species being domesticated by women were basically leafy vegetables while those linked to men were essentially fruit species (Table 6) and the species being domesticated by both men and women were medicinal plants.

## 4. Discussion

*4.1. Diversity, Availability, and Utilisation of the Species.* The process of plant domestication is very active in the rural areas of Benin. The great diversity of the species under domestication recorded in this study is a tangible proof. These results are in support of those reported earlier on yam [11, 12, 31] and on traditional leafy vegetables in Benin [16]. For the 69 species inventoried, 138 vernacular names were recorded. Many names (one to five) were known for each species, and these vary among and within ethnic areas (Table 1). In the study of folk nomenclature in plant, such variation is now well known and documented [16, 32, 33]. The higher numbers of species under domestication were found in the forest zones and most of species recorded (56.47%) were herbaceous. Herbaceous are annual and are not available at the same place all the years and searching for an important wild herb species within the forest when needed is not secure (frequent snakebites, risks of lost). The species inventoried do not have the same ecogeographical distributions, and moreover the indigenous knowledge related to the utilization of the species varies from one area to the other. One understands, therefore, why some species were found under domestication in all the villages surveyed while many others were restricted to only one or two sites.

The ecogeographical consideration also remains the main justification of the partition (based on the species found under domestication) of the five villages surveyed into two clusters corresponding to the arid zone of the north and to the humid zone of the south. The communities interviewed have a good knowledge of the status of the plant species they are domesticating. They reported tree species (*C. bonduc*, *L. taraxacifolia*, and *L. multiflora*) under threat due to overexploitation by people. This is true for *L. taraxacifolia* following Dansi et al. [16] and also for *C. bonduc* and *L. multiflora*, which are even already in the Benin red list of threatened species [20]. The great majority of the species was used for food and/or medicine, the two most important

Table 6: Classification of the species under domestication according to the gender and to their specific utilization.

| Group of species | Total | Ethnic groups | | | | Type of plant | | | | |
|---|---|---|---|---|---|---|---|---|---|---|
| | | NA | MA | GO | LD | LV | NV | Fr | Tb | Md |
| Species being domesticated by women | 31 | 20 | 09 | 04 | 04 | 22 | 02 | 02 | 00 | 05 |
| Species being domesticated by men | 18 | 06 | 03 | 09 | 16 | 02 | 01 | 09 | 02 | 04 |
| Species being domesticated by both men and women | 20 | 10 | 06 | 08 | 07 | 04 | 03 | 03 | 01 | 09 |
| Total | 69 | 36 | 18 | 21 | 27 | 28 | 06 | 14 | 03 | 18 |

*NA: Nago, MA: Mahi, GO: Gourmantché, LD: Lamba/Ditamari, LV: leafy vegetable, NV: nonleafy vegetable, Fr: fruit, Tb: Tuber crop, Md: medicinal plant.*

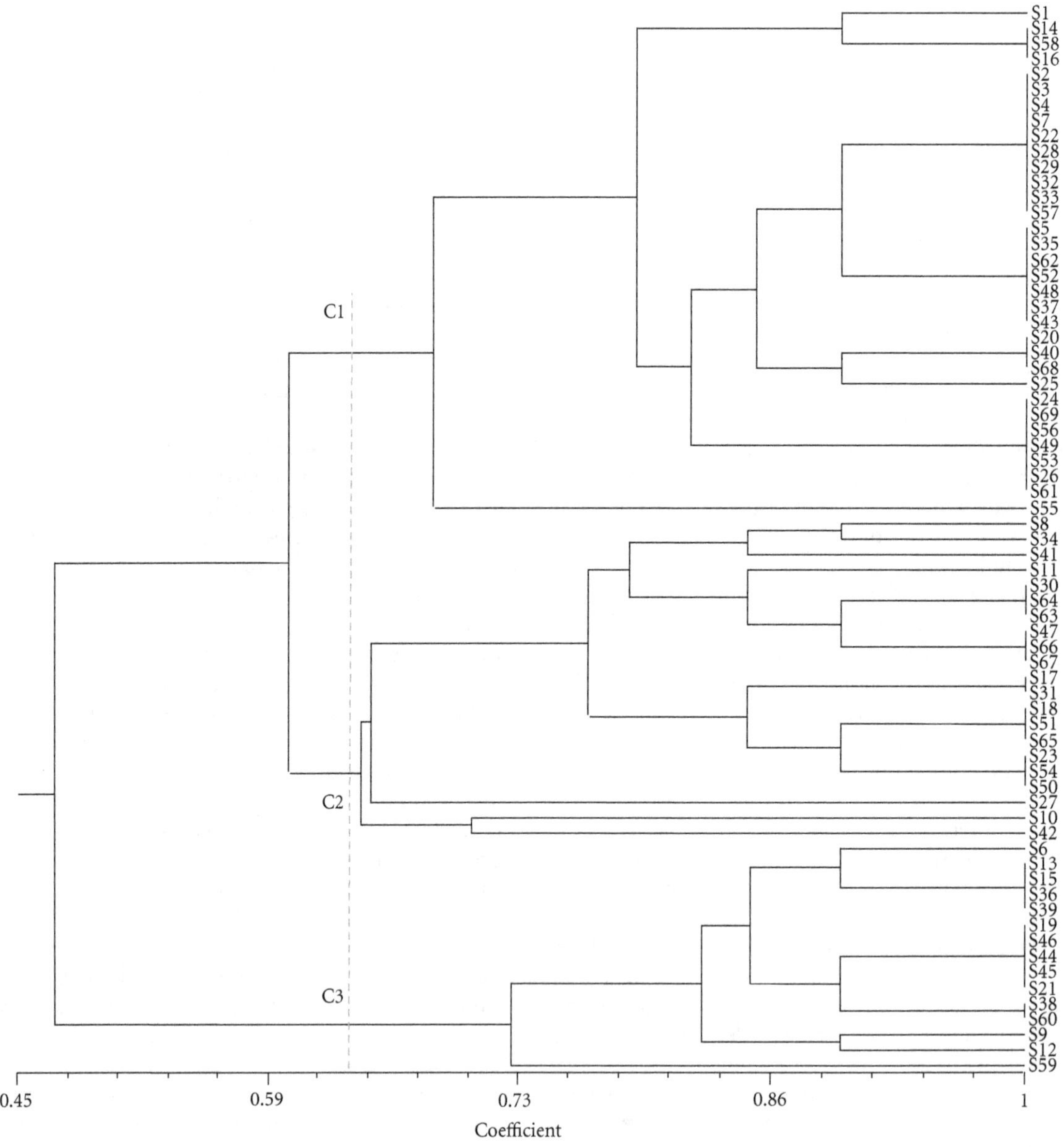


Figure 5: Dendrogram showing the classification of the species base on their habitat and their domestication levels.

vital needs of human being. Similar results were reported by Hildebrand [34] in southwest Ethiopia and by Casas et al. [6] in Mesoamerica. In all the villages surveyed apart from Batia, most of the species are being domesticated for their leaves besides available for use mainly in rainy season. This result is expected as most of the species domesticated for their leaves are leafy vegetables of daily used [16]. At Batia, bordering village of the national park of Pendjari inhabited by the Gourmantché, fruit species are most numerous and the plants whose useful parts are available only in dry season were preferred. The richness of savannah woodland in fruit trees and preference for fruit species by the ethnic groups living in the area may be the explanations of this finding.

*4.2. Motivations behind the Plant Domestication and Domestication Levels.* Farmers reported that plant domestication seeks to bring out the maximum human benefit within a species. It is generally done for dietary, medicinal, economic, and cultural reasons or by simple curiosity. This result is in agreement with those reported by Hildebrand [34] and Casas et al. [6]. Not surprisingly, the number of species domesticated per household is affected by several factors dominated by the education level of the head of the household and the number of food shortages experienced the last ten years. The negative influence noted for the first factor follows the actual general tendency by intellectuals to abandon traditional practices. On the other hand, the negative correlation observed with the number of food shortages experienced the last ten years was unexpected and could be tentatively explained as follow: a species being domesticated for food purposes is rarely cultivated or present on a large area in a short period of time. Consequently, it cannot produce sufficient quantity of food needed to meet the requirements of the households which are generally important. It is therefore normal that the more a household experienced food shortages, the more they will abandon domestication in favour of a more strengthened production of staple crops (cereals, root, and tubers, etc.).

Most of the species were found at low levels of domestication apart from yam where domestication was well studied and understood at both ethnobotanical and molecular levels [11, 12, 31, 35]. Normally species with high economic value should be prioritised for domestication by the households. Unfortunately, *C. tinctorium*, *L. taraxacifolia*, and *P. biglobosa* although having a relatively high economic value are still at very low domestication levels. For the farmers interviewed, *C. tinctorium* is still plenty in the wild and not very far from the villages; therefore, there is no urgent need to cultivate it. On the other hand, collecting its roots from the bush is laborious and grinding them later on into powder after drying is very difficult. They recognize however that *L. taraxacifolia* is becoming rare, but its domestication cannot go further than the "let standing" (practices directed to maintain within human-made environments useful plants that occurred in those areas before the environments were transformed by humans) described by several authors [6, 36–39] due to its reproductive biology (rapid loss of viability of the seeds during storage) not yet understood. For *P. biglobosa*, the reasons are not clear enough. The long time needed for the plant to start producing fruits could be the major handicap. Shortening the growth cycles for most fruit trees will facilitate their domestication process.

The results of the multivariate analysis (PCA and Cluster analysis) indicates that the seven steps (Step 1 to 7) initially defined in the domestication process could be visibly reorganized into three. The first one corresponds to Step 1 , the second to the combination of Steps 2 and 3, and the third one associates Steps 4 to 7. These three newly defined steps correspond to the three different practices (systematic gathering, let standing, encouraging growing) defined by many authors [6, 40–44].

*4.3. Gender Issue and Role of Domestication in Conserving Plant Diversity on Farmlands.* Variation was noted on the number of species found under domestication according to the gender. In the south, female-headed households domesticate more species than male-headed households. In the north, the opposite situation was observed. In both cases, species being domesticated by women were basically leafy vegetables and medicinal plants while those under the control of men were mainly fruits. The cultural division of tasks at household level generally devotes women to food preparation and children care taking, and men to hunting and farming. Richness of savannah woodland in wild fruit trees and the fruit harvest which is typically men activity because of the physical skill and energy it requires could be a comprehensive explanation of these results which are in agreement with those published by Msuya et al. [9] in Tanzania.

The great diversity (69 species) of plant recorded indicates that domestication is a traditional practice for conserving biodiversity. Domestication contributes to increasing plant genetic diversity and to conservation on farm of the agricultural biodiversity. It is a dynamic system which links genetic diversity development, use, and conservation. This observation is in agreement with publications of many scientists [9, 45–50] who studied plant domestication in many parts of the world. Many species that are on the red list of Benin, threatened species like *C. bonduc*, would have completely disappeared, if they have not been domesticated by local communities. Similar results were reported in Cameroon and Madagascar, where domestication of *Prunus africana* Hook. f. has protected the species against extinction because of excessive bark harvesting for export for medicinal use [48, 51].

## 5. Conclusion

This study showed that domestication is actively being carried out in the rural areas of Benin and appears as a one of the most appropriate practices for developing the diversity, increasing its use and conserving agricultural biodiversity *in situ*. The process follows different steps which can be deliberately organised into three, four, or six steps. The results highlighted the role that gender (men and women) plays in plant domestication and revealed that food security and health, two vital needs of human being, are the main motives behind adoption and cultivation of wild species. Thanks to local communities' efforts, experiences, and innovations, plant genetic diversity is being developed, preserved, and sustainably used. Unfortunately, several factors limit full success of farmers' initiatives: limited knowledge of plant reproductive biology, plant diseases and pests' complex, climate variability and its impact on biodiversity, and so forth. Scientific investigations on major constraints to plant domestication are needed. We recommend that multidisciplinary research focusing on individual plant species (leafy vegetables, herbs, fruits, etc.) be conducted to better understand the influence of the domestication on the evolution of the species. Further baseline studies are needed on the uses and values of the species under domestication by the local communities throughout west Africa.

## Acknowledgments

This research was sponsored by Bioversity International. The authors are grateful to Dr. Mauricio Bellon, Director of Diversity for Livelihood Programme at Bioversity International, who has approved the project concept note and has provided useful scientific guidance for the implementation of the work. They are also grateful to Dr. H. Yédomonhan (National Herbarium, Department of Botany and Plant biology, University of Abomey-Calavi) and to Mr. Ph. Akodji (research assistants at IRDCAM) for their technical assistance during the survey. They express their sincere thanks to all the farmers they met or were interviewed during the surveys.

## References

[1] M. Pourkheirandish and T. Komatsuda, "The importance of barley genetics and domestication in a global perspective," *Annals of Botany*, vol. 100, no. 5, pp. 999–1008, 2007.

[2] J. R. Harlan, *Crops and Man*, American Society of Agronomy, Crop Science Society of America, Madison, Wis, USA, 2nd edition, 1992.

[3] D. Zohary and M. Hopf, *Domestication of Plants in the Old World*, Clarendon Press, Oxford, UK, 1993.

[4] M. Sweeney and S. McCouch, "The complex history of the domestication of rice," *Annals of Botany*, vol. 100, no. 5, pp. 951–957, 2007.

[5] J. Ross-Ibarra, P. L. Morrell, and B. S. Gaut, "Plant domestication, a unique opportunity to identify the genetic basis of adaptation," *Proceedings of the National Academy of Sciences of the United States of America*, vol. 104, no. 1, pp. 8641–8648, 2007.

[6] A. Casas, A. Otero-Arnaiz, E. Pérez-Negrón, and A. Valiente-Banuet, "In situ management and domestication of plants in Mesoamerica," *Annals of Botany*, vol. 100, no. 5, pp. 1101–1115, 2007.

[7] B. Pickersgill, "Domestication of plants in the Americas: insights from Mendelian and molecular genetics," *Annals of Botany*, vol. 100, no. 5, pp. 925–940, 2007.

[8] D. A. Vaughan, E. Balázs, and J. S. Heslop-Harrison, "From crop domestication to super-domestication," *Annals of Botany*, vol. 100, no. 5, pp. 893–901, 2007.

[9] T. S. Msuya, M. A. Mndolwa, and C. Kapinga, "Domestication: an indigenous method in conserving plant diversity on farmlands in west Usambara Mountains, Tanzania," *African Journal of Ecology*, vol. 46, supplement 1, pp. 74–78, 2008.

[10] S. J. Midgley and J. W. Turnbull, "Domestication and use of Australian acacias: case studies of five important species," *Australian Systematic Botany*, vol. 16, no. 1, pp. 89–102, 2003.

[11] H. D. Mignouna and A. Dansi, "Yam (Dioscorea ssp.) domestication by the Nago and Fon ethnic groups in Benin," *Genetic Resources and Crop Evolution*, vol. 50, no. 5, pp. 519–528, 2003.

[12] P. Vernier, G. C. Orkwor, and A. R. Dossou, "Studies on yam domestication and farmers' practices in Benin and Nigeria," *Outlook on Agriculture*, vol. 32, no. 1, pp. 35–41, 2003.

[13] Y. Bai and P. Lindhout, "Domestication and breeding of tomatoes: what have we gained and what can we gain in the future?" *Annals of Botany*, vol. 100, no. 5, pp. 1085–1094, 2007.

[14] T. Sang and S. Ge, "The puzzle of rice domestication," *Journal of Integrative Plant Biology*, vol. 49, no. 6, pp. 760–768, 2007.

[15] E. De Caluwé, S. De Smedt, A. E. Assogbadjo, R. Samson, B. Sinsin, and P. Van Damme, "Ethnic differences in use value and use patterns of baobab (Adansonia digitata L.) in northern Benin," *African Journal of Ecology*, vol. 47, no. 3, pp. 433–440, 2009.

[16] A. Dansi, A. Adjatin, H. Adoukonou-Sagbadja et al., "Traditional leafy vegetables in Benin: folk nomenclature, species under threat and domestication," *Acta Botanica Gallica*, vol. 156, no. 2, pp. 183–199, 2009.

[17] H. Adoukonou-Sagbadja, A. Dansi, R. Vodouhe, and K. Akpagana, "Indigenous knowledge and traditional conservation of Fonio millet (*Digitaria* exilis Stapf, *Digitaria* iburua Stapf) in Togo," *Biodiversity and Conservation*, vol. 15, pp. 2379–2395, 2006.

[18] A. Dansi, H. Adoukonou-Sagbadja, and R. Vodouhè, "Diversity, conservation and related wild species of Fonio millet (*Digitaria* spp.) in the northwest of Benin," *Genetic Resources and Crop Evolution*, vol. 57, no. 6, pp. 827–839, 2010.

[19] S. Adam and M. Boko, Editions of Bright / EDICEF, Benin, p. 96, 1993.

[20] A.C. Adomou, *Vegetation patterns and environmental gradients in Benin: implications for biogeography and conservation*, Ph.D. thesis, Wageningen University, Wageningen, The Netherlands, 2005.

[21] A. Akoègninou, W. J. van der Burg, and L. J. G. van der Maesen, Eds., *Flore Analytique de Bénin*, Backhuys Publishers, Leiden, The Netherlands, 2006.

[22] C. Almekinders and W. de Boef, *Encouraging Diversity. The Conservation and Development of Plant Genetic Resources*, Intermediate Technology Publication, London, UK, 2000.

[23] P. Howard-Borjas and W. Cuijpers, "Gender relations in local plant genetic resource management and conservation," in *Biotechnology, in Encyclopedia for Life Support Systems*, H. W. Doelle and E. DaSilva, Eds., EOLSS Publishers, Cambridge, UK, 2002.

[24] H. D. Tuan, N. N. Hue, B. R. Sthapit, and D. I. Jarvis, *On-Farm Management of Agricultural Biodiversity in Vietnam*, International Plant Genetic Resources Institute, Rome, Italy, 2003.

[25] M. K. Anderson, "Pre-agricultural plant gathering and management," in *Encyclopaedia of Plant and Crop Science*, R. Goodman, Ed., pp. 1055–1060, Marcel Dekker, New York, 2004.

[26] SAS, *SAS User's Guide: Statistics. Release 8.02*, SAS Institute, Cary, NC, USA, 1996.

[27] F. J. Rohlf, *NTSYS-pc Version 2.2: Numerical Taxonomy and Multivariate Analysis System*, Exeter Software, New York, NY, USA, 2000.

[28] P. Jaccard, "Nouvelles recherches sur la distribution florale," *Bulletin De La Société Vaudoise Des Sciences Naturelles*, vol. 44, pp. 223–270, 1908.

[29] P. H. A. Sneath and R. O. Sokal, *Numerical Taxonomy*, Freeman, San Francisco, Calif, USA, 1973.

[30] D. L. Swofford and G. J. Olsen, "Phylogeny reconstruction," in *Molecular systematics*, D. M. Hillis and C. Moritz, Eds., pp. 411–501, Sinauer Associates, Sunderland, Mass, USA, 1990.

[31] R. Dumont, A. Dansi, P. Vernier, and J. Zoundjihékpon, *Biodiversity and Domestication of Yams in West Africa. Traditional Practices Leading to Dioscorea Rotundata Poir. Edité par Collection Repères*, CIRAD, 2005.

[32] S. Appa Rao, C. Bounphanousay, J. M. Schiller, A. P. Alcantara, and M. T. Jackson, "Naming of traditional rice varieties by farmers in the Lao PDR," *Genetic Resources and Crop Evolution*, vol. 49, no. 1, pp. 83–88, 2002.

[33] F. Mekbib, "Infra-specific folk taxonomy in sorghum (*Sorghum bicolor* (L.) Moench) in Ethiopia: folk nomenclature, classification, and criteria," *Journal of Ethnobiology and Ethnomedicine*, vol. 3, article 38, 2007.

[34] E. A. Hildebrand, "Motives and opportunities for domestication: an ethno-archaeological study in southwest Ethiopia," *Journal of Anthropological Archaeology*, vol. 22, pp. 358–375, 2003.

[35] N. Scarcelli, S. Tostain, C. Mariac et al., "Genetic nature of yams (Dioscorea sp.) domesticated by farmers in Benin (West Africa)," *Genetic Resources and Crop Evolution*, vol. 53, no. 1, pp. 121–130, 2006.

[36] T. Davis and R. Bye, "Ethnobotany and progressive domestication of Jaltomata spp (Solanaceae) in Mexico and Central America," *Economic Botany*, vol. 36, pp. 225–241, 1982.

[37] S. Zarate, "Ethnobotany and domestication process of Leucaena in Mexico," *Journal of Ethnobiology*, vol. 19, pp. 1–23, 1999.

[38] E. Arellano and A. Casas, "Morphological variation and domestication of *Escontria chiotilla* (Cactaceae) under silvicultural management in the Tehuacán Valley, Central Mexico," *Genetic Resources and Crop Evolution*, vol. 50, no. 4, pp. 439–453, 2003.

[39] P. Gepts, "Crop domestication as a long-term selection experiment," *Plant Breeding Reviews*, vol. 24, pp. 1–44, 2004.

[40] C. Mapes, J. Caballero, E. Espitia, and R. A. Bye, "Morphophysiological variation in some Mexican species of vegetable Amaranthus: evolutionary tendencies under domestication," *Genetic Resources and Crop Evolution*, vol. 43, no. 3, pp. 283–290, 1996.

[41] P. Colunga-Garcíamarín, E. Estrada-Loera, and F. May-Pat, "Patterns of morphological variation, diversity, and domestication of wild and cultivated populations of Agave in Yucatan, Mexico," *American Journal of Botany*, vol. 83, no. 8, pp. 1069–1082, 1996.

[42] S. Zarate, N. Perez-Nasser, and A. Casas, "Genetics of wild and managed populations of *Leucaena esculenta* subsp. Esculenta (Fabaceae: Mimosoideae) in La Montana of Guerrero, Mexico," *Genetic Resources and Crop Evolution*, vol. 52, pp. 941–957, 2005.

[43] A. Carmona and A. Casas, "Management, phenotypic patterns and domestication of Polaskia chichipe (Cactaceae) in the Tehuacán Valley, Central Mexico," *Journal of Arid Environments*, vol. 60, no. 1, pp. 115–132, 2005.

[44] D. Q. Fuller, "Contrasting patterns in crop domestication and domestication rates: recent archaeobotanical insights from the old world," *Annals of Botany*, vol. 100, no. 5, pp. 903–924, 2007.

[45] A. R. S. Kaoneka, *Land use in the West Usambara Mountains: analysis of ecological and socio-economic aspects with special reference to forestry*, Ph.D. thesis, Department of Forestry, Agricultural University of Norway, Oslo, Norway, 1993.

[46] G. C. Kajembe, "Indigenous management systems as a basis for community forestry in Tanzania. A case study of Dodoma Urban, and Lushoto Districts," Tropical Resource Management Series 6, Wageningen Agricultural University, Wageningen, The Netherlands, 1994.

[47] A. Zemede and N. Ayele, "Home gardens in Ethiopia: characteristics and plant diversity," *Ethiopian Journal of Science*, vol. 18, pp. 235–266, 1995.

[48] I. Dawson, "Prunus africana: how agroforestry can help save an endangered medicinal tree," *ICRAF, Agroforestry Today*, vol. 9, pp. 6–9, 1997.

[49] J. F. Kessy, "Conservation and utilization of natural resources in the East Usambara forest reserves: conservational views and local perspectives," Tropical Resource Management Papers 18, Wageningen Agricultural University, Wageningen, The Netherlands, 1998.

[50] O. Tibe, D. M. Modise, and K. K. Mogotsi, "Potential for domestication and commercialization of Hoodia and Opuntia species in Bots," *African Journal of Biotechnology*, vol. 7, no. 9, pp. 1199–1203, 2008, Proceedings of the International Plant Genetic Resources Institute Symposium, Hanoi, Vietnam, December 2001.

[51] A. B. Cunningham and F. T. Mbenkum, "Sustainability of harvesting prunus Africana Bark in Cameroon: a medicinal plant in domestication," *Biodiversity and Conservation*, vol. 6, pp. 1409–1412, 1993.

# The c.1460C>T Polymorphism of *MAO-A* Is Associated with the Risk of Depression in Postmenopausal Women

**R. Słopień,[1] A. Słopień,[2] A. Różycka,[3] A. Warenik-Szymankiewicz,[1] M. Lianeri,[3] and P. P. Jagodziński[3]**

[1] *Department of Gynecological Endocrinology, Poznan University of Medical Sciences, Ul. Polna 33, 60-535 Poznan, Poland*
[2] *Department of Child and Adolescent Psychiatry, Poznan University of Medical Sciences, Ul. Szpitalna 27/33, 60-572 Poznan, Poland*
[3] *Department of Biochemistry and Molecular Biology, Poznan University of Medical Sciences, Ul. Święcickiego 6, 60-781 Poznan, Poland*

Correspondence should be addressed to R. Słopień, asrs@wp.pl and P. P. Jagodziński, pjagodzi@am.poznan.pl

Academic Editors: S. Mastana and J. H. Zhao

*Objective.* The aim of the study was an evaluation of possible relationships between polymorphisms of serotoninergic system genes and the risk of depression in postmenopausal women. *Methods.* We studied 332 women admitted to our department because of climacteric symptoms. The study group included 113 women with a diagnosis of depressive disorder according to the Hamilton rating scale for depression; the controls consisted of 219 women without depression. Serum 17$\beta$-estradiol concentrations were evaluated using radioimmunoassay, while polymorphisms in serotoninergic system genes: serotonin receptors 2A (*HTR2A*), 1B (*HTR1B*), and 2C (*HTR2C*); tryptophan hydroxylase 1 (*TPH1*) and 2 (*TPH2*), and monoamine oxidase A (*MAO-A*) were evaluated using polymerase chain reaction-restriction. *Results.* We found that the 1460T allele of *MAO-A* c.1460C>T (SNP 1137070) appeared with a significantly higher frequency in depressed female patients than in the control group ($P = 0.011$) and the combined c.1460CT + TT genotypes were associated with a higher risk of depression ($P = 0.0198$). Patients with the 1460TT genotype had a significantly higher 17$\beta$-estradiol concentration than patients with the 1460CT genotype ($P = 0.0065$) and 1460CC genotype ($P = 0.0018$). *Conclusions.* We concluded that depression in postmenopausal women is closely related to the genetic contribution of *MAO-A*.

## 1. Introduction

The central serotoninergic system has been implicated in the pathophysiology of a number of neuropsychiatric disorders such as mood disorders, substance abuse, or alcoholism [1, 2]. Well-studied components of this system are the serotonin (5-HT) receptors 2A (5-HT2A), 1B (5-HT1B), and 2C (5-HT2C), and the key regulators of 5-HT metabolism: tryptophan hydroxylase 1 and 2 (TPH1 and TPH2) and monoamine oxidase A (*MAO-A*) (Figure 1). The receptors 5-HT2A, 5-HT2C, and 5-HT1B are members of a family of receptors linked to guanine-nucleotide-binding proteins (G-protein-coupled receptor; GPCR) expressed on the cell body and dendrites of serotoninergic neurons in the brain [3]. The receptors 5-HT2A and 5-HT2C are the main excitatory receptor subtypes among the GPCRs for 5-HT, whereas the 5-HT1B receptor is thought to act as a nerve terminal autoreceptor, inhibiting the release of 5-HT. After acting at its receptor, 5-HT is metabolized by *MAO-A* (Figure 1), and therefore MAO activity may play a critical role in the regulation of the serotoninergic system and in the pathogenesis of depressive disorders [4].

The gene encoding 5-HT2A (*HTR2A*) is considered to be a candidate gene for depression. Genetic association has been reported between the c.102C>T polymorphism in the *HTR2A* gene and depression, as well as suicidal behavior in patients with mood disorders and schizophrenia [5–7], although several studies have failed to replicate these findings [8]. It is likely that other serotoninergic genes are involved in gene-environment interactions related to depression. Among them, the gene encoding the 5-HT1B receptor (*HTR1B*) has been correlated to attempted suicide in patients with major depression, because altered postmortem 5-HT1B receptor binding was found to be associated with suicide in some of

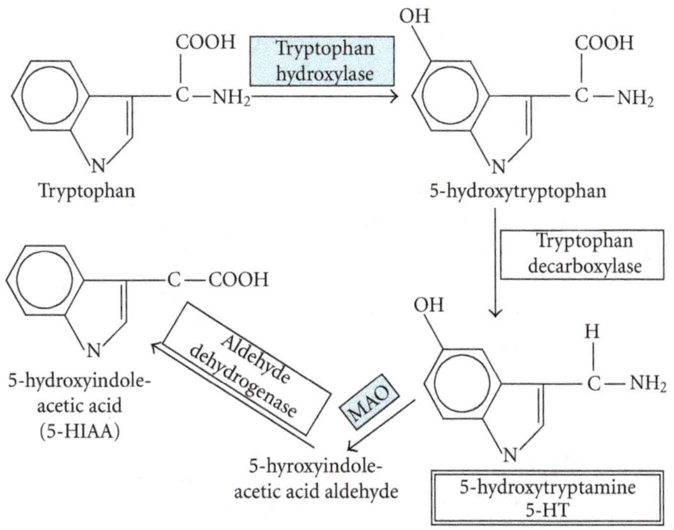

FIGURE 1: Contribution of tryptophan hydroxylase and MAO in serotonin metabolism. Tryptophan hydroxylase catalyzes the monooxygenation of tryptophan to 5-hydroxytryptophan, which is subsequently decarboxylated to form serotonin (5-hydroxytryptamine; 5-HT). Monoamine oxidase (MAO) catalyzes the oxidative deamination of 5-HT to the corresponding aldehyde. This is followed by oxidation by aldehyde dehydrogenase to 5-HIAA, the indole acetic acid derivative.

the studies [9]. A common c.861 G>C polymorphism of the *HTR1B* gene was identified in the coding region of the gene, and major depression appears to be associated with this *locus* [10]. A strong association between suicide and receptor genes of the serotoninergic system has presented evidence for yet another 5-HT receptor gene located on human chromosome Xq24, the 5-HT2C receptor gene (*HTR2C*) [11]. This receptor mediates the release of dopamine (DA) in the brain and can cause anxiety, depression, and compulsive behaviors in human subjects due to the mechanism of rapid downregulation by serotonin. The structural variant c.68 G>C of the *HTR2C* gene that gives rise to a cysteine-to-serine substitution in the *N* terminal extracellular domain of the receptor protein (Cys23Ser) has recently been found to be significantly associated with female suicide victims [12].

Tryptophan hydroxylase (TPH) is the rate-limiting enzyme in the biosynthesis of 5-HT and thus has a major function in regulating the serotoninergic system [13]. Studies of the human brain have revealed two isoforms of the gene coding for tryptophan hydroxylase 1 and 2, termed *TPH1* and *TPH2*. One of them, the *TPH2* gene, is of interest because this tryptophan hydroxylase 2 isoform regulates the biosynthetic pathway of 5-HT in the serotoninergic neurons of *raphe nuclei* and has been implicated in the pathogenesis of major depressive disorder and the mechanism of antidepressant action [14].

Investigations of the functional effects of genetic variations in the *TPH2* gene have demonstrated that a haplotype of 3 SNPs within the gene promoter (−703G>T, −473T>A, and 90 A>G) influenced transcription in human cell lines. Similarly, the *TPH2* c.1077G>A polymorphism within the coding region (Pro312Pro; SNP 7305115) increased expression levels of the gene in the human pons, containing the dorsal and median *raphe nuclei* [15]. This SNP has previously demonstrated an association with major depression and suicide [16].

Monoamine oxidase (MAO) is an important enzyme associated with the metabolism of biogenic amines and neurotransmitters, including norepinephrine (NE), DA, and 5-HT. Two forms of the enzyme, MAOA and MAOB, are found in the human brain [17]. Of the two, MAOA exhibits a higher affinity for 5-HT and NE, whereas DA is preferentially metabolized by both forms of the enzyme. The MAOA gene (*MAO-A*) has been mapped to the short arm of the *X* chromosome; thus, functional polymorphisms of this *locus* are expected to manifest in a sex-specific fashion. The human *MAO-A* contains a variable-number tandem repeat (VNTR) polymorphism in its promoter region that may alter the transcriptional efficiency of MAOA expression [18]. Most of the other known *MAO-A* polymorphisms either affect intronic sequences or introduce a silent change in the open-reading frame (i.e., the *Eco*RV polymorphism and *Fnu*4HI polymorphism, SNPs: 1137070 and 6323, resp.). These variants are unlikely to affect MAO function, although they may be in disequilibrium with other, as yet unidentified, functional variants. The primary role of MAOA in regulating monoamine turnover, and hence ultimately influencing levels of NE, DA, and 5-HT, indicates that its gene is a highly plausible candidate for affecting individual differences in the manifestation of psychological traits and psychiatric disorders [19, 20]. For example, several studies indicate that the *MAO-A* gene may be involved in the pathogenesis of depression and major depressive disorder [21, 22].

However, genetic factors can modulate the risk for depression by influencing monoaminergic activity in a sexually dimorphic manner. Because the *MAO-A* gene is X-linked, males are hemizygous at this *locus*, whereas females are homozygous or heterozygous. Moreover, the *MAO-A Eco*RV polymorphism was found to be associated with depression in males but not in females [23, 24]. A sexually dimorphic pattern of genetic susceptibility to obsessive-compulsive disorder (OCD) may also be present [25–27]. Statistically significant associations were observed between the alleles for the *MAO-A EcoRV* polymorphism and levels of MAO activity in human male fibroblast lines [28]. The functional analyses of this genetic variation have revealed the existence of the high-activity T allele and the low-activity C allele of the *MAO-A EcoRV* polymorphism.

Individuals with a high-activity *MAO-A* genotype would be expected to have greater serotonin turnover (Figure 1). High plasma MAO activity, however, has been found to be significantly correlated with testosterone levels in men and with 17$\beta$-estradiol levels in women, with high testosterone or 17$\beta$-estradiol levels leading to low plasma MAO activity [29]. Women are more likely than men to develop affective disorders, including depression, and this risk increases after menopause, when estrogen production in the ovaries ends [30–33]. We hypothesized that the high-activity *MAO-A* genotype would be associated with depression in postmenopausal women. To test this hypothesis, we evaluated serum 17$\beta$-estradiol concentrations and genotyped the *MAO-A EcoRV* polymorphism in a group of healthy ($n = 219$) and depressed ($n = 113$) postmenopausal women. We also used a case-control study design in the same individuals to investigate a possible association with a susceptibility to depression of the SNPs of five other serotoninergic system genes: *5HTR2A* (SNP 6313), *5HTR1B* (SNP 6296), *HTR2C* (SNP6318), *TPH1* (SNP 1800532), and *TPH2* (SNP 7305115).

## 2. Patients and Methods

*2.1. Patients.* We studied three hundred and thirty-two postmenopausal women, aged 42–67, who were admitted to the Department of Gynecological Endocrinology, Poznan University of Medical Sciences, because of climacteric complaints. All postmenopausal women had their last menstrual flow more than 1 year before the study.

All patients were assessed with the Hamilton rating scale for depression (HRSD) and divided into two groups: diagnosed of depressive disorder (113 women) and without depression (219 women served as the control group). According to HRSD, mild depression was defined as a score more than 7 and less or equal to 17, and moderate depression was defined as a score more than 17 and less or equal to 25. The 113 women with depression included 82 women with mild depression and 31 women with moderate depression. None of the examined women were on hormone replacement therapy (HRT) or on psychotropic drugs.

All experiments were carried out after obtaining informed consent from all the participating women. The local Ethic Review Committee of Poznan University of Medical Sciences approved the study protocol.

*2.2. Blood Samples and Measurements of 17β-Estradiol.* A blood sample was collected from each study participant. 17$\beta$-estradiol serum concentrations were quantified using radioimmunoassay (RIAs). The intraassay and interassay coefficients of variation (CV) were 1.2–3.3%, and 2.0–5.6%, respectively.

*2.3. Genotyping by RFLP.* Genomic DNA was prepared from sodium-versenate-($EDTANa_2$-) treated blood samples. Genotyping for polymorphisms in *5HTR2A* c.102C>T (SNP 6313), *5HTR1B* c.861G>C (SNP 6296), *HTR2C* c.68G>C (SNP6318), *MAO-A* c.1460C>T (SNP 1137070), *TPH1* 218C>A (SNP 1800532), and *TPH2* c.1077A>G (SNP 7305115) was determined by polymerase chain reaction-restriction fragment length polymorphism (PCR-RFLP) assay, using the appropriate restriction enzymes. The digested PCR products were resolved on a 2% agarose gel and stained with ethidium bromide for visualization under UV light.

*2.4. Statistical Analysis.* The genotype and allele frequencies of all analyzed polymorphisms were compared between the group of postmenopausal women with depression and without depression using a case-control study design. Significance was evaluated by the Fisher exact test. Odds ratios (ORs) and 95% confidence intervals (CIs) were estimated using the GraphPad (Instant, USA) program. An online (http://ihg2.helmholtz-muenchen.de/cgi-bin/hw/hwa1.pl) program for deviation from the Hardy-Weinberg equilibrium was applied. Comparison of the 17$\beta$-estradiol serum concentrations between the different genotype groups was performed with the use of the Kruskal-Wallis test.

In all cases, $P < 0.05$ was considered statistically significant.

## 3. Results

In the present study, 113 women with diagnosed mild or moderate depression (patients) and 219 women without depression (healthy controls) were genotyped for polymorphisms in six serotonergic candidate genes: *HTR2A*, *HTR1B*, *HTR2C*, *TPH1*, *TPH2*, and *MAO-A*. The data for allele frequencies and genotype distribution of these polymorphisms for the patients and the controls are presented in Table 1. Distribution of these polymorphisms was consistent with the Hardy-Weinberg equilibrium in the group with depressive disorders, as well as in the control group.

No significant differences were observed in the frequency of either the *5HTR2A* c.102C>T, *5HTR1B* c.861G>C, *HTR2C* c.68G>C, *TPH1* 218C>A, or *TPH2* c.1077A>G genotypes or alleles between the patients and the controls (Table 1).

The *MAO-A* c.1460C>T polymorphism in the patient group demonstrated a significant difference when compared to the control group. The frequency of the T allele in the group of women with depression was higher than that in the control group ($P = 0.011$) (Table 1). The frequency of the homozygous c.1460TT genotype in these groups reached 14% and 8%, respectively (Table 1). Although the frequency of the heterozygous c.1460CT was higher in women with depression (50%) than in women without depression (42%), this was not significant (Table 1).

We also undertook the test for association, using c.1460T as a risk allele (Table 1). The OR and CI were calculated for

Table 1: The genotype distribution between postmenopausal women with (*w*) and without (*w/o*) depression.

| Polymorphism | *n* | Genotype and allele distribution absolute number (frequency) | | | | | Allele *P* Value | Genotype Odds ratio (95% CI); *P* Value |
|---|---|---|---|---|---|---|---|---|
| | | CC | CT | *TT* | C | T | | |
| *5HTR2A* c. 102C>T (SNP 6313) | w depression | 50 | 44 | 19 | 144 | 82 | $P = 0.735$ | 0.769 (0.485–1.219)[a]; $P = 0.2634$[b] |
| | Total 113 | (0.44) | (0.39) | (0.17) | (0.64) | (0.36) | | |
| | w/o depression | 83 | 107 | 29 | 273 | 165 | | |
| | Total 219 | (0.38) | (0.49) | (0.13) | (0.62) | (0.38) | | |
| | | GG | GC | **CC** | G | C | | |
| *5HTR1B* c.861G>C (SNP 6296) | w depression | 62 | 44 | 7 | 168 | 58 | $P = 0.635$ | 1.094 (0.693–1.728)[a]; $P = 0.700$[b] |
| | Total 113 | (0.55) | (0.39) | (0.06) | (0.74) | (0.26) | | |
| | w/o depression | 125 | 83 | 11 | 333 | 105 | | |
| | Total 219 | (0.57) | (0.38) | (0.05) | (0.76) | (0.24) | | |
| | | GG | GC | **CC** | G | C | | |
| *5HTR2C* c. 68G>C (SNP 6318) | w depression | 78 | 30 | 5 | 186 | 40 | $P = 1.000$ | 0.996 (0.610–1.628)[a]; $P = 0.989$[b] |
| | Total 113 | (0.69) | (0.27) | (0.04) | (0.82) | (0.18) | | |
| | w/o depression | 151 | 57 | 11 | 359 | 79 | | |
| | Total 219 | (0.69) | (0.26) | (0.05) | (0.82) | (0.18) | | |
| | | CC | CT | *TT* | C | T | | |
| *MAO-A* c.1460C>T (SNP 1137070) | w depression | 41 | 56 | 16 | 138 | 88 | $P = 0.011$* | 1.772 (1.112–2.825)[a]; $P = 0.0198$*[b] |
| | Total 113 | (0.36) | (0.50) | (0.14) | (0.61) | (0.39) | | |
| | w/o depression | 110 | 91 | 18 | 311 | 127 | | |
| | Total 219 | (0.50) | (0.42) | (0.08) | (0.71) | (0.29) | | |
| | | CC | CA | *AA* | C | A | | |
| *TPH1* 218C>A (SNP 1800532) | w depression | 44 | 49 | 20 | 137 | 89 | $P = 0.801$ | 1.014 (0.637–1.615)[a]; $P = 0.9533$[b] |
| | Total 113 | (0.39) | (0.43) | (0.18) | (0.61) | (0.39) | | |
| | w/o depression | 86 | 99 | 34 | 271 | 167 | | |
| | Total 219 | (0.39) | (0.45) | (0.16) | (0.62) | (0.38) | | |
| | | GG | GA | *AA* | G | A | | |
| *TPH2* c.1077G>A (SNP 7305115) | w depression | 45 | 51 | 17 | 141 | 85 | $P = 0.737$ | 0.870 (0.546–1.386)[a]; $P = 0.5573$[b] |
| | Total 113 | (0.40) | (0.45) | (0.15) | (0.62) | (0.38) | | |
| | w/o depression | 80 | 106 | 33 | 266 | 172 | | |
| | Total 219 | (0.37) | (0.48) | (0.15) | (0.61) | (0.39) | | |

The women's groups were classified based on the severity of depression assessed by the Hamilton rating scale for depression: w/o depression (0–7), mild depression (8–17), and moderate depression (18–25). The 113 women with depression included 82 women with mild depression and 31 women with moderate depression.

[a]The Odds Ratio was calculated for patients homozygous or heterozygous carrying risk allele *vs.* homozygous.

[b]Fisher exact test was used for comparison of patients with depression versus patients without depression.

each genotype and compared with the homozygous values for the genotype of higher frequency among controls, which was set as the reference genotype. When the c.1460CC genotype was used as the reference, the combined c.1460CT + TT genotypes were associated with higher risk of depression (OR = 1.772; CI = 1.112–2.825, $P = 0.0198$).

17$\beta$-estradiol serum concentration was not associated with depression in postmenopausal women. No significant differences in 17$\beta$-estradiol serum concentration were observed between the healthy group of postmenopausal women without depression ($r = 108.1$ pg/mL), the mild depression ($r = 102,1$ pg/mL), or the moderate depression ($r = 74.2$ pg/mL) patient subgroups.

Comparison of the 17$\beta$-estradiol serum concentration between patients with different genotypes of the *MAO-A* c.1460C>T polymorphism yielded a significant difference between the patients with wild-type or heterozygous genotypes (1460CC or 1460CT, resp.) and those with the homozygous genotype variant (1460TT) (Kruskall-Wallis, $\chi^2 = 11.960$, $P = 0.0025$). In the subgroup of the homozygous 1460TT genotype, a significantly higher 17$\beta$- estradiol concentration was observed ($r = 130.03$ pg/mL) as compared to the 1460CT ($r$ = 88.64 pg/mL; $P$ = 0.0065) or 1460CC ($r$ = 83.173 pg/mL; $P$ = 0.0018) genotypes.

No significant correlations between plasma levels of 17$\beta$-estradiol and the other SNP genotype variants were observed.

## 4. Discussion

We found a significant contribution of the *MAO-A* c.1460C>T polymorphic variant to depression in postmenopausal women. An association between the *MAO-A* CT and TT genotypes and depression in postmenopausal women has been evidenced. A C-to-T substitution at the third base of codon470 in exon14 (Asp470Asp) of *MAO-A* results in *Eco*RV restriction length polymorphism. To date, significant associations have been observed between the T allele for the *MAO-A EcoRV* polymorphism and high MAOA activity, with levels ranging more than 50-fold among control subjects, as measured in cultured skin fibroblasts [28]. We provide evidence for an association between depression in postmenopausal women and the T allele of the *MAO-A* gene, previously linked to high MAOA enzymatic activity [26–28]. These findings are in agreement with the well-established action of MAOA inhibitors as antidepressants.

High-activity variants of the *MAO-A* gene have been also associated with OCD. This association, however, has been detected among OCD males with comorbid major depressive disorder, more likely having the high-activity T allele of the *MAO-A* gene than controls [23–25]. On the other hand, females with OCD were more frequently homozygous for the low-activity C allele of the *MAO-A EcoRV* variant compared to controls, with this allele also more frequent in female patients than in controls [26]. Considering that the *MAO-A* gene is localized to the X chromosome, our results may support a sexually dimorphic association between the gender groups as reported by others and the occurrence of alleles of the *MAO-A* gene polymorphism. These discrepancies, however, might also be due to the different populations that have been studied.

The fairly low relative risk of depression observed in postmenopausal women having the CT or TT genotypes (OR = 1.772) may be explained by a predisposition of patients in the postmenopausal period to depression by a variant that may not be in the *MAO-A* gene but within a closely linked, yet-to-be-discovered susceptibility gene. The most likely explanation, however, is that the linked, functional, *MAO-A* high-activity variant has low penetrance or imposes a risk on only a subset of postmenopausal women with depression. This theory is consistent with previous reports of a beneficial effect of MAOA inhibitors in certain OCD cases, although, in general, treatment of OCD with MAOA inhibitors does not seem to be as effective as treatment with selective serotonin reuptake inhibitors (SSRIs) [34]. An association between high-activity variants of the *MAO-A* gene and depression in postmenopausal women suggests, however, that some of them will also respond well to MAO inhibitors.

Estrogen likely promotes serotoninergic neurotransmission by influencing release, metabolism, reuptake, or synthesis [35]. It has been reported that women with a high-activity *MAO-A* genotype differ from men and from women with a low-activity genotype, which suggests that women with a high-activity *MAO-A* genotype may drive some previously reported sex differences in serotoninergic mechanisms [36]. Among women, but not among men, the concentration of the major serotonin metabolite, 5-hydroxyindoleacetic acid (Figure 1), was greater in those with a high-activity genotype than in those with a low-activity genotype [36]. For healthy women, plasma MAO activity is lowest at the time of ovulation, when 17$\beta$-estradiol production is greatest, but MAO activity increases during the luteal phase, when progesterone is secreted [30]. The estrogen deficiency of postmenopausal women who have not used HRT results in higher MAO activity. This may have an influence on the risk of depressive disorders in postmenopausal women [31–33]. Nevertheless, the genetic association between the *MAO-A EcoRV* polymorphism and depression has never been analyzed in postmenopausal women. Several lines of evidence support the hypothesis of an antidepressant effect of estrogens exerted via inhibition of the MAO pathway in women [37]. High-dose estrogen significantly decreased MAOA activity in the hypothalamus and amygdala in adult female rats, with no significant changes in MAOB activity in these areas of the brain [38].

In a human neuroblastoma cell line with transfected cDNA of the estrogen receptor (SK-ER3), estrogen receptor activation by a physiological concentration of 17$\beta$-estradiol was correlated with a marked decrease in MAOA activity [39]. We did not determine MAOA activity in the blood samples studied, and although concentrations of 5-HT and 17$\beta$-estradiol were measured, plasma levels of either 5-HT or 17$\beta$-estradiol did not differ between particular groups of postmenopausal women. However, the Kruskal-Wallis test revealed that the high-activity T allele of the *MAO-A EcoRV* polymorphism is markedly associated with a higher 17$\beta$-estradiol concentration, thus favoring the hypothesis of the presence of a functional link between estrogen and MAOA activity in human cells of neural origin. In fact, a reverse causality in the relationship between high 17$\beta$-estradiol levels and depression disorders may have a protective role in the homozygous 1460TT *MAO-A EcoRV* subgroup of the studied postmenopausal women. On the other hand, the concentration of 17$\beta$-estradiol may not be sufficient in preventing depression in postmenopausal women that are not taking HRT [40].

We concluded that the 1460T allele of *MAO-A* appeared with a significantly higher frequency in depressed female patients than in the control group, and the patients with the 1460CT + TT genotypes showed an increased risk of depression, which indicates that the 1460T allele of *MAO-A* may be a risk factor for depression in postmenopausal women.

## Acknowledgment

This paper is supported by Grant no. 50305-01109136-12261-08039 from the Polish Ministry of Scientific Research and Information Technology.

## References

[1] S. N. Young and M. Leyton, "The role of serotonin in human mood and social interaction: insight from altered tryptophan levels," *Pharmacology Biochemistry and Behavior*, vol. 71, no. 4, pp. 857–865, 2002.

[2] S. Y. Huang, W. W. Lin, F. J. Wan et al., "Monoamine oxidase-A polymorphisms might modify the association between the dopamine D2 receptor gene and alcohol dependence," *Journal of Psychiatry and Neuroscience*, vol. 32, no. 3, pp. 185–192, 2007.

[3] H. S. Jørgensen, "Studies on the neuroendocrine role of serotonin," *Danish Medical Bulletin*, vol. 54, no. 4, pp. 266–288, 2007.

[4] X. Ni, T. Sicard, N. Bulgin et al., "Monoamine oxidase A gene is associated with borderline personality disorder," *Psychiatric Genetics*, vol. 17, no. 3, pp. 153–157, 2007.

[5] V. D. Khait, Y. Y. Huang, G. Zalsman et al., "Association of Serotonin 5-$HT_{2A}$ receptor binding and the T102C polymorphism in depressed and healthy caucasian subjects," *Neuropsychopharmacology*, vol. 30, no. 1, pp. 166–172, 2005.

[6] G. Zalsman, M. Patya, A. Frisch et al., "Association of polymorphisms of the serotonergic pathways with clinical traits of impulsive-aggression and suicidality in adolescents: a multicenter study," *World Journal of Biological Psychiatry*, vol. 12, no. 1, pp. 33–41, 2011.

[7] E. M. Peñas-Lledó, P. Dorado, M. C. Cáceres, A. de la Rubia, and A. Llerena, "Association between T102C and A-1438G polymorphisms in the serotonin receptor 2A (*5-HT2A*) gene and schizophrenia: relevance for treatment with antipsychotic drugs," *Clinical Chemistry and Laboratory Medicine*, vol. 45, no. 7, pp. 835–838, 2007.

[8] R. Y. Chen, P. Sham, E. Y. Chen et al., "No association between T102C polymorphism of serotonin-2A receptor gene and clinical phenotypes of Chinese schizophrenic patients," *Psychiatry Research*, vol. 105, no. 3, pp. 175–185, 2001.

[9] Y. Y. Huang, M. A. Oquendo, J. M. Friedman et al., "Substance abuse disorder and major depression are associated with the human 5-HT1B receptor gene (HTR1B) G861C polymorphism," *Neuropsychopharmacology*, vol. 28, no. 1, pp. 163–169, 2003.

[10] C. Fehr, N. Grintschuk, A. Szegedi et al., "The HTR1B 861G>C receptor polymorphism among patients suffering from alcoholism, major depression, anxiety disorders and narcolepsy," *Psychiatry Research*, vol. 97, no. 1, pp. 1–10, 2000.

[11] B. Lerer, F. Macciardi, R. H. Segman et al., "Variability of 5-HT2C receptor cys23ser polymorphism among European populations and vulnerability to affective disorder," *Molecular Psychiatry*, vol. 6, no. 5, pp. 579–585, 2001.

[12] A. Videtič, T. T. Peternelj, T. Zupanc, J. Balažic, and R. Komel, "Promoter and functional polymorphisms of HTR2C and suicide victims," *Genes, Brain and Behavior*, vol. 8, no. 5, pp. 541–545, 2009.

[13] D. J. Walther, J. U. Peter, S. Bashammakh et al., "Synthesis of serotonin by a second tryptophan hydroxylase isoform," *Science*, vol. 299, no. 5603, p. 76, 2003.

[14] H. Bach-Mizrachi, M. D. Underwood, S. A. Kassir et al., "Neuronal tryptophan hydroxylase mRNA expression in the human dorsal and median raphe nuclei: major depression and suicide," *Neuropsychopharmacology*, vol. 31, no. 4, pp. 814–824, 2006.

[15] F. Haghighi, H. Bach-Mizrachi, Y. Y. Huang et al., "Genetic architecture of the human tryptophan hydroxylase 2 Gene: existence of neural isoforms and relevance for major depression," *Molecular Psychiatry*, vol. 13, no. 8, pp. 813–820, 2008.

[16] L. Ke, Z. Y. Qi, Y. Ping, and C. Y. Ren, "Effect of SNP at position 40237 in exon 7 of the TPH2 gene on susceptibility to suicide," *Brain Research*, vol. 1122, no. 1, pp. 24–26, 2006.

[17] J. C. Shih and R. F. Thompson, "Monoamine oxidase in neuropsychiatry and behavior," *The American Journal of Human Genetics*, vol. 65, no. 3, pp. 593–598, 1999.

[18] B. H. Brummett, A. D. Krystal, I. C. Siegler et al., "Associations of a regulatory polymorphism of monoamine oxidase-A gene promoter (MAOA-uVNTR) with symptoms of depression and sleep quality," *Psychosomatic Medicine*, vol. 69, no. 5, pp. 396–401, 2007.

[19] J. Li, C. Kang, H. Zhang et al., "Monoamine oxidase A gene polymorphism predicts adolescent outcome of attention-deficit/hyperactivity disorder," *American Journal of Medical Genetics B*, vol. 144, no. 4, pp. 430–433, 2007.

[20] M. Preisig, F. Bellivier, B. T. Fenton et al., "Association between bipolar disorder and monoarnine oxidase a gene polymorphisms: results of a multicenter study," *The American Journal of Psychiatry*, vol. 157, no. 6, pp. 948–955, 2000.

[21] B. Gutiérrez, B. Arias, C. Gastó et al., "Association analysis between a functional polymorphism in the monoamine oxidase A gene promoter and severe mood disorders," *Psychiatric Genetics*, vol. 14, no. 4, pp. 203–208, 2004.

[22] T. G. Schulze, D. J. Müller, H. Krauss et al., "Association between a functional polymorphism in the monoamine oxidase A gene promoter and major depressive disorder," *American Journal of Medical Genetics B*, vol. 96, no. 6, pp. 801–803, 2000.

[23] L. Du, D. Bakish, A. Ravindran, and P. D. Hrdina, "MAO-A gene polymorphisms are associated with major depression and sleep disturbance in males," *NeuroReport*, vol. 15, no. 13, pp. 2097–2101, 2004.

[24] E. A. Tivol, C. Shalish, D. E. Schuback, Y. P. Hsu, and X. O. Breakefield, "Mutational analysis of the human MAOA gene," *American Journal of Medical Genetics C*, vol. 67, no. 1, pp. 92–97, 1996.

[25] M. Karayiorgou, C. Sobin, M. L. Blundell et al., "Family-based association studies support a sexually dimorphic effect of COMT and MAOA on genetic susceptibility to obsessive-compulsive disorder," *Biological Psychiatry*, vol. 45, no. 9, pp. 1178–1189, 1999.

[26] B. Camarena, C. Cruz, J. R. de la Fuente, and H. Nicolini, "A higher frequency of a low activity-related allele of the MAO-A gene in females with obsessive-compulsive disorder," *Psychiatric Genetics*, vol. 8, no. 4, pp. 255–257, 1998.

[27] C. Lochner, S. M. Hemmings, C. J. Kinnear et al., "Gender in obsessive-compulsive disorder: clinical and genetic findings," *European Neuropsychopharmacology*, vol. 14, no. 2, pp. 105–113, 2004.

[28] G. S. Hotamisligil and X. O. Breakefield, "Human monoamine oxidase A gene determines levels of enzyme activity," *The American Journal of Human Genetics*, vol. 49, no. 2, pp. 383–392, 1991.

[29] R. L. Sjöberg, F. Ducci, C. S. Barr et al., "A non-additive interaction of a functional MAO-A VNTR and testosterone predicts antisocial behavior," *Neuropsychopharmacology*, vol. 33, no. 2, pp. 425–430, 2008.

[30] C. L. Bethea, N. Z. Lu, C. Gundlah, and J. M. Streicher, "Diverse actions of ovarian steroids in the serotonin neural system," *Frontiers in Neuroendocrinology*, vol. 23, no. 1, pp. 41–100, 2002.

[31] B. de Lignieres and M. Vincens, "Differential effects of exogenous oestradiol and progesterone on mood in post-menopausal women: individual dose/effect relationship," *Maturitas*, vol. 4, no. 1, pp. 67–72, 1982.

[32] R. Słopien, K. Jasniewicz, B. Meczekalski, A. Warenik-Szymankiewicz, M. Lianeri, and P. P. Jagodziński, "Polymorphic variants of genes encoding MTHFR, MTR, and MTHFD1 and the risk of depression in postmenopausal women in Poland," *Maturitas*, vol. 61, no. 3, pp. 252–255, 2008.

[33] D. Deecher, T. H. Andree, D. Sloan, and L. E. Schechter, "From menarche to menopause: exploring the underlying biology of depression in women experiencing hormonal changes," *Psychoneuroendocrinology*, vol. 33, no. 1, pp. 3–17, 2008.

[34] M. Weber, S. Talmon, I. Schulze et al., "Running wheel activity is sensitive to acute treatment with selective inhibitors for either serotonin or norepinephrine reuptake," *Psychopharmacology*, vol. 203, no. 4, pp. 753–762, 2009.

[35] D. B. Imwalle, J. A. Gustafsson, and E. F. Rissman, "Lack of functional estrogen receptor $\beta$ influences anxiety behavior and serotonin content in female mice," *Physiology and Behavior*, vol. 84, no. 1, pp. 157–163, 2005.

[36] B. J. Mickey, F. Ducci, C. A. Hodgkinson, S. A. Langenecker, D. Goldman, and J. K. Zubieta, "Monoamine oxidase A genotype predicts human serotonin 1A receptor availability in vivo," *Journal of Neuroscience*, vol. 28, no. 44, pp. 11354–11359, 2008.

[37] E. L. Klaiber, D. M. Broverman, W. Vogel, L. G. Peterson, and M. B. Snyder, "Relationships of serum estradiol levels, menopausal duration, and mood during hormonal replacement therapy," *Psychoneuroendocrinology*, vol. 22, no. 7, pp. 549–558, 1997.

[38] D. P. Holschneider, T. Kumazawa, K. Chen, and J. C. Shih, "Tissue-specific effects of estrogen on monoamine oxidase A and B in the rat," *Life Sciences*, vol. 63, no. 3, pp. 155–160, 1998.

[39] Z. Q. Ma, E. Violani, F. Villa, G. B. Picotti, and A. Maggi, "Estrogenic control of monoamine oxidase A activity in human neuroblastoma cells expressing physiological concentrations of estrogen receptor," *European Journal of Pharmacology*, vol. 284, no. 1-2, pp. 171–176, 1995.

[40] J. H. Morrison, R. D. Brinton, P. J. Schmidt, and A. C. Gore, "Estrogen, menopause, and the aging brain: how basic neuroscience can inform hormone therapy in women," *Journal of Neuroscience*, vol. 26, no. 41, pp. 10332–10348, 2006.

3

# Mouse Models of Aneuploidy

**Olivia Sheppard,[1] Frances K. Wiseman,[1] Aarti Ruparelia,[1] Victor L. J. Tybulewicz,[2] and Elizabeth M. C. Fisher[1]**

[1] *Department of Neurodegenerative Disease, UCL Institute of Neurology, Queen Square, London WC1N 3BG, UK*
[2] *Division of Immune Cell Biology, MRC National Institute for Medical Research, The Ridgeway, Mill Hill, London NW7 1AA, UK*

Correspondence should be addressed to Frances K. Wiseman, f.wiseman@prion.ucl.ac.uk

Academic Editor: Adele Murrell

Abnormalities of chromosome copy number are called aneuploidies and make up a large health load on the human population. Many aneuploidies are lethal because the resulting abnormal gene dosage is highly deleterious. Nevertheless, some whole chromosome aneuploidies can lead to live births. Alterations in the copy number of sections of chromosomes, which are also known as segmental aneuploidies, are also associated with deleterious effects. Here we examine how aneuploidy of whole chromosomes and segmental aneuploidy of chromosomal regions are modeled in the mouse. These models provide a whole animal system in which we aim to investigate the complex phenotype-genotype interactions that arise from alteration in the copy number of genes. Although our understanding of this subject is still in its infancy, already research in mouse models is highlighting possible therapies that might help alleviate the cognitive effects associated with changes in gene number. Thus, creating and studying mouse models of aneuploidy and copy number variation is important for understanding what it is to be human, in both the normal and genomically altered states.

## 1. Introduction

Traditionally, aneuploidy was defined as a deletion or duplication of a whole chromosome. This genomic abnormality is thought to occur in at least 5% of all clinically recognized pregnancies, usually resulting in spontaneous abortion [1]. Aneuploidy is thought to be usually highly deleterious because many genes are "dosage-sensitive" in that their expression is affected by their copy number in the genome, and changes in gene expression levels may result in altered phenotypes that can be lethal [2]. As well as whole chromosome aneuploidy, deletion of a few kilobases or megabases of DNA (microdeletion) or similarly a duplicated region (microduplication) within a chromosome can also result in changes in gene copy number. Recent advances in genomic technologies have revealed the association of many of these segmental aneuploidies (microdeletions and duplications) with specific genetic syndromes and diseases [3–5].

The most frequently occurring full autosomal aneuploidy is that of trisomy of human chromosome 21 (Hsa21), which causes Down syndrome (DS). DS is the most common cause of genetic intellectual disability, occurring in ~1 in 750 live births in all populations. People with DS have an increased risk of developing cardiac defects, certain leukemias, and early onset Alzheimer's disease as well as many other phenotypes [6]. Trisomies of chromosomes 18 (Hsa18) (Edwards syndrome) and 13 (Hsa13) (Patau syndrome) occur at lower frequency than DS (1 in 4300 and 1 in 7100 live births, respectively), and infants with these conditions have a very short life expectancy, typically less than 1 year and less than 5 years, respectively [7]. Aneuploidy of the sex chromosomes can occur with multiple copies of the X or Y chromosome, or loss of the X or Y chromosome. Relatively common sex chromosome aneuploidies include Klinefelter syndrome (KS) (47 XXY, males with an additional copy of the X chromosome), ~1 in 500–1000 males [8], and Turner syndrome (TS) (45,X, females with monosomy of the X chromosome), 1 in ~4000 live births [9].

Segmental aneuploidies, otherwise known as partial aneuploidies or segmental aneusomies may be more compatible with life than whole chromosomal aneuploidies, and result in a large number of well-defined syndromes (Table 1). Many of these conditions are associated with neurodevelopmental and growth problems that result in epilepsy, intellectual disability, and autism.

The challenge facing scientists and clinicians from the aneuploidy syndromes is how to unravel the interaction between abnormal gene dosage and abnormal gene expression that leads to the specific phenotypes of each syndrome, and then to find therapies for intervention for these phenotypes.

## 2. Mouse Models of Aneuploidy

The use of mouse models of aneuploidy allows scientists to study the direct effects of abnormal gene dosage on specific syndromes, at the molecular, cellular, physiological, and behavioural level. Many technologies exist to manipulate the mouse genome to mutate, overexpress, and knock-out specific genes of interest, and help to define which dosage sensitive genes are causative for any given phenotype (reviewed in [10]). Such technologies now include chromosome engineering whereby large regions of the mouse genome can be deleted or duplicated corresponding to the partial aneuploidies found in humans (reviewed in [11, 12]). However, one confounding factor is that each human chromosome has syntenic regions to two or more mouse chromosomes. An alternative approach has been to transfer an entire human chromosome into a mouse, to overcome this problem [13]. Here, we discuss the contribution of mouse models of whole chromosome and segmental aneuploidy to our biological understanding and highlight possible future models and how they may further our knowledge.

## 3. Mouse Models of Whole Chromosome Aneuploidies

*3.1. Down Syndrome.* A number of mouse models have been developed to study the most frequently occurring autosomal aneuploidy, Down syndrome (Figure 1). The Tc1 transchromosomic mouse model contains a freely segregating maternally inherited copy of Hsa21 and is trisomic for approximately 75% of Hsa21 genes [13]. This mouse has altered learning and memory, synaptic plasticity, a reduced cerebellar neuronal number, heart anomalies, reduced solid tumor development, and defects in angiogenesis and megakaryopoiesis [13–18]. Other mouse models of DS contain an additional copy of regions of mouse chromosomes 16, 17, and 10, which are syntenic with Hsa21. The Ts65Dn mouse model is the most widely used; it contains an extra copy of a segment of mouse chromosome 16 (Mmu16) and is trisomic for about 50% of the genes found on Hsa21 [19]. This model shows impaired learning and motor deficits [19], neuronal degeneration similar to that observed in people with Alzheimer's disease (which is part of the DS phenotype) and heart and angiogenesis defects [20–22]. Another commonly used model is the Ts1Cje mouse which contains a smaller segmental trisomy of Mmu16 including approximately 68 genes; it also exhibits learning and behavioral deficits, but does not exhibit neuronal degeneration [23]. The newest model of DS, developed by Yu and colleagues, contains three copies of all Hsa21 homologs on mouse chromosomes 16, 17, and 10 and shows learning and memory deficits that may be similar to some of the cognitive problems that people with DS experience [24, 25].

To determine the identity of trisomic genes that cause specific phenotypes, aneuploid mouse models of DS can be crossed with mouse models of segmental Hsa21 monosomy (Ms1Yah and Ms4Yah) [26–29] or to gene knockouts to alter dosage of individual genes within a region of trisomy. These techniques have been recently used to identify the genes responsible for trisomy-21-related protection against the tumour formation [30], furthering our understanding of the biology that underlies these important processes. Mouse models of Hsa21 trisomy have been used also for demonstrating the potential for cognitive enhancement therapies for people who have DS [21, 31, 32]. A number of the drugs tested in studies of DS mouse models for their effects on learning and memory are currently in small-scale clinical trials, demonstrating the utility of these mice to combat the deleterious effects of DS.

*3.2. Edwards Syndrome and Patau Syndrome.* A mouse model of Edwards syndrome has yet to be developed. Hsa18 is 78 Mb in length and has conserved synteny with 5 principal regions encoded on three mouse chromosome (Mmu 1, 17, and 18). Similarly, no animal model of Patau syndrome has been reported; Hsa13 has conserved synteny with six mouse chromosome segments. Thus, although technically challenging, it would be possible to generate models of these syndromes by duplication of the mouse syntenic regions. These models could be used to further our understanding of the biology of this devastating conditions.

*3.3. Turner Syndrome and Klinefelter Syndrome.* Mouse models with both paternal and maternally inherited 45,X karyotypes exhibit behavioural changes including reduced attention, growth retardation, and hearing defects, which resemble aspects of human Turner syndrome (TS) (reviewed by [33]). These models have been useful for understanding the X-parent-of-origin-effect on TS-associated phenotypes. However, 39,X mice do not manifest some TS-associated phenotypes, such as motor deficits; this may reflect differences in X-inactivation between mouse and humans. Mouse models of Klinefelter syndrome (XXY male) develop hypogonadism and cognitive problems and have impaired fertility (reviewed by [34]), phenotypes that resemble aspects of KS. Molecular studies undertaken in XXY mouse models have shed light on the possible chemical alterations in the brain that cause cognitive problems observed in KS [34], and this knowledge may lead to the development of therapeutic strategies.

## 4. Mouse Models of Segmental Aneuploidies

Mouse models of segmental aneuploidy are invaluable for our understanding of which dosage-sensitive genes result in

Table 1: Examples of mouse models of segmental aneuploidies.

| Human syndrome | Associated genetic change | Aneuploid mouse models |
|---|---|---|
| Angelman syndrome | deletion of maternal 15q11–13 | *PatDp* [37]<br>*MatDf(Ube3a-Gabrb3)* [38] |
| Prader-Willi syndrome | deletion of paternal 15q11–13 | *MatDp* [37] |
| Autism risk factor | Duplication 15q11–13 | *matDp; pat Dp* [57] |
| Smith-Magenis syndrome | deletion of 17p11/17p11.2 | *Df(11)17* [58]<br>*Df(11)17-1; Df(11)17-2; Df(11)17-3* [59] |
| Potocki-Lupski syndrome | duplication of 17p11/17p11.2 | *Dp(11)17* [58] |
| DiGeorge syndrome | deletion of 22q11.2 | *Df1* [44]<br>*Idd-Ctp* [45]<br>*Idd-Arvcf* [46]<br>*Df2; Df3; Df4; Df5* [60] |
| Williams-Beuren | deletion of 7q11 | *PD* and *DD* [53] |
| — | deletion/duplication of 17q21 | *Df11[1]* and *Dp11[1]* [35] |

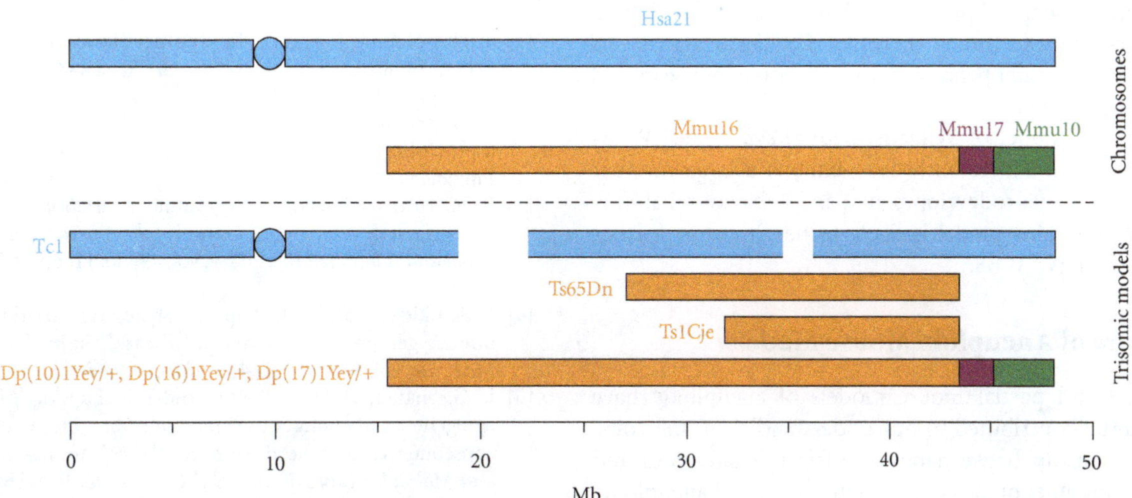


Figure 1: Mouse models of Down syndrome. Hsa21 (in blue) and the syntenic mouse chromosomes (Mmu 16, orange, Mmu 17, purple, Mmu10, green). The trisomic regions of several of the well-established mouse models of DS, the Tc1 mouse, Ts65Dn, Ts1Cje, and Dp(10)1Yey/+, Dp(16)1Yey/+, Dp(17)1Yey/+ are aligned to the corresponding parts of the human and mouse genome.

the deleterious phenotypes that are associated with these genomic changes in humans. Moreover, mouse models of segmental aneuploidy, not associated with a specific human syndrome, can also be used to understand the relationship between gene and phenotype. For example, mouse models with 0.8 Mb reciprocal chromosomal deletions and duplications have been used to identify the role of *Stat5* in immune-hypersensitivity and metabolic syndrome [35].

A number of models of Prader-Willi syndrome (PWS) (deletion of paternal 15q11–13) and Angelman syndrome (AS) (deletion of maternal 15q11–13) have been reported [36–38]. PWS is also associated with chromosome 15 maternal disomy and AS with paternal chromosome 15 disomy. Mouse models of these genetic changes have been reported and both exhibit reduced viability and neonatal growth retardation [36, 37]. Deficits in learning and memory have also been observed in a mouse model with a maternally inherited segmental deletions (*Ube3a-Gabrb3*) corresponding to part of the region lost in AS [38]. Mouse models deficient in the *Ube3a* and *Gabrb3* PWS/AS candidate genes exhibit neurodevelopment and behavior changes, highlighting the key role these genes play in the syndromes [39–41].

Interestingly, maternal duplications of the PWS/AS associated region, 15q11–13, are associated with autism [42]. A mouse model of the duplication of the mouse syntenic region of chromosome 7 exhibits some features that resemble autism, but only when the duplication is paternally inherited in contrast to the inheritance pattern observed in humans [43]. These models will help give insight into the genetic and biochemical abnormalities causing autism.

Mouse models of DiGeorge syndrome (deletion of 1.5–3 Mb at 22q11) have been crucial to the molecular understanding of this condition. A series of complementary mouse models with full or partial deletions of the region syntenic with 22q11 identified the key deleted gene, *Tbx1*, responsible for the syndrome's deleterious phenotypes [44–46]. The 22q11 deletion is also the largest known genetic risk factors for schizophrenia [47, 48], and the DiGeorge mouse models may also be useful to further understanding of this condition [49]. Similarly, mouse models of the complete 3.7 Mb deletion and duplication associated with Smith-Magenis (SMS) and Potocki-Lupski (PTLS) syndromes have helped identify one of the key dosage sensitive genes, *Rai1* [50]. Moreover, these models have also been used to investigate the relative effect of genomic rearrangement versus gene copy number change on gene expression [51]. Work in this field has also highlighted the very complex interactions of genes both within the copy number altered region and those elsewhere in the genome [52], as the penetrance of some PTLS-like features in the mouse models vary with the size of the region disrupted and the genetic background of the model. The effect of genetic background on the penetrance of aneuploidy-associated phenotypes has also been highlighted in the Tc1 mouse model in which DS-like heart defects appear with a greater penetrance on a C57BL/6 mouse inbred line background [15].

A mouse model of Williams-Beuren syndrome (WBS) has been developed recently that exhibits a large number of informative neurodevelopmental and behavioural abnormalities [53]. This model is likely to be crucial to further understanding of WBS.

## 5. Future of Aneuploid Mouse Models

Complete and partial mouse models of aneuploidy have significantly contributed to our understanding of the complex relationship between dosage of individual genes and the resulting phenotypes that arise in individual aneuploidy syndromes. Unfortunately, even for the most widely studied aneuploidy disorders such as DS, we are a long way from understanding much of the molecular basis of the pathology. However, the rate of progress in understanding the effects of gene copy number and expression levels is increasing, and we now know that there is a considerable variation of small genomic regions, copy number variation (CNV), across the entire human genome in normal individuals. These regions can be up to a megabase in size and affect much of normal human phenotypic variation, including susceptibility or resistance to common disorders (e.g., see [54–56]). New mouse models of CNVs will be beneficial to study not only the effects of gene dosage but also to dissect the effects of altering copy number for the regulatory elements found in these regions of the genome.

Advances in our understanding of the human genome will present new opportunities for the development of novel mouse aneuploid models. Equally, findings from existing mouse models will continue to influence human genetic studies. Thus, complementary human and mouse genetic studies are key to unraveling the links between gene copy number and phenotype.

## Acknowledgments

The authors thank Ray Young for assistance with preparation of the figure. F. K. Wiseman, O. Sheppard, A. Ruparelia, and E. M. C. Fisher are funded by the UK Medical Research Council, the Wellcome Trust, the AnEUploidy Grant from Framework Programme 6 of the European Union Commission, and the Alzheimer's Research Trust, the Brain Research Trust. V. L. J. Tybulewicz is funded by the UK Medical Research Council, the AnEUploidy grant from Framework Programme 6 of the European Union Commission, and the Wellcome Trust.

## Authors' Contribution

O. Sheppard and F. K. Wiseman contributed equally to this publication.

## References

[1] T. Hassold and P. Hunt, "To err (meiotically) is human: the genesis of human aneuploidy," *Nature Reviews Genetics*, vol. 2, no. 4, pp. 280–291, 2001.

[2] M. Dierssen, Y. Herault, and X. Estivill, "Aneuploidy: from a physiological mechanism of variance to Down syndrome," *Physiological Reviews*, vol. 89, no. 3, pp. 887–920, 2009.

[3] E. M. Morrow, "Genomic copy number variation in disorders of cognitive development," *Journal of the American Academy of Child and Adolescent Psychiatry*, vol. 49, no. 11, pp. 1091–1104, 2010.

[4] P. Stankiewicz and J. R. Lupski, "Structural variation in the human genome and its role in disease," *Annual Review of Medicine*, vol. 61, pp. 437–455, 2010.

[5] L. G. Shaffer, D. H. Ledbetter, and J. R. Lupski, "Molecular cytogenetics of contiguous gene syndromes: mechanisms and consequences of gene dosage imbalance," in *The Metabolic and Molecular Bases of Inherited Diseases*, C. R. Scriver, A. L. Beaudet, W. S. Sly, D. Valle, B. Vogelstein, and B. Childs, Eds., pp. 1291–1326, McGraw–Hill, New York, NY, USA, 2001.

[6] F. K. Wiseman, K. A. Alford, V. L. J. Tybulewicz, and E. M. C. Fisher, "Down syndrome—recent progress and future prospects," *Human Molecular Genetics*, vol. 18, no. 1, pp. R75–R83, 2009.

[7] G. M. Savva, K. Walker, and J. K. Morris, "The maternal age-specific live birth prevalence of trisomies 13 and 18 compared to trisomy 21 (Down syndrome)," *Prenatal Diagnosis*, vol. 30, no. 1, pp. 57–64, 2010.

[8] J. C. Giltay and M. C. Maiburg, "Klinefelter syndrome: clinical and molecular aspects," *Expert Review of Molecular Diagnostics*, vol. 10, no. 6, pp. 765–776, 2010.

[9] M. L. Davenport, "Approach to the patient with Turner syndrome," *Journal of Clinical Endocrinology and Metabolism*, vol. 95, no. 4, pp. 1487–1495, 2010.

[10] D. Nguyen and X. Tian, "The expanding role of mouse genetics for understanding human biology and disease," *Disease Models and Mechanisms*, vol. 1, no. 1, pp. 56–66, 2008.

[11] V. L. J. Tybulewicz and E. M. C. Fisher, "New techniques to understand chromosome dosage: mouse models of aneuploidy," *Human Molecular Genetics*, vol. 15, no. 2, pp. R103–R109, 2006.

[12] R. Ramírez-Solis, P. Liu, and A. Bradley, "Chromosome engineering in mice," *Nature*, vol. 378, no. 6558, pp. 720–724, 1995.

[13] A. O'Doherty, S. Ruf, C. Mulligan et al., "Genetics: an aneuploid mouse strain carrying human chromosome 21 with Down syndrome phenotypes," *Science*, vol. 309, no. 5743, pp. 2033–2037, 2005.

[14] K. A. Alford, A. Slender, L. Vanes et al., "Perturbed hematopoiesis in the Tc1 mouse model of Down syndrome," *Blood*, vol. 115, no. 14, pp. 2928–2937, 2010.

[15] L. Dunlevy, M. Bennett, A. Slender et al., "Down's syndrome-like cardiac developmental defects in embryos of the transchromosomic Tc1 mouse," *Cardiovascular Research*, vol. 88, no. 2, pp. 287–295, 2010.

[16] M. Galante, H. Jani, L. Vanes et al., "Impairments in motor coordination without major changes in cerebellar plasticity in the Tc1 mouse model of Down syndrome," *Human Molecular Genetics*, vol. 18, no. 8, pp. 1449–1463, 2009.

[17] E. Morice, L. C. Andreae, S. F. Cooke et al., "Preservation of long-term memory and synaptic plasticity despite short-term impairments in the Tc1 mouse model of down syndrome," *Learning and Memory*, vol. 15, no. 7, pp. 492–500, 2008.

[18] L. E. Reynolds, A. R. Watson, M. Baker et al., "Tumour angiogenesis is reduced in the Tc1 mouse model of Downs syndrome," *Nature*, vol. 465, no. 7299, pp. 813–817, 2010.

[19] R. H. Reeves, N. G. Irving, T. H. Moran et al., "A mouse model for Down syndrome exhibits learning and behaviour deficits," *Nature Genetics*, vol. 11, no. 2, pp. 177–184, 1995.

[20] A. Salehi, J. D. Delcroix, P. V. Belichenko et al., "Increased App expression in a mouse model of Down's syndrome disrupts NGF transport and causes cholinergic neuron degeneration," *Neuron*, vol. 51, no. 1, pp. 29–42, 2006.

[21] A. Salehi, M. Faizi, D. Colas et al., "Restoration of norepinephrine-modulated contextual memory in a mouse model of Down syndrome," *Science translational medicine*, vol. 1, no. 7, pp. 7–ra17, 2009.

[22] J. D. Cooper, A. Salehi, J. D. Delcroix et al., "Failed retrograde transport of NGF in a mouse model of Down's syndrome: reversal of cholinergic neurodegenerative phenotypes following NGF infusion," *Proceedings of the National Academy of Sciences of the United States of America*, vol. 98, no. 18, pp. 10439–10444, 2001.

[23] H. Sago, E. J. Carlson, D. J. Smith et al., "Ts1Cje, a partial trisomy 16 mouse model for Down syndrome, exhibits learning and behavioral abnormalities," *Proceedings of the National Academy of Sciences of the United States of America*, vol. 95, no. 11, pp. 6256–6261, 1998.

[24] T. Yu, C. Liu, P. Belichenko et al., "Effects of individual segmental trisomies of human chromosome 21 syntenic regions on hippocampal long-term potentiation and cognitive behaviors in mice," *Brain Research*, vol. 1366, pp. 162–171, 2010.

[25] T. Yu, Z. Li, Z. Jia et al., "A mouse model of Down syndrome trisomic for all human chromosome 21 syntenic regions," *Human Molecular Genetics*, vol. 19, no. 14, Article ID ddq179, pp. 2780–2791, 2010.

[26] V. Besson, V. Brault, A. Duchon et al., "Modeling the monosomy for the telomeric part of human chromosome 21 reveals haploinsufficient genes modulating the inflammatory and airway responses," *Human Molecular Genetics*, vol. 16, no. 17, pp. 2040–2052, 2007.

[27] A. Duchon, S. Pothion, V. Brault et al., "The telomeric part of the human chromosome 21 from Cstb to Prmt2 is not necessary for the locomotor and short-term memory deficits observed in the Tc1 mouse model of Down syndrome," *Behavioural Brain Research*, vol. 217, no. 2, pp. 271–281, 2011.

[28] L. E. Olson, J. T. Richtsmeier, J. Leszl, and R. H. Reeves, "A chromosome 21 critical region does not cause specific down syndrome phenotypes," *Science*, vol. 306, no. 5696, pp. 687–690, 2004.

[29] T. Yu, S. J. Clapcote, Z. Li et al., "Deficiencies in the region syntenic to human 21q22.3 cause cognitive deficits in mice," *Mammalian Genome*, vol. 21, no. 5-6, pp. 258–267, 2010.

[30] T. E. Sussan, A. Yang, F. Li, M. C. Ostrowski, and R. H. Reeves, "Trisomy represses ApcMin-mediated tumours in mouse models of Down's syndrome," *Nature*, vol. 451, no. 7174, pp. 73–75, 2008.

[31] A. C. S. Costa, J. J. Scott-McKean, and M. R. Stasko, "Acute injections of the NMDA receptor antagonist memantine rescue performance deficits of the Ts65Dn mouse model of Down syndrome on a fear conditioning test," *Neuropsychopharmacology*, vol. 33, no. 7, pp. 1624–1632, 2008.

[32] F. Fernandez, W. Morishita, E. Zuniga et al., "Pharmacotherapy for cognitive impairment in a mouse model of Down syndrome," *Nature Neuroscience*, vol. 10, no. 4, pp. 411–413, 2007.

[33] P. M. Y. Lynn and W. Davies, "The 39,XO mouse as a model for the neurobiology of Turner syndrome and sex-biased neuropsychiatric disorders," *Behavioural Brain Research*, vol. 179, no. 2, pp. 173–182, 2007.

[34] J. Wistuba, "Animal models for Klinefelter's syndrome and their relevance for the clinic," *Molecular Human Reproduction*, vol. 16, no. 6, Article ID gaq024, pp. 375–385, 2010.

[35] O. Ermakova, L. Piszczek, L. Luciani et al., "Sensitized phenotypic screening identifies gene dosage sensitive region on chromosome 11 that predisposes to disease in mice," *EMBO Molecular Medicine*, vol. 3, no. 1, pp. 50–66, 2011.

[36] B. M. Cattanach, J. A. Barr, C. V. Beechey, J. Martin, J. Noebels, and J. Jones, "A candidate model for Angelman syndrome in the mouse," *Mammalian Genome*, vol. 8, no. 7, pp. 472–478, 1997.

[37] B. M. Cattanach, J. A. Barr, E. P. Evans et al., "A candidate mouse model for Prader-Willi syndrome which shows an absence of Snrpn expression," *Nature Genetics*, vol. 2, no. 4, pp. 270–274, 1992.

[38] Y. H. Jiang, Y. Pan, L. Zhu et al., "Altered ultrasonic vocalization and impaired learning and memory in Angelman syndrome mouse model with a large maternal deletion from Ube3a to Gabrb3," *PLoS One*, vol. 5, no. 8, Article ID e12278, 2010.

[39] T. M. DeLorey, A. Handforth, G. E. Homanics, and R. W. Olsen, "Mice lacking the gabrb3 gene have epilepsy and behavioral characteristics of Angelman syndrome," *Brain Research*, vol. 809, p. A29, 1998.

[40] T. M. DeLorey, A. Handforth, A. Asatourian et al., "Mice lacking the GABA(A) receptor beta(3) subunit gene have some of the characteristics of Angelmann syndrome," *Journal of Neurochemistry*, vol. 69, p. S236, 1997.

[41] Y. H. Jiang, D. Armstrong, U. Albrecht et al., "Mutation of the Angelman ubiquitin ligase in mice causes increased cytoplasmic p53 and deficits of contextual learning and long-term potentiation," *Neuron*, vol. 21, no. 4, pp. 799–811, 1998.

[42] E. H. Cook, V. Lindgren, B. L. Leventhal et al., "Autism or atypical autism in maternally but not paternally derived proximal 15q duplication," *American Journal of Human Genetics*, vol. 60, no. 4, pp. 928–934, 1997.

[43] T. Takumi, "A humanoid mouse model for autism by a chromosome engineering," *Neuroscience Research*, vol. 65, p. S27, 2009.

[44] E. A. Lindsay, A. Botta, V. Jurecic et al., "Congenital heart disease in mice deficient for the DiGeorge syndrome region," *Nature*, vol. 401, no. 6751, pp. 379–383, 1999.

[45] W. L. Kimber, P. Hsieh, S. Hirotsune et al., "Deletion of 150 kb in the minimal DiGeorge/velocardiofacial syndrome critical region in mouse," *Human Molecular Genetics*, vol. 8, no. 12, pp. 2229–2237, 1999.

[46] A. Puech, B. Saint-Jore, S. Merscher et al., "Normal cardiovascular development in mice deficient for 16 genes in 550 kb of the velocardiofacial/DiGeorge syndrome region," *Proceedings of the National Academy of Sciences of the United States of America*, vol. 97, no. 18, pp. 10090–10095, 2000.

[47] T. Sigurdsson, K. L. Stark, M. Karayiorgou, J. A. Gogos, and J. A. Gordon, "Impaired hippocampal-prefrontal synchrony in a genetic mouse model of schizophrenia," *Nature*, vol. 464, no. 7289, pp. 763–767, 2010.

[48] M. Karayiorgou, T. J. Simon, and J. A. Gogos, "22q11.2 microdeletions: linking DNA structural variation to brain dysfunction and schizophrenia," *Nature Reviews Neuroscience*, vol. 11, no. 6, pp. 402–416, 2010.

[49] D. W. Meechan, E. S. Tucker, T. M. Maynard, and A. S. LaMantia, "Diminished dosage of 22q11 genes disrupts neurogenesis and cortical development in a mouse model of 22q11 deletion/DiGeorge syndrome," *Proceedings of the National Academy of Sciences of the United States of America*, vol. 106, no. 38, pp. 16434–16439, 2009.

[50] K. Walz, R. Paylor, J. Yan, W. Bi, and J. R. Lupski, "Rai1 duplication causes physical and behavioral phenotypes in a mouse model of dup(17)(p11.2p11.2)," *Journal of Clinical Investigation*, vol. 116, no. 11, pp. 3035–3041, 2006.

[51] G. Ricard, J. Molina, J. Chrast et al., "Phenotypic consequences of copy number variation: insights from smith-magenis and Potocki-Lupski syndrome mouse models," *PLoS Biology*, vol. 8, no. 11, Article ID e1000543, 2010.

[52] J. Yan, W. Bi, and J. R. Lupski, "Penetrance of craniofacial anomalies in mouse models of Smith-Magenis syndrome is modified by genomic sequence surrounding Rai1: not all null alleles are alike," *American Journal of Human Genetics*, vol. 80, no. 3, pp. 518–525, 2007.

[53] H. H. Li, M. Roy, U. Kuscuoglu et al., "Induced chromosome deletions cause hypersociability and other features of Williams-Beuren syndrome in mice," *EMBO Molecular Medicine*, vol. 1, no. 1, pp. 50–65, 2009.

[54] S. Girirajan and E. E. Eichler, "Phenotypic variability and genetic susceptibility to genomic disorders," *Human Molecular Genetics*, vol. 19, no. R2, pp. R176–187, 2010.

[55] A. C. Need and D. B. Goldstein, "Whole genome association studies in complex diseases: where do we stand?" *Dialogues in Clinical Neuroscience*, vol. 12, no. 1, pp. 37–46, 2010.

[56] C. Lee and S. W. Scherer, "The clinical context of copy number variation in the human genome," *Expert Reviews in Molecular Medicine*, vol. 12, p. e8, 2010.

[57] J. Nakatani, K. Tamada, F. Hatanaka et al., "Abnormal behavior in a chromosome- engineered mouse model for human 15q11-13 duplication seen in Autism," *Cell*, vol. 137, no. 7, pp. 1235–1246, 2009.

[58] K. Walz, S. Caratini-Rivera, W. Bi et al., "Modeling del(17)(p11.2p11.2) and dup(17)(p11.2p11.2) contiguous gene syndromes by chromosome engineering in mice: phenotypic consequences of gene dosage imbalance," *Molecular and Cellular Biology*, vol. 23, no. 10, pp. 3646–3655, 2003.

[59] J. Yan, V. W. Keener, W. Bi et al., "Reduced penetrance of craniofacial anomalies as a function of deletion size and genetic background in a chromosome engineered partial mouse model for Smith-Magenis syndrome," *Human Molecular Genetics*, vol. 13, no. 21, pp. 2613–2624, 2004.

[60] E. A. Lindsay and A. Baldini, "Recovery from arterial growth delay reduces penetrance of cardiovascular defects in mice deleted for the DiGeorge syndrome region," *Human Molecular Genetics*, vol. 10, no. 9, pp. 997–1002, 2001.

# 4

# PTEN Gene: A Model for Genetic Diseases in Dermatology

**Corrado Romano[1] and Carmelo Schepis[2]**

[1] *Unit of Pediatrics and Medical Genetics, I.R.C.C.S. Associazione Oasi Maria Santissima, 94018 Troina, Italy*
[2] *Unit of Dermatology, I.R.C.C.S. Associazione Oasi Maria Santissima, 94018 Troina, Italy*

Correspondence should be addressed to Carmelo Schepis, cschepis@oasi.en.it

Academic Editors: G. Vecchio and H. Zitzelsberger

PTEN gene is considered one of the most mutated tumor suppressor genes in human cancer, and it's likely to become the first one in the near future. Since 1997, its involvement in tumor suppression has smoothly increased, up to the current importance. Germline mutations of PTEN cause the PTEN hamartoma tumor syndrome (PHTS), which include the past-called Cowden, Bannayan-Riley-Ruvalcaba, Proteus, Proteus-like, and Lhermitte-Duclos syndromes. Somatic mutations of PTEN have been observed in glioblastoma, prostate cancer, and brest cancer cell lines, quoting only the first tissues where the involvement has been proven. The negative regulation of cell interactions with the extracellular matrix could be the way PTEN phosphatase acts as a tumor suppressor. PTEN gene plays an essential role in human development. A recent model sees PTEN function as a stepwise gradation, which can be impaired not only by heterozygous mutations and homozygous losses, but also by other molecular mechanisms, such as transcriptional regression, epigenetic silencing, regulation by microRNAs, posttranslational modification, and aberrant localization. The involvement of PTEN function in melanoma and multistage skin carcinogenesis, with its implication in cancer treatment, and the role of front office in diagnosing PHTS are the main reasons why the dermatologist should know about PTEN.

## 1. PTEN Gene: What It Is and How It Works

PTEN stands for phosphatase and tensin homolog deleted in chromosome 10, and it is considered one of the most mutated tumor suppressor genes in human cancer. In the near future, it is likely to become the first one overcoming the current leader, p53 gene [1]. The involvement of PTEN's alteration in tumorigenesis has been first suspected and subsequently proven in 1997 [2], when high frequency of loss of heterozygosity (LOH) at 10q23 chromosome band was observed in several human tumors. Furthermore, the suppression of tumorigenesis in glioblastoma murine cells by the wildtype chromosome 10 led to envision a tumor suppressor gene mapping in 10q23. Such gene was eventually isolated by the above-mentioned authors and called PTEN. They detected homozygous deletions, frame shift, or nonsense mutations in PTEN in 63% (5/8) of glioblastoma cell lines, 100% (4/4) of prostate cancer cell lines, and 10% (2/20) of breast cancer cell lines. Steck et al. [3] independently isolated the same gene and called it mutated in multiple advanced cancers-1 (MMAC-1). Indeed, a common feature of PTEN somatic mutations, already presented in 10q LOH, is the association with advanced-stage tumors (mainly glial and prostate cancers), whereas this is not true for endometrial cancer, being affected equally at all the stages. This has led to the suggestion that the activation of PTEN is at an early stage in endometrial carcinogenesis, but later on in glial and prostatic carcinogenesis. This mechanism is the cornerstone of the classical two-hit Knudson' hypothesis [4]: a single mutation in one homolog of a tumor-suppressor gene is not sufficient to initiate tumor growth; however, deletion or disabling of the allele on the homologous chromosome results in unregulated cell growth. Both sporadic and hereditary tumors can be explained by such mechanism. In sporadic tumors, both alleles are normal at conception; subsequently, a postzygotic mutation (first hit) in one cell creates the heterozygosity (one mutant and one normal allele); thereafter, a deletion or a new mutation (second hit) in the other allele of that cell provokes the LOH, starting the uncontrolled tumor growth. In hereditary tumors, the heterozygosity for mutant allele (first hit) is present at conception, and is sufficient that a postzygotic mutation (second hit) during life creates the LOH for the onset of uncontrolled tumor growth.

Liaw et al. [5] found germline mutations of PTEN gene in families with Cowden syndrome [6] (CS), showing the function of tumor suppressor gene also in the germline. Furthermore, germline PTEN mutations lead to increased breast cancer incidence, but do not frequently cause familial breast cancer [7], notwithstanding 10% of breast cancer cell lines have inactivated PTEN [2, 3]. Recently it has been shown that PTEN loss is a common event in breast cancers caused by BRCA1 deficiency [8]. Marsh et al. [9] defined PTEN hamartoma tumor syndrome (PHTS) as a syndromic condition including one or more hamartomas which has its biological basis in a germline mutation of the PTEN gene. Following such assumption, PHTS includes patients with the previous diagnosis of CS, Bannayan-Riley-Ruvalcaba syndrome [10] (BRRS), Proteus syndrome [11] (PS), Proteus-like syndrome [12] (PLS), and Lhermitte-Duclos syndrome [13] (LDS).

Li et al. [2] have shown that PTEN gene is a human cdc14 homolog, like CDC14A and CDC14B. The cdc14 gene is a key point for the progression of cell cycle in *Saccharomyces cerevisiae*. Its protein acts in late nuclear division preparing for subsequent DNA replication. The human PTEN gene spans 103,207 bases, is made up of 9 exons, and codes for a 1212-bp transcript and a 403-amino-acid protein.

The PTEN product has the kinetic properties of dual-specific phosphatases [14] and acts on G1 cell cycle progression through negative regulation of the PI3-kinase/Akt or PKB signalling pathway [15]. PTEN is a member of the protein-tyrosine phosphatase (PTP) gene superfamily [2, 3]. These are genes consisting of conserved catalytic domains, flanked by noncatalytic regulatory sequences [16]. The PTP catalytic domains show a "signature motif," which is the canonical sequence HCXXGXXRS/T. Among the PTP superfamily genes, a further split is made in "classic" PTP (acting only towards phosphothyrosine residues) and dual-specificity phosphatase families (dephosphorylating phosphotyrosine, phosphoserine and/or phosphotreonine). PTEN is a dual-specificity phosphatase. The catalytic domain of PTEN has been proven to be essential for its function, which is lost following any mutation within the signature motif [17].

PTEN gene in glioblastoma-derived cell lines regulates hypoxia- and IGF-1-induced angiogenic gene expression by regulating Akt activation of HIF-1 activity [18]. Restoration of wild-type PTEN to glioblastoma cell lines lacking functional PTEN ablates hypoxia and IGF-1 induction of HIF-1-regulated genes. In addition, Akt activation leads to HIF-1$\alpha$ stabilization, whereas PTEN attenuates hypoxia-mediated HIF-1 $\alpha$ stabilization. Loss of PTEN during malignant progression contributes to tumor expansion through the deregulation of Akt activity and HIF-1-regulated gene expression. PTEN abnormalities have been found also in primary acute leukemias and non-Hodgkin's lymphomas [19].

PTEN and phosphorylated Akt levels are inversely correlated in the large majority of the examined samples, suggesting that PTEN regulates phosphatidylinositol 3,4,5-triphosphates and may play a role in apoptosis. Overexpression of PTEN inhibits cell migration, whereas antisense PTEN enhances migration [20]. The phosphatase domain of PTEN is essential because its inactivation does not allow the downregulation of integrin-mediated cell spreading and formation of focal adhesions, peculiar of wild-type PTEN. Overexpression of focal adhesion kinase (FAK) partially antagonizes the effects of PTEN. Thus, the negative regulation of cell interactions with the extracellular matrix could be the way PTEN phosphatase acts as a tumor suppressor.

PTEN gene plays an essential role in human development. Indeed, the additional effect of three homozygotic mutations together (e.g., mutations in both alleles) produces early embryonic lethality in mice, whereas heterozygosis (e.g., mutations in one allele) increases tumor incidence [21–23]. Furthermore, PTEN antagonizes growth factor-induced Shc phosphorylation and inhibits the MAP kinase (MAPK) signalling pathway [24]. The way this inhibition is accomplished is currently understood as a suppression by the PTEN protein phosphatase activity [25].

The function of PTEN gene and the way its mutation can cause a disease can barely be understood if one does not see the PTEN protein inside the PI3K/Akt/mTOR signalling pathway [26] (Figure 1). The activation of phosphoinositide 3′ kinase (PI3K) is the primary event in this pathway [27]. This can occur from several growth factor receptors (GFRs), such as PDGFR, EGFR, FGFR, IGF-1R, VEGFR, IL-R, interferon receptors (IF-Rs), integrin receptors, and the Ras pathway [28–30]. The major role of PI3K is the phosphorylation of phosphatidylinositol (4,5) P (PIP2) to phosphatidylinositol (3,4,5) P (PIP3). PIP3 binds and translocates Akt near the cell membrane where it can be phosphorylated and activated by phosphatidylinositol (3,4,5) P-dependent kinase 1 (PDK1) and phosphatidylinositol (3,4,5) P-dependent kinase 2 (PDK2). Akt has several downstream effectors which mediate its ability to promote cell survival and growth. Then, activation of PI3K/Akt pathway is observed in several human cancers. PTEN is the antagonist of PI3K because it dephosphorylates PIP3 to PIP2. The PI3K/PTEN imbalance, caused for instance by a mutation of PTEN, is then responsible for the progression to human cancer.

Salmena et al. [1] have recently proposed a model for the causes and consequences of PTEN loss. While retinoblastoma (RB) gene has been the foundation of the Knudson's hypothesis [4], showing that only the homozygous loss of such gene can start the retinoblastoma initiation, PTEN behaves in a different way. There is compelling evidence in mice confirming PTEN as a haploinsufficient tumor suppressor gene: loss of one allele leads to the progression of a lethal polyclonal autoimmune disorder [31]; epithelial cancers, such as prostate cancer, are driven by PTEN heterozygosity [32]; cellular levels of PTEN protein inversely correlate with the occurrence of invasive prostate cancer [1]. Consequently, functional loss of one PTEN allele is critical for the onset of cancer in mice. Things in humans are little less compelling, but the evidence is growing. The association between PTEN heterozygous germline mutations and the so-called PHTS

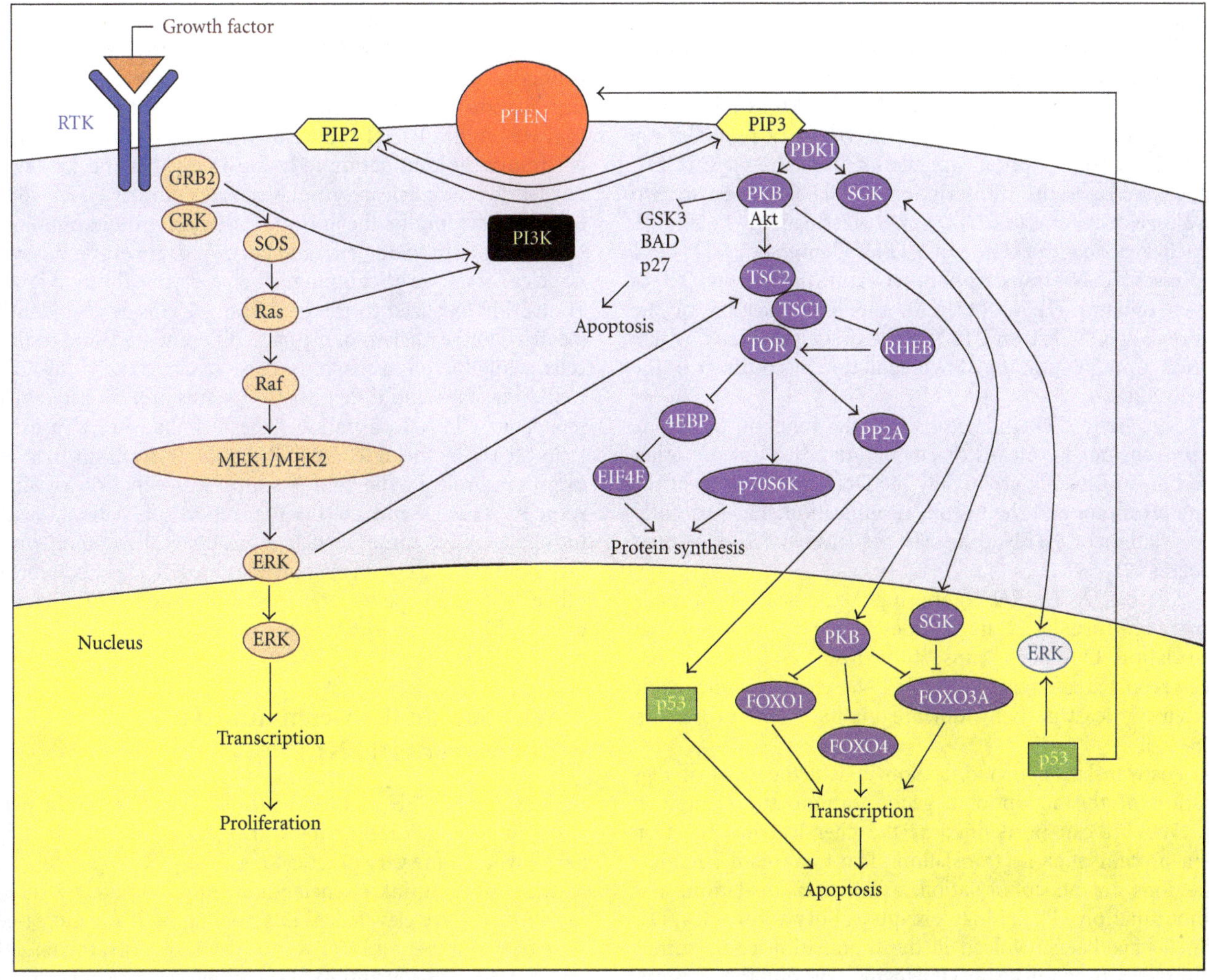


Figure 1: Pathway involving the PTEN protein.

is the first proof. Further evidence for haploinsufficiency is supported by the following observations: some tumors arisen in patients with CS do not show biallelic mutations of PTEN gene [33]; primary prostate cancers are associated with loss or alteration of one PTEN allele in 70% of cases [34], whereas homozygous deletion is present in 10% of cases [35]; the occurrence of monoallelic mutation of PTEN in breast cancer is much more frequent (30–40% versus 5%) than that of biallelic loss [36–38]. Besides the above-mentioned proofs of the contribution of PTEN haploinsufficiency to tumor progression and cancer syndromes, the identification of CS and tumor-derived PTEN mutations preserving partial or full PTEN lipid phosphatase function [39] allows the notion that even minor impairments of PTEN function can lead to cancer. The model of Santena et al. [1] springs from the above results: the loss of PTEN function, caused not only by classical genetic mutations leading to heterozygous (50% of function) or homozygous (0% of function) loss, but also by other molecular mechanisms, such as transcriptional regression, epigenetic silencing, regulation by microRNAs, posttranslational modification, and aberrant localization, can lead to subtle and/or dramatic losses of PTEN function, behaving as a stepwise gradation of function. An unexpected result has been put forward by Chen et al. [40], who studied the relationship between PTEN dose and tumor progression in mouse models of prostate specific loss of PTEN: complete acute loss of PTEN promoted a strong senescence response opposing tumor progression. Campisi and d'Adda di Fagagna [41] interpreted senescence as an antitumor mechanism set off by tumor suppressor genes in response to triggers including DNA damage and oncogene activation.

While heterozygous and homozygous mutations are now widely understood, the other molecular mechanisms deserve some clarification. We will do so, showing that all these mechanisms have a significant impact in the way PTEN functions, as a model of tumor suppressor gene.

While transforming growth factor $\beta$ (TGF$\beta$) was considered the single transcriptional regulator of PTEN gene in 1997 [42], acting in a negative way, today the story is much more complicated. Several factors have been discovered to upregulate, such as early growth-regulated transcription factor-1 (EGR-1) [43] which is a downstream effector of Insulin-like growth factor 2 (IGF-2) [44], and peroxisome proliferation-activated receptor-$\gamma$ (PPAR-$\gamma$) [45], p53 [46], and MYC [47], or downregulate PTEN, such as c-Jun [48], NF$\kappa$B [49], and HES-1 [47]. Furthermore, a peculiar way of PTEN's transcriptional regulation is that of NOTCH1, which increases the transcription of PTEN at least in two fashions: activating MYC [47] or repressing CBF-1 [50, 51], which is a downregulator of PTEN. Conversely, NOTCH1 represses PTEN's transcription, activating the known PTEN's downregulator HES-1 [47]. Another complication of the action of NOTCH1 on PTEN is the tissue specificity, which allows upregulation or downregulation according to the involved tissues.

Epigenetic silencing means that the function of a gene is broken, not by means of DNA mutations, but for other mechanisms, which are mainly the DNA methylation of the gene promoter and the histone modification. The promoter methylation of PTEN gene has been associated to several cancers [52–54].

MicroRNAs (miRNAs) are single-stranded RNAs, made up of a few (usually 22) nucleotides, which repress the mRNA translation. One of the most studied miRNAs, miR-21, has been reported as a repressor of PTEN, exerting its oncogenic activity, at least partially, downregulating PTEN expression [55–59].

Posttranslational modification is a further way of regulation of the action of a gene, without any change in its DNA. It can be defined as the chemical modification of a protein after its translation. The four main chemical reactions are phosphorylation, acetylation, oxidation, and ubiquitination. PTEN has six phosphorylation sites [1], which have been involved in the modulation of its tumor suppressor functions, subcellular distribution, and stability. They are Threonine 366 (Thr366), Serine 370 (Ser370), Ser385, Ser380, Thr382, and Thr383. The last three are collectively referred to as the STT cluster, which has the most important effects on PTEN function. The wild-type or phosphorylated STT cluster is maintained to stabilize PTEN in a closed state, whereas the mutation at these phosphorylation sites opens the protein conformation, making it less stable [60]. A dephosphorylation of the STT cluster is currently considered a common way leading to cancer activation [61]. PTEN interacts with the nuclear histone acetyltransferase-associated PCAF protein, promoting PTEN acetylation at Lysine 125 (Lys125) and Lys128 sites, which in turn negatively regulates PTEN catalytic activity [62]. What is now called reactive oxygen species (ROS), which is made of oxygen ions, free radical, and peroxides, are very small molecules sharing a high reactivity due to presence of unpaired valence shell electrons. ROS modulates PTEN catalytic activity by oxidative-stress-induced formation of a disulfide bond between the active site cysteine 124 (Cys124) and Cys71 [63]. Ubiquitin [64] is a highly conserved regulatory protein that is widely (ubiquitously) expressed in eukaryotes. Ubiquitination refers to the posttranslational modification of a protein by the covalent attachment (via an isopeptide bond) of one or more ubiquitin monomers. The most prominent function of ubiquitin is labeling proteins for proteasomal degradation. Proteasome inhibition increases the half-life of PTEN [65], and the exposure of human bronchial cells to zinc ions promotes ubiquitin-dependent degradation of PTEN [66], but inhibitors of the proteasome may destabilize PTEN [67]. Notwithstanding such incoherent results, ubiquitination to lysine 13 (Lys13) and Lys289 is needed for the nuclear-cytoplasmic shuttling of PTEN [68]. Such last fact implies the notion that PTEN protein can move from cytoplasm to nucleus, and backward. However, this was not clear at the very beginning, when everyone thought it was exclusively localized to the cytoplasm. PTEN is abundantly localized in the nucleus of primary, differentiated, and resting cells, while there is a sharp decrease in cancer cells [69–72]. Cell cycle stage and differentiation status are, consequently, related to PTEN localization. The fact that PTEN protein is localized in the nucleus, besides the cytoplasm, has an important role in the tumor suppressor function of such protein. This is proven by the report of patients with more aggressive tumors, such as esophageal squamous cell carcinoma [73], cutaneous melanoma [74, 75], colorectal cancer [76], and pancreatic islet cell tumors [72], having their PTEN protein absent in the nucleus.

## 2. Why Should the Dermatologist Know about PTEN?

Besides using PTEN as a model for genetic disease, the dermatologist should deepen his knowledge of such fascinating pathway for some very practical reasons.

First of all, think of melanoma. Approximately 70% of melanomas have elevated Akt3 signaling both for increased gene copy number and PTEN loss. Consequently, the targeting of (V600E)B-Raf and Akt3 signalling can prevent or treat cutaneous melanocytic lesions. The development of agents specifically targeting these proteins would be very useful, because they have fewer side effects than those inhibiting both normal and mutant B-Raf protein or targeting all three Akt isoforms. Recently [77], a nanoliposomal-ultrasound-mediated approach reported for delivering small interfering RNA (siRNA) specifically targeting (V600E)B-Raf and Akt3 into melanocytic tumors present in skin to retard melanoma development. Novel cationic nanoliposomes stably encapsulate siRNA targeting (V600E)B-Raf or Akt3, providing protection from degradation and facilitating entry into melanoma cells to decrease expression of these proteins. Low-frequency ultrasound using a lightweight four-cymbal transducer array enables penetration of nanoliposomal-siRNA complex throughout the epidermal and dermal layers of laboratory-generated or animal skin. Nanoliposomal-mediated siRNA targeting of (V600E)B-Raf and Akt3 led to a cooperatively acting approximately 65% decrease in early or invasive cutaneous melanoma compared with inhibition of each singly with negligible associated systemic toxicity. Thus, cationic nanoliposomes loaded with siRNA targeting (V600E)B-Raf and Akt3 provide an effective approach for targeted inhibition of early or invasive cutaneous melanomas. Furthermore, it is currently thought that the progression of human cutaneous melanomas behaves in a stepwise fashion, due to accumulating genetic and epigenetic alterations. The combination of PTEN deficiency and Braf activation induces a melanoma *in-situ*-like phenotype without dermal invasion. Further addition of cell autonomous TGF-$\beta$ activation in the context of PTEN deficiency and Braf activation promotes dermal invasion in skin cultures without significantly promoting proliferation *in vitro* and *in vivo*. This proinvasive phenotype of cell autonomous TGF-$\beta$ activation is genetic context dependent, as hyperactivating the TGF-$\beta$ type I receptor without PTEN deficiency and Braf activation failed to induce an invasive behavior. Evidence of genetic interactions among PTEN deficiency, Braf activation, and cell autonomous TGF-$\beta$ activation shows that distinct stages of human melanoma are genetically tractable in the proper tissue architecture [78].

Secondly, multistage skin carcinogenesis is prone to gene synergism. The ablation of PTEN function in mouse epidermis expressing activated Fos leads to hyperplasia, hyperkeratosis, and tumors that move forward to highly differentiated keratoacanthomas, rather than to carcinomas [79].

Thirdly, the dermatologist can be the front office in diagnosing a PHTS. Acral papular neuromatosis [80] and mucocutaneous neuromas [81] can be the first sign of a disease caused by a PTEN mutation. Katona et al. [82] analysed a series of patients affected by mycosis fungoides, showing that LOH studies are a robust method for evaluating genetic abnormalities in mycosis fungoides, and several loci associated with the PTEN appear to be associated with progression from plaque to tumor stage. It has been shown that PTEN was significantly higher in depigmented epidermis, implying that vitiliginous keratinocytes may be more susceptible to TNF-alpha-mediated apoptosis through impaired Akt and NF-kappaB activation. Keratinocytes showing impaired Akt activation demonstrated increased apoptosis with less activation of NF-kappaB. Thus, reduced activation of NF-kappaB via impaired PI3K/Akt activation under increased TNF-alpha levels could result in increased apoptosis of vitiliginous keratinocytes [83].

Finally, the pathogenic role of mutated PTEN involves some conditions collectively called PTEN hamartoma tumor syndrome (PHTS) [84, 85], such as Cowden syndrome [86], Bannayan-Riley-Ruvalcaba syndrome (BRRS) [87], and Proteus syndrome (PS) [88].

Such rare diseases highlight tumoral degeneration in many body organs. Affected patients show usually variable degrees of intellective disability. The cutaneous surface of PHTS patients leads one to suspect the disease being present, many times in a peculiar way. Main examples are speckled penis in BRRS [87] (Figure 2), plantar cerebriform hyperplasia in 70–80% of PS patients [88, 89] (Figure 3), and three or more trichilemmomas in CS [86].

The spectrum of clinical findings associated with PTEN tumor suppressor gene germline mutations includes also mucocutaneous neuromas, as reported by Shaffer et al., who stated that this is an underrecognized manifestation of the gene [81]. These are only some reasons why the knowledge of PTEN is very much needed for the dermatologist, not only for his/her culture, but also for practical issues, such as a targeted therapy in the near future.

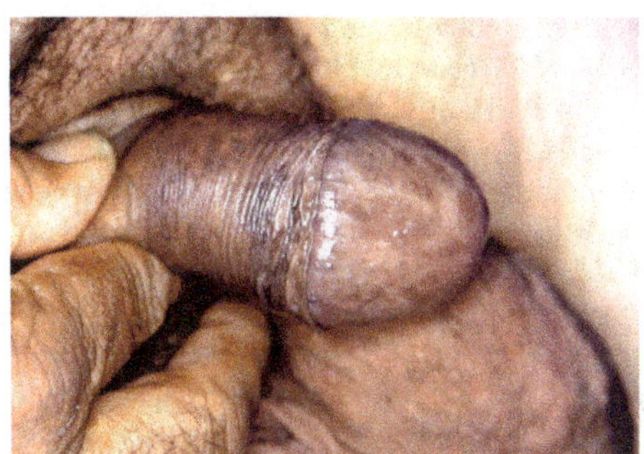

Figure 2: Speckled penis in a patient affected by Bannayan-Riley-Ruvalcaba syndrome.

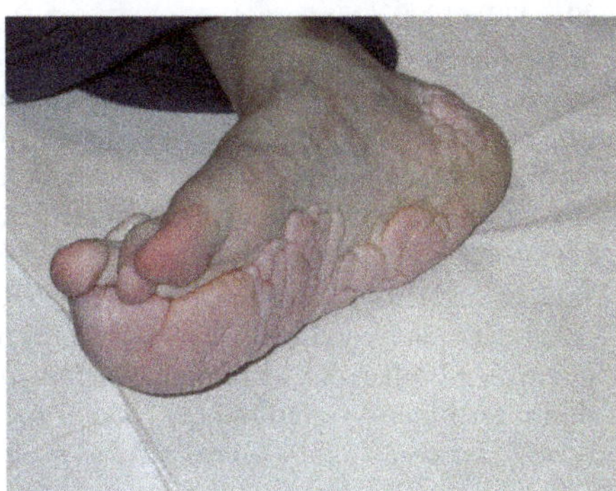

Figure 3: Cerebriform plantar hyperplasia in a girl affected by Proteus syndrome.

## Conflict of Interests

The authors have no conflict of interests to declare.

## References


[1] L. Salmena, A. Carracedo, and P. P. Pandolfi, "Tenets of PTEN tumor suppression," *Cell*, vol. 133, no. 3, pp. 403–414, 2008.

[2] J. Li, C. Yen, D. Liaw et al., "PTEN, a putative protein tyrosine phosphatase gene mutated in human brain, breast, and prostate cancer," *Science*, vol. 275, no. 5308, pp. 1943–1947, 1997.

[3] P. A. Steck, M. A. Pershouse, S. A. Jasser et al., "Identification of a candidate tumour suppressor gene, MMAC1, at chromosome 10q23.3 that is mutated in multiple advanced cancers," *Nature Genetics*, vol. 15, no. 4, pp. 356–362, 1997.

[4] A. G. Knudson, "Mutation and cancer: statistical study of retinoblastoma," *Proceedings of the National Academy of Sciences of the United States of America*, vol. 68, no. 4, pp. 820–823, 1971.

[5] D. Liaw, D. J. Marsh, J. Li et al., "Germline mutations of the PTEN gene in Cowden disease, an inherited breast and thyroid cancer syndrome," *Nature Genetics*, vol. 16, no. 1, pp. 64–67, 1997.

[6] P. E. Weary, R. J. Gorlin, W. C. Gentry, J. E. Comer, and K. E. Greer, "Multiple hamartoma syndrome (Cowden's disease)," *Archives of Dermatology*, vol. 106, no. 5, pp. 682–690, 1972.

[7] J. Chen, P. Lindblom, and A. Lindblom, "A study of the PTEN/MMAC1 gene in 136 breast cancer families," *Human Genetics*, vol. 102, no. 1, pp. 124–125, 1998.

[8] L. H. Saal, S. K. Gruvberger-Saal, C. Persson et al., "Recurrent gross mutations of the PTEN tumor suppressor gene in breast cancers with deficient DSB repair," *Nature Genetics*, vol. 40, no. 1, pp. 102–107, 2008.

[9] D. J. Marsh, J. B. Kum, K. L. Lunetta et al., "PTEN mutation spectrum and genotype-phenotype correlations in Bannayan-Riley-Ruvalcaba syndrome suggest a single entity with Cowden syndrome," *Human Molecular Genetics*, vol. 8, no. 8, pp. 1461–1472, 1999.

[10] M. M. Cohen Jr., "Bannayan-Riley-Ruvalcaba Syndrome: renaming three formerly recognized syndromes as one etiologic entity," *American Journal of Medical Genetics*, vol. 35, no. 2, pp. 291–2292, 1990.

[11] H. R. Wiedemann, G. R. Burgio, P. Aldenhoff, J. Kunze, H. J. Kaufmann, and E. Schirg, "The proteus syndrome. Partial gigantism of the hands and/or feet, nevi, hemihypertrophy, subcutaneous tumors, microcephaly and other skull anomalies and possible accelerated growth and visceral affections," *European Journal of Pediatrics*, vol. 140, pp. 5–12, 1983.

[12] X. P. Zhou, D. J. Marsh, H. Hampel, J. B. Mulliken, O. Gimm, and C. Eng, "Germline and germline mosaic PTEN mutations associated with a Proteus-like syndrome of hemihypertrophy, lower limb asymmetry, arteriovenous malformations and lipomatosis," *Human Molecular Genetics*, vol. 9, no. 5, pp. 765–768, 2000.

[13] H. M. Dastur, S. K. Pandya, and D. H. Deshpande, "Diffuse cerebellar hypertrophy. (Lhermitte Duclos disease)," *Neurology India*, vol. 23, no. 1, pp. 53–56, 1975.

[14] M. P. Myers, J. P. Stolarov, C. Eng et al., "P-TEN, the tumor suppressor from human chromosome 10q23, is a dual-specificity phosphatase," *Proceedings of the National Academy of Sciences of the United States of America*, vol. 94, no. 17, pp. 9052–9057, 1997.

[15] D. M. Li and H. Sun, "PTEN/MMAC1/TEP1 suppresses the tumorigenicity and induces G1 cell cycle arrest in human glioblastoma cells," *Proceedings of the National Academy of Sciences of the United States of America*, vol. 95, no. 26, pp. 15406–15411, 1998.

[16] J. M. Denu, J. A. Stuckey, M. A. Saper, and J. E. Dixon, "Form and function in protein dephosphorylation," *Cell*, vol. 87, no. 3, pp. 361–364, 1996.

[17] M. P. Myers and N. K. Tonks, "PTEN: sometimes taking it off can be better than putting it on," *American Journal of Human Genetics*, vol. 61, no. 6, pp. 1234–1238, 1997.

[18] W. Zundel, C. Schindler, D. Haas-Kogan et al., "Loss of PTEN facilitates HIF-1-mediated gene expression," *Genes and Development*, vol. 14, no. 4, pp. 391–396, 2000.

[19] P. L. M. Dahia, R. C. T. Aguiar, J. Alberta et al., "PTEN is inversely correlated with the cell survival factor Akt/PKB and is inactivated via multiple mechanisms in haematological malignancies," *Human Molecular Genetics*, vol. 8, no. 2, pp. 185–193, 1999.

[20] M. Tamura, J. Gu, K. Matsumoto, S. I. Aota, R. Parsons, and K. M. Yamada, "Inhibition of cell migration, spreading, and focal adhesions by tumor suppressor PTEN," *Science*, vol. 280, no. 5369, pp. 1614–1617, 1998.

[21] A. Di Cristofano, B. Pesce, C. Cordon-Cardo, and P. P. Pandolfi, "Pten is essential for embryonic development and tumour suppression," *Nature Genetics*, vol. 19, no. 4, pp. 348–355, 1998.

[22] A. Suzuki, J. L. de La Pompa, V. Stambolic et al., "High cancer susceptibility and embryonic lethality associated with mutation of the PTEN tumor suppressor gene in mice," *Current Biology*, vol. 8, no. 21, pp. 1169–1178, 1998.

[23] K. Podsypanina, L. H. Ellenson, A. Nemes et al., "Mutation of Pten/Mmacl in mice causes neoplasia in multiple organ systems," *Proceedings of the National Academy of Sciences of the United States of America*, vol. 96, no. 4, pp. 1563–1568, 1999.

[24] L. P. Weng, W. M. Smith, J. L. Brown, and C. Eng, "PTEN inhibits insulin-stimulated MEK/MAPK activation and cell growth by blocking IRS-1 phosphorylation and IRS-1/Grb-2/Sos complex formation in a breast cancer model," *Human Molecular Genetics*, vol. 10, no. 6, pp. 605–616, 2001.

[25] L. P. Weng, J. L. Brown, K. M. Baker, M. C. Ostrowski, and C. Eng, "PTEN blocks insulin-mediated ETS-2 phosphorylation through MAP kinase, independently of the phosphoinositide 3-kinase pathway," *Human Molecular Genetics*, vol. 11, no. 15, pp. 1687–1696, 2002.

[26] H. B. Newton, "Molecular neuro-oncology and development of targeted therapeutic strategies for brain tumors. Part 2: PI3K/Akt/PTEN, mTOR, SHH/PTCH and angiogenesis," *Expert Review of Anticancer Therapy*, vol. 4, no. 1, pp. 105–128, 2004.

[27] I. Vivanco and C. L. Sawyers, "The phosphatidylinositol 3-kinase-AKT pathway in human cancer," *Nature Reviews Cancer*, vol. 2, no. 7, pp. 489–501, 2002.

[28] K. A. Martin and J. Blenis, "Coordinate regulation of translation by the PI 3-kinase and mTOR pathways," *Advances in Cancer Research*, vol. 86, pp. 1–39, 2002.

[29] M. P. Wymann and L. Pirola, "Structure and function of phosphoinositide 3-kinases," *Biochimica et Biophysica Acta*, vol. 1436, no. 1-2, pp. 127–150, 1998.

[30] L. E. Rameh and L. C. Cantley, "The role of phosphoinositide 3-kinase lipid products in cell function," *The Journal of Biological Chemistry*, vol. 274, no. 13, pp. 8347–8350, 1999.

[31] A. Di Cristofano, P. Kotsi, Y. F. Peng, C. Cordon-Cardo, K. B. Elkon, and P. P. Pandolfi, "Impaired Fas response and autoimmunity in Pten(+/−) mice," *Science*, vol. 285, no. 5436, pp. 2122–2125, 1999.

[32] A. Di Cristofano, M. De Acetis, A. Koff, C. Cordon-Cardo, and P. P Pandolfi, "Pten and p27KIP1 cooperate in prostate cancer tumor suppression in the mouse," *Nature Genetics*, vol. 27, no. 2, pp. 222–224, 2001.

[33] P. L. M. Dahia, "PTEN, a unique tumor suppressor gene," *Endocrine-Related Cancer*, vol. 7, no. 2, pp. 115–129, 2000.

[34] I. C. Gray, L. M. D. Stewart, S. M. A. Phillips et al., "Mutation and expression analysis of the putative prostate tumour-suppressor gene PTEN," *British Journal of Cancer*, vol. 78, no. 10, pp. 1296–1300, 1998.

[35] Y. E. Whang, X. Wu, H. Suzuki et al., "Inactivation of the tumor suppressor PTEN/MMAC1 in advanced human prostate cancer through loss of expression," *Proceedings of the National Academy of Sciences of the United States of America*, vol. 95, no. 9, pp. 5246–5250, 1998.

[36] I. U. Ali, L. M. Schriml, and M. Dean, "Mutational spectra of PTEN/MMAC1 gene: a tumor suppressor with lipid phosphatase activity," *Journal of the National Cancer Institute*, vol. 91, no. 22, pp. 1922–1932, 1999.

[37] S. Bose, S. I. Wang, M. B. Terry, H. Hibshoosh, and R. Parsons, "Allelic loss of chromosome 10q23 is associated with tumor progression in breast carcinomas," *Oncogene*, vol. 17, no. 1, pp. 123–127, 1998.

[38] H. E. Feilotter, V. Coulon, J. L. McVeigh et al., "Analysis of the 10q23 chromosomal region and the PTEN gene in human

sporadic breast carcinoma," *British Journal of Cancer*, vol. 79, no. 5-6, pp. 718–723, 1999.

[39] K. A. Waite and C. Eng, "Protean PTEN: form and function," *American Journal of Human Genetics*, vol. 70, no. 4, pp. 829–844, 2002.

[40] Z. Chen, L. C. Trotman, D. Shaffer et al., "Crucial role of p53-dependent cellular senescence in suppression of Pten-deficient tumorigenesis," *Nature*, vol. 436, no. 7051, pp. 725–730, 2005.

[41] J. Campisi and F. d'Adda Di Fagagna, "Cellular senescence: when bad things happen to good cells," *Nature Reviews Molecular Cell Biology*, vol. 8, no. 9, pp. 729–740, 2007.

[42] D. M. Li and H. Sun, "TEP1, encoded by a candidate tumor suppressor locus, is a novel protein tyrosine phosphatase regulated by transforming growth factor $\beta$," *Cancer Research*, vol. 57, no. 11, pp. 2124–2129, 1997.

[43] T. Virolle, E. D. Adamson, V. Baron et al., "The Egr-1 transcription factor directly activates PTEN during irradiation-induced signalling," *Nature Cell Biology*, vol. 3, no. 12, pp. 1124–1128, 2001.

[44] R. A. Moorehead, C. V. Hojilla, I. De Belle et al., "Insulin-like growth factor-II regulates PTEN expression in the mammary gland," *The Journal of Biological Chemistry*, vol. 278, no. 50, pp. 50422–50427, 2003.

[45] L. Patel, I. Pass, P. Coxon, C. P. Downes, S. A. Smith, and C. H. Macphee, "Tumor suppressor and anti-inflammatory actions of PPAR$\gamma$ agonists are mediated via upregulation of PTEN," *Current Biology*, vol. 11, no. 10, pp. 764–768, 2001.

[46] V. Stambolic, D. MacPherson, D. Sas et al., "Regulation of PTEN transcription by p53," *Molecular Cell*, vol. 8, no. 2, pp. 317–325, 2001.

[47] T. Palomero, M. L. Sulis, M. Cortina et al., "Mutational loss of PTEN induces resistance to NOTCH1 inhibition in T-cell leukemia," *Nature Medicine*, vol. 13, no. 10, pp. 1203–1210, 2007.

[48] K. Hettinger, F. Vikhanskaya, M. K. Poh et al., "c-Jun promotes cellular survival by suppression of PTEN," *Cell Death and Differentiation*, vol. 14, no. 2, pp. 218–229, 2007.

[49] D. Xia, H. Srinivas, Y. H. Ahn et al., "Mitogen-activated protein kinase kinase-4 promotes cell survival by decreasing PTEN expression through an NF$\kappa$B-dependent pathway," *The Journal of Biological Chemistry*, vol. 282, no. 6, pp. 3507–3519, 2007.

[50] W. H. Chappell, T. D. Green, J. D. Spengeman, J. A. McCubrey, S. M. Akula, and F. E. Bertrand, "Increased protein expression of the PTEN tumor suppressor in the presence of constitutively active notch-1," *Cell Cycle*, vol. 4, no. 10, pp. 1389–1395, 2005.

[51] J. T. Whelan, S. L. Forbes, and F. E. Bertrand, "CBF-1 (RBP-J$\kappa$) binds to the PTEN promoter and regulates PTEN gene expression," *Cell Cycle*, vol. 6, no. 1, pp. 80–84, 2007.

[52] J. M. García, J. Silva, C. Peña et al., "Promoter methylation of the PTEN gene is a common molecular change in breast cancer," *Genes Chromosomes and Cancer*, vol. 41, no. 2, pp. 117–124, 2004.

[53] A. Goel, C. N. Arnold, D. Niedzwiecki et al., "Frequent inactivation of PTEN by promoter hypermethylation in microsatellite instability-high sporadic colorectal cancers," *Cancer Research*, vol. 64, no. 9, pp. 3014–3021, 2004.

[54] Y. H. Kang, S. L. Hye, and H. K. Woo, "Promoter methylation and silencing of PTEN in gastric carcinoma," *Laboratory Investigation*, vol. 82, no. 3, pp. 285–291, 2002.

[55] F. Meng, R. Henson, M. Lang et al., "Involvement of human micro-RNA in growth and response to chemotherapy in human cholangiocarcinoma cell lines," *Gastroenterology*, vol. 130, no. 7, pp. 2113–2129, 2006.

[56] F. Meng, R. Henson, H. Wehbe-Janek, K. Ghoshal, S. T. Jacob, and T. Patel, "MicroRNA-21 regulates expression of the PTEN tumor suppressor gene in human hepatocellular cancer," *Gastroenterology*, vol. 133, no. 2, pp. 647–658, 2007.

[57] J. A. Chan, A. M. Krichevsky, and K. S. Kosik, "MicroRNA-21 is an antiapoptotic factor in human glioblastoma cells," *Cancer Research*, vol. 65, no. 14, pp. 6029–6033, 2005.

[58] M. L. Si, S. Zhu, H. Wu, Z. Lu, F. Wu, and Y. Y. Mo, "miR-21-mediated tumor growth," *Oncogene*, vol. 26, no. 19, pp. 2799–2803, 2007.

[59] S. Volinia, G. A. Calin, C. G. Liu et al., "A microRNA expression signature of human solid tumors defines cancer gene targets," *Proceedings of the National Academy of Sciences of the United States of America*, vol. 103, no. 7, pp. 2257–2261, 2006.

[60] N. R. Leslie and C. P. Downes, "PTEN function: how normal cells control it and tumour cells lose it," *Biochemical Journal*, vol. 382, no. 1, pp. 1–11, 2004.

[61] F. Vazquez, S. R. Grossman, Y. Takahashi, M. V. Rokas, N. Nakamura, and W. R. Sellers, "Phosphorylation of the PTEN tail acts as an inhibitory switch by preventing its recruitment into a protein complex," *The Journal of Biological Chemistry*, vol. 276, no. 52, pp. 48627–48630, 2001.

[62] K. Okumura, M. Mendoza, R. M. Bachoo, R. A. DePinho, W. K. Cavenee, and F. B. Furnari, "PCAF modulates PTEN activity," *The Journal of Biological Chemistry*, vol. 281, no. 36, pp. 26562–26568, 2006.

[63] S. R. Lee, K. S. Yang, J. Kwon, C. Lee, W. Jeong, and S. G. Rhee, "Reversible inactivation of the tumor suppressor PTEN by $H_2O_2$," *The Journal of Biological Chemistry*, vol. 277, no. 23, pp. 20336–20342, 2002.

[64] D. Nandi, P. Tahiliani, A. Kumar, and D. Chandu, "The ubiquitin-proteasome system," *Journal of Biosciences*, vol. 31, no. 1, pp. 137–155, 2006.

[65] J. Torres and R. Pulido, "The tumor suppressor PTEN is phosphorylated by the protein kinase CK2 at its C terminus. Implications for PTEN stability to proteasome-mediated degradation," *The Journal of Biological Chemistry*, vol. 276, no. 2, pp. 993–998, 2001.

[66] W. Wu, X. Wang, W. Zhang et al., "Zinc-induced PTEN protein degradation through the proteasome pathway in human airway epithelial cells," *The Journal of Biological Chemistry*, vol. 278, no. 30, pp. 28258–28263, 2003.

[67] Y. Tang and C. Eng, "p53 down-regulates phosphatase and tensin homologue deleted on chromosome 10 protein stability partially through caspase-mediated degradation in cells with proteasome dysfunction," *Cancer Research*, vol. 66, no. 12, pp. 6139–6148, 2006.

[68] L. C. Trotman, X. Wang, A. Alimonti et al., "Ubiquitination regulates PTEN nuclear import and tumor suppression," *Cell*, vol. 128, no. 1, pp. 141–156, 2007.

[69] O. Gimm, A. Perren, L. P. Weng et al., "Differential nuclear and cytoplasmic expression of PTEN in normal thyroid tissue, and benign and malignant epithelial thyroid tumors," *American Journal of Pathology*, vol. 156, no. 5, pp. 1693–1700, 2000.

[70] M. E. Ginn-Pease and C. Eng, "Increased nuclear phosphatase and tensin homologue deleted on chromosome 10 is associated with G0-G1 in MCF-7 cells," *Cancer Research*, vol. 63, no. 2, pp. 282–286, 2003.

[71] M. B. Lachyankar, N. Sultana, C. M. Schonhoff et al., "A role for nuclear PTEN in neuronal differentiation," *Journal of Neuroscience*, vol. 20, no. 4, pp. 1404–1413, 2000.

[72] A. Perren, P. Komminoth, P. Saremaslani et al., "Mutation and expression analyses reveal differential subcellular compartmentalization of PTEN in endocrine pancreatic tumors compared to normal islet cells," *American Journal of Pathology*, vol. 157, no. 4, pp. 1097–1103, 2000.

[73] M. Tachibana, M. Shibakita, S. Ohno et al., "Expression and prognostic significance of PTEN product protein in patients with esophageal squamous cell carcinoma," *Cancer*, vol. 94, no. 7, pp. 1955–1960, 2002.

[74] D. C. Whiteman, X. P. Zhou, M. C. Cummings, S. Pavey, N. K. Hayward, and C. Eng, "Nuclear PTEN expression and clinicopathologic features in a population-based series of primary cutaneous melanoma," *International Journal of Cancer*, vol. 99, no. 1, pp. 63–67, 2002.

[75] X. P. Zhou, O. Gimm, H. Hampel, T. Niemann, M. J. Walker, and C. Eng, "Epigenetic PTEN silencing in malignant melanomas without PTEN mutation," *American Journal of Pathology*, vol. 157, no. 4, pp. 1123–1128, 2000.

[76] X. P. Zhou, A. Loukola, R. Salovaara et al., "PTEN mutational spectra, expression levels, and subcellular localization in microsatellite stable and unstable colorectal cancers," *American Journal of Pathology*, vol. 161, no. 2, pp. 439–447, 2002.

[77] M. A. Tran, R. Gowda, A. Sharma et al., "Targeting V600EB-Raf and Akt3 using nanoliposomal-small interfering RNA inhibits cutaneous melanocytic lesion development," *Cancer Research*, vol. 68, no. 18, pp. 7638–7649, 2008.

[78] R. S. Lo and O. N. Witte, "Transforming growth factor-$\beta$ activation promotes genetic context-dependent invasion of immortalized melanocytes," *Cancer Research*, vol. 68, no. 11, pp. 4248–4257, 2008.

[79] D. Yao, C. L. Alexander, J. A. Quinn, W. C. Chan, H. Wu, and D. A. Greenhalgh, "Fos cooperation with PTEN loss elicits keratoacanthoma not carcinoma, owing to p53/p21WAF-induced differentiation triggered by GSK3$\beta$ inactivation and reduced AKT activity," *Journal of Cell Science*, vol. 121, no. 10, pp. 1758–1769, 2008.

[80] M. Ferran, E. Bussaglia, C. Lazaro, X. Matias-Guiu, and R. M. Pujol, "Acral papular neuromatosis: an early manifestation of Cowden syndrome," *British Journal of Dermatology*, vol. 158, no. 1, pp. 174–176, 2008.

[81] J. V. Schaffer, H. Kamino, A. Witkiewicz, J. M. McNiff, and S. J. Orlow, "Mucocutaneous neuromas: an underrecognized manifestation of PTEN hamartoma-tumor syndrome," *Archives of Dermatology*, vol. 142, no. 5, pp. 625–632, 2006.

[82] T. M. Katona, D. P. O'Malley, L. Cheng et al., "Loss of heterozygosity analysis identifies genetic abnormalities in mycosis fungoides and specific loci associated with disease progression," *American Journal of Surgical Pathology*, vol. 31, no. 10, pp. 1552–1556, 2007.

[83] N. H. Kim, S. Jeon, H. J. Lee, and A. Y. Lee, "Impaired PI3K/Akt activation-mediated NF-$\kappa$B inactivation under elevated TNF-$\alpha$ is more vulnerable to apoptosis in vitiliginous keratinocytes," *Journal of Investigative Dermatology*, vol. 127, no. 11, pp. 2612–2617, 2007.

[84] C. Eng, "PTEN: one gene, Many syndromes," *Human Mutation*, vol. 22, no. 3, pp. 183–198, 2003.

[85] C. Romano, "Genetics of PTEN hamatoma tumor syndrome (PHTS)," in *Neurocutaneous Disorders. Phakomatoses and Hamartoneoplastis Syndromes*, M. Ruggieri, I. Pascual Castroviejo, and C. Di Rocco, Eds., Springer, 2008.

[86] C. Hildenbrand, W. H. C. Burgdorf, and S. Lautenschlager, "Cowden syndrome—diagnostic skin signs," *Dermatology*, vol. 202, no. 4, pp. 362–366, 2001.

[87] Y. M. C. Hendriks, J. T. C. M. Verhallen, J. J. Van der Smagt et al., "Bannayan-Riley-Ruvalcaba syndrome: further delineation of the phenotype and management of PTEN mutation-positive cases," *Familial Cancer*, vol. 2, no. 2, pp. 79–85, 2003.

[88] E. Satter, "Proteus syndrome: 2 Case reports and a review or the literature," *Cutis*, vol. 80, no. 4, pp. 297–302, 2007.

[89] C. Schepis, D. Greco, M. Siragusa, and C. Romano, "Cerebriform plantar hyperplasia: The major cutaneous feature of Proteus syndrome," *International Journal of Dermatology*, vol. 47, no. 4, pp. 374–376, 2008.

# 5

# Population Genetic Analysis of *Lobelia rhynchopetalum* Hemsl. (Campanulaceae) Using DNA Sequences from *ITS* and Eight Chloroplast DNA Regions

**Mulatu Geleta and Tomas Bryngelsson**

*Department of Plant Breeding and Biotechnology, Swedish University of Agricultural Sciences, P.O. Box 101, 230 53 Alnarp, Sweden*

Correspondence should be addressed to Mulatu Geleta, mulatu.geleta.dida@slu.se

Academic Editors: N. Kouprina and S. Mastana

DNA sequence data from the internal transcribed spacer of nuclear ribosomal DNA and eight chloroplast DNA regions were used to investigate haplotypic variation and population genetic structure of the Afroalpine giant lobelia, *Lobelia rhynchopetalum*. The study was based on eight populations sampled from two mountain systems in Ethiopia. A total of 20 variable sites were obtained, which resulted in 13 unique haplotypes and an overall nucleotide diversity (ND) of 0.281 ± 0.15 and gene diversity (GD) of 0.85 ± 0.04. Analysis of molecular variance (AMOVA) revealed a highly significant variation ($P < 0.001$) among populations ($F_{ST}$), and phylogenetic analysis revealed that populations from the two mountain systems formed their own distinct clade with >90% bootstrap support. Each population should be regarded as a significant unit for conservation of this species. The primers designed for this study can be applied to any *Lobelia* and other closely related species for population genetics and phylogenetic studies.

## 1. Introduction

*Lobelia* is the largest genus within the subfamily Lobelioideae of Campanulaceae family comprising over 350 species that range from small herbs to giant woody plants. *Lobelia rhynchopetalum* Hemsl., which belongs to the subgenus Tupa, section Rhynchopetalum [1], is one of the 21 giant lobelia species of eastern Africa that represent a premier botanical example of spectacular evolutionary radiations [2, 3]. It belongs to the most famous giant lobelia group exhibiting a giant-rosette growthform [3]. Giant lobelias predated the formation of tall mountains in eastern Africa, and most evolution occurred in parallel up the mountains [4]. These authors also suggested that an extinct forest species gave rise to several alpine giant lobelias, possibly from which *L. rhynchopetalum* has evolved. *L. rhynchopetalum* and other giant lobelias of eastern Africa have the same chromosome number ($2n = 28$) as that of their progenitor [5, 6] and, thus, their evolution has occurred without a change in chromosome number.

*L. rhynchopetalum* is a monocarpic perennial species endemic to the Ethiopian drained sites of the Afroalpine ecoregion (e.g. [1, 3]). The plant is frost tolerant and has an up to two-meter-tall unbranched stem with a large pith and thick and leathery leaves (e.g. [1]), which are suggested to be adaptations to the high altitude tropical environment [7]. Once the plant has flowered and set seeds, it dies, leaving a tall hollow and dried-out stem. The seed capsules contain a huge number of tiny yellow seeds that can be easily dispersed by wind [8]. *L. rhynchopetalum* is most prominent and noticeable in the Afroalpine part of the Bale and Simien mountain systems, commonly within an altitudinal range of 3600–4500 m asl, where it serves as a tourist attraction. The significance of this species is, therefore, not only ecological but also recreational and economic. Despite its significance, it is one of the least studied lobelia species at a molecular level and, to our knowledge, no DNA sequence data from this species is available in GenBank. Little is known about its population genetics, which makes it difficult to conserve at its full range of genetic diversity in the presence of threats from fire and overgrazing.

Fast evolving regions from nuclear and chloroplast genomes have been used to generate intraspecific DNA sequence data for plant population genetic studies [9–12], as it enables us to reveal the distribution of haplotypes, both within and among populations, and to identify species genetic diversity hotspots.

In the present study, DNA sequence data from the internal transcribed spacers (*ITS*) of nuclear ribosomal DNA (rDNA) and eight chloroplast DNA (cpDNA) regions were generated from *L. rhynchopetalum* with the objectives of (1) population genetic and phylogenetic analyses for its conservation and evolutionary significance, (2) evaluating the utility of these DNA regions for population genetic and phylogenetic analyses, particularly within the genus *Lobelia*, and (3) contributing DNA sequence data from this species to Genbank so that it can be used for broader phylogenetic and phylogeographic analyses in combination with DNA sequence data from other *Lobelia* species.

## 2. Materials and Methods

*2.1. Plant Material.* A total of eight populations of *L. rhynchopetalum* collected from the Bale (6°48′N–7°08′N and 39°45′E–39°57′E) and Simien (13°05′N–13°25′N and 37°50′E-38°30′E) mountain systems were used in this study (Table 1). Goba-1, -2, and -3 populations were from the Bale mountains whereas Debark-1-5 populations were from the Simien mountains. Each population was represented by five individual plants. *Lobelia erinus* L. was included as an outgroup species for the phylogenetic analysis of *L. rhynchopetalum* populations and for comparative assessment of the DNA regions used in this study (see Table 2).

*2.2. DNA Extraction.* DNA was extracted from silica-gel-dried young leaves using a modified CTAB procedure as described in [13] except that 100 mg of fine powder of leaf material was used instead of 300 mg. DNA quality and concentration were measured using a Nanodrop ND-1000 spectrophotometer (Saveen Werner, Sweden).

*2.3. PCR and Sequencing.* Target DNA regions (Table 2) were amplified using a GeneAMP PCR system 9700 thermocycler with the following temperature profiles: initial 3 min denaturing at 94°C and final 7 min extension at 72°C with the intervening 30 cycles of 1 min denaturing at 94°C, 1 min primer annealing at 48°C, and 2 min primer extension at 72°C. The *ITS* was amplified and sequenced using *ITS5F* and *ITS4R* primers [14, Table 2]. Eighteen new primers were designed to amplify and sequence the *trnT-trnL*, *trnfM-trnS*, *petN-trnC*, *trnG-trnR*, *psbT-psbB*, *clpP intron-2*, *3′trnK-matK*, and *psbD-trnT* regions of cpDNA (Table 2) using the primer3 primer designing program [15]. The primers were designed to the conserved regions based on the aligned DNA sequences of *Trachelium caeruleum* L. (Campanulaceae) and *Helianthus annuus* L. (Asteraceae) (accession numbers *EU090187* and *DQ383815*, resp.). These primers were named (Table 2) based on their 5′ position (forward primers) and 3′ position (reverse primers) in the *T. caeruleum* complete chloroplast genome sequence.

The *trn*T-*trn*L intergenic spacer was amplified using primers *37258F* and *37820R*. The reverse primer, *37820R*, was used to sequence this region. The *trnfM-trnS* region contains the *trnfM* gene, *trnfM-trnG* intergenic spacer, *trnG* gene, *trnG-psbZ* intergenic spacer, *psbZ* gene, *psbZ-trnS* integenic spacer, and *trnS* gene in that order. This region was amplified in two segments with a combination of four primers (Table 2). The first pair of primers (*48992F* and *49584R*) amplified part of the *trnfM* gene and the full length of the *trnG* gene and the *trnfM-trnG* and the *trnG-psbZ* intergenic spacers. This part of the *trnfM-trnS* region was sequenced using primer *49584R*. The second pair of primers (*49595F* and *50079R*) amplified part of the *psbZ* gene and the *psbZ-trnS* intergenic spacer that were sequenced using primer *50079R*. Similarly, part of the *petN-trnC* intergenic spacer was amplified and sequenced using primers *58099F* and *58955R*.

Primers *10002F* and *10226R* were used to amplify the *trnG-trnR* intergenic spacer (Table 2). The complete sequence of this spacer was obtained using the forward primer, *10002F*. The *psbT-psbB* intergenic spacer was amplified using primers *26837F* and *27102R* and sequenced using the reverse primer, *27102R*. Primers *111104F* and *111454R* were used to amplify *intron-2* (the intron between *exon-2* and *exon-3*) of the *clpP* gene, whose partial sequence was obtained by using primer *111454R*. The *3′trnK-matK* portion of the *trnK* intron was amplified using primers *1825F* and *2195R*, and the amplified fragment was sequenced using primer *1825F*. Similarly, the *psbD-trnT* intergenic spacer was amplified using primers *53562F* and *54107R*. The reverse primer, *54107R*, was used to obtain the partial sequence of this spacer (Table 2). The PCR products were purified by QIAquick PCR purification kit (Qiagen GmbH, Germany) using a microcentrifuge as recommended by the manufacturer. Eight microlitre of purified PCR product (50–100 ng) was mixed with 2 $\mu$L of 5 $\mu$M sequencing primer and sent to the sequencing facility at the University of Oslo (http://www.bio.uio.no/ABI-lab/), where DNA sequencing was carried out. The representative nucleotide sequences of *ITS* and the eight cpDNA regions of *L. rhynchopetalum* were submitted to nucleotide sequence database (NCBI GenBank), and their accession numbers are given in Table 2.

*2.4. Sequence Alignment and Data Analyses.* DNA sequences were edited using BIOEDIT version 7.0.5 [16], and the quality of the sequences was visually inspected using Sequence Scanner version 1.0 (Applied Biosystems). Sequences were aligned using Clustal X version 1.81 [17]. PAUP* 4.0 Beta 10 [18] was used to construct a bootstrap 50% majority rule consensus tree based on Kimura distance coefficient [19]. Trees were generated using heuristic search with the tree-bisection-reconnection (TBR) branch swapping algorithm, and clade support was estimated using 1000 bootstrap replicates (starting trees were obtained via neighbor-joining, and initial Maxtree was set to 1000). Various population genetic analyses including gene and nucleotide diversity, analysis

Table 1: Genetic analysis of *L. rhynchopetalum* populations from the Bale mountains (BM) and Simien mountains (SM) at different hierarchical levels.

| Population/group | Location-mountains | Altitude (m asl) | NGC /SS | NH | NPS | GD | ND | MIHD ± SD | $\pi$ | $\theta_\pi$ | $\theta_S$ | $\theta_K$* | $D$ |
|---|---|---|---|---|---|---|---|---|---|---|---|---|---|
| Goba-1[ad] | 15 km from Goba to Mena-BM | 4000–4300 | 5 | 4 | 5 | 0.90 ± 0.16 | 0.071 ± −0.052 | 0.081 ± 0.045 | 2.99 ± 1.87 | 2.99 ± 2.19 | 2.40 ± 1.51 | 7.11 [1.54, 33.08] | 1.12 |
| Goba-2[a] | 16.8 km from Goba to Mena-BM | 4000–4300 | 5 | 1 | 0 | 0.00 ± 0.00 | 0.000 ± −0.000 | 0.000 ± 0.000 | 0.00 ± 0.00 | 0.00 ± 0.00 | 0.00 ± 0.00 | 0.00 [0.00, 0.00] | 0.00 |
| Goba-3 | 18.5 km from Goba to Mena-BM | 4000–4300 | 5 | 1 | 0 | 0.00 ± 0.00 | 0.000 ± 0.000 | 0.000 ± 0.000 | 0.00 ± 0.00 | 0.00 ± 0.00 | 0.00 ± 0.00 | 0.00 [0.00, 0.00] | 0.00 |
| Debark-1[bcdef] | 47 km from Debark to Sankaber-SM | 3600–3800 | 5 | 5 | 6 | 1.00 ± 0.12 | 0.038 ± 0.031 | 0.042 ± 0.031 | 1.68 ± 1.17 | 1.68 ± 1.37 | 1.92 ± 1.27 | — | −1.09 |
| Debark-2[b] | 48 km from Debark to Sankaber-SM | 3600–3800 | 5 | 1 | 0 | 0.00 ± 0.00 | 0.000 ± 0.000 | 0.000 ± 0.000 | 0.00 ± 0.00 | 0.00 ± 0.00 | 0.00 ± 0.00 | 0.00 [0.00, 0.00] | 0.00 |
| Debark-3[be] | 51 km from Debark to Sankaber-SM | 3600–3800 | 5 | 2 | 1 | 0.60 ± 0.17 | 0.014 ± 0.015 | 0.024 ± 0.024 | 0.61 ± 0.57 | 0.61 ± 0.67 | 0.48 ± 0.48 | 0.69 [0.15, 3.19] | 1.22 |
| Debark-4[b] | 53 km from Debark to Sankaber-SM | 3600–3800 | 5 | 2 | 2 | 0.40 ± 0.24 | 0.019 ± 0.019 | 0.048 ± 0.034 | 0.83 ± 0.70 | 0.83 ± 0.81 | 0.96 ± 0.76 | 0.69 [0.15, 3.18] | −0.97 |
| Debark-5[cf] | 54.5 km from Debark to Sankaber-SM | 3600–3800 | 5 | 2 | 1 | 0.40 ± 0.24 | 0.009 ± 0.012 | 0.024 ± 0.024 | 0.41 ± 0.44 | 0.41 ± 0.52 | 0.48 ± 0.48 | 0.69 [0.15, 3.18] | −0.82 |
| Bale | | | 15 | 5 | 5 | 0.70 ± 0.08 | 0.035 ± 0.025 | 0.071 ± 0.042 | 1.48 ± 0.95 | 1.48 ± 1.06 | 1.54 ± 0.85 | 2.20 [0.77, 5.97] | −0.28 |
| Simien | | | 25 | 8 | 9 | 0.70 ± 0.08 | 0.025 ± 0.018 | 0.046 ± 0.032 | 1.11 ± 0.75 | 1.11 ± 0.83 | 1.85 ± 0.89 | 3.67 [1.57, 8.21] | −1.33 |
| All | | | 40 | 13 | 20 | 0.85 ± 0.04 | 0.28 ± 0.15 | 0.201 ± 0.062 | 8.23 ± 4.29 | 8.23 ± 4.32 | 3.99 ± 2.16 | 6.29 [3.19, 12.06] | 0.81 |

NGC/SS: number of gene copies/sample size; NH: number of haplotypes; NPS: number of polymorphic sites; GD: gene diversity; ND: nucleotide diversity (based on polymorphic loci only); MIHD: mean interhaplotypic distance (based on Kimura 2P method); SD: standard deviation (for both the sampling and the stochastic processes); $\pi$: mean number of pairwise difference (based on polymorphic loci only). Populations sharing the superscript a, b, or c have some haplotypes in common. Populations sharing the superscript d, e, or f were not significantly different from each other when tested using global test of population differentiation, based on Markov chain length of 10000 (significance level = 0.05). $D$: Tajima's test of selective neutrality. $\theta_\pi$, $\theta_S$, and $\theta_K$ are different estimators of theta ($\theta$). *: Values in the square brackets are 95% confidence interval limits around $\theta_K$. Note: for analysis at mountain-system level, all individuals of the same category were pooled together as a single unit.

Table 2: (1) Name and sequence of primers used for the amplification and sequencing of target DNA regions; (2) Genbank accession numbers of representative sequences and some sequence characteristics of each DNA region.

| Target region | Primer name | Primer sequence | GBAN | NVSWLr | TAL[e] | NVS[e] | %VS[e] | NPIS[e] | %PIS[e] | %PIVS[e] |
|---|---|---|---|---|---|---|---|---|---|---|
| *ITS*[a] | *ITS5F*[c]<br>*ITS4R*[d] | 5′-GGAAGGAGAAGTCGTAACAAGG-3′<br>5′-TCCTCCGCTTATTGATATGC-3′ | FJ664108-9 | 11 | 719 | 198 | 27.5 | 51 | 7.1 | 25.8 |
| *trn*T$_{UGU}$-*trn* L$_{UAA}$[b] | 37258F[d]<br>37820R[c] | 5′-TGCAATGCTCTAACCTCTGA-3′<br>5′-CGATTTTATCATTTATCTATCTCCAA-3′ | FJ664110-1 | 2 | 525 | 47 | 9.0 | 15 | 2.9 | 31.9 |
| *trnf*M$_{CAU}$-*trn* S$_{UGA}$[b] | 48992F[d]<br>49584R[c]<br>49595F[d]<br>50079R[c] | 5′-GTAGCTCGCAAGGCTCATAAC-3′<br>5′-TTCTGGTGGGTATCCTTAATTCTC-3′<br>5′-ATACCCACCAGAAAGACTAATCCA-3′<br>5′-CCATCTCTCCGAAAGACAATTTTA-3′ | FJ664119-20 | 1 | 976 | 134 | 13.7 | 31 | 3.2 | 23.1 |
| *pet*N-*trn* C$_{GCA}$[b] | 58099F[c]<br>58955R[c] | 5′-CCCAAGCGAGACTTACTATATCCA-3′<br>5′-AAATCCTTTTTCCCCAGTTCAA-3′ | FJ664115-6 | 3 | 291 | 20 | 6.9 | 12 | 4.1 | 60.0 |
| *trn*G$_{UCC}$-*trn* R$_{UCU}$[b] | 10002F[c]<br>10226R[d] | 5′-CTAGCCTTCCAAGCTAACGATG-3′<br>5′-GACCTCTGTCCTATCCATTAGACAAT-3′ | FJ664121-2 | 2 | 195 | 4 | 2.1 | 4 | 2.1 | 100.0 |
| *psb*T-*psb*B[b] | 26837F[d]<br>27102R[c] | 5′-GAATGTATAAACCAATGCTTCC-3′<br>5′-GAATTTGGAGCATTCCAAAAACT-3′ | FJ664118 | 1 | 205 | 18 | 8.8 | 8 | 3.9 | 44.4 |
| *ClpP intron 2*[b] | 111104F[d]<br>111454R[c] | 5′-GCCTTCGCCATATGAAA-3′<br>5′-ATGATGGCTCCGTTGCT-3′ | FJ664112 | 1 | 311 | 36 | 11.6 | 11 | 3.5 | 30.6 |
| 3′*trn*K$_{UUU}$-*mat*K[b] | 1825F[c]<br>2195R[d] | 5′-CGGAACTAGTCGGATGGAGT-3′<br>5′-GCTTCTTCTATTTCGCGTAGGT-3′ | FJ664113 | 2 | 307 | 26 | 8.5 | 16 | 5.2 | 61.5 |
| *psb*D-*trn* T$_{GGU}$[b] | 53562F[d]<br>54107R[c] | 5′-TGATCTGTAATCAAAGCAAGATAGTGA-3′<br>5′-CGGTAGAGTAAGCCCATGGTA-3′ | FJ664117 | 1 | 468 | 43 | 9.2 | 16 | 3.4 | 37.2 |

[a]Primers' original reference is White et al. [14]; [b]primers were designed for this study; [c]primers were used both for amplification and sequencing; [d]primers were used for amplification only; [e]the values were calculated based on aligned sequences of *Lobelia rhynchopetalum* and *Lobelia erinus*. GBAN: gene bank accession numbers of *L. rhynchopetalum* sequences. Note: in cases when there is more than one accession number for a given DNA region, the accession numbers are given in range. For example, FJ664108-9 represents two accession numbers (FJ664108 and FJ664109). NVSWLr: number of variable sites within *L. rhynchopetalum*; TAL: total aligned length; NVS: number of variable sites; %VS: percent variable sites; NPIS: number of parsimony informative sites; %PIS: percent parsimony informative sites; %PIVS: percent parsimony informative variable sites.

of molecular variance (AMOVA), haplotype distribution and interhaplotypic distance were conducted using Arlequin version 2 [20]. The minimum spanning tree (MST) of haplotypes was also generated using Arlequin.

## 3. Results

*3.1. Some Sequence Characteristics of L. rhynchopetalum.* In this study, full sequence length was obtained for the *ITSs* (*ITS-1*, *5.8S* and *ITS-2*), *trnG*$_{GCC}$ gene, and *trnfM-trnG*$_{GCC}$, *trnG*$_{GCC}$*-psbZ*, and *trnG*$_{UCC}$*-trnR* intergenic spacers. Partial sequences from the *trnT-trnL*, *psbT-psbB*, *psbD-trnT*, *psbZ-trnS*, *petN-trnC* intergenic spacers, *trnfM*, *psbZ*, and *matK* genes, and *clpP intron-2* and *3′trnK-matK* introns were also obtained. A total of 20 variable sites were obtained within *L. rhynchopetalum*, of which 3 are indel positions and the remaining 17 are substitutions. Indels were only obtained in the *ITS* and *trn*T-*trn*L regions. The *trnT-trnL* intergenic spacer sequences of all individuals from the Bale mountains were shorter by one nucleotide as compared to those from the Simien mountains. This single nucleotide long indel has clearly differentiated the populations according to their mountain system of origin.

The number of variable sites for each region of *L. rhynchopetalum* determined after coding indels according to the simple indel coding method of Simmons and Ochoterena [21] is given in the 5th column of Table 2. The highest number of variable sites within *L. rhynchopetalum* was obtained from *ITS*, as expected. Eleven variable sites were obtained in the entire *ITS* region, nine within *ITS-1* and two within *ITS-2*. The *psbT-psbB*, *psbD-trnT*, *trnfM-trnS*, and *clpP intron-2* sequences were the least variable regions, as only a single variable site per region was obtained within *L. rhynchopetalum*. In the case of the *trnfM-trnS* region, the variable site was located within the *psbZ-trnS* intergenic spacer. The average percent variable sites (*%VS*) of cpDNA regions and *ITS* within *L. rhynchopetalum* were 0.4% and 1.4%, respectively. These values were increased to 9% and 28%, in that order when the *L. rhynchopetalum* sequence was aligned with that of *Lobelia erinus*.

To evaluate the usefulness of the DNA regions used in this study for phylogenetic analysis, the DNA sequences of *L. rhynchopetalum* and *L. erinus* were aligned. Based on the aligned sequences of these species, different parameters of sequence variation of each DNA region were generated (Table 2). The %VS (without including indels) ranged from 2.1 (*trnG-trnR*) to 27.5 (*ITS*), of which 100% and 25.8% were parsimony informative, respectively. This analysis revealed a large number of parsimony informative variable sites in all DNA regions investigated in this study, with the exception of the *trnG-trnR* intergenic spacer (Table 2), suggesting their usefulness for phylogenetic analysis of the genus *Lobelia* and its various subgenera and sections.

*3.2. Intrapopulation Genetic Analysis of L. rhynchopetalum.* Genetic analyses of *L. rhynchopetalum* at the intra-population level were based on 20 polymorphic loci, which include both substitutions and indels. The number of haplotypes per population ranged from 1 to 5. No DNA sequence variation was obtained in the Goba-2, Goba-3, and Debark-2 populations. Since each of these populations carried a single haplotype, the estimates for their intra-population gene diversity and other related parameters were zero (see Table 1). Of the eight populations investigated in this study, Debark-1 was the most diverse, as the haplotype from each individual in the population was different. The haplotypes from the eight populations were in total 18, of which eight haplotypes were found in more than one population. Overall, 13 unique haplotypes were identified from the 40 individuals (Table 1). Five and eight of these haplotypes were unique to the Bale and Simien mountains, respectively.

Genetic diversity was estimated for each population and mountain system as gene diversity (GD; [22]) and nucleotide diversity (ND; [22, 23]). Nucleotide substitutions (transitions and transversions) are the major source of gene diversity within populations. The highest gene diversity (GD = 1.00) was obtained in the Debark-1 population with five haplotypes and six polymorphic sites, followed by Goba-1 (GD = 0.90) with 4 haplotypes and six polymorphic sites (Table 1). Population Goba-1 stood first in terms of nucleotide diversity (ND = 0.071), which is the average gene diversity over all loci under consideration (Table 1). Among the Simien mountain populations, the highest ND (0.038) was recorded in Debark-1. The estimates of ND were higher in the Bale mountains than in the Simien mountains (Table 1) while the estimates for GD were the same. Overall, GD and ND in *L. rhynchopetalum* were estimated to be 0.85 and 0.281, respectively.

The mean interhaplotypic distance (MIHD) was calculated based on the Kimura distance method [19]. The highest estimate (0.081) was recorded in the Goba-1 population, which is almost twofold higher than the highest MIHD among the haplotypes in the populations from the Simien mountains. The mean number of pairwise differences ($\pi$) within populations ranged from 0.00 to 2.99, with the highest obtained in the Goba-1 population. The estimates for MIHD and $\pi$ were higher within the Bale mountains than within the Simien mountains (Table 1). Tajima's test of selective neutrality ($D$; [24]) was also applied to each population and mountain system. The estimates of this parameter ($D$) ranged from −1.33 to 1.22 (Table 1), which is not significantly different from zero. We obtained a similar insignificant deviation from zero when each DNA region was considered separately (data not shown). Theta ($\theta$), a central parameter in population genetic models, summarizes the rate at which mutation and random genetic drift generate and maintain variation within a given DNA region. The estimates of three $\theta$ estimators ($\theta_\pi$, $\theta_S$ and $\theta_K$) are given in Table 1. In this analysis, the estimate of $\theta_\pi$ was almost the same as that of $\pi$ except the slight differences in standard deviation. The highest estimates for the three $\theta$ estimators were obtained in the Goba-1 population.

*3.3. Interpopulation Genetic Analysis of L. rhynchopetalum.* We quantified the population differentiation of haplotypes by using the analysis of molecular variance model [25] for all populations and for populations from each mountain system. The differentiation between the Bale populations and

TABLE 3: Analysis of molecular variance (AMOVA) at different levels based on the Kimura K2P distance method.

| Group | Source of variations | df | Sum of squares | Variance components | % variations | Fixation index |
|---|---|---|---|---|---|---|
| The Bale populations | AP | 2 | 4.38 | 0.34Va | 40.42 | $F_{ST}$: 0.40* |
| | WP | 12 | 5.98 | 0.49Vb | 59.58 | |
| | Total | 14 | 10.36 | 0.84 | | |
| The Simien populations | AP | 4 | 6.16 | 0.24Va | 39.89 | $F_{ST}$: 0.40* |
| | WP | 20 | 7.13 | 0.36Vb | 60.11 | |
| | Total | 24 | 13.29 | 0.59 | | |
| All populations | AP | 7 | 95.20 | 2.63Va | 85.39 | $F_{ST}$: 0.85* |
| | WP | 32 | 14.40 | 0.45Vb | 14.61 | |
| | Total | 39 | 109.60 | 3.08 | | |
| Two mountains[a] | AM | 1 | 84.3 | 4.46Va | 87.01 | $F_{ST}$: 0.87* |
| | WM | 38 | 25.3 | 0.66Vb | 12.99 | |
| | Total | 39 | 109.6 | 5.12 | | |
| Two geographic groups | AG | 1 | 84.29 | 4.39Va | 85.87 | FSC: 0.37* |
| | APWG | 6 | 10.91 | 0.27Vb | 5.34 | FST: 0.91* |
| | WP | 32 | 14.40 | 0.45Vc | 8.79 | FCT: 0.86* |
| | Total | 39 | 109.60 | 5.12 | | |

* $P$-value < 0.001 (significance test at 10000 permutations). AP: among populations; WP: within populations; AM: among mountains; WM: within mountains; AG: among groups; APWG: among populations within groups. [a]Analysis was based on sequences pooled from individuals within the same mountain system.

between the Simien populations was similar ($F_{ST} = 0.40$). Overall, the analysis revealed that 85% of the variance in the distance matrix was accounted for by differences among populations and only 15% by diversity within populations (Table 3). When the significance of population differentiation was tested by 10000 permutations, populations were differentiated at a highly significant level ($P < 0.001$) within geographic locations and overall. The hierarchical AMOVA also revealed a significant geographic differentiation of *L. rhynchopetalum* populations. In addition, the minimum spanning tree (Figure 1) and the distance-based 50% bootstrap majority rule consensus tree (Figure 2) obtained from the analyses of the sequence data demonstrated a clear geographic differentiation of this species.

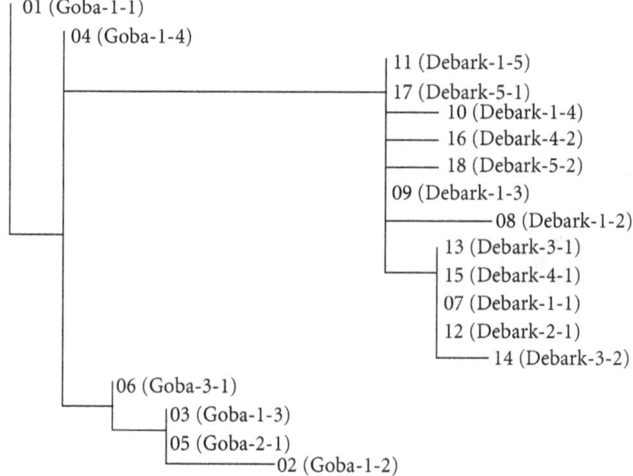


FIGURE 1: Minimum spanning tree (MST) of the 18 haplotypes generated from the ten *L. rhynchopetalum* populations. The MST was based on the K2p distance method. Numbers 01–18 are the haplotypes whereas the text in parenthesis is the population from which the haplotypes were generated. Note: haplotypes 03 and 05 are the same but in different populations; haplotypes 11 and 17 are the same but in different populations; haplotypes 07, 12, 13, and 15 are the same but in different populations.

## 4. Discussion

*4.1. The Utility of the DNA Regions for Population Genetic and Phylogenetics Studies.* The presence of intraspecific variation in nuclear rDNA (e.g., [9, 11, 12]) and cpDNA (e.g., [10, 11, 26]) is well documented. Here, we used the internal transcribed spacers (*ITSs*) of nuclear rDNA and eight cpDNA regions for intra- and interpopulations genetic analyses of *L. rhynchopetalum*. In addition to information from published reports, our choice of the cpDNA regions used in this study was based on the aligned DNA sequences of *Trachelium caeruleum* and *Helianthus annuus*. The alignment of the sequences of these two species revealed mononucleotide repeat microsatellites and a large number of variable sites within the *trnT-trnL*, *trnG-trnR*, *psbT-psbB*, and *psbD-trnT* intergenic spacers, the 3′*trnK-matK* portion of the *trnK* intron, and the *clpP intron-2*. These regions were, therefore, targeted to identify polymorphic microsatellites within *L. rhynchopetalum*. However, no polymorphic microsatellites were found, and thus only variable sites due to nucleotide substitutions and indels were used.

The *ITS* region has been commonly used for plant molecular systematics at lower taxonomic levels and for intraspecific genetic studies since it was first used in phylogenetic inference [27]. It has already become obvious that this region is much more variable than the fastest evolving regions of cpDNA. The result of this study is a further proof to this general understanding, as the *ITS* was over threefold more variable than the average variation obtained within the cpDNA regions in *L. rhynchopetalum*. The *psbD-trnT* region has been considered as one of the highly polymorphic regions of cpDNA [28] and proved to show some degree of polymorphism at the intraspecific level (e.g., [29]). The *trnT-trnL*, *3′trnK-matK*, *petN-trnC*, *psbT-psbB* (as part of *psbB-psbH*), and *trnfM-trnS* regions are also among the fast evolving cpDNA regions [28]. For example, the *trnfM-trnS* region was proved to be informative at the intraspecific level in *Eritrichium nanum* [30] and *Vigna angularis* [29]. The *trnG-trnR* intergenic spacer and the *clpP* intron were not part of the 34 fast evolving cpDNA regions reported in [28] and may not have been used for systematics and population genetic studies.

The *trnG-trnR* region was revealed to be the least variable and parsimony informative in the aligned sequences of *L. rhynchopetalum* and *L. erinus* regardless of the fact that two variable sites were obtained within *L. rhynchopetalum* (see Table 2). On the other hand, when percent variable sites (%*VS*) and percent parsimony informative sites (%*PIS*) were considered, the *clpP intron 2* was found to be as informative as previously reported fast evolving cpDNA regions [28] and hence can be safely used for low taxonomic level phylogenetic studies. Generally, this analysis revealed that all DNA regions included in this study, with the exception of the *trnG-trnR* intergenic spacer, have a large number of parsimony informative variable sites, which can be used for phylogenetic analysis of the genus *Lobelia* and its various subgenera and sections, as exemplified using *L. erinus*.

*L. rhynchopetalum* is more closely related to *Lobelia aberdarica* R. E. Fr. & T. C. E. Fr. than to most other east African giant lobelias, including *Lobelia gibberoa* Hemsl. [3]. Thirty polymorphic sites were obtained within a 650 bp *ITS* aligned sequence of *L. rhynchopetalum* and *L. aberdarica* (accession number, AF163435). The alignment of the *3′trnK-matK* intron partial sequences (309 bp) of *L. rhynchopetalum* and *L. aberdarica* (accession number, AF176898) revealed ten potentially informative sites. Similarly a 489 bp *trnT-trnL* aligned partial sequence of *L. rhynchopetalum* and *L. giberroa* (accession number, DQ285239) revealed 18 potentially informative sites. Hence, the combination of these cpDNA regions could sufficiently resolve the phylogenetic relationships between the east African giant lobelias and beyond.

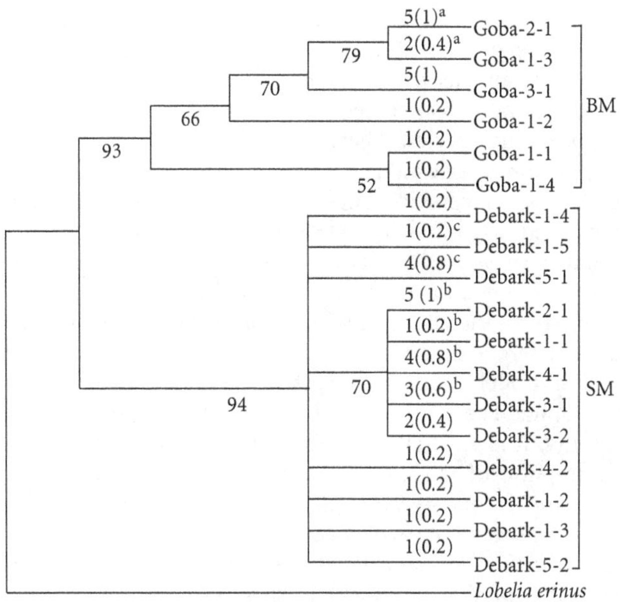


Figure 2: The bootstrap 50% majority rule consensus tree of 18 haplotypes from the ten *L. rhynchopetalum* populations. Bootstrap values greater than 50 are given below the branches. The frequency and relative frequency of each haplotype in each population are given above the branches outside and inside parenthesis, respectively. Haplotypes shared by more than one population are indicated with the same superscript.

*4.2. Intra- and Interpopulation Genetic Analysis of L. rhynchopetalum.* To develop conservation strategies that preserve maximum levels of genetic diversity of *L. rhynchopetalum in situ* and make reasonable decisions about sampling procedures of germplasm for their *ex situ* conservation, one should know how its genetic variation is distributed within the species and what their population genetic structure looks like. The DNA sequence data from *ITS* and eight cpDNA regions proved to be a useful tool for this purpose and successfully applied to reveal the genetic diversity and population genetic structure of this endemic giant lobelia, regardless of limited number of variable sites obtained. Of the eight populations, two populations showed high genetic variation while three populations showed no variation within population. These populations were from both mountain systems, suggesting that the extent of within-population diversity is not limited to geographic regions. However, Simien mountains appeared to be a more favorable habitat for *L. rhynchopetalum* than the Bale mountains, as four of its five populations showed intrapopulation haplotypic diversity.

Many natural ecosystems are subject to habitat fragmentation, which results in smaller and more isolated populations. Plant species remaining in fragmented habitats are of conservation concern due to impacts of decreased population size and increased isolation that threaten their viability due to genetic drift and consequently lower genetic diversity (e.g. [31]). The consequences of such isolation likely increase inbreeding, and in turn greater exposure to genetic drift, resulting in loss of genetic diversity. The absence of genetic variation in some *L. rhynchopetalum* populations may be best explained by genetic drift in relation to their small population size and limited gene flow between populations, as small isolated populations are highly likely to diverge from each other due to genetic drift, which causes fixation of alleles.

Population differentiation is driven by various evolutionary forces such as mutation, gene flow, genetic drift, and selection, and its extent depends on the relative strength of these individual forces in interaction with life history traits of the species. Theta ($\theta$) is a central parameter in population genetic models for the balance between mutation and random genetic drift. For haplotypic data, theta is measured as follow $\theta = 2N_e\mu$, where $N_e$ is effective population size and $\mu$ is a mutation rate per nucleotide site under neutral evolution theory model [32]. Thus, $\theta$ summarizes the rate at which mutations and random genetic drift generate and maintain variation within a given DNA region, under conditions in which natural selection is not operating. There were only two variants per polymorphic site in our aligned DNA sequence data set, which fits to the assumption of the infinite-site model. Pi ($\pi$) is the other measure of sequence variability, which measures the pairwise differences between sequences. In this study, the estimates for $\theta_\pi$ and $\pi$ are similar. This suggests that the data fits with the infinite-site model, as these two parameters have equal expectation under this model (see [32]). Tajima's $D$ test is the test for selective neutrality, and the parameter ($D$) is considered to be zero for neutral loci [24]. In our analysis, $D$ was not significantly different from zero in all populations and groups. Therefore, we conclude that the neutral mutation hypothesis explains the DNA sequence variation obtained in this study.

Thirteen unique haplotypes were obtained by analyzing 40 individual plants. Only three of these, haplotypes were shared among populations of the same mountain systems. Analysis of molecular variance revealed that a high proportion (85%) of the genetic variation was found among populations, and within-population variation only accounted for 15% of the total variation. The 9 variable sites obtained from cpDNA regions were the major source of the highly significant population differentiation at all hierarchical levels, as only three of these sites were variable at the intrapopulation level. Such high population differentiation is not uncommon in endemic species with limited distribution and small population sizes, as a consequence of the pronounced effects of genetic drift (e.g., [33]). When populations are differentiated at a highly significant level, it is recommended to conserve representative samples *ex situ* from each population to reduce the risk of losing unique genetic variants. For *in situ* conservation, priority should be given to populations with relatively high genetic diversity. The results of this study and the ISSR-based study [34] are not in complete agreement as to which of these populations have high diversity, which makes it difficult to prioritize specific populations for *in situ* conservation. However, simultaneous consideration of the two data sets and environmental factors suggests Goba-1 and Debark-1 as good candidates for *in situ* conservation.

AMOVA revealed a highly significant differentiation not only between populations but also among the two mountain systems (Table 3). The differentiation of populations according to geographic areas was also clearly revealed in the 50% bootstrap majority rule consensus tree (Figure 2). Two major clades supported by high bootstrap values (>90%) were formed, in which the haplotypes from the two mountain systems were clearly separated. The result is in agreement with the ISSR-based study [34] in grouping populations according to mountain system of origin. Generally, all analyses revealed a significant differentiation of *L. rhynchopetalum* populations at various hierarchical levels. Such a high population differentiation can be partly explained by limited gene flow. Several factors, such as geographic distance between populations, pollen and seed dispersal mechanisms, and mode of reproduction, have a direct impact on the extent of gene flow between populations. For example, the significant differentiation between the two mountain systems can be partly explained by the Rift Valley as a barrier to gene flow. Similar results were previously reported in *L. giberroa* between these mountain systems [35]. Such a limited gene flow between populations allows further population differentiation, which could lead to speciation.

Knox and Palmer [3] suggested that giant lobelias appear to have initially colonized the ancient upland in East Africa and then moved onto the tall mountains as they arose. *Lobelia acrochila* (E. Wimm.) Knox is the most closely related species to *L. rhynchopetalum* [3, 36]. The inclusion of *L. acrochila* in such type of studies will shed more light to the evolutionary radiation of giant lobelias in eastern Africa. The fact that the Simien mountains are the northernmost end of the distribution of giant lobelias in eastern Africa and the suggestion that giant lobelias were expanding from south towards north [35] may give more weight to the Bale mountains, than to the Simien mountains, as a likely place for the origin of this species. However, since populations from other mountains where this species may be found (though to a lesser extent) were not included in this study, further studies by including these populations are needed to support this suggestion.

This work has revealed the existing haplotypic variation and population genetic structure of *L. rhynchopetalum*. The combination of deterministic and stochastic factors and factors affecting gene flow seems to have played a significant role for the highly significant differentiations of the species at different hierarchical levels. With about 85% of the total genetic variation residing in between populations, each population should be regarded as an important contributor to the overall amount of genetic variation and a significant unit for conservation efforts of this species. Our recommendation is that representative populations from different altitudes and geographic locations should be targeted for conservation purposes, as it reduces the risk of losing unique genetic variants due to several factors. The DNA sequence data generated and submitted to GenBank and the nine pairs of new primers that can be applied to any *Lobelia* and other closely related species, both for population genetic and phylogenetic studies, are also a significant contribution of this work. Furthermore, since the cpDNA primers were designed based on the aligned sequences of *Trachelium caeruleum* (Campanulaceae) and *Helianthus annuus* (Asteraceae), they may be useful for similar studies not only in lobelias but also in various Campanulaceae and Asteraceae species.

## Acknowledgments

This work was financed by the Swedish University of Agricultural Sciences to which the authors are highly grateful. The authors thank Mrs. Ann-Charlotte Strömdahl for her assistance in the laboratory-related work.

## References

[1] D. J. Mabberley, "The Pachycaul lobelias of Africa and St. Helena," *Kew Bulletin*, vol. 29, no. 3, pp. 535–584, 1974.

[2] E. B. Knox, S. R. Downie, and J. D. Palmer, "Chloroplast genome rearrangements and the evolution of giant lobelias from herbaceous ancestors," *Molecular Biology and Evolution*, vol. 10, no. 2, pp. 414–430, 1993.

[3] E. B. Knox and J. D. Palmer, "Chloroplast DNA evidence on the origin and radiation of the giant lobelias in eastern Africa," *Systematic Botany*, vol. 23, no. 2, pp. 109–149, 1998.

[4] R. E. Fries and T. C. E. Fries, "Die riesen-lobelien Africas," *Svensk Botanisk Tidskrift*, vol. 16, pp. 383–416, 1922.

[5] M. Thulin, "Some tropical African Lobeliaceae. Chromosome numbers, new taxa and comments on the taxonomy and nomenclature," *Nordic Journal of Botany*, vol. 3, no. 3, pp. 371–382, 1983.

[6] E. B. Knox and R. R. Kowal, "Chromosome numbers of the East African giant senecios and giant lobelias and their evolutionary significance," *Amercan Journal of Botany*, vol. 80, no. 7, pp. 847–853, 1993.

[7] E. B. Knox and J. D. Palmer, "Chloroplast DNA variation and the recent radiation of the giant senecios (Asteraceae) on the tall mountains of eastern Africa," *Proceedings of the National Academy of Sciences of the United States of America*, vol. 92, no. 22, pp. 10349–10353, 1995.

[8] D. Teketay, "Germination ecology of three endemic species (*Inula confertiflora, Hypericum quartinianum* and *Lobelia rhynchopetalum*) from Ethiopia," *Tropical Ecology*, vol. 39, no. 1, pp. 69–77, 1998.

[9] R. A. Jorgensenand and P. D. Cluster, "Modes and tempos in the evolution of nuclear ribosomal DNA: new characters for evolutionary studies and new markers for genetic and population studies," *Annals of the Missouri Botanical Garden*, vol. 75, no. 4, pp. 1238–1247, 1988.

[10] J. F. Bain and R. K. Jansen, "Numerous chloroplast DNA polymorphisms are shared among different populations and species in the aureoid *Senecio* (Packera) complex," *Canadian Journal of Botany*, vol. 74, no. 11, pp. 1719–1728, 1996.

[11] R. G. Terry, R. S. Nowak, and R. J. Tausch, "Genetic variation in chloroplast and nuclear ribosomal DNA in Utah juniper (*Juniperus osteosperma*, Cupressaceae): evidence for interspecific gene flow," *American Journal of Botany*, vol. 87, no. 2, pp. 250–258, 2000.

[12] G. Baraket, O. Saddoud, K. Chatti et al., "Sequence analysis of the internal transcribed spacers (ITSs) region of the nuclear ribosomal DNA (nrDNA) in fig cultivars (*Ficus carica* L.)," *Scientia Horticulturae*, vol. 120, no. 1, pp. 34–40, 2009.

[13] E. Bekele, M. Geleta, K. Dagne et al., "Molecular phylogeny of genus *Guizotia* (Asteraceae) using DNA sequences derived from ITS," *Genetic Resources and Crop Evolution*, vol. 54, no. 7, pp. 1419–1427, 2007.

[14] T. J. White, T. Bruns, S. Lee et al., "Amplification and direct sequencing of fungal ribosomal RNA genes for phylogenetics," in *PCR Protocols: A Guide to Methods and Applications*, pp. 315–322, Academic Press, 1990.

[15] S. Rozen and H. J. Skaletsky, "Primer3 on the WWW for general users and for biologist programmers," in *Bioinformatics Methods and Protocols: Methods in Molecular Biology*, pp. 365–386, Humana Press, 2000.

[16] T. Hall, BioEdit v. 7.0.5: Biological sequence alignment editor for Windows. Ibis Therapeutics a division of Isis pharmaceuticals, 2005, http://www.mbio.ncsu.edu/BioEdit/bioedit.html.

[17] J. D. Thompson, T. J. Gibson, F. Plewniak, F. Jeanmougin, and D. G. Higgins, "The CLUSTAL X windows interface: flexible strategies for multiple sequence alignment aided by quality analysis tools," *Nucleic Acids Research*, vol. 25, no. 24, pp. 4876–4882, 1997.

[18] D. L. Swofford, *PAUP**: Phylogenetic Analysis Using Parsimony, Version 4.0, Beta*, Sinauer Associates Inc, Sunderland, Mass, USA, 2000.

[19] M. Kimura, "A simple method for estimating evolutionary rates of base substitutions through comparative studies of nucleotide sequences," *Journal of Molecular Evolution*, vol. 16, no. 2, pp. 111–120, 1980.

[20] S. Schneider, D. Roessli, and L. Excoffier, *Arlequin: A Software for Population Genetics Data Analysis, Version 2.000*, Genetics and Biometry Laboratory Department of Anthropology: University of Geneva, 2000.

[21] M. P. Simmons and H. Ochoterena, "Gaps as characters in sequence-based phylogenetic analyses," *Systematic Biology*, vol. 49, no. 2, pp. 369–381, 2000.

[22] M. Nei, *Molecular Evolutionary Genetics*, Columbia University Press, New York, NY, USA, 1987.

[23] F. Tajima, "Evolutionary relationship of DNA sequences in finite populations," *Genetics*, vol. 105, no. 2, pp. 437–460, 1983.

[24] F. Tajima, "Statistical method for testing the neutral mutation hypothesis by DNA polymorphism," *Genetics*, vol. 123, no. 3, pp. 585–595, 1989.

[25] L. Excoffier, P. Smouse, and J. Quattro, "Analysis of molecular variance inferred from metric distances among DNA haplotypes: application to human mitochondrial DNA restriction data," *Genetics*, vol. 131, no. 2, pp. 479–491, 1992.

[26] R. J. Mason-Gamer, K. E. Holsinger, and R. K. Jansen, "Chloroplast DNA haplotype variation within and among populations of *Coreopsis grandiflora* (Asteraceae)," *Molecular Biology and Evolution*, vol. 12, no. 3, pp. 371–381, 1995.

[27] B. G. Baldwin, "Phylogenetic utility of the internal transcribed spacers of nuclear ribosomal DNA in plants: An example from the compositae," *Molecular Phylogenetics and Evolution*, vol. 1, no. 1, pp. 3–16, 1992.

[28] J. Shaw, E. B. Lickey, E. E. Schilling, and R. L. Small, "Comparison of whole chloroplast genome sequences to choose noncoding regions for phylogenetic studies in angiosperms: the Tortoise and the hare III," *American Journal of Botany*, vol. 94, no. 3, pp. 275–288, 2007.

[29] T. T. Ye and H. Yamaguchi, "Sequence variation of four chloroplast non-coding regions among wild, weedy and cultivated *Vigna angularis* accessions," *Breeding Science*, vol. 58, no. 3, pp. 325–330, 2008.

[30] I. Stehlik, F. R. Blattner, R. Holderegger, and K. Bachmann, "Nunatak survival of the high Alpine plant *Eritrichium nanum* (L.) Gaudin in the central Alps during the ice ages," *Molecular Ecology*, vol. 11, no. 10, pp. 2027–2036, 2002.

[31] N. C. Ellstrand and D. R. Elam, "Population genetic conseqences of small population size: implications for plant conservation," *Annual Review of Ecology and Systematics*, vol. 24, pp. 217–242, 1993.

[32] A. G. Clark, K. M. Weiss, D. A. Nickerson et al., "Haplotype structure and population genetic inferences from nucleotide-sequence variation in human lipoprotein lipase," *American Journal of Human Genetics*, vol. 63, no. 2, pp. 595–612, 1998.

[33] C. T. Cole, "Genetic variation in rare and common plants," *Annual Review of Ecology, Evolution, and Systematics*, vol. 34, pp. 213–237, 2003.

[34] M. Geleta and T. Bryngelsson, "Inter simple sequence repeat (ISSR) based analysis of genetic diversity of *Lobelia rhynchopetalum* (Campanulaceae)," *Hereditas*, vol. 146, no. 3, pp. 122–130, 2009.

[35] M. Kebede, D. Ehrich, P. Taberlet, S. Nemomissa, and C. Brochmann, "Phylogeography and conservation genetics of a giant lobelia (*Lobelia giberroa*) in Ethiopian and Tropical East African mountains," *Molecular Ecology*, vol. 16, no. 6, pp. 1233–1243, 2007.

[36] E. B. Knox, "The species of giant senecio (Compositae) and giant lobelia (Lobeliaceae) in eastern Africa," *Contributions from the University of Michigan Herbarium*, vol. 19, pp. 241–257, 1993.

# 6

# Flavonoid-Deficient Mutants in Grass Pea (*Lathyrus sativus* L.): Genetic Control, Linkage Relationships, and Mapping with Aconitase and S-Nitrosoglutathione Reductase Isozyme Loci

**Dibyendu Talukdar**

*Department of Botany, R.P.M. College, University of Calcutta, Uttarpara, West Bengal, Hooghly 712 258, India*

Correspondence should be addressed to Dibyendu Talukdar, dibyendutalukdar9@gmail.com

Academic Editors: K. Chakravarty, E. Olmos, and K. Shoji

Two flavonoid-deficient mutants, designated as *fldL-1* and *fldL-2*, were isolated in EMS-mutagenized (0.15%, 10 h) $M_2$ progeny of grass pea (*Lathyrus sativus* L.). Both the mutants contained total leaf flavonoid content only 20% of their mother varieties. Genetic analysis revealed monogenic recessive inheritance of the trait, controlled by two different nonallelic loci. The two mutants differed significantly in banding patterns of leaf aconitase (ACO) and S-nitrosoglutathione reductase (GSNOR) isozymes, possessing unique bands in *Aco 1, Aco 2,* and *Gsnor 2* loci. Isozyme loci inherited monogenically showing codominant expression in $F_2$ (1 : 2 : 1) and backcross (1 : 1) segregations. Linkage studies and primary trisomic analysis mapped *Aco 1* and *fld 1* loci on extra chromosome of trisomic-I and *Aco 2, fld 2,* and *Gsnor 2* on extra chromosome of trisomic-IV in linked associations.

## 1. Introduction

Flavonoids are secondary metabolites derived from phenylalanine and acetyl CoA that perform a variety of important functions in plant growth, reproduction, and survival and also serve as important micronutrients in human and animal diets [1, 2]. The pigmented flavonoid metabolites have been used as phenotypic markers in many model plant species [3, 4] and have proven to be an excellent tool to study the genetic, molecular, and biochemical processes [4, 5]. One of the functional tools in this regard is the genetic characterization of mutants, exhibiting significantly altered flavonoid compounds. A good number of mutants with altered flavonoid levels have been utilized in *Arabidopsis*, maize, grape, and *Petunia* to reveal biosynthetic pathway of different flavonoids and their diverse roles [6–10]. Although very rich in flavonoid components [11], no reports are available regarding the genetic analysis of flavonoid mutant in leguminous plants.

Plant flavonoids play pivotal role in protection/tolerance against different types of abiotic stress [12]. Evidences are accumulating about functional interplay between flavonoid metabolism and thiol-based (glutathione/thioredoxin) antioxidant defense system in plants during stress response [13, 14], where nitric oxide (NO) functions as signaling molecule [15]. The enzyme aconitase (ACO) is known to be responsible in iron homeostasis and in regulating resistance to oxidative stress [16], but its activity is inhibited by NO [17]. On the other hand, the S-nitrosoglutathione reductase (GSNOR) activity has been described to be associated with the enzyme glutathione-dependent formaldehyde dehydrogenase [18]. GSNOR uses GSNO as its substrate which is formed by the reaction of reduced glutathione with NO molecule. GSNOR is extremely important in maintenance and turnover of cellular NO pool and modulation of hormonal response such as jasmonic acid and salicylic acid, responsible for alteration of stress-induced phenylpropanoid pathway [14, 19].

Grass pea (*Lathyrus sativus* L.), an annual winter legume crop, possesses high level of bioactive compounds including flavonoids [20]. The potential of this hardy crop has been extensively utilized in recent years through isolation and genetic analysis of novel mutants for plant habit [21], flower and seed coat colour [22, 23], pod indehiscence [24], seed size [25], and so forth. Some of these mutant lines are now being tested for their fitness to different abiotic stresses including salinity [26, 27] and arsenic [28], and very recently, a novel ascorbate-deficient mutant has been detected [29]. Linkage mapping and chromosomal assignment of desirable mutations are now being accomplished through establishment of a functional cytogenetic stocks including aneuploids [30, 31], polyploids [32], and translocation lines [33, 34]. Perusal of literature cites only limited information regarding inheritance and linkage association of morphological, biochemical (isozyme), and other molecular markers in grass pea [35]. Although isozyme markers are widely used in gene mapping of different crops and have advantages over other markers due to their codominant expression, lack of sufficient number of polymorphic isozymes loci possesses problems in existing germplasms of grass pea [36]. Creation of additional variability in esterase and root peroxidase isozyme systems through induced mutagenesis has recently been successfully explored in dwarf mutant population of this crop, and genetic control of their allozyme variants has been studied [37]. During screening of desirable mutations in EMS-mutagenized population, two variant plants with white flower color was isolated. The mutants were later found to be highly deficient in total flavonoid content in their leaves. Despite immense importance of flavonoids in legume crops and its relation with enzymes involved in stress responses, no reports in these regards are available in grass pea.

Keeping all these in mind, a genetic approach has been taken to investigate the basis of flavonoid deficiency in the preset materials of grass pea and its association with isozymes of ACO and GSNOR enzymes. The main objectives of the present work are to (1) trace the mode of inheritance of flavonoid deficiency and the zymogram phenotypes of both enzymes, (2) investigate the segregation pattern and linkage associations between different isozymes loci and loci controlling flavonoid deficiency, and (3) ascertain their possible chromosome location through primary trisomic analysis.

## 2. Methods

*2.1. Plant Materials.* Altogether eleven parents have been used in the present study of which four were diploid ($2n = 14$) and rest seven were primary trisomic ($2n+1 = 15$) types. Among the diploid parents, two varieties "BioL-212" and "Hooghly Local" were used as mother control throughout the experiment. Fresh and healthy seeds of these two varieties presoaked with water (6 h) were treated with freshly prepared 0.15% aqueous solution of EMS (Sigma-Aldrich) for 10 h with intermediate shaking at 25 ± 2°C. $M_1$ seeds were sown treatment-wise in completely randomized block design as reported earlier [25]. Two variant plants showing white flowers and absence or modified stipule morphology were distinguished from usual occurrence of blue flower and typical papilionaceous stipules in EMS-treated $M_2$ progeny. During screening of antioxidant activities of different mutant lines, these two plants exhibited abnormally low foliar flavonoid contents. The levels were again confirmed at $M_3$ generation, and on the basis of stipule characters the progeny of the two plants was primarily designated as *fld L-1* (flavonoid-deficient *Lathyrus* type 1 mutant, white flower, estipulate) and *fld L-2* (flavonoid deficient *Lathyrus* type 2 mutant, white flower, linear-acicular stipule). Both the mutants bred true for their phenotypes, and no significant change in leaf flavonoid content was found in $M_3$ generation. Chromosome location of different loci was performed by utilizing a set of primary trisomics, isolated and characterized earlier in grass pea [30, 38].

Table 1: Total foliar flavonoid contents (mg g$^{-1}$ extract) in aqueous and ethanol extract of grass pea (*Lathyrus sativus* L.) mutants (*fld L-1* and *fld L-2*) and mother plants (BioL-212 and Hooghly Local).

| Genotype | Aqueous extracts | Ethanol extract |
|---|---|---|
| BioL-212 | 160.55 ± 3.6 | 354.37 ± 3.9 |
| Hooghly Local | 148.59 ± 3.2 | 346.40 ± 3.1 |
| *fld L-1* | 150.53 ± 3.2 | 70.13 ± 2.2* |
| *fld L-2* | 30.77 ± 1.4* | 337.27 ± 2.9 |

*Significantly different from mother plants at $P < 0.05$.

*2.2. Determination of Total Flavonoid Content.* Total flavonoid content from leaves of mutants and their control varieties were determined spectrophotometrically in both ethanol and aqueous extracts, based on the formation of a flavonoid-aluminium complex [39]. An amount of 2% ethanolic $AlCl_3$ solution (0.5 mL) was added to 0.5 mL of sample. After 1 h at room temperature, the absorbance was measured at 420 nm. A yellow color indicated the presence of flavonoids. Extract samples were evaluated at a final concentration of 0.1 mg mL$^{-1}$. Total flavonoid contents were calculated as rutin (mg g$^{-1}$ of extract) (Table 1).

*2.3. Isozyme Analysis: Gel Electrophoresis and Nomenclature.* Horizontal 10% starch-gel (Sigma) electrophoresis was carried out to analyse the banding profile of aconitase (ACO, EC 4.2.1.3) in mutants ($M_4$) and control varieties and trisomic lines. Crude extracts were prepared by macerating young leaf tissues of 4-d-old seedlings in ice-cold extraction buffer containing 20% sucrose, 5% PVP-40, 0.1 M $KH_2PO_4$, 0.05% triton X-100 (Sigma), and 14 mM 2-mercaptoehanol (Sigma) at pH 7.0. Triton X-100 and 2-mercaptoehanol were added just before use. After extraction, sample was stored at −20°C for future use. ACO isozymes were separated using the electrode and gel buffer system (pH 6.5) of Cardy et al. [40]. Bands of ACO systems were stained according to the recipes (0.1 M Tris-HCL, pH 8.0, *cis*-Aconitic acid, $MgCl_2$, Isocitrate dehydrogenase, MTT, PMS, and NADP) of Cardy and Beversdorf [41]. For GSNOR (EC 1.2.1.1) activity, native PAGE was done using 6% acrylamide gels in TRIS-boric-EDTA buffer (pH 8.0). For staining of GSNOR activity, gels were soaked in 0.1 M sodium phosphate, pH 7.4, containing 2 mM NADH for 15 min in an ice bath. Excess buffer was drained, and gels were covered with filter paper strips soaked in freshly prepared 3 mM GSNO. After 10 min, the filter paper was removed, and gels were exposed to UV-light and analysed for the disappearance of the NADH fluorescence, indicating GSNOR activity [42].

Based on the observed variations, isozyme bands were assigned to putative loci following the principles of Weeden [43]. The isozymes were designated as all letters capitals (ACO and GSNOR) but the loci controlling these two isozymes had only the first letter capitalized and presented in italics (*Aco* and *Gsnor*). When two or more isozymes, coded by different loci in an enzyme, were visualized on gel, they were numbered sequentially according to their mobility relative to the anode with the most anodal isozyme being number one, and subsequent isozymes were assigned sequentially higher numbers. Likewise, the most anodal allele producing allozyme (fastest variant) of a particular locus was termed as "a" and progressively slower forms "b", "c", and so on. Only clearly visible bands for both enzyme systems were scored in the present study.

*2.4. Inheritance and Linkage Analysis.* Inheritance and linkage of loci controlling flavonoid deficiency and different isozymes were traced in segregating populations of $F_2$ and backcross generations derived from single locus as well as joint segregation of two loci in different cross-combinations. Following two generations of selfing, intercrosses including reciprocals were made among control varieties and the mutant lines ($M_4$) to raise $F_1$ and, subsequently, backcross ($BC_1$) and $F_2$ progenies (Table 2). Measures were taken at every stage from sowing to harvesting to prevent any type of outcrossing pollination and intermixing. For allelism test, intercrosses were made among *fld L-1* and *fld L-2*. Chi-square test was employed to test the goodness of fit between observed and expected values for all crosses (Table 2). Zymogram phenotypes of both ACO and GSNOR were studied in selfed and intercrossed ($F_2$ and backcross) progenies of different parents.

Linkage associations of the segregating isozyme markers along with flavonoid deficiency trait were examined for pair-wise combinations of different isozyme loci and also between pairs of isozyme loci and loci controlling flavonoid deficiency for the expected ratio of 1 : 2 : 1 : 2 : 4 : 2 : 1 : 2 : 1 and 3 : 1 : 6 : 2 : 3 : 1, respectively, in the $F_2$ progeny. Testcross population was raised by crossing $F_1$ plant with the parent showing comparatively slow moving allozymes in case of isozyme loci and with recessive lines in segregation of flavonoid deficiency. Chi-square test was employed to test the goodness of fit, and significant deviation from the expected ratio was considered as linkage between the markers. Recombination fraction ($r$) was calculated from testcross data and was converted to map distance in centiMorgans (cM) through Kosambi's mapping function [44]. Data from different families was pooled when homogeneous for analysis (Table 3).

*2.5. Mapping Flavonoid-Deficient Mutant and Isozyme Loci by Primary Trisomic Analysis.* The seven primary trisomic types were crossed as female parent with the four different homozygous diploid genotypes (two controls, two mutants), and $F_1$ population was obtained in each case. The trisomic $F_1$ plants could be readily identified at early seedling stage on the basis of their specific leaflet phenotypes [30]. Trisomic $F_1$ plant was self-fertile and subsequently selfed to obtain $F_2$ progeny and also backcrossed to the respective diploid parent to produce $BC_1$ population. In the segregating $F_2$ progeny, banding patterns were analysed by means of chi-square test for a fit to a normal disomic ratio. Significant deviation from the expected disomic ratio of 1 : 2 : 1 in $F_2$ and 1 : 1 in $BC_1$ was further tested with the expected trisomic ratios of 4 : 4 : 1 in $F_2$, 2 : 1 in $BC_1$ in diploid portion and 2 : 7 : 0 ($F_2$) in trisomic portion of the progeny to locate possible chromosome/s, bearing gene/s of different isozyme loci (Table 4). Necessary cytological confirmation of trisomy was performed at meiosis-I following Talukdar and Biswas [30]. To save space, only segregation of trisomics carrying concerned loci has been presented in Table 4.

*2.6. Statistical Analysis.* Total flavonoid contents in leaves of mother and mutants are presented as mean ± standard error (SE) with 20 plants in each of the four genotypes. Significant differences between mother and mutant plants for total flavonoids were determined by simple "$t$-test." A probability of $P < 0.05$ was considered significant.

## 3. Results

*3.1. Total Flavonoid Contents and Morphology of Mutant and Mother Plants.* Total flavonoid contents as determined in aqueous and ethanol extract in leaves of mutant and mother leaves were significantly ($P < 0.05$) different. Leaves of both the mutants contained total content ($mg\,g^{-1}$ extract) only 20% of mother plants (Table 1). However, flavonoid content was nearly normal in aqueous extract of *fldL-1* mutant, but reduced by about 5-fold in ethanol extract. In contrast, flavonoid content reduced marginally in ethanol extract of *fldL-2* leaves but had reduced by nearly 5-fold in aqueous extract (Table 1).

Both the mutants produced characteristic white flower and modification in stipule characters. While *fldL-1* was completely estipulate, a linear-acicular type of stipule was observed in *fldL-2* plants. Furthermore, *fldL-1* showed normal pollen fertility (98.77%) like mother plants, while it reduced (66%) in *fldL-2* plants. Root formation in both mutants, however, was quite normal.

*3.2. Inheritance and Allelic Relationship of Gene/s Controlling Flavonoid Deficiency.* Reciprocal crosses between *fldL-1* as well as *fldL-2* and mother control varieties yielded $F_1$ plants with normal level of flavonoids (Table 2). Segregation of normal and flavonoid-deficient plant type showed good fit to 3 : 1 in $F_2$ and 1 : 1 in backcross (Table 2). Flavonoid content from leaves of every genotype was tested and verified with parents. The recessive mutants recovered in $F_2$ generation of the above two crosses were also self-pollinated, and in $F_3$ all the 210 plants exhibited only marginal variations in total flavonoid content compared with their respective parents (data not in table).

Table 2: Single locus segregation of *fld 1* and *fld 2* mutations, two aconitase (*Aco1 & 2*), and S-nitrosoglutathione reductase 2 (*Gsnor 2*) isozyme loci in $F_2$ and backcross ($BC_1$) populations of different intercrosses among four parents in *Lathyrus sativus* L. [a]FF-Homozygote of fast alleles, SS-Homozygote of slow allele, FS-Heterozygotes. *, **, and *** consistent with 1 : 2 : 1, 1 : 1, and 3 : 1 ratios, respectively, at 5% level of significance, ++parent/s showing slow allozyme used in testcross with $F_1$ and pooled data of several crosses presented.

| Cross++ | Locus | Phenotype ($F_1$) | $F_2/BC_1$ phenotype[a] Normal | | Deficient | N | $\chi^2$ (3 : 1/1 : 2 : 1/1 : 1) |
|---|---|---|---|---|---|---|---|
| *Flavonoid mutant* | | | | | | | |
| BioL-212 × *fldL-1* | *fld 1* | Normal flavonoid | 61 | — | 23 | 84 | 0.25*** |
| $F_1$ × fldL-1 | *fld 1* | — | 54 | — | 43 | 97 | 1.25** |
| HL × *fldL-1* | *fld 1* | Normal flavonoid | 81 | — | 28 | 109 | 0.02*** |
| $F_1$ × *fldL-1* | *fld 1* | — | 37 | — | 30 | 67 | 0.72** |
| BioL-212 × *fldL-2* | *fld 2* | Normal flavonoid | 118 | — | 41 | 159 | 0.05*** |
| $F_1$ × *fldL-2* | *fld 2* | — | 33 | — | 27 | 60 | 0.60** |
| HL × *fldL-2* | *fld 2* | Normal flavonoid | 120 | — | 43 | 163 | 0.16*** |
| $F_1$ × *fldL-2* | *fld 2* | — | 48 | — | 42 | 90 | 0.04** |
| | | | Normal | *fldL-1* type | *fldL-2* type | Double recessive | $\chi^2$ (9 : 3 : 3 : 1) |
| *fldL-1* × *fldL-2* | *fld 1/fld 2* | Normal flavonoid | 153 | 50 | 57 | 20 | 0.96 |
| Isozyme loci | Locus | Alleles | FF | FS | SS | N | $\chi^2$ (3:1/1:2:1/1:1) |
| BioL-212/HL × *fldL-2* | *Gsnor 2* | ab | 47 | 101 | 51 | 199 | 0.20* |
| $F_1$ × *fldL-2* | *Gsnor 2* | ab | — | 54 | 61 | 115 | 0.43** |
| *fldL-1* × *fldL-2* | *Gsnor 2* | ab | 60 | 118 | 54 | 232 | 0.36* |
| $F_1$ × *fldL-2* | *Gsnor 2* | ab | — | 50 | 46 | 96 | 0.16** |
| *fldL-1/fldL-2* × BioL-212/HL | *Aco 1* | ab | 25 | 54 | 25 | 104 | 0.15* |
| $F_1$ × BioL-212/HL | *Aco 1* | ab | — | 37 | 44 | 81 | 0.60** |
| *fldL-1* × *fldL-2* | *Aco 2* | ab | 51 | 92 | 44 | 187 | 0.57* |
| $F_1$ × *fldL-2* | *Aco 2* | ab | — | 44 | 38 | 82 | 0.44** |
| BioL-212/HL × *fldL-2* | *Aco 2* | bc | 46 | 88 | 39 | 173 | 0.62* |
| $F_1$ × BioL-212/HL | *Aco 2* | bc | — | 40 | 34 | 74 | 0.49** |
| BioL-212/HL × *fldL-1* | *Aco 2* | ac | 23 | 38 | 20 | 81 | 0.53* |
| $F_1$ × BioL-212/HL | *Aco 2* | ac | — | 17 | 13 | 30 | 0.53** |

In order to study the allelic relationships of genes governing flavonoid deficiency in grass pea, *fldL-1* and *fldL-2* were reciprocally crossed. All the $F_1$ plants derived from the crosses contained normal flavonoid level like mother control plants. In $F_2$, four types of plants: normal type, *fldL-1*, *fldL-2,* and a variant type appeared in the progeny showing good fit to 9 : 3 : 3 : 1 ratio (Table 2). Gene symbols of *Fld* for normal type and *fld 1* and *fld 2* for *fldL-1* and *fldL-2* were assigned, respectively. The normal plant type, thus recovered, manifested usual phenotypes such as blue flower, papilionaceous stipules, and normal level of foliar flavonoids. The variant plant type exhibited extreme reduction in total flavonoid contents, containing only 10% of that in mother control, and this feature was accompanied with reduced root length, absence of stipules, abnormal elongation of leafless stem, and much higher pollen sterility (79.33%) than either of its parents.

### 3.3. *Inheritance of Isozyme-Banding Pattern in Selfed and Intercrossed Progenies*

*3.3.1. ACO.* Two mutant lines and the control varieties bred true for their respective single-banded phenotypes in successive selfed generations ($M_2$-$M_4$). Two zones of enzyme activity were conspicuous of which the most anodal

TABLE 3: Joint segregation of pairs of four isozyme loci and *fld1 and fld 2* genes exhibiting significant deviations from expected $F_2$ and backcross ($BC_1$) ratios of random assortment in *Lathyrus sativus* L. [a]$H_1$-Heterozygous for alleles at "X" locus, H-heterozygous for alleles at "Y" locus. *r*-recombinant value. *, **, and *** significant at 5% level for 3 : 1 : 6 : 2 : 3 : 1, 1 : 1 : 1 : 1, and 1 : 2 : 1 : 2 : 4 : 2 : 1 : 2 : 1, respectively.

| Loci (X)-(Y) | Progeny | Number of progeny with designated phenotypes[a] ($F_2/BC_1$ generation) XY | XH | Xy | $H_1Y$ | $H_1H$ | $H_1y$ | xY | xH | xy | Total | $\chi^2$ | *r* | Map distance (cM.) |
|---|---|---|---|---|---|---|---|---|---|---|---|---|---|---|
| *Aco 1-fld 1* | $F_2$ | 27 | — | 11 | 07 | — | 28 | 02 | — | 39 | 114 | 205.55* | — | — |
| *Aco 1-fld 1* | $BC_1$ | — | — | — | — | 49 | 08 | — | 05 | 73 | 135 | 96.68** | 0.0963 | 9.75 |
| *Aco 2-fld 2* | $F_2$ | 22 | — | 13 | 10 | — | 03 | 02 | — | 17 | 67 | 86.39* | — | — |
| *Aco 2-fld 2* | $BC_1$ | — | — | — | — | 57 | 06 | — | 10 | 65 | 138 | 82.56** | 0.116 | 11.80 |
| *Gsnor 2-fld 2* | $F_2$ | 39 | — | 09 | 07 | — | 15 | 11 | — | 30 | 111 | 126.78* | | — |
| *Gsnor 2-fld 2* | $BC_1$ | — | — | — | — | 62 | 11 | — | 07 | 40 | 120 | 67.12** | 0.150 | 15.48 |
| *Gsnor2-Aco-2* | $F_2$ | 18 | 01 | 02 | 03 | 10 | 07 | 03 | 05 | 21 | 70 | 137.61*** | — | — |
| *Gsnor2-Aco-2* | $BC_1$ | — | — | — | — | 52 | 17 | — | 14 | 46 | 129 | 35.49** | 0.240 | 26.19 |

TABLE 4: Segregations of *fld 1* and *fld 2* along with *Aco 1, Aco 2,* and *Gsnor 2* isozyme loci in $F_2$ and $BC_1$ generations obtained from several crosses (data pooled) between two different primary trisomics (Tr I and IV) and four diploid parents. Data of only critical trisomics carrying isozyme loci presented here. [a]F- Homozygous for fast/dominant allele, S-Homozygous for slow/recessive allele, H-Heterozygous; $(2n+1)^b$-consistent at 5% level; *, **, and *** consistent with 4 : 4 : 1, 8 : 1 in $F_2$, and 2 : 1 in $BC_1$ at 5% level of significance, respectively.

| Trisomic types | Progeny | Loci | F2 and BC1 phenotypes[a] 2n FF | FS | SS | Total | $\chi^2$ (1 : 2 : 1/3 : 1) | (1:1) | (4 : 4 : 1/8 : 1) | (2: 1) | $(2n+1)^b$ $X^2$ (2 : 7 : 0) |
|---|---|---|---|---|---|---|---|---|---|---|---|
| Tr-I | $F_2$ | *fld 1* | 129 | — | 17 | 146 | 13.89 | — | 0.04** | — | — |
| Tr-I | $BC_1$ | *fld 1* | — | 68 | 37 | 105 | — | 9.15 | — | 0.17*** | — |
| Tr-IV | $F_2$ | *fld 2* | 118 | — | 14 | 132 | 14.08 | — | 0.03** | — | |
| Tr-IV | $BC_1$ | *fld 2* | — | 45 | 25 | 70 | — | 5.71 | — | 0.18*** | — |
| Tr-I | $F_2$ | *Aco-1* | 50 | 60 | 14 | 124 | 21.04 | — | 0.912* | — | 2.53 |
| Tr-I | $BC_1$ | *Aco-1* | — | 31 | 25 | 56 | — | 15.68 | — | 0.51*** | — |
| Tr-IV | $F_2$ | *Aco-2* | 47 | 52 | 11 | 110 | 23.90 | — | 0.39* | — | 0.51 |
| Tr-IV | $BC_1$ | *Aco-2* | — | 71 | 30 | 101 | — | 16.64 | — | 0.60*** | — |
| Tr-IV | $F_2$ | *Gsnor 2* | 35 | 29 | 08 | 72 | 22.98 | — | 0.56* | — | 0.008 |
| Tr-IV | $BC_1$ | *Gsnor 2* | — | 44 | 19 | 63 | — | 9.92 | — | 0.29*** | — |

one, designated as ACO-1 contained a total of three bands exhibiting two different types of migration (Figures 1 and 4). The fast moving allozyme (ACO-1a) was unique to bothmutants (lane 1 and 3), whereas relatively slower variant (ACO-1b) was common in mother control variety (lane 2). In the ACO-2 zone, three different types of mobility were manifested by a total of three bands; one of them was fast moving (ACO-2a) and developed only in *fldL-1* (lane 1). It was followed by a unique slower band (ACO-2b) generated in *fldL-2* mutant (lane 3). The slowest band in this zone (ACO-2c) was found specific to mother variety, BioL-212 (Figures 1 and 4, lane 2). The other variety "Hooghly Local" produced identical zymograms of var. BioL-212 (not shown in figure).

Two types of allozyme activity in ACO-1 zone have been confirmed in segregating populations of different $F_2$ and backcross (Figures 2 and 5). Segregation of *Aco-1a/Aco-1b* was found in mutant (*Aco 1a/Aco 1a*) × control variety (*Aco 1b/Aco 1b*) (Table 2). In the *Aco 2* locus, heterozygous individuals for the alleles *Aco 2a/Aco 2b* were detected in the $F_2$ progeny of crosses involving *fld L-1* (*Aco 2a/Aco 2a*) and *fld L-2* (*Aco 2b/Aco 2b*) mutants. Similarly, crosses between *fld L-2* (*Aco 2b/Aco 2b*) and two control varieties (*Aco 2c/Aco 2c*) and between *fld L-1* (*Aco 2a/Aco 2a*) and control varieties yielded homozygous individuals parental types and heterozygous individuals for *Aco 2b/Aco 2c* in former cross and for *Aco 2a/Aco 2c* in case of latter. $F_1$ phenotype and one parental phenotype were observed in backcross progeny. In case of both loci, three types of phenotypes-two single-banded homozygous parental and one double-banded heterozygous phenotypes of fast and slow allozymes in $F_2$ and two phenotypes: one heterozygous and one respective parental type in corresponding backcrosses segregated and conformed well with 1 : 2 : 1 in $F_2$ and 1 : 1 in backcrosses, respectively (Table 2).

*3.3.2. GSNOR.* Two control varieties and two induced mutant lines together generated 4 bands which could be

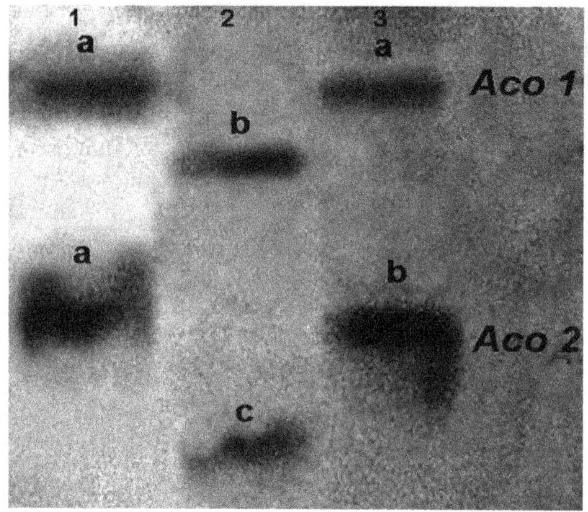


FIGURE 1: Zymogram phenotype of aconitase isozymes; lane 1-*fldL-1*, lane 2-mother variety BioL-212, lane 3-*fldL-2* mutant, small letters indicate alleles of respective loci in *Lathyrus sativus* L.

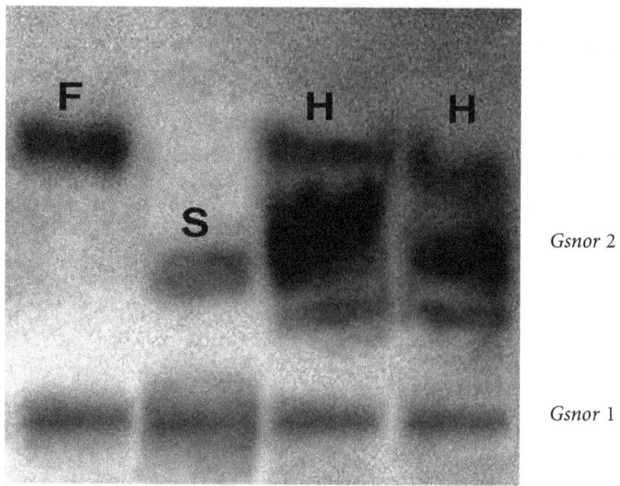


FIGURE 3: Segregation of S-nitrosoglutathione reductase loci, *Gsnor 2* in F2 generation of *fldL-2* × mother plant; F-Fast allele, S-slow allele, H-heterozygote. No segregation was observed in *Gsnor 1* locus in *Lathyrus sativus* L.

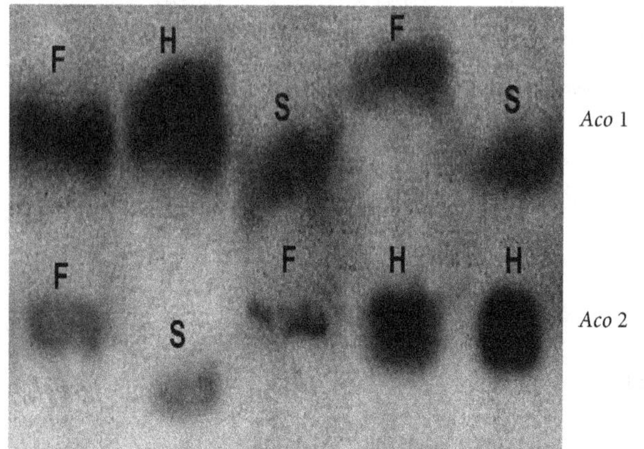


FIGURE 2: Segregation of phenotypes in $F_2$ derived from crosses between BioL-212 × *fldL-1* in *Aco 1* and *Aco 2* loci; F-Fast allele, S-slow allele, H-heterozygotes in different lanes, in *Lathyrus sativus* L.

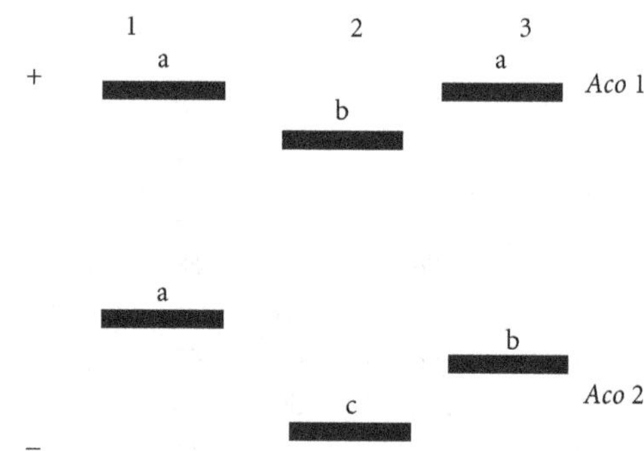


FIGURE 4: Zymogram phenotype of aconitase isozymes; lane 1-*fldL-1*, lane 2-mother variety BioL-212, and lane 3-*fldL-2* mutant. Small letters indicate alleles of respective loci in *Lathyrus sativus* L.

clearly resolved in two separate zones of enzyme activity tentatively designated as GSNOR1 and GSNOR 2 from anodal side of the gel in the present material. All the four parents bred true in successive selfed ($M_2$-$M_4$) generations for the single-banded pattern corresponding to different allozymes in these two zones, and only representative zymograms showing differences in banding pattern have been shown. The mutant line *fld L-2* was conspicuously different from control and also from *fld L-1* by possessing a unique band in zymogram (lane 2). The GSNOR 1 zone was monomorphic with same mobility and intensity of bands in all four parents. By contrast, in GSNOR 2 zone, the fastest band at lane 1 (GSNOR 2a) was present in control variety and *fld L-1* mutant, while the slower one at lane 2 (GSNOR 2b) was visualized as unique band in *fldL-2* line (Figure 3).

$F_2$ progeny revealed allelic segregation (single locus) in GSNOR 2 zones of enzyme activity in crosses between *fld L-2* mutant and other three parents (Figures 3 and 6). Allozymes in this zone segregated into three phenotypic classes: two homozygotes for respective parental alleles (lanes 1 and 2) and one heterozygote of these two alleles (lanes 3 and 4) showing good agreement with the expected 1 : 2 : 1 ratio in $F_2$ generation (Figures 3 and 6; Table 2), and no segregation distortion was found. The $F_1$ hybrid was backcrossed to parents slowing slow allozyme and zymogram phenotypes agreed well with 1 parental : 1 hybrid ratio in each cross (Table 2). Segregation of allozymes could not be detected in $F_2$ progeny of two control varieties in this zone. No segregation of banding pattern was observed in GSNOR 1 zone also.

*3.4. Linkage Analysis between Isozyme Loci, fld 1, and fld 2* . Genetic linkage relationship was analyzed on the basis of joint segregation of zymogram phenotype and flavonoid level in $F_2$ and backcrosses (Table 3). Individual locus in *Aco* and *Gsnor* loci (except *Gsnor 1*) as well as *fld 1 and fld 2* exhibited normal mendelian segregation (1 : 2 : 1/3 : 1 in $F_2$ and 1 : 1 in backcrosses) of alleles, but their joint segregation

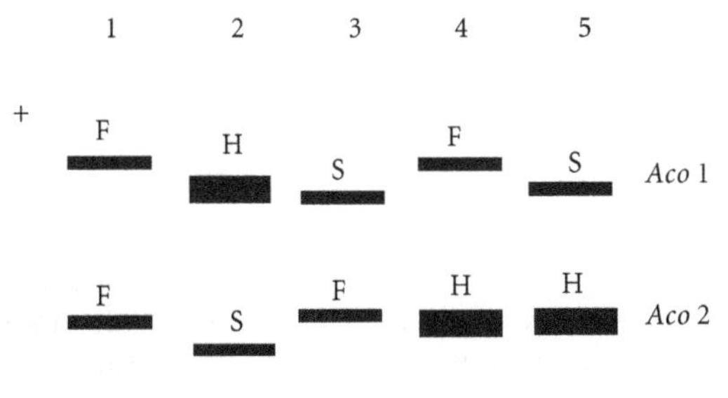


Figure 5: Segregation of phenotypes in $F_2$ derived from crosses between BioL-212 × *fldL-1* in *Aco 1* and *Aco 2* loci; F-Fast allele, S-slow allele, H-heterozygotes in different lanes, in *Lathyrus sativus* L.

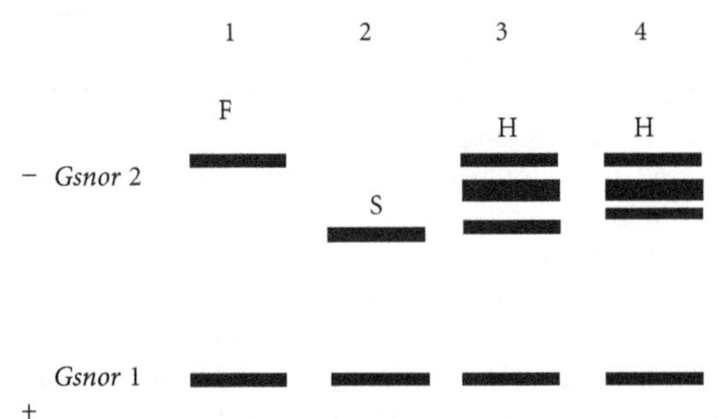


Figure 6: Segregation of S-nitrosoglutathione reductase loci, *Gsnor 2* in $F_2$ generation of *fldL-2* × mother plant; F-Fast allele, S-slow allele, H-heterozygote. No segregation was observed in *Gsnor 1* locus in *Lathyrus sativus* L.

in different cross-combinations showed significant deviations ($P < 0.05$) from the expected ratios of independent assortment in different $F_2$ families and respective backcross progenies (Table 3). In each case, recombination fraction ($r$) calculated from backcross data was put into Kosambi's mapping function, and map distance between loci was estimated. *Aco 1* and *fld1* was linked with a map distances of 9.75 cM, whereas *fld 2* and *Aco 2* were mapped 11.80 cM apart. A linked association with 15.48 cM and 26.19 cM map distance was found between *fld 2* and *Gsnor 2* and between *Aco 2* and *Gsnor 2*, respectively (Table 3). *Gsnor1* could not be mapped due to absence of segregating alleles.

*3.5. Chromosome Location of fld1, fld2, and Isozyme Loci by Primary Trisomic Analysis.* Chromosomal association of *fld1 and fld2* loci controlling different phenotypes of flavonoid deficiency in grass pea was traced in crosses between seven trisomics as female parents and the diploid mutant lines as their male counterpart. For trisomic-I and IV, the segregation of leaf flavonoid content as normal, recessive mutant phenotype derived from *fldL-1* × trisomic-I and from *fldL-2* × trisomic-IV, respectively, exhibited a large and significant $\mathcal{X}^2$ value ($P < 0.05$) for 3:1 in $F_2$ and 1:1 in backcrosses but agreed well with expected trisomic ratios of 8:1 in $F_2$ and 2:1 in testcross progenies (Table 4). On the other hand, segregation of normal and recessive mutant type in rest of the crosses involving other trisomics showed good fit with expected disomic ratio of 3:1 ratio in F2 and 1:1 ratio in corresponding testcrosses in diploid population, and a good number of recessive homozygotes in $2n + 1$ portion of these crosses were cytologically detected as trisomic plants (data not in table).

Among the four isozyme loci visualized in gel, linkage was detected only between *Aco 2* and *Gsnor 2*. Presumably, these two isozyme loci were on the same chromosome. To confirm this assumption and to localize them on chromosomes, the trisomics were crossed as female parent with diploid control and two mutant lines, and $F_1$ progeny in each case was raised. The rationale of the trisomic analysis in the present material involved trisomic segregation of different phenotypes in diploid portion of $F_2$ and $BC_1$. A significant ($P < 0.05$) departure of allozyme segregation coded by *Aco 1* from normal disomic ratios in $F_2$ (1:2:1) as well as backcross (1:1) ratios was manifested in the progenies involving only trisomic-I. Similar situation was encountered for *Aco 2* and *Gsnor 2* loci in trisomic-IV (Table 4). For both cases, segregation of allozymes in respective trisomics agreed well with the expected trisomic ratio of 4:4:1 in $F_2$ and to 2:1 in $BC_1$ generations of diploid portion and to 2:7:0 ($F_2$) in trisomic portion of the progeny (Table 4). Segregation was disomic for all other trisomics in $F_2$ and corresponding backcrosses (data not presented).

Linkage studies and trisomic segregation pattern in $F_2$ as well as $BC_1$ generations revealed that *Aco1 and fld1loci* were linked with each other on extra chromosome of trisomic-I, whereas *fld 2* and two isozyme loci *Aco 2 and Gsnor 2* were carried by extra chromosome of trisomic-IV in linked conditions. Based on the result, the map positions (in cM) among different loci are as shown in Figure 7.

## 4. Discussion

Both *fldL-1* and *fldL-2* mutants, isolated in EMS-treated $M_2$ progeny, exhibited huge deficiency in total foliar flavonoid contents, containing only 20% of normal level as measured in mother controls. However, the mutants differed from each other in the type of extract, where the flavonoid content reduced, ethanol extract for *fldL-1* and aqueous extract for *fldL-2* leaves. Flavonoid deficiency was found associated with modification of usual blue colour of flower into white and stipule characters in both type of mutants, and also rising level of pollen sterility in *fldL-2* plants. The modification of flower colour was ascribed to the deficiency of anthocyanin biosynthesis and is of considerable assistance in plant breeding [45]. In *Petunia*, flavonoid deficiency resulted in male sterility [46], while in maize male fertility was not affected at all [47]. Both the phenomena, however, were found in the present mutants, supporting differential behavior of the two mutants and deficiency of total flavonoids might be due to reduction of ethanol-dissolved and water-soluble compounds. Flavonoid deficiency in *Arabidopsis* was found associated with modifications in seed testa colour [7] and UV-sensitivity [6]. Both *fldL-1* and *fldL-2* in the present study

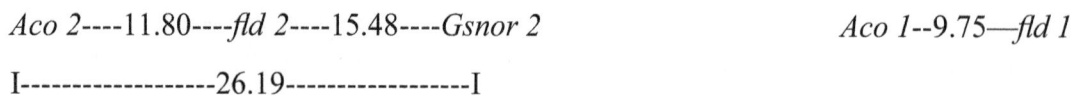

*Aco 2*----11.80---*fld 2*----15.48----*Gsnor 2* *Aco 1*--9.75—*fld 1*

I------------------26.19-----------------I

FIGURE 7

are nonlethal and provided easily detectable phenotypes such as flower color. No significant variation of flavonoid content, however, was found in two mother plants, suggesting lack of its variation in common genotypes of grass pea.

Mode of inheritance of flavonoid deficiency was traced in self-pollinated as well as in intercrossed population, involving two mutants and two mother controls. In all four parents, marginal variation in flavonoid content was found in advanced generations, indicating true breeding nature of the mutant traits. Inheritance studies in intercrossed progeny obtained from control × mutant plants revealed monogenic recessive nature of the low flavonoid content in both the mutant types with dominant allele that was always with mother plants. The result is in agreement with monogenic recessive nature of different flavonoid mutants in plants including *Arabidopsis* [7]. Interestingly, flower colour and stipule characteristics appeared unmodified in the respective recessive mutant type, confirming their true breeding nature in the present material.

A completely different result, however, was obtained when the two mutants were crossed reciprocally. Occurrence of $F_1$ plants with normal level of flavonoids and usual presence of blue flower and papilionaceous stipules and its segregations into four different plant types: normal, *fldL-1*, *fldL-2*, and a double-mutant type, consistent with 9 : 3 : 3 : 1 ratio in $F_2$, suggested involvement of two independent nonallelic loci *Fld1*/*fld 1* (for *fldL-1* mutant) and *Fld2*/*fld 2* (for *fld L-2* mutant) in controlling flavonoid deficiency in two mutant types under study. Both the genes (*Fld 1* and *Fld 2*) exhibited dominance over their respective recessive alleles (*fld1* and *fld 2*). In presence of both the genes in dominant form (*Fld 1*–*Fld 2*-), normal phenotype appeared whereas presence of *fld 2* gene in double recessive form (*fld 2 fld 2 Fld 2*-) produced phenotypes characteristic of *fldL-2* type. On the other hand, *fldL-1* type occurred in the presence of double recessive nature of *fld 1* gene (*Fld1-fld 1 fld1*). In homozygous recessive condition of both the genes (*fld1 fld1 fld2 fld2*) variant plant type showing leaf flavonoid content only 10% of mother control plants and high pollen sterility (79.33%) resulted in the $F_2$ progeny. This type bred true in advanced generations and tentatively designated as "flavonoid-deficient double mutant type" in grass pea. Recovery of *fldL-1* and *fldL-2* phenotypes in $F_2$ and occurrence of the double mutant type strongly indicated possibility of multiple blockages in flavonoid biosynthesis pathway which was different in two mutant types, but combined in double mutant plants, leading to further depletion of its flavonoid content in relation to *fldL-1* and *fldL-2* levels. The double mutants have immense significance as it provides valuable clues in functional biology of glutathione, NO and thioredoxin-mediated redox signaling in plants [14, 48].

The differences in genetic constitution of flavonoid deficiency between two mutant plant types were also manifested by banding profiles of aconitase and S-glutathione reductase isozymes. Inheritance pattern in the present study revealed that *fldL-1 and fldL-2* were not only different from control varieties but also differed from each other due to variant banding profiles that were heritable and bred true for all the four loci resolved here. The distinct zones of enzyme activity are mostly coded by different loci, and the variants within a particular zone are usually due to presence of different alleles or their interaction as heterozygotes [43]. In the present material, consistency in two zones of enzyme activity for both ACO and GSNOR enzymes was confirmed in successive self-pollinated and intercrossed populations of four parents.

Quite remarkably, the "loss-of-function" mutation in flavonoid content led to gain of isozyme functions in the present mutants. Allozyme variation is essential for construction of saturated linkage map with other markers in grass pea [35, 36]. Although control varieties showed monomorphic banding pattern, consistent occurrences of mutant specific bands in the present *Aco* and *Gsnor* loci indicated evolution of variant alleles which inherited as recessive gene mutations in the present material of grass pea. Both the mutant lines possessed some unique bands coded by specific alleles: *Aco 2a* in *fld L-1* and *Gsnor 2b* and *Aco 2b* in *fld L-2*. Obviously, *Aco 2* was triple allelic while *Aco 1* and *Gsnor 2* both were double allelic, resulting in increased polymorphism in the present mutants over their control plants. Presence of more than two alleles was also reported in *Aco* loci of grass pea [35] and lima bean [49]. Like *Gsnor 1*, single zone of activity was reported in GSNOR enzyme of *Pisum sativum* L. [15].

Segregation pattern of different allozymes in the present $F_2$ and backcross-population indicated involvement of codominant alleles in monogenic segregation of *Aco 1*, *Aco 2* loci of ACO system, and *Gsnor 2* locus of GSNOR enzyme. Presence of double-banded phenotypes in heterozygotes suggested monomeric nature of aconitase in grass pea, and no distorted segregation was apparent in $F_2$ generation. The GSNOR, on the other hand, was functionally dimeric as confirmed by the presence of four-banded phenotypes in the heterozygotes. However, single-banded phenotype was exhibited in heterozygotes of $F_2$ and backcross-populations obtained from crosses involving mutant and control parents for *Gsnor 1* locus, confirming its monomorphic nature in the present material. In $F_2$ population of crosses between different subaccessions of *Lathyrus sativus* L., Chowdhury and Slinkard [35] also detected polymorphism in both *Aco 1* and *Aco 2* loci with occurrence of codominant alleles, while, in regenerated plants of soybean, a rare mutation in *Aco2b* locus was detected as a null allele [50]. Polymorphisms displaying segregation ratios close to those expected for single locus traits suggested involvement of different alleles at the structural loci in generation of variation in different isozyme loci [51]. Among the closely related genera of *Lathyrus*, *Aco 1* was monomorphic, but *Aco 2* was polymorphic in *Vicia faba* L. [52]. Polymorphic *Aco* loci showing codominant expression of different alleles were also studied in pea [43], lens [53], and *Cicer arietinum* L. [54, 55].

Absence of distortion in $F_2$ single locus and joint segregation was another interesting feature in the present study consisting of four true breeding parental lines of *Lathyrus sativus* L. for the concerned traits. The result was in contrast with earlier reports of distorted segregation of other isozyme loci in grass pea [35]. Helentjaris et al. [56] explained that intraspecific cross-minimized genetic distortion and other errors than wide crosses to establish linkage maps. The true breeding nature of isozyme phenotypes in selfed progeny and simple segregation in $F_2$ population of different intercrossed progenies confirmed their stability in the present material.

Mutation has been identified as one of the main sources of isozyme variation in higher plants [57]. For the first time, allelic variations in *Aco* and *Gsnor* loci have been generated in two stable mutant lines, deficient in flavonoid contents, of grass pea through induced mutagenesis. Consistent presence of polymorphism in isozymes of both enzymes indicated origin of different molecular forms of allozymes. Induction of variant allele in leaf isozyme system has been reported in different legumes including *Glycine max* [50] and *Trifolium resupinatum* [58]. However, in some accessions of *Lathyrus sativus* L. and *Centrosema* occurrences of higher number of alleles per locus have been attributed to heterozygosity induced by significant outcrossing rate in these crops [36]. In the present study, in addition to using outcrossing preventive measure during hybridization, effective isolation between lines and populations has been maintained throughout the experiment to prevent intermixing, and inheritance studies were carried out in advanced selfed generation ($M_4$) of different true breeding parental lines. It seemed likely that the occurrences of new alleles in *Aco* and *Gsnor* loci resulted from the action of the recessive genes induced by EMS treatments in the present materials.

Linkage analysis involving *Aco 1*, *Aco 2*, and *Gsnor 2* isozyme loci, and *fld 1* and *fld 2* mutations revealed independent assortment between two *Aco* loci, of which *Aco 1* was linked tightly with *fld 1* whereas *Aco 2* was mapped with *fld 2* and *Gsnor 2* loci in linked states showing a distance of 11.80 cM and 26.19 cM, respectively. Absence of linkage between different *Aco* loci was also reported in different genotypes of grass pea, soybean, and *Phaseolus vulgaris* [35, 59], and this was confirmed in the present study also. However, for the first time, a *Gsnor* locus was mapped in linked association with a flavonoid-deficient locus and also with an *Aco* locus in any leguminous crop. The aconitase is exquisitely sensitive to NO and other ROS [17], while Gsnor reportedly showed reduced band intensity in cadmium-treated *Pisum sativum* L. [15]. Altered expression of the present *Aco* and *Gsnor* loci indicated modulation of enzyme activities under flavonoid-deficient conditions, and the mapping of their isozyme loci with *fld 1* and *fld 2* genes in closely linked state confirmed this assumption.

The importance of any mutant trait as a potential tool in functional biology enhanced once it was assigned to a particular chromosome. Primary trisomic has been used as an excellent tool in legume crops to confirm possible chromosomal location of various traits [60]. When the loci under study were located on a particular chromosome in trisomy, the normal disomic segregation ratio was modified due to presence of an extra chromosome. Trisomic segregation of electrophoretic phenotypes of different isozymes in the present zymogram strongly indicated possible location of *Aco 1* on extra chromosome of trisomic-I and *Aco 2 and Gsnor 2* on extra chromosome of trisomic-IV. Similarly, a good fit of *fld 1* and *fld 2* to trisomic segregation strongly indicated possible location of *fld 1* gene on extra chromosome of trisomic I and that of *fld 2* gene on extra chromosome of trisomic IV. The deviations from the normal segregation ratio are ascribed to the phenomenon of primary trisomy. Furthermore, no recessive homozygote plant in trisomic portion was recovered in these crosses, and all the recessive homozygotes in population were cytologically confirmed as diploids (data not presented). Segregating phenotypes in $F_2$ and $BC_1$ generations involving other trisomic types in respective crosses were consistent with normal mendelian disomic ratios and confirmed the above observation. In grass pea, primary trisomic has been successfully utilized to assign genes of agronomic interest on specific chromosomes [21, 24, 61] and to study gene-dosage effect of aneuploidy on antioxidant defense enzymes [62]. The isolation of two different flavonoid-deficient mutants and their mapping with closely linked isozyme markers of two prominent enzymatic systems on specific chromosomes may provide vital clues in understanding the role of flavonoid in integrated antioxidant defense system and their genetic basis in grass pea.

## Conflict of Interests

No conflict of interests is involved in any way with the present work.

## References

[1] L. P. Taylor and E. Grotewold, "Flavonoids as developmental regulators," *Current Opinion in Plant Biology*, vol. 8, no. 3, pp. 317–323, 2005.

[2] Y. Miyagi, A. S. Om, K. M. Chee, and M. R. Bennink, "Inhibition of azoxymethane-induced colon cancer by orange juice," *Nutrition and Cancer*, vol. 36, no. 2, pp. 224–229, 2000.

[3] T. A. Holton and E. C. Cornish, "Genetics and biochemistry of anthocyanin biosynthesis," *Plant Cell*, vol. 7, no. 7, pp. 1071–1083, 1995.

[4] S. Chopra, A. Hoshino, J. Boddu, and S. Iida, "Flavonoid pigments as tools in molecular genetics," in *The Science of Flavonoids*, E. Grotewold, Ed., pp. 147–173, The Ohio State University, Columbus, Ohio, USA, 2006.

[5] R. Koes, W. Verweij, and F. Quattrocchio, "Flavonoids: a colorful model for the regulation and evolution of biochemical pathways," *Trends in Plant Science*, vol. 10, no. 5, pp. 236–242, 2005.

[6] J. Li, T. M. Ou-Lee, R. Raba, R. G. Amundson, and R. L. Last, "Arabidopsis flavonoid mutants are hypersensitive to UV-B irradiation," *Plant Cell*, vol. 5, no. 2, pp. 171–179, 1993.

[7] B. W. Shirley, W. L. Kubasek, G. Storz et al., "Analysis of Arabidopsis mutants deficient in flavonoid biosynthesis," *Plant Journal*, vol. 8, no. 5, pp. 659–671, 1995.

[8] S. Albert, M. Delseny, and M. Devie, "Banyuls, a novel negative regulator of flavonoid biosynthesis in the Arabidopsis seed coat," *Plant Journal*, vol. 11, no. 2, pp. 289–299, 1997.

[9] K. Bieza and R. Lois, "An *Arabidopsis* mutant tolerant to lethal ultraviolet-B levels shows constitutively elevated accumulation of flavonoids and other phenolics," *Plant Physiology*, vol. 126, no. 3, pp. 1105–1115, 2001.

[10] M. Sharma, M. Cortes-Cruz, K. R. Ahern, M. McMullen, T. P. Brutnell, and S. Chopra, "Identification of the Pr1 gene product completes the anthocyanin biosynthesis pathway of maize," *Genetics*, vol. 188, no. 1, pp. 69–79, 2011.

[11] R. A. Dixon and L. W. Sumner, "Legume natural products: understanding and manipulating complex pathways for human and animal health," *Plant Physiology*, vol. 131, no. 3, pp. 878–885, 2003.

[12] J. Filkowski, O. Kovalchuk, and I. Kovalchuk, "Genome stability of vtc1, tt4, and tt5 *Arabidopsis thaliana* mutants impaired in protection against oxidative stress," *Plant Journal*, vol. 38, no. 1, pp. 60–69, 2004.

[13] M. R. Alfenito, E. Souer, C. D. Goodman et al., "Functional complementation of anthocyanin sequestration in the vacuole by widely divergent glutathione *S*-transferases," *Plant Cell*, vol. 10, no. 7, pp. 1135–1149, 1998.

[14] T. Bashandy, L. Taconnat, J.-P. Renou, Y. Meyer, and J.-P. Reichheld, "Accumulation of flavonoids in an ntra ntrb mutant leads to tolerance to UV-C," *Molecular Plant*, vol. 2, no. 2, pp. 249–258, 2009.

[15] J. B. Barroso, F. J. Corpas, A. Carreras et al., "Localization of *S*-nitrosoglutathione and expression of *S*-nitrosoglutathione reductase in pea plants under cadmium stress," *Journal of Experimental Botany*, vol. 57, no. 8, pp. 1785–1793, 2006.

[16] W. Moeder, O. Del Pozo, D. A. Navarre, G. B. Martin, and D. F. Klessig, "Aconitase plays a role in regulating resistance to oxidative stress and cell death in *Arabidopsis* and *Nicotiana benthamiana*," *Plant Molecular Biology*, vol. 63, no. 2, pp. 273–287, 2007.

[17] D. A. Navarre, D. Wendehenne, J. Durner, R. Noad, and D. F. Klessig, "Nitric oxide modulates the activity of tobacco aconitase," *Plant Physiology*, vol. 122, no. 2, pp. 573–582, 2000.

[18] A. Sakamoto, M. Ueda, and H. Morikawa, "*Arabidopsis* glutathione-dependent formaldehyde dehydrogenase is an S-nitrosoglutathione reductase," *FEBS Letters*, vol. 515, no. 1–3, pp. 20–24, 2002.

[19] M. Díaz, H. Achkor, E. Titarenko, and M. C. Martínez, "The gene encoding glutathione-dependent formaldehyde dehydrogenase/GSNO reductase is responsive to wounding, jasmonic acid and salicylic acid," *FEBS Letters*, vol. 543, no. 1–3, pp. 136–139, 2003.

[20] E. Pastor-Cavada, R. Juan, J.E. Pastor, J. Girón-Calle, M. Alaiz, and J. Vioque, "Antioxidant activity in Lathyrus species," *Grain Legumes*, vol. 54, pp. 10–11, 2009.

[21] D. Talukdar, "Dwarf mutations in grass pea (*Lathyrus sativus* L.): Origin, morphology, inheritance and linkage studies," *Journal of Genetics*, vol. 88, no. 2, pp. 165–175, 2009.

[22] D. Talukdar and A. K. Biswas, "Inheritance of flower and stipule characters in different induced mutant lines of grass pea (*Lathyrus sativus* L.)," *Indian Journal of Genetics and Plant Breeding*, vol. 67, pp. 396–400, 2007.

[23] D. Talukdar and A. K. Biswas, "Induced seed coat colour mutations and their inheritance in grass pea (*Lathyrus sativus* L.)," *Indian Journal of Genetics and Plant Breeding*, vol. 65, pp. 135–136, 2005.

[24] D. Talukdar, "Genetics of pod indehiscence in *Lathyrus sativus* L," *Journal of Crop Improvement*, vol. 25, pp. 1–15, 2011.

[25] D. Talukdar, "Bold-seeded and seed coat colour mutations in grass pea (*Lathyrus sativus* L.): origin, morphology, genetic control and linkage analysis," *International Journal of Current Research*, vol. 3, pp. 104–112, 2011.

[26] D. Talukdar, "Flower and pod production, abortion, leaf injury, yield and seed neurotoxin levels in stable dwarf mutant lines of grass pea (*Lathyrus sativus* L.) differing in salt stress responses," *International Journal of Current Research*, vol. 2, pp. 46–54, 2011.

[27] D. Talukdar, "Isolation and characterization of NaCl-tolerant mutations in two important legumes, *Clitoria ternatea* L. and *Lathyrus sativus* L.: Induced mutagenesis and selection by salt stress," *Journal of Medicinal Plant Research*, vol. 5, no. 16, pp. 3619–3628, 2011.

[28] D. Talukdar, "Effect of arsenic-induced toxicity on morphological traits of *Trigonella foenum-graecum* L. and *Lathyrus sativus* L during germination and early seedling growth," *Current Research Journal of Biological Sciences*, vol. 3, pp. 116–123, 2011.

[29] D. Talukdar, "Ascorbate deficient semi-dwarf *asfL1* mutant of grass pea (*Lathyrus sativus* L.) exhibits alterations in antioxidant defense," *Biologia Plantarum*. In press.

[30] D. Talukdar and A. K. Biswas, "Seven different primary trisomics in grass pea (*Lathyrus sativus* L.). I. Cytogenetic characterisation," *Cytologia*, vol. 72, no. 4, pp. 385–396, 2007.

[31] D. Talukdar, "Cytogenetic characterization of seven different primary tetrasomics in grass pea (*Lathyrus sativus* L.)," *Caryologia*, vol. 61, no. 4, pp. 402–410, 2008.

[32] D. Talukdar, "Cytogenetic characterization of induced autotetraploids in grass pea (*Lathyrus sativus* L.)," *Caryologia*, vol. 63, no. 1, pp. 62–72, 2010.

[33] D. Talukdar, "Reciprocal translocations in grass pea (*Lathyrus sativus* l.): pattern of transmission, detection of multiple interchanges and their independence," *Journal of Heredity*, vol. 101, no. 2, pp. 169–176, 2010.

[34] D. Talukdar, "Cytogenetic analysis of a novel yellow flower mutant carrying a reciprocal translocation in grass pea (*Lathyrus sativus* L.)," *Journal of Biological Research-Thessaloniki*, vol. 15, pp. 123–134, 2011.

[35] M. A. Chowdhury and A. E. Slinkard, "Genetics of isozymes in grasspea," *Journal of Heredity*, vol. 91, no. 2, pp. 142–145, 2000.

[36] J. F. Gutiérrez, V. Francisca, and J. V. Francisco, "Genetic mapping of isozyme loci in *Lathyrus sativus* L," *Lathyrus Lathyrism Newsletter*, vol. 2, pp. 74–78, 2001.

[37] D. Talukdar, "Allozyme variations in leaf esterase and root peroxidase isozymes and linkage with dwarfing genes in induced dwarf mutants of grass pea (*Lathyrus sativus* L.)," *International Journal of Genetics and Molecular Biology*, vol. 2, no. 6, pp. 112–120, 2010.

[38] D. Talukdar and A. K. Biswas, "Seven different primary trisomics in grass pea (*Lathyrus sativus* L.). II. Pattern of transmission," *Cytologia*, vol. 73, no. 2, pp. 129–136, 2008.

[39] N. Wu, K. Fu, Y.-J. Fu et al., "Antioxidant activities of extracts and main components of pigeonpea [*Cajanus cajan* (L.) Millsp.] leaves," *Molecules*, vol. 14, no. 3, pp. 1032–1043, 2009.

[40] B. J. Cardy, C. W. Stuber, and M. M. Goodman, *Techniques for starch gel electrophoresis of enzymes from maize (Zea mays L.),*

Institute of Statistics, Mimeograph Series No. 1317, N.C. State University, Raleigh, NC, USA, 1980.

[41] B. J. Cardy and W. D. Beversdorf, "A procedure for the starch gel electrophoretic detection of isozymes of soybean (*Glycine max* [L] *Merr.*)," *Technical Bulletin*, no. 119/8401, Department of Crop Science, University of Guelph, Guelph, Canada, 1984.

[42] M. R. Fernández, J. A. Biosca, and X. Parés, "S-nitrosoglutathione reductase activity of human and yeast glutathione-dependent formaldehyde dehydrogenase and its nuclear and cytoplasmic localisation," *Cellular and Molecular Life Sciences*, vol. 60, no. 5, pp. 1013–1018, 2003.

[43] N. F. Weeden, "A suggestion for the nomenclature of isozyme loci," *Pisum Newsletter*, vol. 20, pp. 44–45, 1988.

[44] D. D. Kosambi, "The estimation of map distance from recombination values," *Annals of Eugenics*, vol. 12, pp. 172–175, 1944.

[45] R. A. Dixon and N. L. Paiva, "Stress-induced phenylpropanoid metabolism," *Plant Cell*, vol. 7, no. 7, pp. 1085–1097, 1995.

[46] I. M. Van Der Meer, M. E. Stam, A. J. Van Tunen, J. N. M. Mol, and A. R. Stuitje, "Antisense inhibition of flavonoid biosynthesis in petunia anthers results in male sterility," *Plant Cell*, vol. 4, no. 3, pp. 253–262, 1992.

[47] Y. Mo, C. Nagel, and L. P. Taylor, "Biochemical complementation of chalcone synthase mutants defines a role for flavonols in functional pollen," *Proceedings of the National Academy of Sciences of the United States of America*, vol. 89, no. 15, pp. 7213–7217, 1992.

[48] A. Shahpiri, B. Svensson, and C. Finnie, "The NADPH-dependent thioredoxin reductase/thioredoxin system in germinating barley seeds: gene expression, protein profiles, and interactions between isoforms of thioredoxin h and thioredoxin reductase," *Plant Physiology*, vol. 146, no. 2, pp. 789–799, 2008.

[49] I. Zoro Bi, A. Maquet, B. Wathelet, and J.-P. Baudoin, "Genetic control of isozymes in the primary gene pool of *Phaseolus lunatus* L," *Biotechnology, Agronomy, Society and Environment*, vol. 3, pp. 10–27, 1999.

[50] L. A. Amberger, R. C. Shoemaker, and R. G. Palmer, "Inheritance of two independent isozyme variants in soybean plants derived from tissue culture," *Theoretical and Applied Genetics*, vol. 84, no. 5-6, pp. 600–607, 1992.

[51] B. Wolko and N. F. Weeden, "Additional markers for chromosome 6," *Pisum Newsletter*, vol. 22, pp. 71–74, 1990.

[52] B. Román, Z. Satovic, D. Pozarkova et al., "Development of a composite map in *Vicia faba*, breeding applications and future prospects," *Theoretical and Applied Genetics*, vol. 108, no. 6, pp. 1079–1088, 2004.

[53] D. Zamir and G. Ladizinsky, "Genetics of allozyme variants and linkage groups in lentil," *Euphytica*, vol. 33, no. 2, pp. 329–336, 1984.

[54] P. M. Gaur and A. E. Slinkard, "Genetic control and linkage relations of additional isozyme markers in chick-pea," *Theoretical and Applied Genetics*, vol. 80, no. 5, pp. 648–656, 1994.

[55] K. Kazan, F. J. Muehlbauer, N. E. Weeden, and G. Ladizinsky, "Inheritance and linkage relationships of morphological and isozyme loci in chickpea (*Cicer arietinum* L.)," *Theoretical and Applied Genetics*, vol. 86, no. 4, pp. 417–426, 1993.

[56] T. Helentjaris, M. Slocum, S. Wright, A. Schaefer, and J. Nienhuis, "Construction of genetic linkage maps in maize and tomato using restriction fragment length polymorphisms," *Theoretical and Applied Genetics*, vol. 72, no. 6, pp. 761–769, 1986.

[57] Z. Bartošová, B. Obert, T. Takáč, A. Kormuták, and A. Pretová, "Using enzyme polymorphism to identify the gametic origin of flax regenerants," *Acta Biologica Cracoviensia Series Botanica*, vol. 47, no. 1, pp. 173–178, 2005.

[58] D. R. Malaviya, A. K. Roy, A. Tiwari, P. Kaushal, and B. Kumar, "In vitro callusing and regeneration in *Trifolium resupinatum*—a fodder legume," *Cytologia*, vol. 71, no. 3, pp. 229–235, 2006.

[59] B. R. Hedges and R. G. Palmer, "Tests of linkage of isozyme loci with five primary trisomics in soybean, *Glycine max* (L.) Merr," *Journal of Heredity*, vol. 82, no. 6, pp. 494–496, 1991.

[60] R. J. Singh, G. H. Chung, and R. L. Nelson, "Landmark research in legumes," *Genome*, vol. 50, no. 6, pp. 525–537, 2007.

[61] D. Talukdar, "Recent progress on genetic analysis of novel mutants and aneuploid research in grass pea (*Lathyrus sativus* L.)," *African Journal of Agricultural Research*, vol. 4, no. 13, pp. 1549–1559, 2009.

[62] D. Talukdar, "The aneuploid switch: Extra-chromosomal effect on antioxidant defense through trisomic shift in *Lathyrus sativus* L.," *Indian Journal of Fundamental and Applied Life Sciences*, vol. 1, no. 4, pp. 263–273, 2011.

7

# A Study of Epstein-Barr Virus BRLF1 Activity in a *Drosophila* Model System

**Amy Adamson and Dennis LaJeunesse**

*Department of Biology, University of North Carolina at Greensboro, Greensboro, NC 27402, USA*

Correspondence should be addressed to Amy Adamson, aladamso@uncg.edu

Academic Editors: I. R. Arkhipova, B. Harrach, and F. Meggetto

Epstein-Barr virus, a member of the herpesvirus family, infects a large majority of the human population and is associated with several diseases, including cancer. We have created *Drosophila* model systems to study the interactions between host cellular proteins and the Epstein-Barr virus (EBV) immediate-early genes BRLF1 and BZLF1. BRLF1 and BZLF1 function as transcription factors for viral transcription and are also potent modifiers of host cell activity. Here we have used our model systems to identify host cell genes whose proteins modulate BRLF1 and BZLF1 functions. Via our *GMR-R* model system, we have found that BRLF1 expression results in overproliferation of fly tissue, unlike BZLF1, and does so through the interaction with known tumor suppressor genes. Through an additional genetic screen, we have identified several *Drosophila* genes, with human homologs, that may offer further insights into the pathways that BRLF1 interacts with in order to promote EBV replication.

## 1. Introduction

Epstein-Barr virus is a human herpesvirus that infects a majority of the human population. In addition to being the causative agent of infectious mononucleosis, Epstein-Barr virus (EBV) is also associated with several different cancers. Such malignancies include Burkitt's lymphoma, Hodgkin's lymphoma, nasopharyngeal carcinoma, breast cancer, and gastric cancer [1].

EBV can exist in a productive (lytic) phase or dormant (latent) phase. The EBV genome encodes more than 85 genes, subsets of which are expressed during the latent phase or during the lytic phase (which is broken down further into immediate-early, early, and late genes). BRLF1 (R) and BZLF1 (Z) are essential transcriptional activators expressed during the lytic phase that activate transcription of the EBV early genes. R and Z also have important roles in modulating the intracellular environment. For instance, R has been shown to interact with and alter the functions of the transcriptional regulators CREB-binding protein (CBP), Rb, and MCAF1 [2–4]. The Ran-binding-protein M (RanBPM) has also been shown to directly bind to R and act as a coactivator of R-mediated transcription [5]. R has been shown to promote cell cycle progression by activating S phase in fibroblast and epithelial cell lines [6], and conversely to promote senescence in an epithelial cell line [7]. Recently, R has been shown to inhibit expression of IRF3 and IRF7, leading to a decrease in the induction of interferon-$\beta$ [8]. All of these findings have been accomplished via cell culture studies.

In order to study viral protein function in a more comprehensive way, we have created *Drosophila* model systems for both R and Z. We previously examined Z protein activity in *Drosophila* and were able to investigate Z's function at both the molecular and genetic level [9]. We identified the *Drosophila* gene *shaven* as a potent modifier of Z activity in fly tissue [9]. The human homolog of *shaven*, Pax5, also interacted with Z in human cells and plays an important role in EBV biology [9].

There are several cellular pathways that are important for cell cycle regulation within *Drosophila*. These pathways contain numerous tumor suppressors that, when mutated, contribute to tissue overgrowth. These tumor suppressors generally fall into one of three classes: hyperplastic (mutations that cause increased cell proliferation with normal tissue structure), such as those in the *target of rapamycin*

TABLE 1: Candidate gene screen of tumor suppressors.

| Allele | Nature of allele* | Effect |
|---|---|---|
| *14-3-3zeta*$^{P1375}$ | P element insertion | Enhancer of R |
| *brat*$^{1}$, *brat* $^{k06028}$ | EMS, P element insertion | Enhancer of R |
| *Csk*$^{c04256}$ | P element insertion | Enhancer of R |
| *GMR-E2F* | Overexpresser | Enhancer of R |
| *InR*$^{E19}$ | EMS | Enhancer of R |
| *Merlin*$^{4}$ | loss of function | Enhancer of R |
| *Rab5*$^{EY10619}$ | P element enhancer upstream, | Enhancer of R |
| *Rab5*$^{KG05684}$ | P element insertion | Enhancer of Z (*Rab5*$^{EY10619}$ only) |
| *ago*$^{EY19092}$ | P element insertion | Enhancer of Z |
| *dlg1*$^{G0276}$ | P element insertion | Enhancer of Z |
| *GMR-p53* | Overexpresser | Enhancer of Z<br>Suppressor of R |
| *GMR-reaper* | Overexpresser | Enhancer of Z<br>Suppressor of R |
| *Ras85*$^{e1B}$ | EMS missense mutation | Enhancer of Z |
| *Tor*$^{DeltaP}$, *Tor*$^{k17004}$ | Deletion, P element insertion | Enhancer of Z<br>Suppressor of R |
| *awd*$^{j2A4}$ | P element insertion | Suppressor of R |
| *chico*$^{KG00336}$ | P element insertion | Suppressor of R |
| *GMR-dacapo* | Overexpresser | Suppressor of R |
| *GMR-Rbf* | Overexpresser | Suppressor of R |
| *scrib*$^{j7b3}$ | Hypomorph | Suppressor of R |
| *hyd*$^{15}$ | EMS | Suppressor of Z |
| *l(2)gd1*$^{EY04750}$ | P element insertion | Suppressor of *Z* |
| *14-3-3epsilon*$^{s-969}$ | antimorph | Suppressor of Z |

* Information about alleles from FlyBase.
EMS: ethyl methanesulfonate; hypomorph: less protein activity; antimorph: dominant negative protein activity.

(*Tor*) or *insulin receptor* pathways, neoplastic (mutations that cause increased cell proliferation with abnormal tissue structure and cause invasiveness) such as those in the *discs large (dlg)* or *Rab5* pathways, and nonautonomous (the overgrowth of wild-type cells due to neighboring cells being mutant) such as those in the *hyperplastic discs* (*hyd*) pathway [10]. Most of these tumor suppressors have human homologs that function in the same manner as in *Drosophila* cells.

Here we have made use of *Drosophila's* powerful genetic system to investigate R function and to investigate the pathways by which R may cause aberrant cell division. Via our *GMR-R* model system, we found that R expression causes overproliferation in fly tissue, as it has done in human cell culture. Through genetic screens, we have identified several

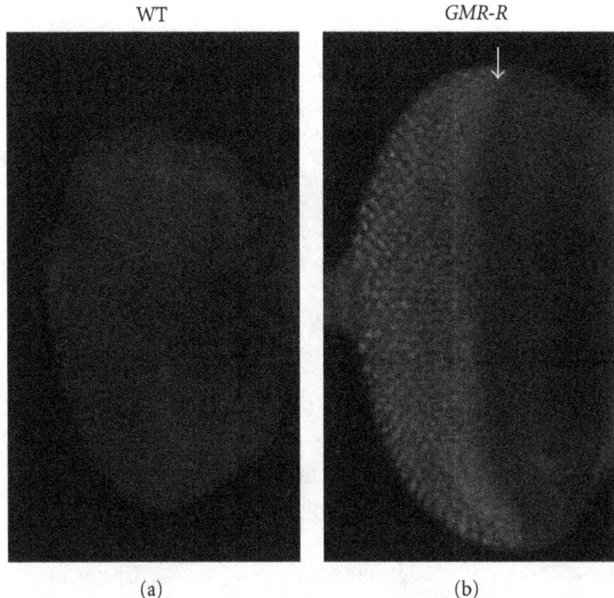

(a) (b)

FIGURE 1: BRLF1 protein expression in the *GMR-R/+* eye imaginal disc. Wild-type (a) and *GMR-R/+* (b) imaginal eye discs were stained with an anti-BRLF1 antibody. Confocal microscopy was used to image the discs. The arrow refers to the morphogenetic furrow.

*Drosophila* genes that are important for this R-mediated phenotype. The genes identified confirm previous findings from human cell culture and offer insights into how R interacts with host cell proteins and pathways to promote EBV replication.

## 2. Materials and Methods

*2.1. Fly Culture.* Flies were maintained at 20°C in plastic vials on a medium of cornmeal, yeast, molasses, and agar with methyl 4-hydroxybenzoate added as a mold inhibitor. $w^{1118}$ was used as the wild-type line. Fly stocks for the genetic screens were purchased from the Bloomington stock center. Crosses were performed at 20°.

*2.2. P-Element-Mediated Transformation.* The BRLF1 cDNA was cloned into the pGMR vector. Germline transformations were performed using the standard P-element protocol [11]. Several *GMR-R* lines were isolated.

*2.3. Scanning Electron Microscopy.* Flies were stored in 95% ethanol until ready to be sputter-coated. Flies were dried briefly, mounted onto stubs, and sputter-coated with gold. Sputter-coated flies were imaged in a Leica scanning electron microscope and images recorded at 2000x and 500x magnifications.

*2.4. Immunostaining of Imaginal Discs.* Eye-antenna imaginal discs were immunostained as described [12]. The anti-R antibody (Argene) was used at a 1 : 50 dilution and the anti-phospho-histone H3 antibody (Upstate) used at a 1 : 1000

Table 2: EP Line (second chromosome) Genetic Screen: Suppressors (S) and Enhancers (E) of the *GMR-R* phenotype.

| Class | *Drosophila* gene | Insertion** | Human homolog | Function |
|---|---|---|---|---|
| S | Beta coatomer protein | Upstream | COPB2 | Vesicle transport from Golgi |
| S | Defective proventriculus | Within | SATB1 | Transcription/chromatin organization |
| S | Paxillin | Within | TGF$\beta$1I1 | Signal transduction/adaptor |
| S | Rack1 | Within | GNB2L1 | Signal transduction |
| S | RanBPM | Upstream | RANBP9 | Scaffolding/proapoptotic |
| E* | cdc14 | Within/up | CDC14A | Phosphatase/cell cycle/p53 |
| E* | Chip | Upstream | LDB2 | LIM domain binding |
| E* | Female sterile (2) ketel | Upstream | KPNB1 | Protein import into nucleus |
| E* | NAT1 | Within/up | EIF4G2 | Repressor of translation initiation |
| E* | Tre oncoprotein related | Upstream | TBC1D3C | Rab GTPase activator |
| E* | Yippee interacting protein 2 | Within/up | ACAA2 | Acetyl Coa-acyltransferase |
| E | 14-3-3 zeta | Within | YWHAZ | Signal transduction/insulin pathway |
| E | Alpha-adaptin | Upstream | AP2A2 | Clathrin-coated vesicle transport |
| E | Bicoid-interacting protein 3 | Upstream | MEPCE | Transcription |
| E | cdGAPr | Within/up | AC108065 | GTPase activator |
| E | CG16896 | Upstream | WDR67 | Rab GTPase activator |
| E | Chickadee | Within/up | PFN4 | Cytokinesis |
| E | Genghis khan | Upstream | CDC42BPA | Phosphorylation/actin polymerization |
| E | Pendulin | Within/up | KPNA | Importin alpha 2/nuclear import |
| E | RhoGEF2 | Within | ARHGF12 | Rho GEF/actin dynamics |
| E | Shroom | Within | SHROOM3 | Actin binding |
| E | TBPH | Upstream | TARDBP | Mrna binding/repressor of transcription |

*Indicates a strong enhancer of the *GMR-R* phenotype.

**Indicates whether the p element insertion, containing the enhancer, lies upstream, where it likely causes overexpression of the gene, or within, where it likely interrupts the coding region. Within/up indicates that the p element lies in the 5′ untranslated region of the gene.

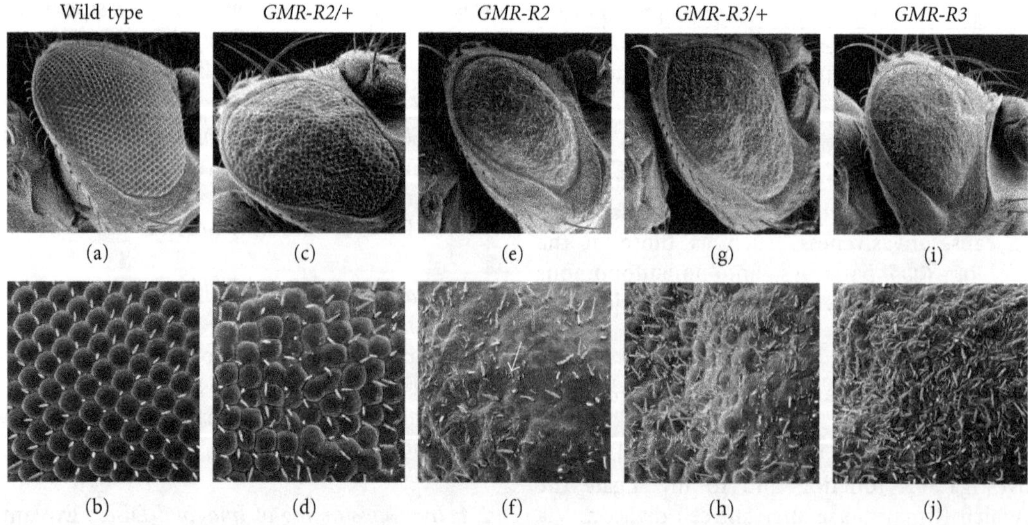


Figure 2: The *GMR-R* phenotype. SEMs of wild-type (a, b), *GMR-R2* heterozygous (c, d), *GMR-R2* homozygous (e, f), *GMR-R3* heterozygous (g, h), and *GMR-R3* homozygous (i, j) eyes are presented. The images are presented as 500x (top panels) and 2000x (bottom panels). Arrow points to hair-like bristles.

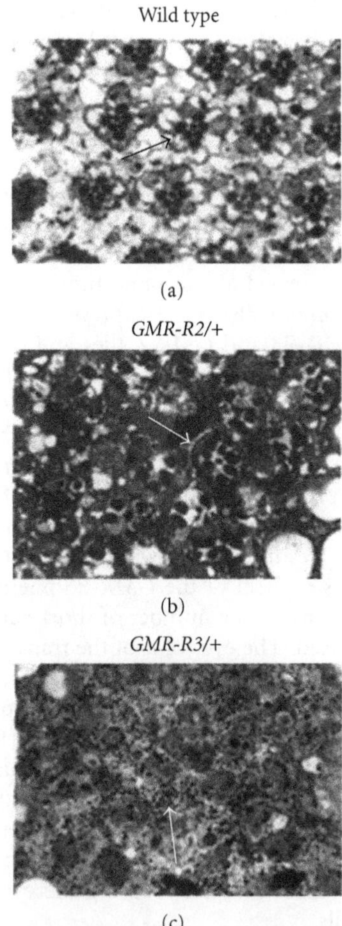


FIGURE 3: Sections of *GMR-R* adult eyes. Adult eyes from wild-type (a), *GMR-R2/+* (b) and *GMR-R3/+* (c) were embedded and sectioned. Arrows in (a) and (b) refer to photoreceptor clusters, arrow in (c) refers to pigment granules. Note that the photoreceptor cluster in (b) contains eight photoreceptors instead of seven.

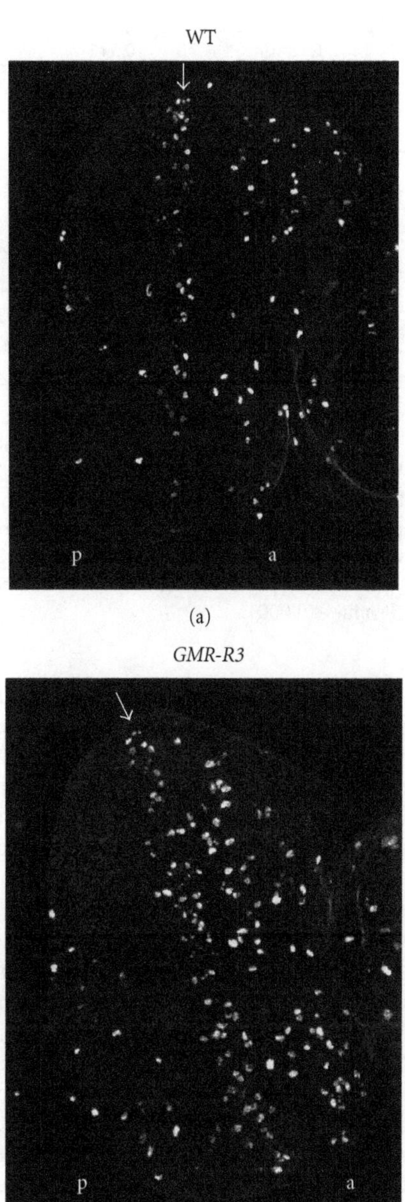


FIGURE 4: Increased cell division in *GMR-R* eyes. Third instar eye imaginal discs from wild-type (a) and *GMR-R3* (b) larvae were stained with anti-phospho-histone-H3 (Ser 10) antibody. Confocal microscopy was used to image the discs. Arrows refer to the morphogenetic furrows. Posterior is to the left of each morphogenetic furrow.

dilution. Each primary antibody was incubated with several (~10) discs overnight. The secondary antibodies (donkey anti-mouse CY3 and donkey anti-rabbit FITC (Jackson Immunoresearch)) were used at a 1 : 2000 dilution, and were incubated with the discs for 2 hr. Discs were mounted in anti-fade media (Dako Cytomation). Images were obtained by confocal microscopy and analyzed with FluoView software and MicroSuite software.

## 3. Results

*3.1. BRLF1 Produces a Dose-Sensitive Rough Eye Phenotype in Drosophila.* BRLF1 transgenic flies were created by cloning the BRLF1 cDNA into the *Drosophila* P-element vector pGMR (Glass-mediated response) [13]. This vector allowed for eye-specific expression of BRLF1. Expression from this construct begins during the larval stage, with a peak during the third larval instar and can be seen in cells posterior to the morphogenetic furrow in eye imaginal discs (Figure 1(b)). Several lines of *GMR-R* were obtained, each with a dose-sensitive phenotype. Expression of BRLF1 in the *Drosophila* eye resulted in a rough adult eye phenotype (Figure 2). While wild-type eyes had an organized pattern of ommatidia (Figures 2(a) and 2(b)), flies heterozygous for *GMR-R2* (*GMR-R2/+*) had an unorganized ommatidial and bristle structure (Figures 2(c) and 2(d)). Flies homozygous for *GMR-R2* had a more severe eye phenotype, including a loss of ommatidia, and included the appearance of short hair-like bristles (Figures 2(e) and 2(f)). The *GMR-R* line *GMR-R3* (Figures 2(g)–2(j)) displayed a more severe phenotype than the *GMR-R2* line (Figure 2(c)–2(f)). The number of short, hair-like bristles was greatly increased, especially in homozygous

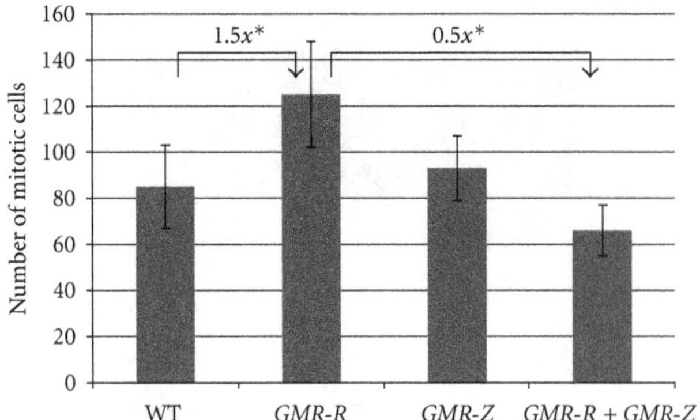


Figure 5: Expression of Z counteracts the overproliferation phenotype within *GMR-R* eye discs. Third instar eye imaginal discs from wild-type (WT), *GMR-R3, GMR-Z, and GMR-R/GMR-Z* larvae were stained with anti-phospho-histone-H3 (Ser 10) antibody. The number of positively stained cells within the furrow or posterior to the furrow was counted. Cells from approximately 12 discs were counted for each genotype. Error bars show the standard error; *indicates a $P$ value < 0.0001.

*GMR-R3* eyes. *GMR-R* flies, and especially *GMR-R3* flies, had a darker red eye color than wild type (data not shown). When adult eyes were sectioned, we found that *GMR-R2/+*, with the more mild phenotype (Figure 2(d)), had disorganized but recognizable ommatidia (Figure 3(b)), while *GMR-R3/+*, with the more severe phenotype (Figure 2(h)), had no recognizable ommatidial structures, but had an overabundance of pigment granules (Figure 3(c)).

*3.2. Expression of BRLF1 Causes Overproliferation of Eye Cells.* As previous work suggested that BRLF1 expression increased entry into the cell cycle in tissue culture cells [6], we tested whether BRLF1 also increased entry into the cell cycle in *GMR-R* discs. To identify mitotic cells in the eye discs, we stained wild-type and homozygous *GMR-R3* discs with an anti-phospho-histone H3 (Ser 10) antibody. As histone H3 is phosphorylated on Ser 10 during mitosis, this antibody identifies mitotic cells only. In wild-type discs, mitotic cells were seen in the morphogenetic furrow, with some mitotic cells anterior to and posterior to the furrow (Figure 4(a)). In *GMR-R3* discs, more mitotic cells were found in the morphogenetic furrow as well as more mitotic cells posterior to the furrow (Figure 4(b)). The number of cells in the furrow and that posterior to the furrow were counted in wild-type and *GMR-R3* discs. Figure 5 shows that *GMR-R3* discs have an average of 1.5 times more cells in mitosis than wild-type discs. Overproliferation was also evident in *GMR-R* adult eyes. Figure 3(b) shows a representative ommatidium that contained an extra photoreceptor cell. In addition, Figure 3(c) shows an overabundance of pigment granules, which contributed to the darker adult eye color in these flies.

*3.3. Genetic Interaction between GMR-R and GMR-Z.* We have previously analyzed the expression of the EBV BZLF1 gene in *Drosophila* [9]. The *GMR-Z/+* phenotype was different than that of *GMR-R* and included a rough eye phenotype and diminished pigment. Homozygosity for *GMR-Z* (referred to as "strong Z") led to a loss of ommatidia (leaving a smooth eye) and a complete loss of pigment. In our studies we found that strong Z inhibited the cell cycle and prevented the differentiation of cone cells in homozygous *GMR-Z* eyes [9].

Although the *GMR-R* and *GMR-Z* phenotypes seem to be antagonistic to each other (as R increases entry into the cell cycle and Z decreases entry into the cell cycle), during the normal EBV infection process, the two proteins R and Z are expressed at the same time and in fact work together to promote EBV gene expression [14]. Therefore we crossed *GMR-R3* and *GMR-Z* flies and examined their progeny (Figure 6). We also counted the number of mitotic cells in *GMR-R3/weak GMR-Z* larval eye discs. We found that while Z alone did not significantly alter the number of mitotic cells in eye discs, expression of Z did reverse *GMR-R3* overproliferation, such that the number of mitotic cells in *GMR-R3/weak GMR-Z* eye discs was half that of those in *GMR-R3* alone (resulting in a lower number than wild type) (Figure 5). This reversal of the *GMR-R3* phenotype can be seen in Figure 6(h), as the number of short hair-like bristles was also decreased. The eye color of the trans-heterozygotes was very similar to the *GMR-Z/+* eye color (orange), suggesting that the pigment cells had not over-proliferated as in *GMR-R3* eyes. However, Figures 6(g) and 6(h) shows that the *GMR-R3/weak GMR-Z* trans-heterozygous flies had a complete loss of ommatidia, although the bristles remained. This suggests that while Z counteracted the overproliferation phenotype of R, the presence of R and Z together elicited a significant cellular response causing the eventual loss of ommatidial cells.

*3.4. GMR-R and GMR-Z Genetically Interact with Growth Regulator Mutants.* Both *R* and *Z* impact cell division, either positively or negatively. However, *GMR-R* and *GMR-Z* genetically interact to produce a more severe phenotype than either alone, suggesting that they participate in different pathways. To elucidate these pathways, we performed a genetic screen with candidate genes—known genes involved in growth regulation. We crossed both *GMR-R3* and *GMR-Z* to flies mutant for or overexpressing genes involved in cell cycle, signal transduction, and apoptosis. In most cases, more than one allele for each gene was tested. Enhancers of the *GMR-R* phenotype were those that increased the roughness/disorganization/pigmentation of the eye tissue, and/or increased the number of short hair-like bristles (e.g., see Figures 7(i) and 7(j)). Suppressors of the *GMR-R* phenotype were those that restored the eye to a more wild-type organization of ommatidia/pigmentation; this was typically accompanied by a reduction in the number of short hair-like bristles (e.g., see Figures 7(e), 7(f), 7(m), and 7(n)). Enhancers of the *GMR-Z* phenotype were those that led to a loss of ommatidia and pigmentation, while suppressors of the *GMR-Z* phenotype were those that restored wild-type ommatidial organization and pigmentation. Of the 51 genes tested, 15 modified the *GMR-R3* phenotype and 10 modified the *GMR-Z* phenotype (Table 1). Of these, there were 4 genes that modified both the *GMR-R* and *GMR-Z* phenotypes: *Rab5, p53, reaper*, and *Tor*.

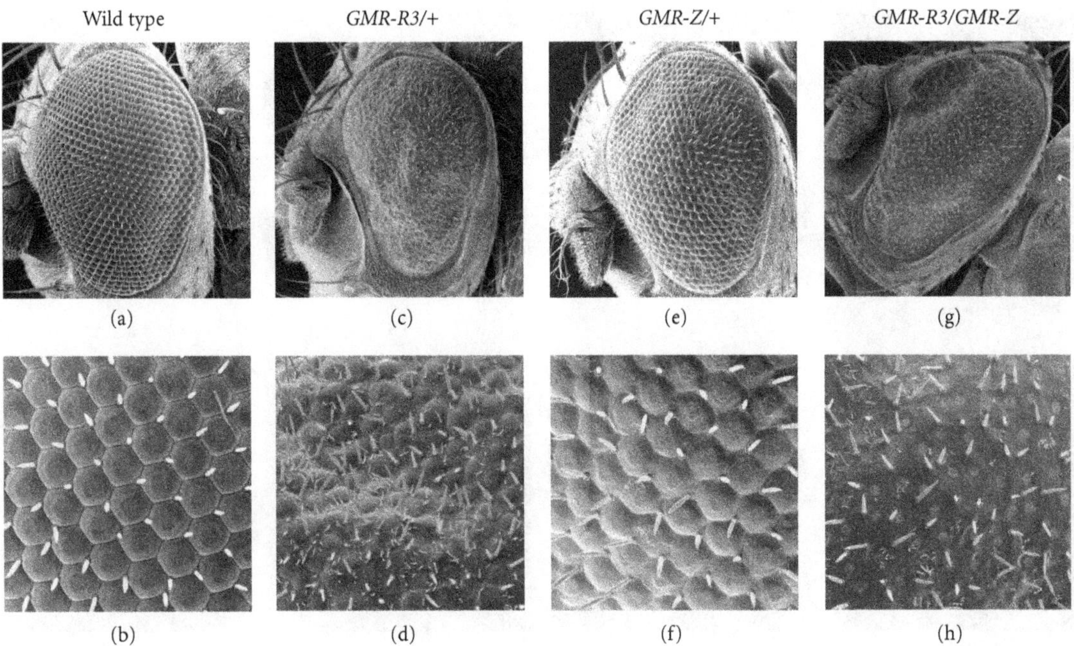


FIGURE 6: BZLF1 expression alters the *GMR-R* phenotype. A cross was performed between *GMR-R* and *GMR-Z* flies. SEMs were taken at 500x (top panels) and 2000x (bottom panels). Wild-type (a, b), *GMR-R3/+* (c, d), *GMR-Z/+* (e, f), and *GMR-R/GMR-Z* (g, h) adult eyes are presented.

Table 1 shows that fly lines that over-expressed the wild-type tumor suppressors *Rbf* (homolog of Rb), *dacapo* (homolog of p21), and *p53* suppressed the R overproliferation phenotype, while the overexpression of the cell cycle promoter *E2F* or a loss of a tumor suppressor (*brat, Csk, Merlin*) enhanced the R overproliferation phenotype. Conversely, fly lines that had a loss of a tumor suppressor (*hyd, l(2)gd1*) suppressed the Z phenotype, while the loss of cell cycle promoters (*Ras85*) enhanced the Z phenotype.

Interestingly, four fly gene mutants affected both the *GMR-R* and *GMR-Z* phenotypes. The *GMR-R* overproliferation phenotype was suppressed by over-expression of the cell cycle inhibitor *p53*, as well as by the proapoptotic gene *reaper*; the *GMR-Z* phenotype was enhanced by over-expression of both *p53* and *reaper*. Decreased levels of Tor suppressed the *GMR-R* phenotype, while the same mutants enhanced the *GMR-Z* phenotype. Furthermore, misexpression of *Rab5*, a GTPase that promotes endocytic vesicle fusion and helps terminate signaling pathways, enhanced both the *GMR-R* and *GMR-Z* phenotypes.

To further define R's biological role in cells, we performed a more objective genetic screen by crossing *GMR-R3* to the second chromosome EPgy2 misexpression line collection [15] (http://flystocks.bio.indiana.edu/Browse/in/misexpression-top.php). Each of these approximately 1000 lines tested contained a P element inserted either upstream of or within the coding region of a specific gene, causing that gene to either be overexpressed (if upstream) or mutant (if within). Lines that caused either enhancement or suppression of the *GMR-R* phenotype were crossed to a control line (*GMR* alone). Enhancers or suppressors that were specific to the *GMR-R* phenotype are presented in Table 2. Forty-nine genes were identified; only genes with human homologs and defined protein functions have been listed (22 of the 49). While some genes identified are involved in cell cycle regulation (*cdc14*), others are involved in signal transduction (*Paxillin, Rack1, 14-3-3 zeta, cdGAPr*) regulating the cytoskeleton (*RhoGEF2, genghis khan, Shroom, chickadee*) and vesicle transport (*beta coatomer protein, alpha-adaptin*) and specifically Rab GTPase activators (*CG16896, TRE oncoprotein related*). Others are transcriptional regulators (*defective proventriculus, bicoid interacting protein 3, TBPH*) or are involved in protein import into the nucleus (*Female sterile (2) ketel, PENdulin*). Interestingly, the overexpression of *RanBPM*, a scaffolding protein involved in many signal transduction pathways [16], suppressed the R phenotype; it has been previously reported that RanBPM physically interacts with R and promotes the transactivation ability of R in human B cells [5]. Also, while not listed in Table 2, the EP line screen identified *brat* as a potent modifier of the *GMR-R* phenotype, confirming our results in Table 1. *brat* does not have a human homolog, but functions as a tumor suppressor in *Drosophila* brain tissue [17].

## 4. Discussion

We have created a model system to investigate the cellular consequences of R expression. We found that R is expressed in the nuclei of eye cells while under the control of the GMR promoter element and that R expression causes more cells to enter the cell cycle. This overproliferation is evident from the *GMR-R* mutant eye phenotype—the ommatidia and bristles become unorganized, an overabundance of pigment leads to a dark red eye color, and short, fine bristles appear. These

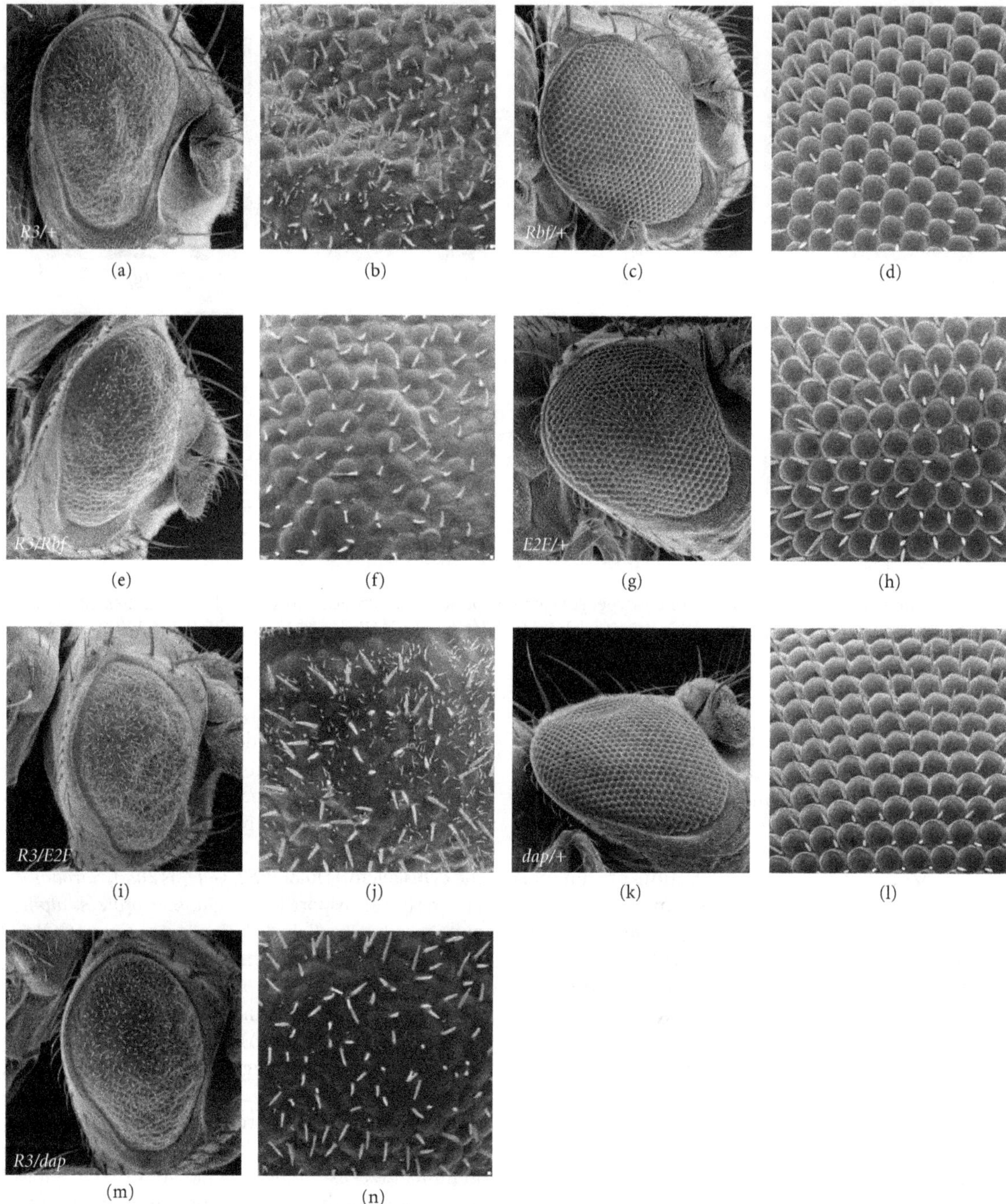


Figure 7: The *GMR-R* phenotype is suppressed by *Drosophila* Rb and p21 and enhanced by E2F. Crosses were performed between *GMR-R3* and *GMR-Rbf*, *GMR-E2F*, and *GMR-dacapo*. SEMs, at 500x and 2000x, are shown for *GMR-R3/+* (a, b), *GMR-Rbf/+* (c, d), *GMR-R3/GMR-Rbf* (e, f), *GMR-E2F/+* (g, h), *GMR-R3/GMR-E2F* (i, j), *GMR-dacapo/+* (k, l), and *GMR-R3/GMR-dacapo* (m, n).

R-mediated effects are dose sensitive—the more R expressed, the more severe the phenotype. The coexpression of Z curbed the overproliferation phenotype in *GMR-R* eyes: fewer cells underwent mitosis in larval eye discs, and there was less pigment and fewer hair-like bristles in adult eyes. However the combination of R and Z did not produce a wild-type eye, as a different mutant phenotype took the place of both the *GMR-R* and *GMR-Z* phenotypes in *GMR-R/GMR-Z* eyes.

In order to determine which cellular genes were mediating the *GMR-R* and *GMR-Z* phenotypes, we performed a candidate gene screen with both the *GMR-R* and *GMR-Z* flies, as well as an additional EP line screen with the *GMR-R* flies (Tables 1 and 2). Via the tumor suppressor candidate gene screen, we were able to determine that R's ability to promote the cell cycle is sensitive to the activities of a variety of cell cycle regulators, especially those falling into the

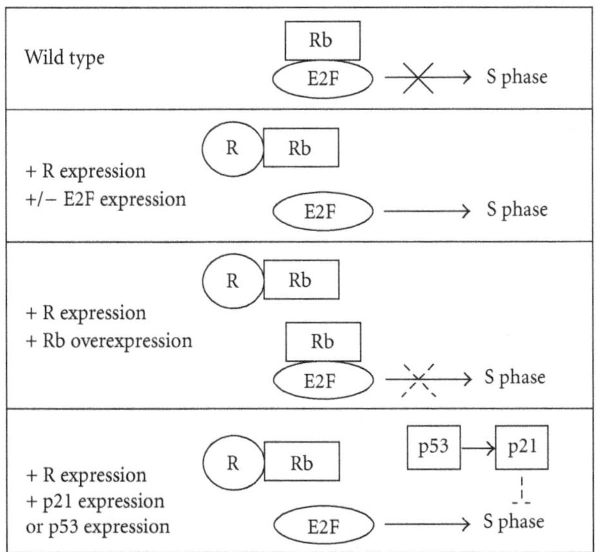


Figure 8: Model of R's effect upon S phase entry. Unless given a signal, cells will have Rb bound to E2F to inhibit E2F activity. R protein binds to Rb, displacing it from E2F, in which case E2F will promote S phase entry. Overexpression of E2F enhances this effect. Overexpression of Rb allows for renewed E2F inhibition. p53 or p21 overexpression inhibits cell cycle progression after it is initiated by active E2F.

*Drosophila* hyperplastic growth category. Our work confirms previous work showing that R interacts with Rb and E2F to promote entry into the cell cycle [3, 6] (for our model, see Figure 8), but also indicates that R activity is influenced by a number of signaling pathways, including the insulin receptor pathway and the Tor pathway. It is interesting to note that from this screen we found an equal number of enhancers for both R and Z (7 each), but more suppressors for R than for Z (8 for R versus 3 for Z). Therefore it appears to be "easier" to suppress R activity than to suppress Z activity.

From the EP line screen with *GMR-R*, we identified 5 suppressors and 17 enhancers of R activity. It is interesting to note that many of these modifiers code for proteins that are involved in protein/vesicle trafficking. Specifically, we identified two genes involved with protein transport into the nucleus: *Female sterile (2) ketel* and *Pendulin*. It appears that the likely overexpression of *Female sterile (2) ketel* and *Pendulin* leads to the enhancement of R activity, perhaps due to an increase of R protein import into the nucleus. Similar to our finding that *Rab5* was a modifier of both R and Z activity in our candidate gene screen, we identified other genes involved in vesicle trafficking, as well as Rab GTPase activators, as modifiers of R activity in this screen. Namely, the likely overexpression of *tre oncoprotein related* and *CG16896*, both Rab GTPase activators, enhanced the *GMR-R* phenotype, just as overexpression of *Rab5* (Table 1) enhanced the *GMR-R* phenotype. Rab proteins are involved in vesicle trafficking, and Rab5 is specifically involved in endocytosis [18].

Another class of genes that modify R activity are signal transduction mediators: *Paxillin, Rack1, RanBPM*, and *14-3-3zeta*. The fact that we found so many adaptors, which are involved in a variety of signal transduction pathways, to modulate R activity, indicates that the R protein exerts its effects via manipulation of several different signaling pathways.

We are very interested in the four genes that we identified in our candidate gene screen that modify both the *GMR-R* and *GMR-Z* phenotypes: *reaper, p53, Rab5*, and *Tor*. As both R and Z are present together in lytically replicating EBV-positive cells and they both contribute to transactivation of EBV early genes as well as to manipulation of their cellular environments, we are interested in cellular pathways that are common to both. We are especially interested in the mammalian mTOR pathway, which controls cell growth and protein translation [19] and how it will affect R and Z functioning during EBV lytic replication, as well as the mammalian Rab5 protein, and how vesicle trafficking may affect or be affected by EBV lytic replication.

Overall, we have established a model of R activity that mimics how the protein functions in human cells and have identified several cellular mediators of R activity that will be interesting foci for future study.

## Acknowledgments

The authors would like to acknowledge the Bloomington Stock Center and FlyBase for fly stocks and *Drosophila* gene-related information.

## References

[1] M. P. Thompson and R. Kurzrock, "Epstein-barr virus and cancer," *Clinical Cancer Research*, vol. 10, no. 3, pp. 803–821, 2004.

[2] J. J. Swenson, E. Holley-Guthrie, and S. C. Kenney, "Epstein-Barr virus immediate-early protein BRLF1 interacts with CBP, promoting enhanced BRLF1 transactivation," *Journal of Virology*, vol. 75, no. 13, pp. 6228–6234, 2001.

[3] V. L. Zacny, J. Wilson, and J. S. Pagano, "The epstein-barr virus immediate-early gene product, BRLF1, interacts with the retinoblastoma protein during the viral lytic cycle," *Journal of Virology*, vol. 72, no. 10, pp. 8043–8051, 1998.

[4] L. K. Chang, J. Y. Chung, Y. R. Hong, T. Ichimura, M. Nakao, and S. T. Liu, "Activation of Sp1-mediated transcription by Rta of Epstein-Barr virus via an interaction with MCAF1," *Nucleic Acids Research*, vol. 33, no. 20, pp. 6528–6539, 2005.

[5] L. K. Chang, S. T. Liu, C. W. Kuo et al., "Enhancement of transactivation activity of Rta of Epstein-Barr virus by RanBPM," *Journal of Molecular Biology*, vol. 379, no. 2, pp. 231–242, 2008.

[6] J. J. Swenson, A. E. Mauser, W. K. Kaufmann, and S. C. Kenney, "The Epstein-Barr virus protein BRLF1 activates S phase entry through E2F1 induction," *Journal of Virology*, vol. 73, no. 8, pp. 6540–6550, 1999.

[7] Y. L. Chen, Y. J. Chen, W. H. Tsai, Y. C. Ko, J. Y. Chen, and S. F. Lin, "The Epstein-Barr virus replication and transcription activator, Rta/BRLF1, induces cellular senescence in epithelial cells," *Cell Cycle*, vol. 8, no. 1, pp. 58–65, 2009.

[8] G. L. Bentz, R. Liu, A. M. Hahn, J. Shackelford, and J. S. Pagano, "Epstein-Barr virus BRLF1 inhibits transcription of

IRF3 and IRF7 and suppresses induction of interferon-$\beta$," *Virology*, vol. 402, no. 1, pp. 121–128, 2010.

[9] A. L. Adamson, N. Wright, and D. R. LaJeunesse, "Modeling early Epstein-Barr virus infection in *Drosophila* melanogaster: the BZLF1 protein," *Genetics*, vol. 171, no. 3, pp. 1125–1135, 2005.

[10] I. K. Hariharan and D. Bilder, "Regulation of imaginal disc growth by tumor-suppressor genes in *Drosophila*," *Annual Review of Genetics*, vol. 40, pp. 335–361, 2006.

[11] I. Rebay, R. G. Fehon, and S. Artavanis-Tsakonas, "Specific truncations of *Drosophila* Notch define dominant activated and dominant negative forms of the receptor," *Cell*, vol. 74, no. 2, pp. 319–329, 1993.

[12] T. Wolff, "Histological techniques for the *Drosophila* eye, part I: larva and pupa," in *Drosophila Protocols*, W. Sullivan, M. Ashburner, and R. S. Hawley, Eds., pp. 201–227, Cold Spring Harbor Press, New York, NY, USA, 2000.

[13] B. A. Hay, T. Wolff, and G. M. Rubin, "Expression of baculovirus P35 prevents cell death in *Drosophila*," *Development*, vol. 120, no. 8, pp. 2121–2129, 1994.

[14] E. Kieff and A. B. Rickinson, "Epstein-Barr virus and its replication," in *Field's Virology*, D. M. Knipe and P. M. Howley, Eds., pp. 2603–2654, Lippincott Williams & Wilkins, Philadelphia, Pa, USA, 2007.

[15] H. J. Bellen, R. W. Levis, G. Liao et al., "The BDGP gene disruption project: single transposon insertions associated with 40% of *Drosophila* genes," *Genetics*, vol. 167, no. 2, pp. 761–781, 2004.

[16] L. C. Murrin and J. N. Talbot, "RanBPM, a scaffolding protein in the immune and nervous systems," *Journal of Neuroimmune Pharmacology*, vol. 2, no. 3, pp. 290–295, 2007.

[17] E. Arama, D. Dickman, Z. Kimchie, A. Shearn, and Z. Lev, "Mutations in the $\beta$-propeller domain of the *Drosophila* brain tumor (brat) protein induce neoplasm in the larval brain," *Oncogene*, vol. 19, no. 33, pp. 3706–3716, 2000.

[18] H. Stenmark and V. M. Olkkonen, "The Rab GTPase family," *Genome Biology*, vol. 2, no. 5, article 3007, 2001.

[19] Y. Mamane, E. Petroulakis, O. LeBacquer, and N. Sonenberg, "mTOR, translation initiation and cancer," *Oncogene*, vol. 25, no. 48, pp. 6416–6422, 2006.

# 8

# Assessment of Tools for Marker-Assisted Selection in a Marine Commercial Species: Significant Association between MSTN-1 Gene Polymorphism and Growth Traits

**Irma Sánchez-Ramos,[1] Ismael Cross,[1] Jaroslav Mácha,[2] Gonzalo Martínez-Rodríguez,[3] Vladimir Krylov,[2] and Laureana Rebordinos[1]**

[1] *Laboratorio de Genética, Universidad de Cádiz, Poligono Río San Pedro s/n, 11510 Puerto Real, Spain*
[2] *Department of Cell Biology, Faculty of Science, Charles University in Prague, Prague 2, Viničná 7, 12843 Prague, Czech Republic*
[3] *Instituto de Ciencias Marinas de Andalucía, Consejo Superior de Investigaciones Científicas, República Saharaui, no. 2, 11510 Puerto Real, Spain*

Correspondence should be addressed to Laureana Rebordinos, laureana.rebordinos@uca.es

Academic Editor: Sardana Fedorova

Growth is a priority trait from the point of view of genetic improvement. Molecular markers linked to quantitative trait loci (QTL) have been regarded as useful for marker-assisted selection in complex traits as growth. Polymorphisms have been studied in five candidate genes influencing growth in gilthead seabream (*Sparus aurata*): the growth hormone (*GH*), insulin-like growth factor-1 (*IGF-1*), myostatin (*MSTN-1*), prolactin (*PRL*), and somatolactin (*SL*) genes. Specimens evaluated were from a commercial broodstock comprising 131 breeders (from which 36 males and 44 females contributed to the progeny). In all samples eleven gene fragments, covering more than 13,000 bp, generated by PCR-RFLP, were analyzed; tests were made for significant associations between these markers and growth traits. ANOVA results showed a significant association between *MSTN-1* gene polymorphism and growth traits. Pairwise tests revealed several RFLPs in the *MSTN-1* gene with significant heterogeneity of genotypes among size groups. *PRL* and *MSTN-1* genes presented linkage disequilibrium. The *MSTN-1* gene was mapped in the centromeric region of a medium-size acrocentric chromosome pair.

## 1. Introduction

The value of aquaculture production was estimated at USD 98.4 billion in 2008 and has continued to show strong growth, increasing at an average annual growth rate of 6.2 percent [1]. Among fishes, the gilthead seabream *Sparus aurata*, from the Sparidae family, is one of the most important fish species cultivated in the Mediterranean region, where the main producer countries are Greece, Turkey, and Spain [2]. The strong competition between producer companies and the steep increase in production have caused the selling prices of the product to fall; this in turn has seriously reduced the profit margins of the companies, which in many cases have been squeezed to unsustainable levels. For this reason improvements in the systems of cultivation have become essential, and programmes of improvement by selection must be put into action to reduce production costs. Nevertheless, from cultivated species, only a few of them have ongoing selective breeding programmes. Traditionally, these have been carried out successfully using pedigree information by selecting individuals based on breeding values using an *animal model*. However, information on individual genes with medium or large effects cannot be used in this manner [3]. Molecular markers that directly affect or are linked to quantitative trait loci (QTL) have been regarded as useful for marker-assisted selection (MAS) or gene-assisted selection (GAS) programmes. The allelic variations at these individual genes have a major influence on overall phenotypic expression, and the number of genotypes managed is smaller, so it may be faster and more efficient to implement these improvement programs rather than others under industrial production conditions. Besides, traits as growth,

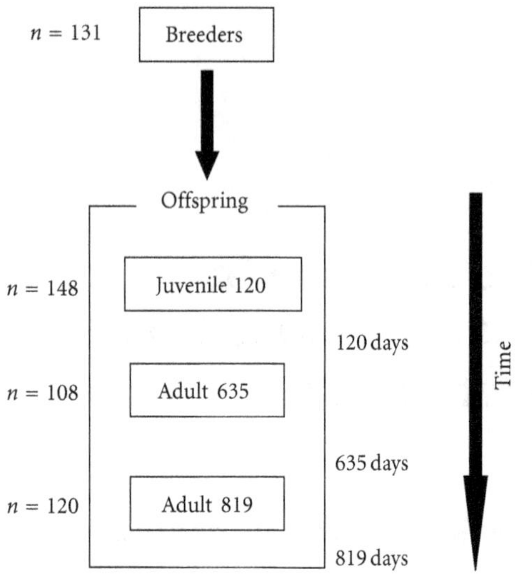


FIGURE 1: Schematic diagram of sampling during the different life stages.

with low heritability and relatively few records per traits measured, are those most benefiting from incorporating marker information [4].

Tests of association between candidate gene polymorphisms and quantitative traits have been widely applied in recent years using genetic markers. In that context, PCR-RFLPs (polymerase chain reaction-restriction fragment length polymorphisms) have been demonstrated to be very useful genetic markers for candidate gene studies; such studies have revealed polymorphisms associated with quantitative traits in Atlantic salmon *Salmo salar* [5], oysters *Crassostrea gigas* [6], cattle [7, 8], chickens [9], and sheep [10].

Growth is a priority trait from the point of view of genetic improvement since economic advantages can be gained from shortening the time required for production and improving the rate of feed conversion. Increasing homogeneity in the rate of growth of individuals is another characteristic of interest, since it would reduce the number of gradings to be carried out until harvest size is reached. In this context, the search for candidate genes influencing growth has been a major focus of research, and several candidate genes influencing growth in finfish have been isolated from the genome, and their effects quantified [11]. The growth is a complex trait from the genetic point of view because its genetic con-trol is not well known and there are dozens of candidate genes acting upon it.

The five candidate genes selected for this study were growth hormone (*GH*), insulin-like growth factor-1 (*IGF-1*), myostatin (*MSTN-1*), prolactin (*PRL*), and somatolactin (*SL*) genes. All of them are known to be related to the somatotropic axis or transforming growth factors. Growth hormone (*GH*) plays a major role in stimulating somatic growth primarily through the induction of insulin-like growth factor I (*IGF*) [12]. Myostatin (*MSTN*) or growth differentiation factor 8 gene (*GDF-8*) seems to be a negative regulator of skeletal muscle growth in mammals [13]. There are two genes in gilthead seabream that code for MSTN: *MSTN-1* [14] and *MSTN-2* [15]; *MSTN-1* is expressed mainly in skeletal muscle, at both adult and juvenile stages, and *MSTN-2* is expressed almost exclusively in the central nervous system, at late larval stages [14, 15]. The insulin-like growth factor I gene (*IGF-I*) appears to be linked to nutritional condition, environmental adaptation, embryonic development, and growth regulation of teleosts [16]. Other candidate genes acting upon growth are the prolactin (*PRL*) gene, which performs several functions including mitogenic, somatotropic, and osmoregulatory activities [17] and the somatolactin (*SL*) gene, involved in a variety of physiological functions in teleosts [18].

Genetic mapping of genes allows a complete identification and the location of a gene and hence can be used in programs of genetic improvement in aquaculture. Physical maps enable the integration of linkage maps and karyotypes and are essential tools for comprehensive comparative genomic studies. In addition, the existence of a well-characterized physical map makes it more feasible to undertake a whole genome sequencing project [19]. Here, we report the localization of *MSTN-1* gene, which presented in this work statistically significant association with growth traits, on *S. aurata* chromosomes.

With the object of gaining new understanding of genes related to growth traits in a marine commercial species, the objectives of the present study were to find and describe polymorphisms in candidate genes from *S. aurata* populations produced under commercial conditions and to test for associations between these polymorphisms and growth traits. Furthermore, the *MSTN-1* gene, whose polymorphism has shown, in this study, a statistically significant correlation with growth traits, was mapped to the chromosomes of this species by means of the tyramide signal amplification (TSA) fluorescence system.

## 2. Materials and Methods

*2.1. Animals and Phenotypic Traits.* The samples were obtained from CUPIMAR S.A., an aquaculture company that operates in the area of the Bay of Cadiz, on the Atlantic coast of southwest Spain. One broodstock and their offspring were analyzed at different stages of production (Figure 1). The broodstock in these installations have mixed origins; wild specimens are added to those from the company's own aquaculture production, because the transformation of males into females after the third year results in a relative lack of males. The broodstock was utilised during several spawning seasons to generate the $F_1$ that are cultivated and grown for subsequent transfer to the on-growing facilities until reaching harvesting size. The breeders were maintained under environmental conditions of temperature, salinity, and oxygen, according to the management practices of the farm. A total of 131 individual breeders were sampled: 48 males and 83 females ($N_B = 131$). Within each gender, the breeders were all of similar age.

Fish were anaesthetized with 0.025% 2-phenoxyethanol solution; fork length ($L$) and body weight ($W$) were recorded, and 1–1.5 mL of blood was extracted for DNA isolation. Maturation was induced by control of the photoperiod. Spawning took place in several tanks under production conditions, and offspring were then grouped together in one tank. The offspring originated from the spawning that took place in one single day.

Total residence time, calculated from the hatching date to the first sampling date, ranged between 112 days and 126 days (J120). A total of 148 juvenile fishes ($N_{J120} = 148$) were measured (fork length and body weight) during this period of fourteen days; specimens were graded, and either fin clips were excised or blood samples were taken, in both cases being preserved in 70% ethanol at 4°C till processing for DNA isolation. Juveniles were kept in nursery tanks and transferred to floating cages, where they were cultivated until reaching adult stage. At this stage, the individuals from two cages were sampled at different dates, representing either 635 days ($N_{A635} = 108$) or 819 days ($N_{A819} = 120$) after the date of spawning, respectively, and the 228 individuals in total were measured for fork length and body weight (Figure 1). Finally, quantities of 1.0 to 1.5 mL of blood were extracted for subsequent DNA analysis. A condition index was used to assess relative fatness and provide an estimate of the energy reserves, morphology, and health of the fish and is the parameter that best reflects the criterion utilised by the producer for its commercial classification of the product. Body weight × fork length$^{-3}$ × 100 is the metric of the condition index ($C$).

*2.2. DNA Extraction, Genotyping, and Parental Assignment.* Genomic DNA from breeders, adults, and some juveniles was extracted from blood according to the protocol described by Martínez et al. [21]. For the rest of the juveniles, genomic DNA was extracted from fin clips using a modification of the salting-out extraction method. In all cases, the DNA was kept at 4°C until utilisation. All 376 offspring and 131 breeders were genetically characterised using microsatellite markers (multiplex) and familial assignments, determined according to Porta et al. [22].

*2.3. Polymerase Chain Reaction (PCR) and Sequencing.* The 11 regions from the five candidate genes that were screened by restriction fragment length polymorphisms (RFLP) are listed in Table 1. Except for growth hormone intron I amplification, which was carried out using oligonucleotides as described in Almuly et al. [20], all other amplification products were performed using forward and reverse primers based on *S. aurata* sequences from GenBank and designed with *Primer3* software [23]. Primers were designed to anneal in the exons and/or promoters of the candidate genes so that we could amplify 0.7–1.9 kb DNA regions including at least one intron or part of the promoter (Table 2 and Figure 2).

The PCR amplification reactions were performed in a Gene Amp PCR System 2700 (Applied Biosystem) thermal cycler that was programmed for denaturation at 94°C for 5 min.; 35 cycles of amplification with 30 s at 94°C, 30 s at 59–63°C (depending on DNA fragment), 1 min. at 72°C; and a final extension at 72°C for 10 min. Reactions were carried out in a final volume of 50 $\mu$L with 200 mM dNTPs, 0.2 $\mu$M of the forward and reverse primers, 2.0-3.0 mM $MgCl_2$, 60–120 ng gilthead seabream genomic DNA, 2–5 U of EuroTaq polymerase (EuroClone Genomics), and 5.0 $\mu$L 10x reaction buffer (EuroClone Genomics). The PCR products were examined by electrophoresis through a 1.5% agarose gel containing 0.5 $\mu$g mL$^{-1}$ ethidium bromide. Amplified products of the predicted size were cut and purified from gels with a nucleoSpin extract II kit (Macherey-Nagel) before carrying out the sequencing reactions. DNA sequencing was performed with fluorescence-labelled terminator (BigDye Terminator 3.1 Cycle Sequencing Kit; Applied Biosystems) in an ABI3100 Genetic Analyzer. Sequence data from this paper have been deposited with the GenBank Data Library under Accession numbers FJ827497 to FJ827506.

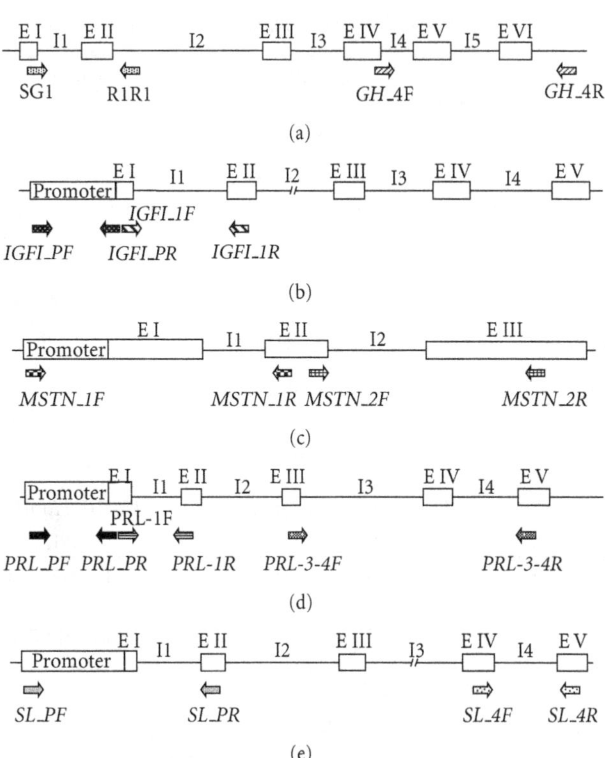


FIGURE 2: Structure of the candidate genes and oligonucleotide primers used for amplification by PCR in this study. Growth hormone (a), Insulin-Like Growth Factor I (b), Myostatin-1 (c), Prolactin (d), and Somatolactin (e) genes. I: Introns; E: Exons; Arrows indicate relative position of the primers.

*2.4. RFLP Detection and Genotyping (PCR-RFLP Alleles).* The selection of the restriction enzymes needed to observe the RFLPs was performed using software available on the Internet, such as *In silico* [24] and *Webcutter 2.0* (http://rna.lundberg.gu.se/cutter2/). Preferential selection was made for those enzymes that produced cuts within the target sequence and with the number of targets not higher than three, to avoid products of small size that might not be detected later in the electrophoresis. Different enzymes were tested for the different PCR products, with the aim of obtaining at least two RFLPs for each sequence amplified.

The digestions were performed by incubation of the samples in a water bath at the optimum temperature for each enzyme according to the manufacturer. The reactions were carried out in a final volume of 50 $\mu$L containing 0.05–1.0 $\mu$L

Table 1: Description of RFLPs in five candidate genes related to growth traits in gilthead seabream analyzed in this work.

| Gene | PCR fragment | Enzyme used to detect polymorphism | Polymorphic region | RFLP alleles |
|---|---|---|---|---|
| *GH* | A: Exon 1 to intron2 | *Hae*III & *Hap*II | Intron I | Size (micros.) |
| *GH* | B: Exon 4 to 3′UTR | *Hae*III, *Mbo*II, *Pst*I & *Taq*I | N.P. | — |
| *IGF-I* | A: Promoter to exon 1 | *Taq*I | Promoter | T → G/C |
| *IGF-I* | B: Exon 1 to exon 2 | *Hae*II, *EcoR*I *Dra*I, *Hap*II & *BspLU11*I | N.P. | — |
| *MSTN-1* | A: Promoter to exon 2 | *Hae*III | Exon 1 | C → T |
| | | *Dra*I | Intron 1 | A → C |
| *MSTN-1* | B: Exon 2 to exon 3 | *Hae*III | Intron 2 | G → T |
| *PRL* | A: Promoter to exon 1 | *SspB*I & *Dra*I | Promoter | Size (micros.) |
| *PRL* | B: Exon 1 to exon 2 | *EcoR*I | Intron 1 | A → G |
| | | *BspLU11*I | Intron 1 | C → T |
| *PRL* | C: Exon 3 to exon 5 | *Hae*III | Intron 3/Exon 4 | Size (indel) |
| | | *Fok*I | Intron 3 & 4/Exon 3 | Size (indel) |
| *SL* | A: Promoter to exon 2 | *Hae*III, *Hap*II, *Afa*I, *Dra*I & *Taq*I | N.P. | — |
| *SL* | B: Exon 4 to exon 5 | *Tru9*I, *Taq*I, *Ssp*I, *Hae*III | N.P. | — |

N.P.: nonpolymorphic; *GH*: growth hormone; *IGF-1*: insulin-like growth factor I; *MSTN*: myostatin; *PRL*: prolactin; *SL*: somatolactin.

Table 2: PCR oligonucleotide primers used for RFLP analysis in the five candidate genes studied in gilthead seabream.

| Amplified fragment | | Primers | | | |
|---|---|---|---|---|---|
| Fragment | Length (bp) | Name | Position | Sequence (5′-3′) | Tm (°C) |
| *GH_A* | 1045–1900 | SG1 [20] | Exon 1 | AGAACCTGAACCAGACATGG | 52.8 |
| | | *R1R1* [20] | Intron 2 | CTGCTGCCAGAGAATTACTG | 54.1 |
| *GH_B* | 1172 | *GH_4F* | Exon 4 | GTTCTCTGTCTGGCGGTTCT | 56.3 |
| | | *GH_4R* | 3′UTR | AGCAACTGGGTCTAATGAATGT | 54.4 |
| *IGF-I_A* | 944/954 | *IGFI_PF* | Promoter | TCATCAGATGTCATTTGCAGAC | 52.9 |
| | | *IGFI_PR* | Exon 1 | TGCCACTGAAAGGAAAGAGC | 56.8 |
| *IGF-I_B* | 1506 | *IGFI-1F* | Exon 1 | AGCGCTCTTTCCTTTCAGTG | 56.8 |
| | | *IGFI-1R* | Exon 2 | GCCTCTCTCTCCACACACAA | 55.1 |
| *MSTN1_A* | 934 | *MSTN_1F* | Promoter | GCCTGTCAGTGTGGGACTTT | 55.0 |
| | | *MSTN_1R* | Exon 2 | ACGATTCGATTGGCTTGAAT | 57.4 |
| *MSTN1_B* | 881/886 | *MSTN_2F* | Exon 2 | AACAAGTGTTGAGCGTGTGG | 55.5 |
| | | *MSTN_2R* | Exon 3 | CTCCGAGATCTTCACCTCCA | 57.2 |
| *PRL_A* | 1014–1063 | *PRL_PF* | Promoter | GACTTTAACATGACCTGGAGGA | 54.4 |
| | | *PRL_PR* | Exon 1 | TGGTTTCTCTGTGAGCCATCT | 57.7 |
| *PRL_B* | 864/765 | *PRL_1F* | Exon 1 | ATGGCTCACAGAGAAACCAA | 54.0 |
| | | *PRL_1R* | Exon 2 | GAGTTGTGCTGAGGGAGTGC | 55.7 |
| *PRL_C* | 1549/1496 | *PRL_3-4F* | Exon 3 | GTAGGCTGGACGATGATGC | 54.8 |
| | | *PRL_3-4R* | Exon 5 | AGCAGGACAACAGGAAATGG | 56.7 |
| *SL_A* | 1650 | *SL_PF* | Promoter | GAACAGTGGTAATGACAGTCTCAA | 53.9 |
| | | *SL_PR* | Exon 2 | GTGAGCAGATAGGGCCAGAG | 56.2 |
| *SL_B* | 704–721 | *SL_4F* | Exon 4 | CATTCTGTGCTGATGCTGGT | 55.9 |
| | | *SL_4R* | Exon 5 | GGGCATCTTTCTTGAAGCAG | 56.1 |

N.P.: nonpolymorphic; *GH*: Growth Hormone; *IGF-1*: Insulin-Like Growth Factor I; *MSTN*: Myostatin; *PRL*: Prolactin; *SL*: Somatolactin. A, B, and C fragments defined by the primer positions.

of enzyme (Roche Molecular Biochemicals, Amersham Biosciences or Fermentas), 1x of the appropriate buffer, and 2–5 $\mu$L of the PCR product.

To determine the size of the products, the samples were loaded into agarose gels (1–1.5%) in 0.5x TBE with ethidium bromide (0.5 $\mu$g mL$^{-1}$). Gels were visualised and analyzed using *Gel Doc XR* Molecular Imager (BioRad) and Quantity One 1-D Analysis programme, respectively.

*2.5. Statistical Analysis.* One-way analysis of variance (ANOVA) was used to test significant associations between the marker genotype (at each of the eight RFLP loci) and

Table 3: Means (standard deviation) for body weight, fork length, and condition index in samples analyzed.

| Group | $N$ | Weight (g) | Length (cm) | Condition index ($g \cdot cm^{-3}$) |
|---|---|---|---|---|
| B-M | 48 | 2,048.42 (±533.75) | 50.97 (±3.85) | 1.50 (±0.13) |
| B-F | 83 | 2,656.22 (±491.51) | 55.86 (±3.00) | 1.52 (±0.15) |
| B-T | 131 | 2,433.51 (±584.64) | 54.07 (±4.08) | 1.51 (±0.14) |
| J120 | 148 | 6.14 (±3.62) | 7.22 (±1.51) | 1.43 (±0.26) |
| A635 | 108 | 394.44 g (±139.56) | 27.73 (±3.13) | 1.77 (±0.17) |
| A819 | 120 | 549.58 g (±183.14) | 29.14 (±2.87) | 2.13 (±0.20) |

B-M, B-F, and B-T: males, females, and total individuals among breeders, respectively; J120: juveniles sampled 120 days after hatching; A635 and A819: adults sampled 635 and 819 days after hatching.

the three phenotypic traits ($L$, $W$, and $C$). The procedures utilised in the comparison of means are not very sensitive to the lack of normality. When there are no atypical observations and the sample distributions are approximately symmetric, ANOVA can safely be used even for very few samples ($n$ = 4-5) [25]. On the other hand, the results of the $F$ test of the ANOVA can be considered reliable, when the largest standard deviation is less than twice as large as the smallest standard deviation [25]. The ANOVA test was applied only in those cases in which the standard deviations complied with this ratio. In those cases in which there were significant values and the sample standard deviations did not comply with this ratio, the nonparametric test of Kruskal-Wallis was applied.

As the gilthead seabream is a protandrous hermaphroditic teleost, in which sex is reversed in males after about 2 years of age, the broodstock presented a marked difference between the means of the males and females. For this reason, we analyzed length, body weight, and condition index variables for each locus by fitting a general linear model (GLM) to the data from breeders that included gender as a factor:

$$Y_{ij} = \mu + \alpha_i + \beta_j + \varepsilon_{ij}, \quad (1)$$

where $Y_{ij}$ is the phenotypic trait (e.g., weight in breeder group), $\mu$ is the mean value of the trait, $\alpha_i$ is the genotype at the RFLP marker, $\beta_j$ is gender, and $\varepsilon_{ij}$ is the random residual.

The proportion of the phenotypic variation explained by each RFLP marker was calculated as $r^2$ = SS effect/SS total.

All these tests were applied using the SPSS 14.0 statistical software programme. In all cases when multiple comparisons were made, the Bonferroni adjustment was applied [26].

To determine the existence of linkage disequilibrium between pairs of loci analyzed, the null hypothesis "*genotypes at one locus are independent from genotypes at the other locus*" was tested using the GENEPOP programme package [27]. Contingency tables were created for all pairs of loci in each age group and across all the population, and Fisher's exact test was performed for each table.

The distribution of the genotypes in each age group, subclassified by sizes, was studied. The null hypothesis tested was "*the genotypic distribution is identical across the size-group*". For each locus, the test was performed on contingency tables. An unbiased estimate of the $P$ value of a log-likelihood (G)-based exact test was performed [28] using the GENEPOP program package [27]. The test was performed for all pairs of samples, for all loci. According to fork length and body weight distribution, adults and breeders were classified in two groups: small and large (adults) and males and females (breeders). Juveniles were classified in two size groups, small (J-S) and large (J-L) which were subclassified in small (S), medium (M), and large (L). When multiple comparisons were made, the Bonferroni correction was applied to adjust the $\alpha$ value for each age group.

*2.6. Cytogenetic Techniques.* Chromosome preparations were made from 1-2-day-old *S. aurata* larvae according to Cross et al. [29]. To make the myostatin probe, first-strand cDNA was obtained from *S. aurata* larvae RNA using a SMART RACE cDNA Amplification Kit (Clontech Laboratories). The cDNA for the *MSTN-1* probe was obtained by PCR from the first-strand cDNA using a forward primer in exon 1: 5′-AGAAGACACGGAGCTGTGC-3′ and a reverse primer in exon 3: 5′-AAGAGCATCCACAACGGTCT-3′. The amplification yielded a fragment of 1025 bp, and its structure was verified by sequencing. Digoxigenin (dig-11-dUTP) labelling of 1 $\mu$g DNA probe was performed by random priming using the Decalabel DNA Labeling Kit (Fermentas) according to the manufacturer's instructions. Following the labelling, the sample was purified on a column of QIAquick Gel Extraction Kit (Qiagen) in a final volume of 50 $\mu$L. FISH-TSA were carried out essentially according to Krylov et al. [30] with minor modifications. Photographs were taken using a fluorescence microscope (Olympus BX-40) and a SONY EXwave HAD black and white camera. Images were processed and coloured using the ACC Program, v.5.0 (SOFO, Brno, Czech Republic).

## 3. Results

*3.1. Biometric Data of Breeders and Progeny.* The analysis of parentage revealed that a total of 80 fishes, 47 males and 33 females, from 131 breeders, contributed to the progeny (exclusion power > 0.999 and an estimated probability of identity of $2.13 \cdot 10^{-31}$). A total of 108 families of full and half-siblings, were represented in the sample of 376 offspring samples, with an average of 3.2 (±3.15) descendents per full siblings family.

Table 3 summarizes biometric data (weight, length, and condition index) from breeder and offspring in the different sampled groups. Because of the biology of *S. aurata*, a clear size difference can be observed between male and female breeders; all individuals are males at the end of their first year of life, and at approximately two years of age, most individuals become females. In our study, all the offspring sampled were male. In the breeder group, the 83 females sampled presented mean weight and length values higher than those of the 48 males. The coefficients of variation (CV) can be deduced from Table 3, and for weight, CV is approximately 3 times the values obtained for *L* and for *C*.

*3.2. Polymorphisms of the Five Candidate Genes.* We studied five candidate genes for growth rate in *S. aurata* and tested for association with three quantitative traits (body weight, fork length, and condition factor) at different stages of development: breeders, juveniles, and adults. To find usable RFLP markers in the proposed candidate genes, 11 gene fragments, covering more than 13000 bp, were amplified and sequenced. Several enzymes were chosen to test for usable polymorphism. Four of the 11 fragments studied (*GH_B*, *IGF-I_B*, *SL_A*, and *SL_B*) were nonpolymorphic, but seven exhibited polymorphism of size and/or restriction sites, as shown in Table 1 (RFLP patterns obtained in *PRL 3_4* for *Hae*III and *Fok*I were identical, and therefore only one of them (*Fok*I) was considered in the ANOVA analysis).

*3.3. One-Tail ANOVA Analysis of the Association between Genotypes of the Different Age-Groups, of the Five Candidate Genes and Phenotypic Traits.* When the ANOVA test was applied, several putative associations between phenotypic traits (*W*, *L*, and *C*) and genetic marker genotypes were found (Table 4). However, as we tested seven independent hypotheses (seven RFLPs) in each age group the corrected value for $\alpha$ was considerably smaller ($\alpha_{\text{Bonferroni}} = 0.05/7 = 0.00714$). Under this $\alpha$-correction for the ANOVA test, six putative associations between markers and traits remained significant: five of the six were associated with the *MSTN1* RFLP (in some of the breeders, adults and juvenile groups) (Table 4). After Bonferroni corrections in the Kruskal-Wallis nonparametric tests (applied to *W* and *C* in the juveniles group, and *C* in the A635 adults), two putative *MSTN1*-RFLP associations (both in breeders for *W* and *L*) remained significant (*MSTN1_A-Hae*III). This marker accounted for 5.8% and 6.2% of the phenotypic variance in *W-B* and *L-B*, respectively.

Size differences between males and females are a characteristic of this species due to its protrandrous hermaphroditism, by which individuals mature first as males and later change to females. Because of this difference, the study of associations between genotypes and quantitative characteristics in breeders may be affected by sex. When a GLM was used to test for association between genotypic marker and the three quantitative traits (*W*, *L* and *C*) in breeders with gender factor (data not shown), the association between *MSTN1_A-Hae*III and weight/length traits did not remain significant ($P_W = 0.084$ and $P_L = 0.061$).

*3.4. Pairwise Analysis of the Association between Genotypes, Five Candidate Genes, and Size Classes in Each Age Group.* To determine the distribution of genotypes between pairs of size classes for each age group, an analysis using contingency tables was made. Pairwise tests revealed several RFLPs with significant heterogeneity of genotypes among size-group pairs in juveniles, again most of them at the *MSTN-1* gene (*P* ranged from 0.002 to 0.017), and one significant *P* value ($P = 0.019$) in the A819 adult group at the *IGF-I* gene (Table 5). As numerous comparisons were performed for each age group, Bonferroni's correction was applied. Under these corrected values of $\alpha$, the comparison of the small-large groups (S-L), in the RFLP *MSTN1_B-Hae*III in the J-S group, remained significant. With this new distribution of individuals, in function of their size, an ANOVA analysis was performed for the small juveniles (S-J120); it was found that, for the weight, there was a significant association ($P = 0.003$) with the RFLP *MSTN1_B-Hae*III, confirming the putative relationship of *MSTN-1* with growth traits obtained in the various statistical treatments reflected in Tables 4 and 5.

To assess whether heterozygosity at the *MSTN-1* gene influences any of the traits studied, the genotypes were grouped by homozygous/heterozygous type, and this new factor was used in an analysis of variance (data not shown). It was observed that again, for weight of both breeders (*MSTN1_A-Hae*III) and juveniles (*MSTN1_B-Hae*III), the homozygous genotypes were larger ($P < 0.05$) than the heterozygous.

When we tested for the existence of linkage disequilibrium between pairs of loci, our results revealed that, curiously, *MSTN1_A-Dra*I were in linkage disequilibrium with both *PRL-1-BspLU11*I ($P \leq 0.05$) and *PRL-1-Eco*RV ($P \leq 0.05$) (Fisher's exact test), which implies a statistically significant linkage disequilibrium between these two genes, *MSTN1* and *PRL*.

*3.5. Chromosome Location of the MSTN-1 Gene.* FISH-TSA experiments on *S. aurata* chromosomes have led us to localize the *MSTN-1* gene of *S. aurata* in the centromeric region of two medium-sized acrocentric chromosomes (Figure 3(a)). Signal position was evaluated in 194 metaphase spreads, of which 65.9% displayed a specific signal. The remaining metaphase spreads did not show any FISH signal, apart from a few metaphase spreads (8.7%) that presented four possible signals; that second signal also appears in a centromeric region (Figure 3(b)).

## 4. Discussion

The infinitesimal model, which assumes that for any particular quantitative trait the individual effects of genes are unknown and usually are only small in their magnitude, does not fully explain the observed patterns of genetic variance. On the contrary, for many traits, allelic variation at individual genes has been shown to have a major influence on overall phenotypic expression [11]. Consequently, there has been substantial investment in livestock sciences directed towards the detection of major/candidate genes differentially influencing trait expression. Taking these factors into account, this study was conducted to obtain more practical results that are closely related to conditions in fish farming and to the biology of the species. The analysis of polymorphism

Table 4: Analysis of variance (ANOVA) for weight, length, and condition index within genotypes for different age groups in *S. aurata*. Breeders are considered without respect to gender.

| Locus | | | Phenotypic variable | | | | | | | | | | | |
|---|---|---|---|---|---|---|---|---|---|---|---|---|---|---|
| | | | *W*-B | *L*-B | *C*-B | *W*-J120 | *L*-J120 | *C*-J120 | *W*-A635 | *L*-A635 | *C*-A635 | *W*-A819 | *L*-A819 | *C*-A819 |
| | | *F* | 0.325 | 0.492 | 0.758 | 0.046 | 0.243 | 4.227 | 3.328 | 3.101 | 1.227 | 1.457 | 1.408 | 1.137 |
| *IGF-I-P* | *Taq*I | *P* | 0.807 | 0.689 | 0.520 | 0.996 | 0.913 | 0.003[b] | 0.040[a] | 0.049[a] | 0.297 | 0.220 | 0.236 | 0.343 |
| | | *N* | 131 | 131 | 131 | 148 | 148 | 148 | 108 | 108 | 108 | 120 | 120 | 120 |
| | | *F* | 8.010 | 8.578 | 0.614 | 0.177 | 0.105 | 0.117 | 0.665 | 0.807 | 0.297 | 3.521 | 2.832 | 2.546 |
| | *Hae*III | *P* | 0.005[b] | 0.004[b] | 0.435 | 0.675 | 0.675 | 0.732 | 0.417 | 0.371 | 0.587 | 0.063 | 0.095 | 0.113 |
| | | *N* | 131 | 131 | 131 | 148 | 148 | 148 | 108 | 108 | 108 | 120 | 120 | 120 |
| *MSTN1_A* | | *F* | 0.303 | 0.712 | 0.290 | 0.012 | 0.003 | 1.430 | 0.422 | 1.859 | 11.918 | — | — | — |
| | *Dra*I | *P* | 0.583 | 0.400 | 0.591 | 0.912 | 0.960 | 0.234 | 0.518 | 0.176 | 0.001[b] | — | — | — |
| | | *N* | 131 | 131 | 131 | 148 | 148 | 148 | 108 | 108 | 108 | 120 | 120 | 120 |
| | | *F* | 4.236 | 3.723 | 0.440 | 5.103 | 3.924 | 0.047 | 4.747 | 3.292 | 4.510 | 0.518 | 0.144 | 2.929 |
| *MSTN1_B* | *Hae*III | *P* | 0.017[a] | 0.027[a] | 0.645 | 0.007[b] | 0.022[a] | 0.954 | 0.011[a] | 0.041[a] | 0.013[a] | 0.597 | 0.866 | 0.057 |
| | | *N* | 131 | 131 | 131 | 148 | 148 | 148 | 108 | 108 | 108 | 118 | 118 | 118 |
| | | *F* | 1.534 | 1.779 | 0.416 | 0.182 | 0.136 | 1.255 | 1.587 | 1.380 | 3.429 | 2.111 | 2.043 | 0.573 |
| | *Eco*RV | *P* | 0.220 | 0.173 | 0.660 | 0.834 | 0.873 | 0.288 | 0.197 | 0.253 | 0.020[a] | 0.126 | 0.134 | 0.565 |
| | | *N* | 131 | 131 | 131 | 148 | 148 | 148 | 108 | 108 | 108 | 120 | 120 | 120 |
| *PRL_1* | | *F* | 1.123 | 1.199 | 0.415 | 0.453 | 0.463 | 0.716 | 2.086 | 1.882 | 1.513 | 0.944 | 1.019 | 0.365 |
| | *BspLU*11I | *P* | 0.352 | 0.313 | 0.838 | 0.810 | 0.803 | 0.613 | 0.061 | 0.102 | 0.181 | 0.467 | 0.417 | 0.900 |
| | | *N* | 131 | 131 | 131 | 148 | 148 | 148 | 108 | 108 | 108 | 120 | 120 | 120 |
| | | *F* | 0.030 | 0.026 | 0.103 | 1.022 | 1.395 | 0.400 | 0.437 | 0.122 | 1.230 | 0.416 | 0.336 | 0.653 |
| *PRL_3_4* | *Fok*I | *P* | 0.97 | 0.974 | 0.902 | 0.362 | 0.251 | 0.671 | 0.647 | 0.885 | 0.297 | 0.661 | 0.715 | 0.522 |
| | | *N* | 131 | 131 | 131 | 147 | 147 | 147 | 107 | 107 | 108 | 119 | 119 | 119 |

*W*-, *L*-, and *C*-: body weight, fork length, and condition index, respectively; B: breeders; J120: juveniles sampled 120 days after hatching; A635: adults sampled 635 days after hatching; A819 Adults sampled 819 days after hatching; *F*: *F* value; *P*: *P* value; *N*: number of individuals analyzed.

[a]Significant at $\alpha = 0.05$.

[b]Significant at $\alpha_{\text{Bonferroni}} = 0.00714$.

Table 5: Probability values (*P*) for exact tests for homogeneity of RFLP genotype frequencies in candidate genes between analyzed groups of *S. aurata*.

| Locus | | | B | | J-S | | | J-L | | A635 | A819 |
|---|---|---|---|---|---|---|---|---|---|---|---|
| | | | F-M | S-M | M-L | S-L | S-M | M-L | S-L | S-L | S-L |
| *IGF-I-P* | *Taq*I | *P* | 0.623 | 1.000 | 0.772 | 1.000 | 0.547 | 0.576 | 0.435 | 0.334 | 0.019[a] |
| | | *N* | 131 | 50 | 51 | 51 | 48 | 47 | 49 | 108 | 120 |
| *MSTN1_A* | *Hae*III | *P* | 0.157 | 0.349 | 0.191 | 1.000 | 1.000 | 1.000 | 1.000 | 0.617 | 0.119 |
| | | *N* | 131 | 50 | 51 | 51 | 48 | 47 | 49 | 108 | 120 |
| | *Dra*I | *P* | 0.355 | 0.609 | 0.349 | 1.000 | 1.000 | 1.000 | 1.000 | 1.000 | — |
| | | *N* | 131 | 50 | 51 | 51 | 48 | 47 | 49 | 108 | 120 |
| *MSTN1_B* | *Hae*III | *P* | 0.309 | 0.616 | 0.005[a] | 0.002[b] | 0.817 | 0.442 | 0.324 | 0.017[a] | 1.000 |
| | | *N* | 131 | 50 | 51 | 51 | 48 | 47 | 49 | 108 | 120 |
| *PRL_1* | *Eco*RV | *P* | 0.144 | 0.566 | 0.248 | 0.727 | 0.725 | 1.000 | 1.000 | 0.327 | 0.092 |
| | | *N* | 131 | 50 | 51 | 51 | 48 | 47 | 49 | 108 | 120 |
| | *BspLU*11I | *P* | 0.281 | 0.443 | 0.404 | 0.762 | 1.000 | 0.732 | 0.719 | 0.054 | 0.196 |
| | | *N* | 131 | 50 | 51 | 51 | 48 | 47 | 49 | 108 | 120 |
| *PRL_3_4* | *Fok*I | *P* | 0.889 | 1.000 | 0.427 | 0.313 | 1.000 | 0.818 | 1.000 | 0.788 | 0.407 |
| | | *N* | 131 | 50 | 51 | 51 | 48 | 47 | 49 | 108 | 119 |

B: breeders; J: juveniles; A635: adults sampled 635 days after hatching; A819: adults sampled 819 days after hatching; S: small; M: medium; L: large; *N*: number of samples analyzed; *P*: *P* value.
[a]Significant at $\alpha = 0.05$.
[b]Significant at $\alpha_{\text{Bonferroni}} = 0.0024$.

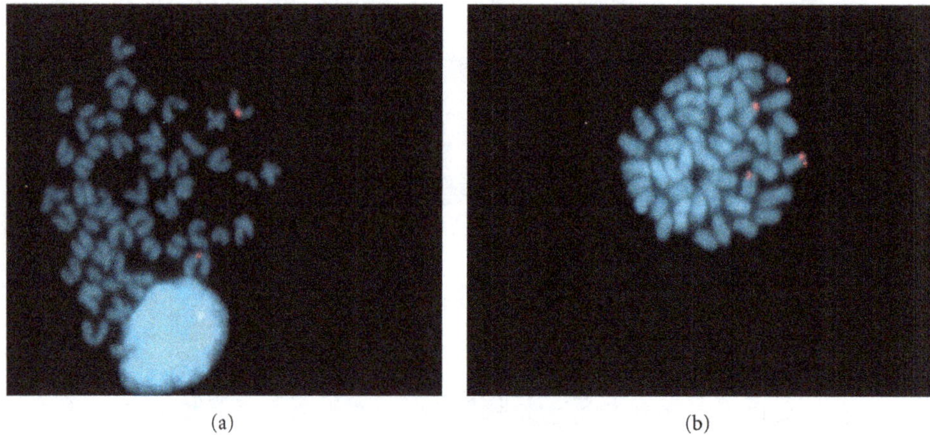

(a) (b)

Figure 3: Localization of the myostatin-1 probe on *S. aurata* chromosomes. (a) *MSTN-1* probe in the centromeric region of acrocentric chromosomes. (b) Localization of *MSTN-1* and putative *MSTN-2* genes both in centromeric regions of two different acrocentric chromosomes.

in eleven DNA fragments, belonging to five candidate genes, allowed us to use seven PCR-RFLPs to test for association with three quantitative traits (*W*, *L*, and *C*). The majority of the associations found, after several corrections and nonparametric tests (when necessary), were associated with the *MSTN-1* gene. In breeders, when ANOVAs were performed considering gender as a factor (data not shown), no significant values were obtained. The considerable weight lost at spawning in females may explain the effect of transition from male to female on length and weight traits. Pairwise tests revealed significant heterogeneity of genotypes among size groups in one RFLP marker (*MSTN1_B-Hae*III) (Table 5) and confirmed the putative relationship of *MSTN-1* with growth traits observed in the ANOVA tests. When we tested for association between homozygous/heterozygous genotypes of the *MSTN-1* gene and phenotype traits, the results suggest selection at the genotype level, with homozygotes being favoured at certain ages for certain alleles present in the population. In order to test rigorously and confirm the putative associations uncovered in this study, cross-test family-based analysis should be applied, under similar conditions of industrial production.

In this study, myostatin polymorphisms were putatively linked with growth traits in *S. aurata* under industrial production conditions. An association between this candidate gene and growth is not surprising because myostatin acts as a negative regulator of skeletal muscle growth [31], and many studies show that myostatin can affect muscle mass in several species [32–34], and the presence in the population of favourable alleles would result in greater growth of those individuals that were homozygous for growth-increasing alleles. Unlike terrestrial vertebrates, piscine MSTN mRNA has been detected in several tissues including muscle, gill, skin, brain, renal, and gonadal tissue [35]; this pattern of expression suggests that MSTN is involved in numerous physiological activities such as osmoregulation, gonadal development, growth, and maturation.

No significant correlations of molecular variants and quantitative traits were revealed in the present study at the other candidate genes analyzed. However, genes within the somatotropic axis and transforming growth factor superfamily have been shown to influence growth variation in terrestrial livestock, as well as in other vertebrates [11].

We tested the null hypothesis that genotypes at one locus are independent of genotypes at the other locus, for each locus pair across the population. A locus pair located at the same gene (e.g., introns) should be in linkage disequilibrium; the results show a significant linkage disequilibrium between *MSTN-1* and *PRL*. However, not all of them revealed this expected disequilibrium across all age groups. This is because several or many families were present in each group, and different genotypic combinations could be observed. The finding is remarkable because it indicates that genotypes of two RFLP markers (located in *MSTN-1* and *PRL* genes) are not independent. If these loci are located on the same chromosome, they can be selected at the same time. If they are not linked, it means that some genotypic combinations of the two loci are favourable (or are selected against) under some specific conditions and they can be used in marker-assisted selection (MAS), too. Sarropoulou et al. [36] localized the *S. aurata* growth hormone and prolactin genes on radiation hybrid group RH24 of the gene-based radiation hybrid map of gilthead seabream. This location is very interesting because our results suggest linkage disequilibrium between *PRL* and *MSTN-1* genes; therefore, if this disequilibrium is because the two genes are linked in the genome, a linkage between three candidate genes, *PRL*, *MSTN-1*, and *GH*, for putative growth-related QTLs would have been revealed.

In relation to cytogenetic mapping, the application of fluorescence *in situ* hybridization (FISH) to fish genetics and genomics has been widely tested (reviewed in [37]). Chromosome mapping of single copy genes is useful in isolating QTL of importance in aquaculture, and it is the first step towards linking physical and linkage maps. In fishes, to date, only one study, on *D. rerio*, has been found in which the *MSTN-1* gene was localized on chromosome 9, close to the marker Z8363 [38]. The fact that some metaphases, in our study, showed four signals may be due to the presence of two *MSTN* genes in *S. aurata*. The relative similarity at the nucleotide level of the two myostatin cDNA genes (72%) could be enough to find some metaphase spreads with four signals. In the light of the results, the two genes (*MSTN-1* and *MSTN-2*) are located on different chromosomes, as was found in zebrafish, where the *MSTN-2* gene was situated on chromosome 22 (Zv7 of the zebrafish genome, NCBI data base). Our results show that the genotypes of *MSTN-1* gene should be considered for selective improvement programmes of *S. aurata* and could be a start point in order to detect candidate genes for growth in other aquatic species of economic interest.

## Acknowledgments

This research was supported by BIOANDALUS 08/5/l2.3 and PETRI (PTR1995-0648-OP-03-03) in collaboration with CUPIMAR S.A., grants of Junta Andalucia to the BIO-219 group, and by Grant MSM0021620858 from the Ministry of Education, Youth and Sports of the Czech Republic (V. Krylov and J. Mácha). I. Sánchez-Ramos was recipient of a fellowship from the University of Cadiz. The authors wish to thank Rosa Vázquez and all the members of the Laboratory of Marine Culture from the University of Cadiz for the larvae supplies.

## References

[1] Food and Agriculture Organization of The United Nations, Overview: Major Trends and Issues. Fisheries and Aquaculture Department, 2011.

[2] Food and Agriculture Organization of The United Nations, Fisheries Department, Fishery Information Data and Statistics unit, 2009.

[3] V. Martinez, "Marker assisted selection in fish and shellfish breeding schemes," in *Marker-Assisted Selection: Current Status and Future Perspectives in Crops, Livestock, Forestry and Fish*, E. Guimarães, J. Ruane, B. Scherf, A. Sonnino, and J. Dargie, Eds., pp. 329–362, Electronic Publishing Policy and Support Branch Communication Division, FAO, Rome, Italy, 2007.

[4] T. H. E. Meuwissen, "Genomic selection the future of marker-assisted selection and animal breeding," in *Marker-Assisted Selection: a Fast Track to Increase Genetic Gain in Plant and Animal Breeding?*, 2003.

[5] R. Gross and J. Nilsson, "Restriction fragment length polymorphism at the growth hormone 1 gene in Atlantic salmon (*Salmo salar* L.) and its association with weight among the offspring of a hatchery stock," *Aquaculture*, vol. 173, no. 1–4, pp. 73–80, 1999.

[6] M. Prudence, J. Moal, P. Boudry et al., "An amylase gene polymorphism is associated with growth differences in the Pacific cupped oyster *Crassostrea gigas*," *Animal Genetics*, vol. 37, no. 4, pp. 348–351, 2006.

[7] C. Zhang, B. Liu, H. Chen et al., "Associations of a HinfI PCR-RFLP of POU1F1 gene with growth traits in Qinchuan cattle," *Animal Biotechnology*, vol. 20, no. 2, pp. 71–74, 2009.

[8] A. Gorbai, R. V. Torshizi, M. Bonyadi, and C. Amirinia, "A MspI PCR-RFLP within bovin growth hormone gene and its association with sperm quality traits in Iranian Holstein bulls," *African Journal of Biotechnology*, vol. 8, no. 19, pp. 4811–4816, 2009.

[9] Q. Wang, H. Li, L. Leng et al., "Polymorphism of heart fatty acid-binding protein gene associatied with fatness traits in the chicken," *Animal Biotechnology*, vol. 18, no. 2, pp. 91–99, 2007.

[10] M. Mohammadi, M. T. B. Nasiri, K. Alami-Saeid, J. Fayazi, M. Mamoee, and A. S. Sadr, "Polymorphism of calpastatin gene in Arabic sheep using PCR- RFLP," *African Journal of Biotechnology*, vol. 7, no. 15, pp. 2682–2684, 2008.

[11] C. de-Santis and D. R. Jerry, "Candidate growth genes in finfish—Where should we be looking?" *Aquaculture*, vol. 272, no. 1–4, pp. 22–38, 2007.

[12] T. T. Chen, M. Shamblott, C. M. Lin et al., "Structure and evolution of fish growth hormone and insulin-like growth factor genes," *Perspective in Comparative Endocrinology*, pp. 352–364, 1994.

[13] A. C. McPherron, A. M. Lawler, and S. J. Lee, "Regulation of skeletal muscle mass in mice by a new TGF-$\beta$ superfamily member," *Nature*, vol. 387, no. 6628, pp. 83–90, 1997.

[14] L. Maccatrozzo, L. Bargelloni, G. Radaelli, F. Mascarello, and T. Patarnello, "Characterization of the myostatin gene in the gilthead seabream (*Sparus aurata*): sequence, genomic structure, and expression pattern," *Marine Biotechnology*, vol. 3, no. 3, pp. 224–230, 2001.

[15] L. Maccatrozzo, L. Bargelloni, B. Cardazzo, G. Rizzo, and T. Patarnello, "A novel second myostatin gene is present in teleost fish," *FEBS Letters*, vol. 509, no. 1, pp. 36–40, 2001.

[16] M. H. C. Chen, G. H. Lin, H. Y. Gong et al., "Cloning and characterization of insulin-like growth factor I cDNA from black seabream (*Acanthopagrus schlegeli*)," *Zoological Studies*, vol. 37, no. 3, pp. 213–221, 1998.

[17] L. A. Manzon, "The role of prolactin in fish osmoregulation: a review," *General and Comparative Endocrinology*, vol. 125, no. 2, pp. 291–310, 2002.

[18] M. Rand-Weaver and H. Kawauchi, "Growth hormone, prolactin and somatolactin: a structural overview," in *Biochemistry and Molecular Biology of Fishes: Molecular Biology Frontiers*, P. W. Hochachka and T. P. Mommsen, Eds., pp. 39–56, Elsevier, Amsterdam, The Netherlands, 1993.

[19] S. H. S. Ng, C. G. Artieri, I. E. Bosdet et al., "A physical map of the genome of Atlantic salmon, *Salmo salar*," *Genomics*, vol. 86, no. 4, pp. 396–404, 2005.

[20] R. Almuly, B. Cavari, H. Ferstman, O. Kolodny, and B. Funkenstein, "Genomic structure and sequence of the gilthead seabream (*Sparus aurata*) growth hormoneencoding gene: identification of minisatellite polymorphism in intron I," *Genome*, vol. 43, no. 5, pp. 836–845, 2000.

[21] G. Martínez, E. M. Shaw, M. Carrillo, and S. Zanuy, "Protein salting-out method applied to genomic DNA isolation from fish whole blood," *Biotechniques*, vol. 24, no. 2, pp. 238–239, 1998.

[22] J. Porta, J. M. Porta, J. Béjar, and M. C. Alvarez, "Development of a microsatellite multiplex genotyping tool for the fish Gilthead seabream (*Sparus aurata*): applicability in population genetics and pedigree analysis," *Aquaculture Research*, vol. 41, no. 10, pp. 1514–1522, 2010.

[23] S. Rozen and H. J. Skaletsky, "Primer3 on the WWW for general users and for biologist programmers," in *Bioinformatics Methods and Protocols: Methods in Molecular Biology*, S. Krawetz and S. Misener, Eds., pp. 365–386, Humana Press, Totowa, NJ, USA, 2000.

[24] J. Bikandi, R. S. Millan, A. Rementeria, and J. Garaizar, "In silico analysis of complete bacterial genomes: PCR, AFLP-PCR and endonuclease restriction," *Bioinformatics*, vol. 20, no. 5, pp. 798–799, 2004.

[25] D. S. Moore, *The Basic Practice of Statistics*, W.H. Freeman, New York, NY, USA, 2003.

[26] W. R. Rice, "Analysing tables of statistical tests," *Evolution*, vol. 43, no. 1, pp. 223–225, 1989.

[27] M. Raymond and F. Rousset, "GENEPOP (version 1.2): population genetics software for exact tests and ecumenicism," *Journal of Heredity*, vol. 86, no. 3, pp. 248–249, 1995.

[28] J. Goudet, M. Raymond, T. de Meeüs, and F. Rousset, "Testing differentiation in diploid populations," *Genetics*, vol. 144, no. 4, pp. 1933–1940, 1996.

[29] I. Cross, A. Merlo, M. Manchado, C. Infante, J. P. Cañavate, and L. Rebordinos, "Cytogenetic characterization of the sole *Solea senegalensis* (Teleostei: Pleuronectiformes: Soleidae): Ag-NOR, $(GATA)_n$, $(TTAGGG)_n$ and ribosomal genes by one-color and two-color FISH," *Genetica*, vol. 128, no. 1–3, pp. 253–259, 2006.

[30] V. Krylov, T. Tlapakova, and J. Macha, "Localization of the single copy gene Mdh2 on *Xenopus tropicalis* chromosomes by FISH-TSA," *Cytogenetic and Genome Research*, vol. 116, no. 1-2, pp. 110–112, 2007.

[31] L. Grobet, L. J. Martin, D. Poncelet et al., "A deletion in the bovine myostatin gene causes the double-muscled phenotype in cattle," *Nature Genetics*, vol. 17, no. 1, pp. 71–74, 1997.

[32] L. Tang, Z. Yan, Y. Wan, W. Han, and Y. Zhang, "Myostatin DNA vaccine increases skeletal muscle mass and endurance in mice," *Muscle and Nerve*, vol. 36, no. 3, pp. 342–348, 2007.

[33] X. Wang, X. Meng, B. Song, X. Qiu, and H. Liu, "SNPs in the myostatin gene of the mollusk *Chlamys farreri*: association with growth traits," *Comparative Biochemistry and Physiology B*, vol. 155, no. 3, pp. 327–330, 2010.

[34] S. Welle, K. Burgess, C. A. Thornton, and R. Tawil, "Relation between extent of myostatin depletion and muscle growth in mature mice," *American Journal of Physiology*, vol. 297, no. 4, pp. E935–E940, 2009.

[35] G. Radaelli, A. Rowlerson, F. Mascarello, M. Patruno, and B. Funkenstein, "Myostatin precursor is present in several tissues in teleost fish: a comparative immunolocalization study," *Cell and Tissue Research*, vol. 311, no. 2, pp. 239–250, 2003.

[36] E. Sarropoulou, R. Franch, B. Louro et al., "A gene-based radiation hybrid map of the gilthead sea bream *Sparus aurata* refines and exploits conserved synteny with *Tetraodon nigroviridis*," *BMC Genomics*, vol. 8, article 44, 2007.

[37] R. B. Phillips, "Application of fluorescence *in situ* hybridization (FISH) to fish genetics and genome mapping," *Marine Biotechnology*, vol. 3, Supplement 1, pp. S145–S152, 2001.

[38] C. Xu, G. Wu, Y. Zohar, and S.-J. Du, "Analysis of myostatin gene structure, expression and function in zebrafish," *Journal of Experimental Biology*, vol. 206, no. 22, pp. 4067–4079, 2003.

# Allelic Variation at the *Rht8* Locus in a 19th Century Wheat Collection

**Linnéa Asplund,[1,2] Matti W. Leino,[3,4] and Jenny Hagenblad[1,4,5]**

[1] *Department of Ecology and Evolution, Uppsala University, SE-752 36 Uppsala, Sweden*
[2] *Department of Crop Production Ecology, Swedish University of Agricultural Sciences, SE-750 07 Uppsala, Sweden*
[3] *Swedish Museum of Cultural History, SE-643 98 Julita, Sweden*
[4] *IFM Molecular Genetics, Linköping University, SE-581 83 Linköping, Sweden*
[5] *Department of Biology, Norwegian University of Science and Technology, NO-7491 Trondheim, Norway*

Correspondence should be addressed to Jenny Hagenblad, jenny.hagenblad@liu.se

Academic Editors: T. Nakazaki and I. Tokatlidis

Wheat breeding during the 20th century has put large efforts into reducing straw length and increasing harvest index. In the 1920s an allele of *Rht8* with dwarfing effects, found in the Japanese cultivar "Akakomugi," was bred into European cultivars and subsequently spread over the world. *Rht8* has not been cloned, but the microsatellite marker WMS261 has been shown to be closely linked to it and is commonly used for genotyping *Rht8*. The "Akakomugi" allele is strongly associated with *WMS261-192bp*. Numerous screens of wheat cultivars with different geographical origin have been performed to study the spread and influence of the *WMS261-192bp* during 20th century plant breeding. However, the allelic diversity of WMS261 in wheat cultivars before modern plant breeding and introduction of the Japanese dwarfing genes is largely unknown. Here, we report a study of WMS261 allelic diversity in a historical wheat collection from 1865 representing worldwide major wheats at the time. The majority carried the previously reported 164 bp or 174 bp allele, but with little geographical correlation. In a few lines, a rare 182 bp fragment was found. Although straw length was recognized as an important character already in the 19th century, *Rht8* probably played a minor role for height variation. The use of WMS261 and other functional markers for analyses of historical specimens and characterization of historic crop traits is discussed.

## 1. Introduction

During the green revolution in 1960's and 1970's the yield of cereal grain increased dramatically and annual production doubled [1]. This was partly due to changed cultivation practices but primarily a result of the development of new varieties of wheat, corn, and rice. One important aspect of the new varieties was the shorter, sturdier straw that could take large amounts of fertilizers without suffering from lodging.

The reduction in straw length was a result of cultivars being insensitive to gibberellin [2]. For example, Peng et al. [3] reported that the mutant alleles of the genes *reduced height-1, (Rht-B1 and Rht-D1)* leading to dwarfism in wheat, as well as the maize gene *dwarf-8* (*d8*), are orthologues of the *Arabidopsis Gibberellin Insensitive (GAI)* gene. Unfortunately the two *Rht* genes, *Rht-B1* and *Rht-D1,* also reduce seedling establishment and coleoptile length under some environmental conditions.

Such negative effects on seedling vigour have not been found in another semidwarfism gene, *Rht8* [13], located on chromosome 2D [14]. Although the molecular identity of *Rht8* is still unknown, Korzun et al. [15] showed a close association with the microsatellite marker WMS261. Several alleles of this marker exist and the 164 bp allele increased height with 3 cm compared to the 174 bp allele, while the 192 bp allele, diagnostic for the semidwarf phenotype, was associated with a reduction in 7-8 cm compared to the 174 bp allele.

Table 1: Studies of WMS261 allelic diversity.

| Reference | Plant material | Number of accessions |
|---|---|---|
| Worland et al. [4] | World-wide cultivars | 118 |
| Chebotar et al. [5] | Ukrainian cultivars and breeding lines | 27 |
| | US and European cultivars and breeding lines | 20 |
| Worland et al. [6] | World-wide cultivars | 870 |
| Ahmad and Sorrells [7] | Mainly US and NZ cultivars | 71 |
| Manifesto and Suárez [8] | Argentinian cultivars | 165 |
| Schmidt et al. [9] | Australian cultivars | 24 |
| Liu et al. [10] | Chinese cultivars and breeding lines | 408 |
| | CIMMYT, US and European cultivars and breeding lines | 98 |
| Ganeva et al. [11] | Bulgarian cultivars | 89 |
| Zhang et al. [12] | Chinese landraces, cultivars, and breeding lines | 220 |

A few exceptions to the linkage between WMS261 and *Rht8* have been reported [16]. Nevertheless, WMS261 has been useful in a large number of screens for *Rht8* polymorphisms in various materials (Table 1). The dwarfing allele of *Rht8* and associated *WMS261-192 bp* was introduced from the Japanese variety "Akakomugi" through Italian breeding programs in the 1920's [4]. After that, it was used in several crossings and spread to the rest of the world [17]. In southern and central Europe, this allele is now very abundant [4, 6] and it is found almost exclusively in certain areas like Ukraine [5] and Bulgaria [11]. Additionally, in China the *WMS261-192 bp* is very common [10]. Interestingly, *WMS261-192 bp* is also found in several Chinese landraces suggesting an alternative source of *Rht8* in Chinese cultivars to the "Akakomugi"-Italian breeding origin [12]. The semidwarf CIMMYT varieties usually carry the *WMS261-164 bp* allele [4]. These lines have reduced height through *Rht-B1b* and *Rht-D1b*. Worland et al. [4] speculate that addition of *Rht8* would lead to a too strong dwarfing phenotype. In varieties from USA, UK, Germany, and France *WMS261-174 bp* is the most common allele. This was suggested to be due to its linkage with the photoperiod sensitive *Ppd-D1b* allele that might be beneficial for northern varieties [4, 6].

In spite of these extensive screenings, the world-wide distribution of WMS261 alleles in the era before modern plant breeding as well as introduction of the "Akakomugi" allele is unknown. The objective of this study is to explore the presence of the different WMS261 alleles in a historic 19th century material. Although no formal plant breeding (i.e., planned crossings and pedigree-based selections) took place in the 19th century, numerous well-characterized wheat cultivars existed [18, 19]. These, more or less, pure lines derived from landraces were multiplied and sold by seed companies and were thus spread and cultivated over large areas.

Several of the most recognized wheat cultivars in the 1860s were displayed at the International Exhibition in London 1862. Seed samples from the exhibition were taken to Stockholm, Sweden where they, together with some German cultivars, were multiplied at the Experimental Field of The Royal Academy of Agriculture during subsequent years [20]. Samples of the harvest of 1865 were saved in glass containers and stored at the academy museum for 100 years before being moved to the Swedish Museum of Cultural History where the samples have been kept since. Here, we report on the WMS261 genotyping of these 147-year-old seeds and the possible influence of *Rht8* in 19th century wheats.

## 2. Material and Methods

*2.1. Historical Plant Material.* Fifty-nine historical wheat varieties, harvested in 1865, were obtained from the seed collection of The Royal Swedish Academy of Forestry and Agriculture (Table 2). The seeds in this seed collection are no longer viable but genetic analysis of the aged DNA is possible [22]. Information regarding sample origin, cultivar origin, and subspecies (Table 2) was gathered from the archives of The Royal Swedish Academy of Forestry and Agriculture and complemented with data from 19th century literature on cereal cultivation [18–21]. Data on straw length (Table 2) and lodging resistance in test cultivations 1865 was taken from Juhlin-Dannfelt [21].

*2.2. Molecular Analysis.* DNA extractions of historical material were made at Linköping University in a laboratory where cereal DNA work is not regularly performed. DNA was extracted from single seeds using the FastDNA SPIN Kit and FastPrep Instrument (MP Biomedicals), with extraction blanks performed in parallel as negative controls.

*Rht8* was genotyped through a seminested PCR for the marker WMS261. The primer pair Rht8f (TGTAAAACC-ACGGCCAGTCTCCCTGTACGC) and Rht8r (CTCGCG-CTACTAGCCATTG) was used for a first round of PCR, followed by a second round using a fluorescently-labelled forward primer, M13f, together with Rht8r. Each PCR reaction of 20 $\mu$L consisted of 0.5 U Taq DNA Polymerase (New England BioLabs), 1X New England BioLabs ThermoPol Reaction Buffer, 0.25 $\mu$M of each dNTP, 0.1 $\mu$M each of the primers, and 1 $\mu$L and 3 $\mu$L of DNA-template for the first and the second PCR, respectively, where PCR product from the first PCR was used as template for the second. PCR amplifications were run at 3 min initial denaturation at 94°C, 30 cycles of 94°C for 20 s, 55°C for 1 min 20 s, and 72°C for 30 s and a final extension step of 72°C for 10 min. In samples failing to amplify the PCR reaction was repeated twice, the second time with an annealing temperature of 51°C to allow for annealing to mutated primer sites. Fragment lengths of PCR products were analyzed using MegaBACE 1000 (Amersham Biosciences) and MegaBACE Fragment Profiler version 1.2

TABLE 2: Historical cultivars screened for WMS261 allelic diversity. Acc.nr refers to the seed collection inventory number in the Swedish Museum of Cultural History. Data on height and lodging are from Juhlin Dannfelt [21].

| Acc.nr | Species[1] | Cultivar name | Country of origin | WMS261 allele/s | Height (cm) | Notes on lodging |
|---|---|---|---|---|---|---|
| NM1080 | *T. ae. ae.* | Hartswood | England | 164 | 122 | Lodging |
| NM1081 | *T. ae. ae.* | West Canada | Canada | 174 | 91 | Little lodging |
| NM1082 | *T. ae. ae.* | Fife | Canada | 164 | 102 | |
| NM1083 | *T. ae. ae.* | Cloves Highland | Holland | 174 | | |
| NM1084 | *T. ae. ae.* | Stevens | Australia | 164 | 102 | Early lodging |
| NM1085 | *T. ae. ae.* | Hunters Winter | Germany | 174 | 114 | Lodging |
| NM1086 | *T. ae. ae.* | Tappahannock | United States | 164 | 119 | Somewhat lodging |
| NM1087 | *T. ae. ae.* | Richmond's Giant | England | 164 | 114 | Late lodging |
| NM1088 | *T. ae. ae.* | Marigold | Germany | 164, 174 | | |
| NM1090 | *T. ae. ae.* | Red Lammas | England | 174 | 117 | Somewhat lodging |
| NM1091 | *T. ae. ae.* | Chiddam | England | 174 | 122 | Somewhat lodging |
| NM1092 | *T. ae. ae.* | Petticoat | Canada | 174 | 122 | |
| NM1093 | *T. ae. ae.* | Hundredfold | England | 164 | 112 | Late lodging |
| NM1094 | *T. ae. ae.* | Victoria | Venezuela | 174 | 102 | Somewhat lodging |
| NM1095 | *T. ae. ae.* | Drewett's | Unknown | 174 | 114 | |
| NM1096 | *T. ae. ae.* | Tuscany | Italy | 174 | 114 | |
| NM1097 | *T. ae. ae.* | Hopetoun | Germany | 174 | 117 | Somewhat lodging |
| NM1098 | *T. ae. ae.* | Southern Australia | Australia | 174 | | |
| NM1099 | *T. ae. ae.* | Long bearded | Unknown | 164 | 112 | Lodging |
| NM1100 | *T. ae. ae.* | Red from Tschernigow | Ukraine | 164 | 94 | Lodging |
| NM1101 | *T. ae.* | Summer wheat | Unknown | 174 | | |
| NM1102 | *T. ae. ae.* | Australia | Australia | 164 | | |
| NM1103 | *T. ae. ae.* | Mummy | England | 164, 174 | 117 | Somewhat lodging |
| NM1104 | *T. ae. ae.* | Ringelblumen | Germany | 182 | | |
| NM1106 | *T. ae. ae.* | Red Essex | England | 174 | 127 | Somewhat lodging |
| NM1108 | *T. ae. ae.* | Canadian | Canada | 182 | | |
| NM1109 | *T. ae. ae.* | Dayton | Unknown | 164, 182 | 117 | Late lodging |
| NM1110 | *T. ae. ae.* | White Belgian | Belgium | 174 | 114 | |
| NM1111 | *T. ae. ae.* | Hungarian | Hungary | 174 | | |
| NM1112 | *T. ae. sp.* | White Schwanen | Unknown | 164 | | |
| NM1113 | *T. ae. co.* | Igel | Switzerland | 174 | 114 | Somewhat lodging |
| NM1115 | *T. ae. ae.* | Galizian | Poland | 174 | 102 | Lodging |
| NM1116 | *T. ae. ae.* | Eley's Giant | Switzerland | 164 | 99 | |
| NM1118 | *T. ae. ae.* | Sixrow | Unknown | | 112 | Somewhat lodging |
| NM1120 | *T. ae. ae.* | Stålvete | Sweden | 174 | | |
| NM1122 | *T. ae. ae.* | Nottingham | England | 174 | 119 | Much lodging |
| NM1123 | *T. ae. ae.* | Hungarian | Hungary | 174 | | |
| NM1125 | *T. ae. ae.* | Three-row Chevalier | Unknown | 174 | | |
| NM1126 | *T. ae. ae.* | Hungarian | Hungary | 174 | | |
| NM1129 | *T. ae. ae.* | Sandomirka from Volhynia | Ukraine | 174 | 117 | Somewhat lodging |
| NM1135 | *T. ae. ae.* | White Essex | England | 164 | 112 | |
| NM1136 | *T. ae. ae.* | Probsteier | Germany | 174 | 122 | Somewhat lodging |
| NM1139 | *T. ae. ae.* | Grano tenero | Italy | 164, 174 | 114 | lodging |
| NM1140 | *T. ae. ae.* | Lammas | England | 164 | 114 | Lodging |
| NM1141 | *T. ae. ae.* | Fenton | Scotland | 174 | 102 | |
| NM1178 | *T. ae. ae.* | Bluestem | Canada | 174 | 109 | |

TABLE 2: Continued.

| Acc.nr | Species[1] | Cultivar name | Country of origin | WMS261 allele/s | Height (cm) | Notes on lodging |
|---|---|---|---|---|---|---|
| NM1179 | *T. ae. ae.* | Talavera | Spain | 174 | 107 | |
| NM1180 | *T. ae. ae.* | Red-chaffed-pearl | United States | 164 | 127 | Lodging |
| NM1181 | *T. ae. ae.* | Southern Australia | Australia | 164 | | |
| NM1182 | *T. ae. ae.* | Italian | Italy | 174 | | |
| NM1186 | *T. ae. ae.* | Swedish (Sammets) | Sweden | 174 | | |
| NM1187 | *T. ae. ae.* | Hopetoun | England | 174 | 119 | Somewhat lodging |
| NM1189 | *T. ae. ae.* | Hickling's prolific | England | 174 | 122 | Lodging |
| NM1800 | *T. ae. sp.* | White winter spelt | Germany | 174 | 109 | Lodging |
| NM1802 | *T. ae. sp.* | Winter spelt | Germany/France | 174 | | |
| NM1803 | *T. ae. sp.* | White-club-shaped spelt | Germany/Switzerland | 174 | | |
| NM1805 | *T. ae. sp.* | Red winter | Germany | 174 | 107 | |
| NM1807 | *T. ae. sp.* | Schlegel's winter | Germany | | 97 | |
| NM1811 | *T. ae. sp.* | White winter emma spelt | Unknown | 174 | 107 | |

[1]T: Triticum, ae: aestivum, co: compactum, sp: spelta.

## 3. Results

We successfully amplified the marker WMS261 in 57 out of 59 seed samples harvested in 1865 and used it as a proxy for genotyping the linked *Rht8* locus. Among the samples yielding a PCR product we found 15 accessions carrying the *WMS261-164 bp* genotype and 36 accessions with the *WMS261-174 bp* allele. Two accessions had an allele of length 182 bp. In addition four accessions were heterozygous, three for the 164, 174 genotype, and one for the 164, 182 genotype (Table 2).

For most accessions the country of origin was known. We were unable to detect any clear pattern with respect to country of origin and *Rht8* genotype. In most countries from which we had more than one accession both the *WMS261-164 bp* and the *WMS261-174 bp* allele were present. The exceptions were Hungary (all three *WMS261-174 bp*), Sweden (both *WMS261-174 bp*) and the US (both *WMS261-164 bp*). All spelt wheats studied, except one, carried the *WMS261-174 bp* allele.

We evaluated data on straw length from the test cultivations performed in 1865, the cultivations from which the seeds were taken. Data was available for 41 cultivars and straw lengths ranged from 91 to 127 cm. We found no correlation between straw length and the two WMS261-genotypes, -164 bp and -174 bp (two sample $t$-test, $df = 34$, $P = 0.78$). The degree of lodging was registered in the cultivation records and we note that several of the tallest accessions suffer from lodging. Evidently, lodging was considered as a serious problem and tall strawwas an undesirable trait.

## 4. Discussion

The genetic diversity at the *WMS261* microsatellite has been an important diagnostic tool for genotyping the *Rht8* locus (Table 1). Previous studies have shown three different alleles, *WMS261-174 bp*, *WMS261-164 bp*, and *WMS261-192 bp*, to be internationally widespread. The majority of the accessions in our sample had either of the first two of these alleles. Some of our PCR products yielded fragments that were sized a few base pairs larger than *WMS261-164 bp* or *WMS261-174 bp*, but in accordance with Schmidt et al. [9] we did not consider them as distinct alleles, but a result of slippage or "stutter".

Our choice of samples is in many ways comparable with those of previous studies [9, 10, 15] in that it comprises of a range of, at the time, widely cultivated and internationally representative wheat accessions. As expected the *WMS261-164 bp* and the *WMS261-174 bp* alleles were the most common ones (28 and 68% of the homozygotes, resp.). Our samples were harvested some 60 years before the first use of "Akakomugi" in crosses and the 1865 test cultivations did not include any Japanese or Chinese accessions. It is therefore not surprising that we do not detect the *WMS261-192 bp* allele.

It has been suggested that the *WMS261-174 bp* allele is linked to the *Ppd-D1b* allele and has been selected for in northern Europe. However, in contrast to screens of extant material [4, 6] we did not see the clear dominance of the *WMS261-174 bp* allele in cultivars from northern Europe and North America. In our material we found both the *WMS261-174 bp* and the *WMS261-164 bp* alleles in wheats from a wide range of countries and in many cases we found both alleles in wheats from the same country. Although all the accessions from the same country in a few cases shared the same allele we could not distinguish any clear geographic pattern in the distribution of the *WMS261-174 bp* and *WMS261-164 bp* alleles. The limited number of accessions restricts the possibility to recognize geographic patterns, but the geographic segregation of allele types [4, 6] might actually have arisen later during modern plant improvement, often based on a few key cultivars. In the cultivation data for the winter wheats flowering time (not shown) was slightly earlier for accessions with the *WMS261-174 bp* than those with the *WMS261-164 bp* allele (298 versus 300 days after sowing) but not significantly (two sample $t$-test, $df = 27$, $P = 0.19$) and did thus not show any clear support for linkage to *Ppd-D1b*.

In addition to the two major alleles (*WMS261-164 bp* and *WMS261-174 bp*), we found a few accessions with a 182 bp allele. Other studies have reported alleles differing from the three main alleles, but an allele in the 182 bp size range has only been reported previously in a single cultivar, "Madison" [7]. The wheats with the 182 bp allele was a Canadian wheat and a German wheat called "Ringelblumen" and it does not seem that they have a shared or limited origin that might otherwise have explained why the allele has been undetected in most previous studies. Unfortunately we lacked cultivation data for both the accessions homozygous for the 182 bp allele. Its correlation with a specific effect on plant height should be worthwhile investigating to further explore the relationship between different alleles at the WMS261 locus and differences in plant height.

In this study, we cannot find any correlation between genotype and plant height. The average effect of the *WMS261-174 bp* allele compared to the *WMS261-164 bp* allele is a reduction of 3 cm [15] and in the limited number of accessions this effect might be too small to detect. The test cultivations, carried out at the experimental fields of the Royal Swedish Agricultural Academy, were also performed in small and nonreplicated test plots, which further limit the possibility to reveal any effects of the *Rht8*. However, it is clear from the cultivation data that straw length and the amount of lodging were traits of concern to the 19th century plant breeders. Although *Rht8* probably contributed little, if at all, to variation in straw length, the set of cultivars of 1865 displayed a large height range.

The cultivars studied here are from the time period when the seed industry first emerged in the Western world. Line selections from landraces with desirable traits were developed and multiplied to give rise to more uniform seed materials with more predictable traits, that is, cultivars. The most popular of these was named and described in the contemporary literature and received both national and international attention [23]. The major wheat cultivars of the 19th century have in some cases survived to the present in genebank collections, and several of the cultivars genotyped in this study can be obtained as extant material from genebanks. Most of the cultivars studied here are, however, long extinct. For these, samples from historical collections provide the only possibility to study the genetic composition of early wheat cultivars. Also for accessions still available in genebanks, there are advantages in using historical material instead. Concerns regarding the integrity of genebank material have been raised [24] and the geographic distribution of functional alleles have been shown to be much more distinct with historical than extant material [25, 26]. The specific nature of the historic specimens used here, that is, large containers with thousands of seeds [22], also permits repeated or complementary experiments.

Molecular identification of genes involved in domestication and plant improvement has recently accelerated [27, 28]. By screening the genetic diversity present and testing for selection the individual importance of different alleles can be explored. In the case of *Rht8* its role in 20th century wheat improvement is well known from extensive screens and documented crossings and pedigrees. Here we can add insight into the genetic diversity of the *Rht8* during the transition from traditional and modern agriculture, a time less well documented and more difficult to study. The use of historical and archaeological plant material [29] in addition to extant plant material can in this way help to reveal a clearer picture to the processes that formed crop plants of today.

## Acknowledgments

Dr. Per Larsson is acknowledged for help with analyses of data. This research was funded by the Lagersberg, Carl Tryggers and Nilsson-Ehle foundations, and the Swedish Research Council for Environment, Agricultural Sciences and Spatial Planning (FORMAS).

## References

[1] G. S. Khush, "Green revolution: the way forward," *Nature Reviews Genetics*, vol. 2, no. 10, pp. 815–822, 2001.

[2] P. Hedden, "The genes of the green revolution," *Trends in Genetics*, vol. 19, no. 1, pp. 5–9, 2003.

[3] J. Peng, D. E. Richards, N. M. Hartley et al., ""Green revolution" genes encode mutant gibberellin response modulators," *Nature*, vol. 400, no. 6741, pp. 256–261, 1999.

[4] A. J. Worland, V. Korzun, M. S. Röder, M. W. Ganal, and C. N. Law, "Genetic analysis of the dwarfing gene *Rht8* in wheat—part II. The distribution and adaptive significance of allelic variants at the *Rht8* locus of wheat as revealed by microsatellite screening," *Theoretical and Applied Genetics*, vol. 96, no. 8, pp. 1110–1120, 1998.

[5] S. V. Chebotar, V. N. Korzun, and Y. M. Sivolap, "Allele distribution at locus *WMS261* marking the dwarfing gene *Rht8* in common wheat cultivars of Southern Ukraine," *Russian Journal of Genetics*, vol. 37, no. 8, pp. 894–898, 2001.

[6] A. J. Worland, E. J. Sayers, and V. Korzun, "Allelic variation at the dwarfing gene *Rht8* locus and its significance in international breeding programmes," *Euphytica*, vol. 119, no. 1-2, pp. 155–159, 2001.

[7] M. Ahmad and M. E. Sorrells, "Distribution of microsatellite alleles linked to *Rht8* dwarfing gene in wheat," *Euphytica*, vol. 123, no. 2, pp. 235–240, 2002.

[8] M. M. Manifesto and E. Y. Suárez, "Microsatellite screening of the *Rht8* dwarfing gene in Argentinian wheat cultivars," *Annual Wheat Newsletter*, vol. 48, pp. 23–24, 2002.

[9] A. L. Schmidt, K. R. Gale, M. H. Ellis, and P. M. Giffard, "Sequence variation at a microsatellite locus (XGWM261) in hexaploid wheat (*Triticum aestivum*) varieties," *Euphytica*, vol. 135, no. 2, pp. 239–246, 2004.

[10] Y. Liu, D. Liu, H. Zhang et al., "Allelic variation, sequence determination and microsatellite screening at the XGWM261 locus in Chinese hexaploid wheat (*Triticum aestivum*) varieties," *Euphytica*, vol. 145, no. 1-2, pp. 103–112, 2005.

[11] G. Ganeva, V. Korzun, S. Landjeva, N. Tsenov, and M. Atanasova, "Identification, distribution and effects on agronomic traits of the semi-dwarfing *Rht* alleles in Bulgarian common wheat cultivars," *Euphytica*, vol. 145, no. 3, pp. 305–315, 2005.

[12] X. Zhang, S. Yang, Y. Zhou, Z. He, and X. Xia, "Distribution of the Rht-B1b, Rht-D1b and *Rht8* reduced height genes in autumn-sown Chinese wheats detected by molecular markers," *Euphytica*, vol. 152, no. 1, pp. 109–116, 2006.

[13] M. H. Ellis, G. J. Rebetzke, P. Chandler, D. Bonnet, W. Spielmeyer, and R. A. Richards, "The effect of different height reducing genes on the early growth of wheat," *Functional Plant Biology*, vol. 31, no. 6, pp. 583–589, 2004.

[14] J. Worland, C. N. Law, and S. Petrovic, "Height reducing genes and their importance to Yugoslavian winter wheat varieties," *Savremena Poljo-Privreda*, vol. 38, pp. 245–258, 1990.

[15] V. Korzun, M. S. Röder, M. W. Ganal, A. J. Worland, and C. N. Law, "Genetic analysis of the dwarfing gene (*Rht8*) in wheat. Part I. Molecular mapping of *Rht8* on the short arm of chromosome 2D of bread wheat (*Triticum aestivum* L.)," *Theoretical and Applied Genetics*, vol. 96, no. 8, pp. 1104–1109, 1998.

[16] M. H. Ellis, D. G. Bonnett, and G. J. Rebetzke, "A 192bp allele at the Xgwm261 locus is not always associated with the *Rht8* dwarfing gene in wheat (*Triticum aestivum* L.)," *Euphytica*, vol. 157, no. 1-2, pp. 209–214, 2007.

[17] K. Borojevic and K. Borojevic, "The transfer and history of "reduced height genes" (Rht) in wheat from Japan to Europe," *Journal of Heredity*, vol. 96, no. 4, pp. 455–459, 2005.

[18] F. Alefeld, *Landwirtschaftliche Flora*, Wiegandt & Hempel, Berlin, Germany, 1866.

[19] F. Körnicke and H. Werner, *Handbuch des Getreidebaues*, Parey, Berlin, Germany, 1885.

[20] C. Juhlin-Dannfelt and A. Müller, *Berättelse öfver verksamheten vid Kongl.Landtbruks-Akademiens försöksfält och agrikulturkemiska laboratorium underåren 1862 och 1863*, P. A. Norstedt och söner, Kongl. Boktryckare, Stockholm, Sweden.

[21] C. Juhlin-Dannfelt, *Berättelse öfver verksamheten vid Kongl. Landtbruks-Akademiens försöksfält under åren 1865 och 1868*, P.A. Norstedt och söner, Kongl. Boktryckare, Stockholm, Sweden, 1870.

[22] M. W. Leino, J. Hagenblad, J. Edqvist, and E. M. K. Strese, "DNA preservation and utility of a historic seed collection," *Seed Science Research*, vol. 19, no. 3, pp. 125–135, 2009.

[23] A. P. Bonjean and W. J. Angus, *The World Wheat Book: A History of Wheat Breeding*, Intercept, London, UK, 2001.

[24] J. Hagenblad, J. Zie, and M. W. Leino, "Exploring the population genetics of genebank and historical landrace accessions," *Genetic Resources and Crop Evolution*. In press.

[25] H. Jones, D. L. Lister, M. A. Bower, F. J. Leigh, L. M. Smith, and M. K. Jones, "Approaches and constraints of using existing landrace and extant plant material to understand agricultural spread in prehistory," *Plant Genetic Resources*, vol. 6, no. 2, pp. 98–112, 2008.

[26] D. L. Lister, S. Thaw, M. A. Bower et al., "Latitudinal variation in a photoperiod response gene in European barley: insight into the dynamics of agricultural spread from 'historic' specimens," *Journal of Archaeological Science*, vol. 36, no. 4, pp. 1092–1098, 2009.

[27] J. F. Doebley, B. S. Gaut, and B. D. Smith, "The molecular genetics of crop domestication," *Cell*, vol. 127, no. 7, pp. 1309–1321, 2006.

[28] J. C. Burger, M. A. Chapman, and J. M. Burke, "Molecular insights into the evolution of crop plants," *American Journal of Botany*, vol. 95, no. 2, pp. 113–122, 2008.

[29] S. A. Palmer, O. Smith, and R. G. Allaby, "The blossoming of plant archaeogenetics," *Annals of Anatomy*, vol. 194, no. 1, pp. 146–156, 2012.

# 10

# Genetic Diversity of Upland Rice Germplasm in Malaysia Based on Quantitative Traits

**M. Sohrabi,[1] M. Y. Rafii,[1,2] M. M. Hanafi,[1] A. Siti Nor Akmar,[1] and M. A. Latif[2,3]**

[1] *Institute of Tropical Agriculture, Universiti Putra Malaysia, 43400 Serdang, Selangor, Malaysia*
[2] *Department of Crop Science, Faculty of Agriculture, Universiti Putra Malaysia, 43400 Serdang, Selangor, Malaysia*
[3] *Bangladesh Rice Research Institute (BRRI), Gazipur 1701, Bangladesh*

Correspondence should be addressed to M. Y. Rafii, mrafii@putra.upm.edu.my

Academic Editors: V. C. Concibido and Y. Yu

Genetic diversity is prerequisite for any crop improvement program as it helps in the development of superior recombinants. Fifty Malaysian upland rice accessions were evaluated for 12 growth traits, yield and yield components. All of the traits were significant and highly significant among the accessions. The higher magnitudes of genotypic and phenotypic coefficients of variation were recorded for flag leaf length-to-width ratio, spikelet fertility, and days to flowering. High heritability along with high genetic advance was registered for yield of plant, days to flowering, and flag leaf length-to-width ratio suggesting preponderance of additive gene action in the gene expression of these characters. Plant height showed highly significant positive correlation with most of the traits. According to UPGMA cluster analysis all accessions were clustered into six groups. Twelve morphological traits provided around 77% of total variation among the accessions.

## 1. Introduction

Around 3 billion people of the world use rice as a critical or basic food that provides 50 to 80% of their daily calories. Rice is cultivated on more than 150 million hectares, and annual world production is around 600 million tons [1–3]. Upland rice comprises eleven percent of global rice production and is cultivated on around 14 million hectares. Upland rice has a small role in total production but is major food in some tropical countries [4]. Bangladesh, Indonesia, and Philippines are the areas that plant the most upland rice, but the yield is so low (about 1 t/ha on average) and highly variable [5, 6].

In Malaysia, two types of rice are cultivated: wetland rice in Peninsular Malaysia (503,184 ha) and upland rice in Sabah and Sarawak (165,888 ha). The average yield of wetland rice is around 3.3 t/ha; in good conditions, however, it can increase to around 10 t/ha. In contrast, the average yield of upland rice ranges from 0.46 to 1.1 t/ha. In 2005, the total national rice production was roughly 2.24 million metric tons. In Malaysia, upland rice is usually cultivated for home consumption by rural people living in Sabah and Sarawak [7].

Genetic diversity is the basis of plant breeding, so understanding and assessing it is important for crop management, crop improvement by selection, use of crop germplasm, detection of genome structure, and transfer of desirable traits to other plants [8, 9]. Rice is one of the best plants for the study of genome structure and genetic diversity because it is diploid and has a small genome size of 430 Mb [10], a significant level of genetic polymorphism [11, 12], and a large amount of well-conserved genetically diverse material.

The breeders are interested to evaluate genetic diversity based on morphological traits because they are inexpensive, rapid, and simple to score. The study of these traits needs neither sophisticated methods nor complicated equipments, and also these traitscan be inherited without either specific biochemical or molecular techniques. Until now scientific classification of plant was based on morphological traits [13, 14]. The rice plant (*Oryza sativa*) shows great morphological variation, especially in vegetative traits such as plant height and leaf length. Therefore, the present study was undertaken

Table 1: Information on locations, seasons of seed collection and local name of the upland rice accessions.

| SL | Accessions | Location | Season | Local name |
|---|---|---|---|---|
| 1 | 6040 | Peninsular Malaysia | Main season (MS) | Bedor |
| 2 | 6041 | Peninsular Malaysia | Off season (OS) | Berjer |
| 3 | 6043 | Peninsular Malaysia | Main season (MS) | Buih |
| 4 | 6044 | Peninsular Malaysia | Main season (MS) | Gemalah |
| 5 | 6045 | Peninsular Malaysia | Off season (OS) | Kura |
| 6 | 6048 | Peninsular Malaysia | Main season (MS) | Piya |
| 7 | 6050 | Peninsular Malaysia | Off season (OS) | Ulat |
| 8 | 6059 | Peninsular Malaysia | Off season (OS) | Rengan bembang |
| 9 | 6067 | Peninsular Malaysia | Main season (MS) | Lumut/Kuku balam |
| 10 | 6068 | Peninsular Malaysia | Main season (MS) | Padi Kuku balam |
| 11 | 6070 | Peninsular Malaysia | Main season (MS) | Selayang |
| 12 | 6071 | Peninsular Malaysia | Main season (MS) | Lalang |
| 13 | 7531 | Sabah | Main season (MS) | Kungkuling A |
| 14 | 7534 | Sabah | Main season (MS) | Bukit |
| 15 | 7535 | Sabah | Main season (MS) | Pagalan |
| 16 | 7537 | Sabah | Main season (MS) | Sibuku |
| 17 | 7538 | Sabah | Main season (MS) | Lapaung |
| 18 | 7539 | Sabah | Main season (MS) | Sanding |
| 19 | 7540 | Sabah | Main season (MS) | Putus tunang |
| 20 | 7541 | Sabah | Main season (MS) | Ruabon |
| 21 | 7543 | Sabah | Main season (MS) | Semiali |
| 22 | 7544 | Sabah | Main season (MS) | Tadaong |
| 23 | 7545 | Sabah | Main season (MS) | Tayakon kecil |
| 24 | 7546 | Sabah | Main season (MS) | Teun |
| 25 | 7597 | Sabah | Main season (MS) | Batangan |
| 26 | 7596 | Sabah | Main season (MS) | Kaca |
| 27 | 7595 | Sabah | Main season (MS) | Turayo |
| 28 | 7594 | Sabah | Main season (MS) | Tarakan |
| 29 | 7590 | Sabah | Main season (MS) | Dinabor |
| 30 | 7589 | Sabah | Main season (MS) | Rangayat |
| 31 | 7588 | Sabah | Main season (MS) | Turakin |
| 32 | 7585 | Sabah | Main season (MS) | Peturu |
| 33 | 7576 | Sabah | Main season (MS) | Pagalan |
| 34 | 7571 | Sabah | Main season (MS) | Turayan |
| 35 | 7574 | Sabah | Main season (MS) | Dedawar |
| 36 | 7575 | Sabah | Main season (MS) | Lelangsat |
| 37 | 3824 | Peninsular Malaysia | Main season (MS) | Huma kuning lenggong |
| 38 | 3825 | Peninsular Malaysia | Main season (MS) | Huma wangi lenggong |
| 39 | 3826 | Peninsular Malaysia | Off season (OS) | Jarom mas |
| 40 | 3828 | Peninsular Malaysia | Main season (MS) | Kunyit |
| 41 | 3830 | Peninsular Malaysia | Main season (MS) | Langsat |
| 42 | 3831 | Peninsular Malaysia | Off season (OS) | Lenggong |
| 43 | 3832 | Peninsular Malaysia | Main season (MS) | Puteh perak |
| 44 | 3834 | Peninsular Malaysia | Off season (OS) | Rambut |
| 45 | 3833 | Peninsular Malaysia | Off season (OS) | Putih |
| 46 | 3837 | Peninsular Malaysia | Main season (MS) | Tangkai langsat |
| 47 | 3838 | Peninsular Malaysia | Main season (MS) | Wangi puteh |
| 48 | 3835 | Peninsular Malaysia | Off season (OS) | Rengan wangi |
| 49 | 7508 | Sabah | Main season (MS) | Beliong |
| 50 | 7509 | Sabah | Main season (MS) | Bedumpok |

Table 2: List of quantitative traits of upland rice.

| Traits | Method of evaluation |
|---|---|
| Plant height (PH, cm) | The average of height from the base to the tip of last leaf (flag leaf) |
| Days to flowering (DF, days) | The number of days from seeding to flowering day |
| Days to maturity (DM, days) | The number of days from seeding to maturing day |
| Flag leaf length to width ratio (FLR, cm) | Dividing the flag leaf length to width |
| Number of tillers per hill (NT, no.) | Counting of the tillers per hill |
| Number of grains per panicle (NG, no.) | Counting the number of grains on per panicle |
| One thousand grain weight (1000 GW, g) | 200 grains were weighted then 1000 weight grains were calculated from these weights |
| Yield of plant per pot (YP, g) | Weighting total grains per pot |
| Number of panicles per hill (NP, no.) | Counting the panicles per hill |
| Panicle length (PL, cm) | From base of the lowest spikelet to the top of latest spikelet on panicle |
| Spikelet per panicle (SP, no.) | Counting the spikelet per panicle |
| Spikelet fertility (SF, %) | Dividing ripped spikelet to all spikelet |

Table 3: ANOVA showing source of variation, degrees of freedom, means square, and error mean square.

| Source of variation | df | MS | EMS |
|---|---|---|---|
| Blocks ($r$) | $r-1$ | MSB | $\sigma_e^2 + g\sigma_r^2$ |
| Accessions ($g$) | $g-1$ | MSG | $\sigma_e^2 + r\sigma_g^2$ |
| Groups ($t$) | $[t-1]$ | MST | $\sigma_e^2 + r\sigma_{g/t}^2 + rg\sigma_t^2$ |
| Accessions/groups | $[t(g-1)]$ | MSG/T | $\sigma_e^2 + r\sigma_{g/t}^2$ |
| Error | $(r-1)(g-1)$ | MSE | $\sigma_e^2$ |

$r$: blocks, $g$: accessions, $t$: groups, $e$: error, df: degree of freedom, MS: mean squares, EMS: expected mean squares.

to assess the genetic diversity of upland rice genotypes in Malaysia.

## 2. Materials and Methods

*2.1. Plant Material and Experimental Design.* Fifty accessions of upland rice were selected from MARDI (24 from Peninsular Malaysia and 26 from Sabah). The accessions were cultivated in experimental field of Universiti Putra Malaysia. Sprouted seeds were sown in the pots (Table 1). Randomized complete block design (RCBD) with three replications was used with 50 pots for each replication.

*2.2. Data Collection.* Twelve quantitative traits were recorded for all accessions at each replication: plant height (cm), days to flowering (day), days to maturity (day), flag-leaf-length-to-width ratio (cm), number of tillers per hill (no.), number of grains per panicle (no.), one thousand grains weight (g), yield of plant per pot (g), number of panicles per hill (no.), panicle length (cm), spikelet per panicle (no.), and spikelet fertility (%) (Table 2).

*2.3. Statistical Analysis.* The analysis of variance (ANOVA) revealed the main interaction effects. Least significant difference (LSD) was calculated using Statistical analysis system software (SAS version 9.1) (Table 3). Genetic parameters were estimated to identify genetic variation among accessions and to determine genetic and environmental effects on various characters. These genetic parameters were calculated by the formula given by Burton [15], Burton and De Vane [16], and Johnson et al. [17]. These parameters include the following.

(a) Genotypic variance:

$$\sigma_g^2 = \frac{\text{MSG} - \text{MSE}}{r}, \quad (1)$$

where MSG is the mean square of accessions, MSE is mean square of error, and $r$ is number of replications.

(b) Phenotypic variance:

$$\sigma_p^2 = \sigma_g^2 + \sigma_e^2, \quad (2)$$

where $\sigma_g^2$ is the genotypic variance and $\sigma_e^2$ is the mean squares of error.

(c) Phenotypic coefficient of variance (PCV):

$$\text{PCV}(\%) = \frac{\sqrt{\sigma_p^2}}{\overline{X}} \times 100, \quad (3)$$

where $\sigma_p^2$ is the phenotypic variance and $\overline{X}$ is the mean of trait.

(d) Genotypic coefficient of variance (GCV):

$$\text{GCV}(\%) = \frac{\sqrt{\sigma_g^2}}{\overline{X}} \times 100, \quad (4)$$

where $\sigma_g^2$ is the genotypic variance and $\overline{X}$ is the mean of character.

(e) Heritability (broad sense):

$$h_B^2 = \frac{\sigma_g^2}{\sigma_p^2}, \quad (5)$$

where $\sigma_g^2$ is the genotypic variance and $\sigma_p^2$ is the phenotypic variance.

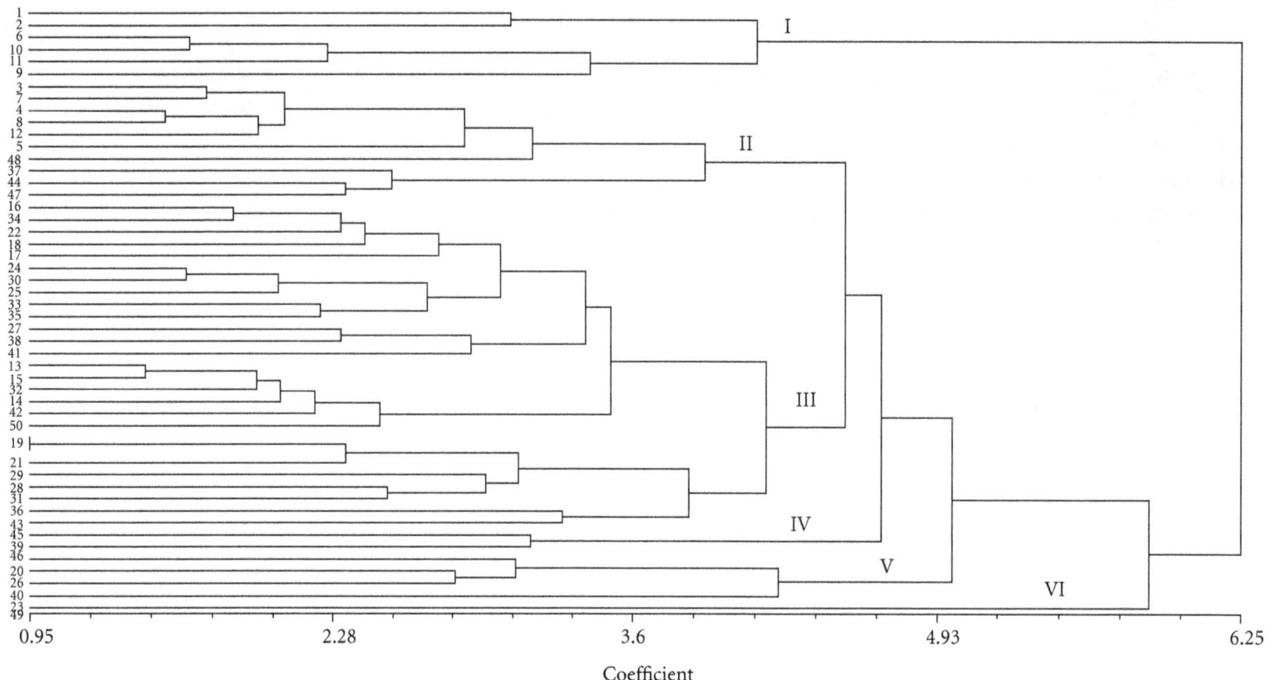


Figure 1: The dendrogram of 50 accessions of upland rice based on 12 quantitative traits.

(f) Expected genetic advance (GA):

$$\mathrm{GA}(\%) = K \times \sigma_p^2 \times h_B^2 \times 100. \quad (6)$$

GA is a percent of the mean assuming selection of the superior 5% of accession:

$$\mathrm{GA}(\%) = K \times \frac{\sqrt{\sigma_p^2}}{\overline{X}} \times h_B^2 \times 100, \quad (7)$$

where $K$ is a constant, $\sqrt{\sigma_p^2/\overline{X}}$ is the phenotypic standard deviation, $h_B^2$ is the heritability, and $\overline{X}$ is the mean of traits.

The correlation coefficient was analyzed to evaluate the relationships among the different variables in the experiment using SAS software (version 9.1). Data were also analyzed based on Jaccard's similarity coefficient by NTSYS-pc software (version 2.1). UPGMA algorithm and SAHN clustering were applied for calculating genetic relationships. The PCA of fifty accessions was calculated by EIGEN and PROJ modules of NTSYS-pc and Minitab software (version 15).

## 3. Result

*3.1. Variation and Genetic Parameters among Accessions.* Eight traits including plant height, days to flowering, flag leaf length-to-width ratio, 1000-GW, yield per pot, panicle length, spikelet per panicle, and spikelet fertility showed highly significant ($P \leq 0.01$) variation and the rest of them such as days to maturity, number of tillers per hill, number of grains per panicle, and number of panicles per hill were significant ($P \leq 0.05$) among all accessions (Table 4).

In this study, most of the growth traits showed higher PCV compared to yield and yield component traits. However, lower PCV belonged to plant height (15.85%) while flag leaf length-to-width ratio (69.63%) was recorded as higher value. Spikelet fertility (47.31%), days to flowering (40.94%), and days to maturity (40.77%) were recorded as higher values of PCV and number of grains per panicles (21.27%), number of panicle (24.54%), and panicle length (24.63%) showed lower values. The higher GCV was recorded at flag leaf length-to-width ratio (66.66%) and the lower was found at plant height (14.92%). GCV value was low in yield and yield components compared to growth characters. Board sense heritability ranged from 60.26 to 99.84%. The highest and the lowest amount of heritability was recorded at yield of plant and number of panicles, respectively. The estimates of heritability were high for 1000 GW (99.76%), spikelet fertility (94.08%), panicle length (91.69%), flag leaf length-to-width ratio (91.63%), plant height (88.57%), days to flowering (85.54%), spikelet per panicle (81.35%), and days to maturity (80.28%) whereas other characters showed relatively low heritability. GA ranged from 28.93% for plant height to 131.45% for flag leaf length-to-width ratio. The average of GA value in growth traits was higher than the average of GA value in yield and yield components (Table 5).

*3.2. Association between Traits.* Pearson's correlation coefficient was computed between 12 quantitative traits among 50 accessions of upland rice (Table 6). Positive correlation was found between most of traits. Plant height was highly significant and positively correlated with most of traits such as days to flowering, days to maturity, flag leaf width-to-

Table 4

(a) Mean squares of analysis of variance for 5 traits among 50 accessions of upland rice.

| Source of variation | df | PH | DF | DM | FLR | NT |
|---|---|---|---|---|---|---|
| Blocks | 2 | $0.01^{ns}$ | $6.61^{ns}$ | $41.33^{ns}$ | $0.01^{ns}$ | $0.0027^{ns}$ |
| Accessions | 49 | 1295.77** | 3236.17** | 5410.90* | 617.37** | 2.12* |
| Groups | [1] | 4977.62** | 15272.12** | 37117.26** | 214.76** | 0.35** |
| Groups/Accessions | [48] | 85.94** | 276.13** | 537.95** | 62.83** | 1.10** |
| Error | 98 | 53.43 | 172.58 | 397.56 | 18.22 | 0.35 |

* Significant at 0.05. **Highly significant at 0.01. PH: plant height, DF: days to flowering, DM: days to maturing, FLR: flag leaf length-to-width ratio, and NT: number of tillers per hill.

(b) Mean squares of analysis of variance for 7 traits among 50 accessions of upland rice.

| Source of variation | df | NG | 1000 GW | YP | NP | PL | SP | SF |
|---|---|---|---|---|---|---|---|---|
| Blocks | 2 | $32.03^{ns}$ | $0.007^{ns}$ | $0.13^{ns}$ | $0.090^{ns}$ | $0.03^{ns}$ | $145.68^{ns}$ | $8.39^{ns}$ |
| Accessions | 49 | 1775.43* | 155.001** | 373.65** | 1.92* | 129.23** | 5644.79** | 4051.39** |
| Groups | [1] | 3998** | 160.58** | 2434.58** | 0.31* | 41.10** | 1607.47** | 351.94** |
| Groups/Accessions | [48] | 454.84** | 154.88** | 330.71** | 1.046** | 13.18** | 1269.94** | 318.59** |
| Error | 98 | 167.46 | 0.115 | 0.199 | 0.346 | 3.78 | 400.53 | 83.21 |

* Significant at 0.05. **Significant at 0.01. NG: number of grains per panicle, 1000 GW: one thousand grain weight, YP: yield per pot, NP: number of panicles per hill, PL: panicle length, SP: spikelet per panicle, and SF: spikelet fertility.

Table 5: Genetic variance of 12 morphological characteristics.

| Traits | MEAN | MSG | MSE | $\sigma_g^2$ | $\sigma_p^2$ | PCV (%) | GCV (%) | $h_B^2$ (%) | GA (%) |
|---|---|---|---|---|---|---|---|---|---|
| PH | 136.36 | 1295.77 | 53.43 | 414.11 | 467.54 | 15.86 | 14.92 | 88.57 | 28.93 |
| DF | 84.39 | 3236.17 | 172.59 | 1021.2 | 1193.78 | 40.94 | 37.87 | 85.54 | 72.15 |
| DM | 111.55 | 5410.90 | 397.57 | 1671.11 | 2068.68 | 40.77 | 36.65 | 80.78 | 67.85 |
| FLR | 21.20 | 617.37 | 18.23 | 199.71 | 217.94 | 69.64 | 66.66 | 91.63 | 131.45 |
| NT | 3.86 | 2.12 | 0.35 | 0.59 | 0.94 | 25.15 | 19.92 | 62.72 | 32.49 |
| NG | 124.65 | 1775.43 | 167.47 | 535.98 | 703.45 | 21.28 | 18.57 | 76.19 | 33.40 |
| 1000 GW | 24.06 | 155.00 | 0.12 | 51.62 | 51.74 | 29.89 | 29.86 | 99.77 | 61.44 |
| YP | 43.45 | 373.65 | 0.20 | 124.48 | 124.68 | 25.70 | 25.68 | 99.84 | 52.85 |
| NP | 3.81 | 1.93 | 0.35 | 0.52 | 0.87 | 24.55 | 19.06 | 60.26 | 30.47 |
| PL | 27.41 | 129.23 | 3.79 | 41.81 | 45.60 | 24.64 | 23.59 | 91.69 | 46.54 |
| SP | 159.14 | 5644.79 | 400.54 | 1748.09 | 2148.62 | 29.13 | 26.27 | 81.35 | 48.82 |
| SF | 79.25 | 4051.39 | 83.21 | 1322.73 | 1405.94 | 47.31 | 45.89 | 94.08 | 91.70 |

PH: plant height, DF: days to flowering, DM: days to maturing, FLR: flag leaf length to width ratio, NT: number of tiller per hill NG: number of grains per panicle, 1000 GW: one thousand grain weight, YP: yield per pot, NP: number of panicles per hill, PL: panicle length, SP: spikelet per Panicle, and SF: spikelet fertility, MSG: mean square of accessions, MSE: mean square of error, PCV: phenotypic coefficient of variation, GCV: genotypic coefficient of variation $h_B^2$: board sense heritability, GA: genetic advance, $\sigma_g^2$: genotypic variance, and $\sigma_p^2$: phenotypic variance.

length ratio, number of grains per panicle, yield of plant, panicle length, and spikelet fertility. Yield of plant had highly significant ($P < 0.01$) and positively correlated with plant height ($r = 0.38$), days to maturity ($r = 0.36$), and number of panicles ($r = 0.48$) at 1% probability level and also significant ($P \leq 0.05$) and positively correlated with days to flowering ($r = 0.31$) and 1000-grain weight ($r = 0.34$).

*3.3. Cluster Analysis.* Fifty accessions of upland rice were clustered into six groups by 12 quantitative traits. As evident from Figure 1 and Table 7 cluster III was the biggest (27 accessions) and cluster VI was the smallest (only one member) group. Cluster I, II, IV, and V consisted of 6, 10, 2, and 4 members, respectively. The first group had the highest average in comparison with the other five groups considering five traits (Table 8) such as plant height (147.9 cm), days to flowering (112.8 days), days to maturity (144 days), flag leaf length-to-width ratio (31.1 cm), and panicle length (30.07 cm). Group VI included the highest average for four traits such as number of tillers (4.7), 1000 GW (33 g), yield of plant (55.1 g), and spikelet fertility (95.8%). On the other hand, accessions having this group (VI) showed the lowest average values in the characters such as plant height, days to maturity, flag leaf length-to width ratio, number of panicles, panicle length, and spikelet per panicle.

*3.4. Principal Component Analysis (PCA).* PCA approximately confirmed the cluster analysis for distant accession,

Table 6: Pearson's correlation coefficient among 12 quantitative traits of upland rice.

| | PH | DF | DM | FLR | NT | NG | 1000 GW | YP | NP | PL | SP | SF |
|---|---|---|---|---|---|---|---|---|---|---|---|---|
| PH | 1.00 | | | | | | | | | | | |
| DF | 0.77** | 1.00 | | | | | | | | | | |
| DM | 0.76** | 0.90** | 1.00 | | | | | | | | | |
| FLR | 0.48** | 0.64** | 0.48** | 1.00 | | | | | | | | |
| NT | 0.12 | 0.22 | 0.22 | 0.27 | 1.00 | | | | | | | |
| NG | 0.77** | 0.93** | 0.99** | 0.74** | 0.28 | 1.00 | | | | | | |
| 1000 GW | 0.28 | 0.19 | 0.15 | 0.02 | −0.02 | 0.13 | 1.00 | | | | | |
| YP | 0.38** | 0.31* | 0.36** | 0.06 | 0.48** | 0.34* | 0.34* | 1.00 | | | | |
| NP | 0.11 | 0.20 | 0.22 | 0.25 | 0.99** | 0.28 | −0.01 | 0.48** | 1.00 | | | |
| PL | 0.46** | 0.52** | 0.45** | 0.53** | 0.14 | 0.51** | 0.25 | 0.20 | 0.15 | 1.00 | | |
| SP | −0.08 | −0.13 | −0.12 | −0.04 | −0.23 | −0.12 | −0.28* | −0.08 | −0.24 | −0.10 | 1.00 | |
| SF | 0.39** | 0.37** | 0.29* | 0.26 | 0.18 | 0.31* | 0.23 | 0.26 | 0.18 | 0.26 | −0.67** | 1.00 |

*Significantly at 0.05. **Significantly at 0.01. PH: plant height, DF: days to flowering, DM: days to maturing, FLR: flag Leaf length to width ratio, NT: number of tillers per hill, NG: number of grains per panicle, 1000 GW: one thousand grain weight, YP: yield per pot, NP: number of panicles per hill, PL: panicle length, SP: spikelet per panicle, and SF: spikelet fertility.

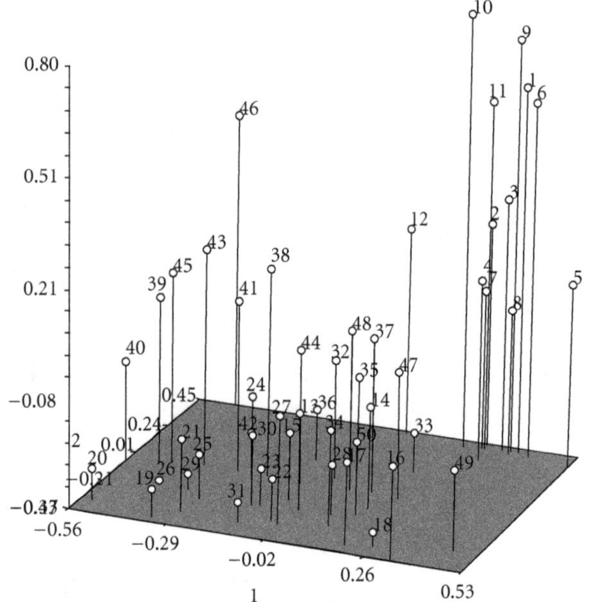

Figure 2: Three-dimensional graph of 50 upland rice accessions based on 12 quantitative traits.

07508, and it was clustered alone in cluster VI (Figure 2). But on the other hand, some accessions were close together in PCA such as accessions 06040, 06041, 06070, 06067, 06043, 06050, and 06059, whereas they were clustered into 2 groups (group I and II) in cluster analysis. According to PCA, the first four principal components accounted for around 76.7% of total variation of all morphological traits. The analysis of eigenvectors indicated the information of morphological traits for percentage of variation to the first four principal components, which were 36.4, 17.9, 12.8, and 9.6%, respectively (Table 9).

Table 7: Groups of upland rice accessions according to cluster analysis.

| Groups | Accessions |
|---|---|
| Group I | 06040, 06041, 06048, 06068, 06070, 06067 |
| Group II | 06043, 06050, 06044, 06059, 06071, 06045, 03835, 03824, 03834, 03838 |
| Group III | 07537, 07571, 07544, 07539, 07538, 07546, 07589, 07597, 07576,07574, 07595, 03825, 03830, 07531, 07535, 07585, 07534, 03831,07509, 07540, 07543, 07590, 07594, 07588, 07575,03832, 03833 |
| Group IV | 03826, 03837 |
| Group V | 07541, 07596, 03828, 07545 |
| Group VI | 07508 |

## 4. Discussion

All traits showed highly significant ($P < 0.01$) and significant ($P < 0.05$) variations among 50 accessions, which originated in Peninsular Malaysia and Sabah. Pandey et al. [18] recorded highly significant difference among 40 genotypes of rice with 12 quantitative traits. Wang et al. [19] observed 95% differentiation among 5 populations of rice by 20 morphological traits. Caldo et al. [20] measured highly significant difference ($P < 0.01$) in 41 morphological characteristics between 81 ancestors of rice and also CV ranged from 2.0% for grain length, grain width, and 1000 grains weight to 22.1% for culm number. Chandra et al. [21] and Abarshahr et al. [22] measured highly significant variation at 0.01 revealed by 14 and 19 quantitative traits among 57 accessions of upland rice and 30 genotypes of rice, respectively.

Correlation between traits is so important because it helps the breeder to select important characters from the studied traits. Most of the traits such as yield and yield component traits are influenced by interaction of genotype and environment, and, therefore, selection based on correlation coefficient makes it easy for plant breeders [23]. As

Table 8: Mean value of 12 quantitative traits for six groups by cluster analysis on 50 upland rice accessions.

| Group | PH | DF | DM | FLR | NT | NG | 1000 GW | YP | NP | PL | SP | SF |
|---|---|---|---|---|---|---|---|---|---|---|---|---|
| I | 147.97 | 112.87 | 144.00 | 31.12 | 4.50 | 134.36 | 25.99 | 52.42 | 4.40 | 30.06 | 148.30 | 90.70 |
| II | 142.81 | 88.73 | 121.85 | 19.26 | 3.53 | 141.97 | 26.32 | 48.40 | 3.49 | 27.79 | 164.55 | 86.84 |
| III | 132.04 | 77.65 | 102.40 | 19.59 | 3.71 | 116.66 | 23.71 | 39.10 | 3.68 | 26.97 | 153.28 | 76.68 |
| IV | 142.10 | 96.64 | 129.58 | 24.96 | 4.00 | 117.32 | 23.32 | 47.92 | 3.92 | 30.07 | 187.43 | 62.62 |
| V | 131.10 | 72.07 | 94.74 | 21.13 | 4.40 | 128.78 | 16.03 | 41.90 | 4.29 | 25.44 | 199.16 | 64.63 |
| VI | 128.06 | 76.89 | 92.28 | 17.31 | 4.75 | 107.03 | 33.03 | 55.13 | 4.67 | 22.20 | 111.70 | 95.86 |

PH: plant height, DF: days to flowering, DM: days to maturing, FLR: flag leaf length-to-width ratio, NT: number of tiller per hill NG: number of grains per panicle, 1000 GW: one thousand grain weight, YP: yield per pot, NP: number of panicles per hill, PL: panicle length, SP: spikelet per panicle, and SF: spikelet fertility.

Table 9: Eigenvectors and eigenvalues of the first four principal components.

| Variable | Eigenvectors | | | |
|---|---|---|---|---|
| | PC 1 | PC 2 | PC 3 | PC 4 |
| Eigenvalue | 4.37 | 2.14 | 1.53 | 1.14 |
| Variation (%) | 36.4 | 17.9 | 12.8 | 9.6 |
| Cumulative (%) | 36.4 | 54.5 | 67.1 | 76.7 |
| PH | 0.388 | −0.227 | −0.004 | −0.081 |
| DF | 0.416 | −0.179 | 0.076 | 0.167 |
| DM | 0.394 | −0.140 | 0.093 | 0.115 |
| FLR | 0.317 | −0.122 | 0.206 | 0.399 |
| NT | 0.227 | 0.541 | 0.269 | −0.014 |
| NG | 0.180 | −0.353 | 0.114 | −0.473 |
| 1000 GW | 0.152 | 0.007 | −0.454 | −0.288 |
| YP | 0.266 | 0.228 | 0.042 | −0.591 |
| NP | 0.224 | 0.549 | 0.259 | −0.017 |
| PL | 0.306 | −0.135 | 0.008 | 0.197 |
| SP | −0.141 | −0.303 | 0.607 | −0.303 |
| SF | 0.281 | 0.038 | −0.463 | −0.071 |

PH: plant height, DF: days to flowering, DM: days to maturing, FLR: flag leaf length to width ratio, NT: number of tillers per hill NG: number of grains per panicle, 1000 GW: one thousand grain weight, YP: yield per pot, NP: number of panicles per hill, PL: panicle length, SP: spikelet per panicle, and SF: spikelet fertility, and PC: principal components.

mentioned, in this assay yield of plant had positive correlation with 12 morphological traits. Lasalita-Zapico et al. [24] evaluated correlation coefficient of 10 quantitative traits for 32 upland rice varieties.In this distinguish significant positive correlation among the majority of the morphological traits was recorded except flag leaf angle that had negative correlation with most of characters such as panicle length, leaf length, leaf width, ligule length, leaf area, and culm length. Zafar et al. [25] recorded positive correlation coefficient of panicle length (yield component) with tillers of plant and 100 grains weight and also significantly positive correlation with grain length (0.278).

The computing of heritability and genetic advance useful for selection on phenotypic expression [17]. Therefore, high amount of heritability and genetic advance can be the base of selection according to morphological traits. In present study, flag leaf length-to-width ratio, spikelet fertility, yield of plant, and days to flowering indicated both high heritability and genetic advance. Thus, selection based on these traits would bring about improvement in the genotypes. In previous studies, Sedeek et al. [26] reported both high heritability and high genetic advance for days to heading, flag leaf area, number of filled grains per panicle, and grain yield per plant. The heritability ranged between 86% and 99.4%, and for genetic advance was ranged from 17.81% for number of panicles per plant to 46.16% for grain yield per plant among 24 of rice varieties. Laxuman et al. [27] recorded high heritability (more than 60%) and high genetic advance (more than 20%) for chlorophyll meter reading, number of productive tillers per plant, panicle weight and number of grains per panicle, and 1000 grain weight. Pandey et al. [18] recorded high broad sense heritability among 40 rice varieties for plant height (99.8%), biological yield (99.6%), harvest index (99%), test weight (98.8 g), number of panicles per hills (98.5%), number of spikelets per panicle (98.3%), and grain yield (98.11 g).

In our study, fifty accessions of the upland rice were clustered into six groups based on 12 quantitative traits. Ahmadikhah et al. [23] clustered 58 rice varieties into four groups based on 18 morphological traits, and genetic distance was around 0.75. Group A composed of only one member and groups B, C, and D contained 14, 20, and 23 members, respectively. Veasey et al. [28] computedclustering for 23 populations of rice by 20 morphological characteristics. So the varietieswere clustered into 10 groups of the last group was the biggest group with seven members and groups 1, 2, 7, and 8 were the smallest groups including only one variety. So, genotypes having distant clusters could be hybridized to get the higher heterotic responses. The similar studies were reported by several authors [29–31].

Principal component analysis indicated diversity among 50 accessions of upland rice by a few eigenvectors. In the present study, the first four principal components indicated 76.4% of total variationfor which PC1 showed 36.4% of the variation PC2, PC3, and PC4 explained 17.9%, 12.8%, and 9.6% of total variation, respectively. Lasalita-Zapico et al. [24] computed approximately 82.7% of total variation among 32 upland rice varieties, 66.9% variation for PC1 and 15.87% for PC2. Caldo et al. [20] recorded the first 10 principal components accounting for 67% of total variation. This suggested a strong correlation among characters being examined. Rajiv et al. [32] reported the first two principal components accounting for 82.1% of total variation in control and 68.6% in the stress-induced genotypes.

## 5. Conclusion

Fifty accessions of upland rice were clustered into six main groups. To achieve a wide spectrum of variation among the segregates, genotypes having distant cluster, group I (accessions 6040, 6041, 6048, 6068, 6070, and 6067) could be hybridized with group V (accessions 7541, 7596, 3828, and 7545) and group VI (accession 7508). Principal component analysis indicated 76.4% of the total variation. PCA and cluster analysis complemented each other with some slight inconsistencies in terms of cluster composition. Heritability is one of the most important factors in statistical analysis. Separation and selection of varieties based on high heritability of traits make it easy for breeders. Most researchers agree that high heritability alone is not enough; both high heritability and high genetic advance are needed. In this experiment, flag leaf length-to-width ratio, spikelet fertility, yield of plant, and days to flowering had high heritability and high genetic advance. Most traits such as plant height, yield of plant, panicle length, number of panicles, and days to flowering had positive correlations among each other, which suggested that utilization of these traitscould improve the genotype by selection of desirable varieties.

## Acknowledgment

The authors greatly acknowledge the Universiti Putra Malaysia for providing research facilities and financial support.

## References

[1] A. K. Tyagi, J. P. Khurana, P. Khurana et al., "Structural and functional analysis of rice genome," *Journal of Genetics*, vol. 83, no. 1, pp. 79–99, 2004.

[2] E. P. Guimarães, "Rice breeding," in *Cereals*, M. J. Carena, Ed., pp. 1–28, Springer, New York, NY, USA, 2009.

[3] M. Kondo, P. P. Pablico, D. V. Aragones et al., "Genotypic and environmental variations in root morphology in rice genotypes under upland field conditions," *Plant and Soil*, vol. 255, no. 1, pp. 189–200, 2003.

[4] N. D. Thanh, H. G. Zheng, N. V. Dong, L. N. Trinh, M. L. Ali, and H. T. Nguyen, "Genetic variation in root morphology and microsatellite DNA loci in upland rice (*Oryza sativa* L.) from Vietnam," *Euphytica*, vol. 105, no. 1, pp. 43–51, 1999.

[5] E. Reuveni, "The genetic background effect on domesticated species: a mouse evolutionary perspective," *TheScientificWorldJournal*, vol. 11, pp. 429–436, 2011.

[6] M. H. Musa, H. Azemi, A. S. Juraimi, and T. M. Mohamed, "Upland rice varieties in Malaysia: agronomic and soil physico-chemical characteristics," *Pertanika Journal of Tropical Agricultural Science*, vol. 32, no. 2, pp. 225–246, 2009.

[7] A. Melchinger, M. Messmer, M. Lee, W. Woodman, and K. Lamkey, "Diversity and relationships among US maize inbreds revealed by restriction fragment length polymorphisms," *Crop Science*, vol. 31, no. 3, pp. 669–678, 1991.

[8] R. K. Varshney, T. Thiel, T. Sretenovic-Rajicic et al., "Identification and validation of a core set of informative genic SSR and SNP markers for assaying functional diversity in barley," *Molecular Breeding*, vol. 22, no. 1, pp. 1–13, 2008.

[9] T. Sasaki, "The map-based sequence of the rice genome," *Nature*, vol. 436, no. 7052, pp. 793–800, 2005.

[10] S. R. McCouch, G. Kochert, Z. H. Yu et al., "Molecular mapping of rice chromosomes," *Theoretical and Applied Genetics*, vol. 76, no. 6, pp. 815–829, 1988.

[11] Z. Y. Wang, G. Second, and S. D. Tanksley, "Polymorphism and phylogenetic relationships among species in the genus Oryza as determined by analysis of nuclear RFLPs," *Theoretical and Applied Genetics*, vol. 83, no. 5, pp. 565–581, 1995.

[12] M. A. Latif, M. M. Rahman, M. S. Kabir, M. A. Ali, M. T. Islam, and M. Y. Rafii, "Genetic diversity analyzed by quantitative traits among rice (*Oryza sativa* L.) genotypes resistant to blast disease," *African Journal of Microbiology Research*, vol. 5, no. 25, pp. 4383–4391, 2011.

[13] L. S. Kumar, "DNA markers in plant improvement: an overview," *Biotechnology Advances*, vol. 17, no. 2-3, pp. 143–182, 1999.

[14] R. Din, M. Y. Khan, M. Akmal et al., "Linkage of morphological markers in Brassica," *Pakistan Journal of Botany*, vol. 42, no. 5, pp. 2995–3000, 2010.

[15] G. W. Burton, "Quantitative inheritance in grasses," in *Proceedings of the International Grassland Congress*, vol. 1, pp. 277–283, 1952.

[16] G. W. Burton and E. De Vane, "Estimating heritability in tall fescue (*Festuca arundinacea*) from replicated clonal material," *Agronomy Journal*, vol. 45, no. 10, pp. 478–481, 1953.

[17] H. W. Johnson, H. Robinson, and R. Comstock, "Estimates of genetic and environmental variability in soybeans," *Agronomy Journal*, vol. 47, pp. 314–318, 1955.

[18] P. Pandey, P. John Anurag, D. K. Tiwari, S. K. Yadav, and B. Kumar, "Genetic variability, diversity and association of quantitative traits with grain yield in rice (*Oryza sativa* L.)," *Journal of Bio-Science*, vol. 17, no. 1, pp. 77–82, 2009.

[19] J. L. Wang, Y. B. Gao, N. X. Zhao et al., "Morphological and RAPD analysis of the dominant species Stipa krylovii Roshev. in Inner Mongolia steppe," *Botanical Studies*, vol. 47, no. 1, pp. 23–35, 2006.

[20] R. Caldo, L. Sebastian, and J. Hernandez, "Morphology-based genetic diversity analysis of ancestral lines of Philippine rice cultivars," *Philippine Journal of Crop Science*, vol. 21, no. 3, pp. 86–92, 1996.

[21] R. Chandra, S. Pradhan, S. Singh, L. Bose, and O. Singh, "Multivariate analysis in upland rice genotypes," *World Journal of Agricultural Sciences*, vol. 3, no. 3, pp. 295–300, 2007.

[22] M. Abarshahr, B. Rabiei, and H. S. Lahigi, "Assessing genetic diversity of rice varieties under drought stress conditions," *Notulae Scientia Biologicae*, vol. 3, no. 1, pp. 114–123, 2011.

[23] A. Ahmadikhah, S. Nasrollanejad, and O. Alishah, "Quantitative studies for investigating variation and its effect on heterosis of rice," *International Journal of Plant Production*, vol. 2, no. 4, pp. 297–308, 2008.

[24] F. C. Lasalita-Zapico, J. A. Namocatcat, and J. L. Cariño-Turner, "Genetic diversity analysis of traditional upland rice cultivars in Kihan, Malapatan, Sarangani Province, Philippines using morphometric markers," *Philippine Journal of Science*, vol. 139, no. 2, pp. 177–180, 2010.

[25] N. Zafar, S. Aziz, and S. Masood, "Phenotypic divergence for agro-morphological traits among landrace genotypes of rice (*Oryza sativa* L.) from Pakistan," *International Journal of Biological Sciences*, vol. 6, no. 2, pp. 335–339, 2006.

[26] S. Sedeek, S. Hammoud, M. Ammar, and T. Metwally, "Genetic variability, heritability, genetic advance and cluster analysis for for some physiological traits and grain yield nad its components in rice (*Oryza sativa* L.)," *Journal of Agricultural Sciences*, vol. 35, no. 3, pp. 858–878, 2009.

[27] L. Laxuman, P. Salimath, H. Shashidhar et al., "Analysis of genetics variability in interspecific backcross inbred lines in rice (*Oryza sativa* L.)," *Karnataka Journal of Agricultural Sciences*, vol. 23, no. 4, pp. 563–565, 2010.

[28] E. A. Veasey, E. F. Da Silva, E. A. Schammass, G. C. X. Oliveira, and A. Ando, "Morphoagronomic genetic diversity in American wild rice species," *Brazilian Archives of Biology and Technology*, vol. 51, no. 1, pp. 95–104, 2008.

[29] M. A. Latif, M. Rafii Yusop, M. Motiur Rahman, and M. R. Bashar Talukdar, "Microsatellite and minisatellite markers based DNA fingerprinting and genetic diversity of blast and ufra resistant genotypes," *Comptes Rendus Biologies*, vol. 334, no. 4, pp. 282–289, 2011.

[30] M. A. Latif, M. M. Rahman, M. S. Kabir, M. A. Ali, M. T. Islam, and M. Y. Rafii, "Genetic diversity analyzed by quantitative traits among rice (*Oryza sativa* L.) genotypes resistant to blast disease," *African Journal of Microbiology Research*, vol. 5, no. 25, pp. 4383–4391, 2011.

[31] N. Abdullah, M. Rafii Yusop, M. Ithnin, G. Saleh, and M. A. Latif, "Genetic variability of oil palm parental genotypes and performance of its' progenies as revealed by molecular markers and quantitative traits," *Comptes Rendus Biologies*, vol. 334, no. 4, pp. 290–299, 2011.

[32] S. Rajiv, P. Thivendran, and S. Deivanai, "Genetic divergence of rice on some morphological and physichemical responces to drought stress," *Pertinaka Journal*, vol. 32, no. 2, pp. 315–328, 2010.

# Evaluation of the Utility of the Random Amplified Polymorphic DNA Method and of the Semi-Specific PCR to Assess the Genetic Diversification of the *Gerbera jamesonii* Bolus Line

**Zbigniew Rusinowski and Olga Domeradzka**

*Department of Ornamental Plants, Warsaw School of Life Sciences, Faculty of Horticulture and Landscape Architecture, Warsaw University of Life Sciences, Nowoursynowska 159, 02776 Warsaw, Poland*

Correspondence should be addressed to Zbigniew Rusinowski, zbyszek@rusinowski.pl

Academic Editor: Kaye Spratt

An attempt was made to evaluate the utility of a method which employs semi-specific PCR using partially specific primers for the coding sequence (ET) at the exon-intron contact and of the RAPD method to identify eight Polish cultivars of gerbera. It was demonstrated that the PCR method which employs semi-specific primers is as simple and economical as the RAPD method, simultaneously the images of the multiplied by means of the semi-specific PCR method DNA fragments are more complex and polymorphic than those obtained through the RAPD method. The studies of the genetic diversification of *Gerbera* cultivars employing the aforementioned methods made it possible to conduct a concentration analysis and evaluation of the genetic distance between the lines, manifesting at the same time the superiority of the semi-random PCR method. Moreover, it transpired that the use of mixtures of RAPD primers not always leads to an increase of the number of generated polymorphic bands.

## 1. Introduction

At the turn of the 21st century, identification of cultivars and protection of breeders' rights became an important issue, especially when it is impossible to tell them apart by their morphologic features such as, for example, the shape of seeds, bulbs, or cuttings [1]. It frequently occurs that cultivars, phenotypically similar, for example, whose flowers are of the same color stem from different breeding companies, and may be confused. Thanks to modern methods based on DNA analysis techniques allow to identify and select plants bearing desired features [2]. The current methods quickly enable us to profoundly analyze of consanguinity and pedigree and to determine the value of breeding material [3]. RAPD is one of the methods recommended to identify the cultivars of *Gerbera jamesonii* [4, 5].

*Gerbera* has the current methods quickly allows for analysis of an important position in the floral market for several years. Each year new cultivars are introduced; they are to attract customers not only with their color but also with their shape and the size of their inflorescence and to attract growers with higher yield and tolerance to growing conditions. The preferences of Polish customers as for cut flowers have changed in recent years, and, after a short-termed stagnation which affected the gerbera market, we may witness a slow but significant rise interest among those who purchase these species.

The present paper attempts to compare the usability of the RAPD method and semi-specific PCR with the use of partially specific primers for the coding sequence lying on the exon-intron junction, never hitherto applied in the case of the gerbera cultivar to identify eight Polish cultivars bred by the Pętoś company.

## 2. Materials and Methods

The experiment was conducted in the years 2004–2006 at the Faculty of Horticulture and Landscape Architecture at the Warsaw University of Life Sciences (*Wydział Ogrodnictwa* i *Architektury Krajobrazu SGGW*) in Warsaw. The material

Table 1: Primers used to evaluate the consanguinity of the *G. jamesonii* cultivars with the nucleotide sequence, total bands number, number of differentiating bands, and the contribution of differentiating bands expressed as a percentage.

| Primer code | Nucleotide sequence of the primers | Total bands number | Contribution of differentiating bands (%) |
|---|---|---|---|
| A2 | 5′-TGCCGAGCTG-3′ | 5 | 40.0% |
| C11 | 5′-AAAGCTGCGG-3′ | 10 | 20.0% |
| D5 | 5′-TGAGCGGACA-3′ | 7 | 28.6% |
| D8 | 5′-GTGTGCCCCA-3′ | 8 | 25.0% |
| D11 | 5′-AGCGCCATTG-3′ | 7 | 28.6% |
| G12 | 5′-CAGCTCACGC-3′ | 9 | 33.3% |

was delivered by the *Pętoś "Specjalistyczny Zakład Ogrodniczy Bartoszyce"* company. Eight new Polish plant cultivars raised by the aforementioned company were used (*Amelia, Bartoszyce, Bolesławiec, Delfin, Kraków, Kreta, Safona, Samuraj*).

To isolate the DNA young leaves were used as source. Two DNA isolation methods were employed. The isolation by means of Genomic Mini AX plant kits from the A&A Biotechnology company and the cTAB procedure after Murray and Thompson [6].

In the RAPD experiments, the amplification program consisted of 40 m cycles; initial denaturation 72°-5′, denaturation 94°-1′, attachment 39°-1′, final temperature 72°-2′ [7]. The composition of the amplification was performed in two-stages mixture PCR whose total volume equaled 25 $\mu$L 15.95 $\mu$L of water dd (miliQ), 2 $\mu$L of $MgCl^2$, 2.5 $\mu$L of buffer 10x, 1.25 $\mu$L of dNTP, 1 $\mu$L of primer, 0.3 $\mu$L of TAQ polymerase DNA, and 2 $\mu$L of tested DNA (approx. 10 ng/$\mu$L).

In the case of semi-specific primers, the DNA amplification was two stage. At the first stage for the 15 nucleotide primers in comparison to the 18 nucleotide primers, the primer attachment temperature was by 10° higher (50–60°) at the second stage it was by 10° lower (64–54°) the attachment time was 1 minute. In both cases this amplification program was used: denaturation 94°-40″, lengthening of the DNA chains in 72°-2′, multiplication completed in 72°-5′, at the second stage next denaturation followed in 94°-40″, lengthening of the DNA chains in 72°-2′, and final multiplication 72°-10 [8]. The composition of the mixture of the semispecific PCR by Rafalski [8], of which volume requaled 20 $\mu$L, 10 $\mu$L of water dd (miliQ), 2 $\mu$L of $MgCl_2$, 2 $\mu$L of buffer 10x with ammonium sulfate, 0.5 $\mu$L of dNTP, 3 $\mu$L of primer, 1 TAO unit of DNA polymerase, and 2 $\mu$L of the tested DNA (approx. 10 ng/$\mu$L). Reagents from MBI Fermentas were used, with the exception of dNTP bought from Invitrogen. The electrophoretic separation was conducted on 1.5% agarose gel in TAE 1x buffer with addition of 10 $\mu$L 0.01% of ethidium bromide. The visualization took place in the UV light. The experiment was documented with the help of a monochromatic camera and software by Biometra BioDell.

The genetic similarity of the analyzed cultivars was computed by means of the Nei and Li formula [9] of the mean UPGMA connections. The concentration analysis was performed with the use of the NTSYS-pc program, v. 2.1.

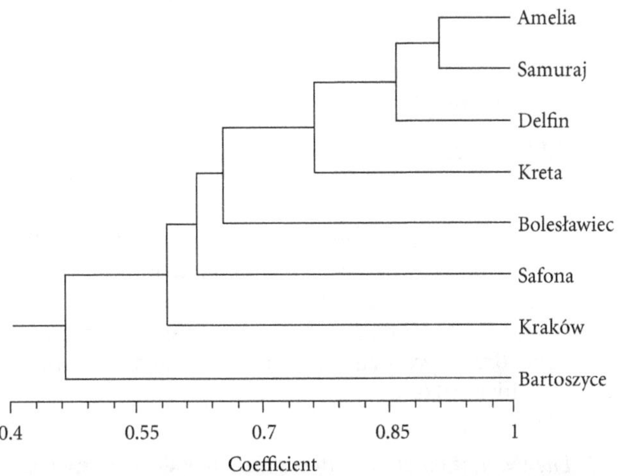


Figure 1: Similarity diagram of the tested *G. jamesonii* cultivars comuted from the mean connections method (UPGMA) and from a RAPD analysis of 19 DNA fragments.

## 3. Results

*3.1. The Diversification Characteristics of Eight G. jamesonii Cultivars by Means of Selected RAPD Primers.* The usability of twenty 10-nucleotide RAPD primers prepared according to the operon nomenclature in the DNA Sequence-forming Laboratory at the Institute of Biochemistry and Biophysics *PAN* was tested.

The RAPD technique was used to differentiate eight gerbera cultivars. For the purpose of the study, 6 out of the 20 tested primers were chosen: those which generated the highest total number of bands and the highest number of differentiating bands (Table 1).

The primer which gave the highest percentage of differentiating bands (40%) generating the lowest number of total bands was the A2 primer. Primer C11, which generated the highest number of bands (10), gave relatively small, equals 20%, contribution of differentiating bands. The other primers rendered on average 28.9% polymorphic bands (Figure 3).

The evaluation of genetic similarity of the cultivars was based on the analysis of 19 DNA fragments obtained from experiments using the RAPD method (Figure 1). The genetic distance coefficients between the tested lines were computed

Table 2: Primers used to evaluate the consanguinity among the cultivars of *G. jamesonii* with the nucleotide sequence, total bands number and the fraction of polymorphic bands in percentage.

| Primer code | Total bands number | Participation of differentiating polymorphic bands (%) |
|---|---|---|
| ET 1/18 | 6 | 66.6% |
| ET 2/18 | 11 | 90.9% |
| ET 3/18 | 4 | 25% |
| ET 4/18 | 4 | 75% |
| ET 5/18 | 9 | 33.3% |
| ET 6/18 | 18 | 100% |
| ET 31/15 | 0 | 0 |
| ET 32/15 | 10 | 80% |
| ET 33/15 | 3 | 0 |
| ET 34/15 | 6 | 50% |
| ET 35/15 | 9 | 55.5% |
| ET 36/15 | 8 | 62.5% |

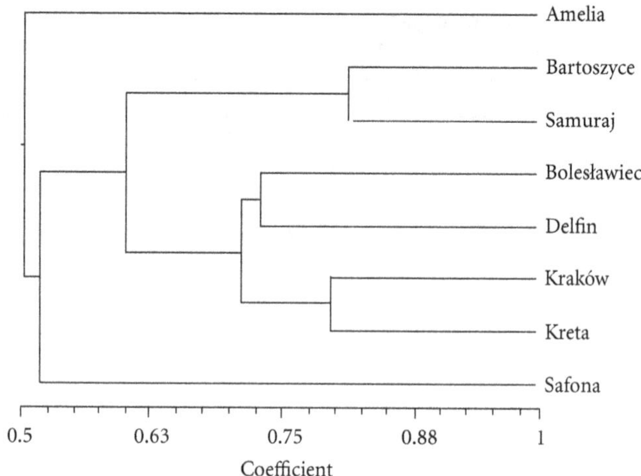


Figure 2: Similarity diagram of the *G. jamesonii* cultivars computed from the mean connections method (UPGMA) and from an analysis of 59 DNA fragments obtained by means of semi-specific PCR.

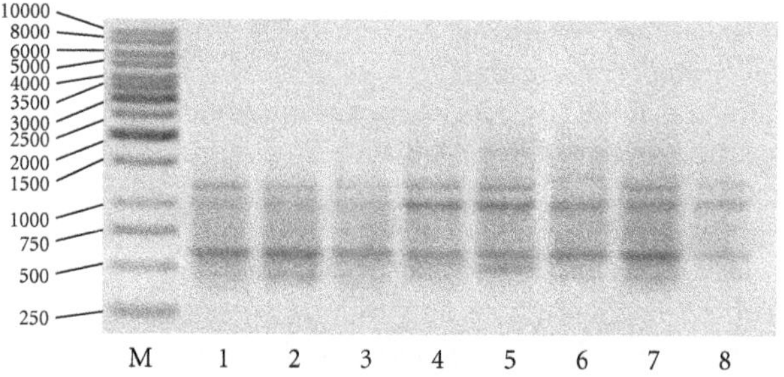


Figure 3: The amplification profile generated with the use of D11, from the left: marker (M) of the size 1 kb, *G. jamesonii* cultivars: 1 *Amelia, 2 Bartoszyce,* 3 *Bolesławiec,* 4 *Delfin,* 5 *Kraków,* 6 *Kreta,* 7 *Safona, and* 8 *Samuraj.*

from the RAPD experiments and ranged from 0.09 to 0.62 (the average 0.37). It had been computed that the minimum genetic distance is between the *Amelia* and *Samuraj* cultivars (0.09), and the maximum distance separates the *Bolesławiec* from the *Bartoszyce* cultivar and the *Bartoszyce* from the *Safona* cultivar (0.62).

*3.2. The Evaluation of the Differentiation of Gerbera sp. Cultivars by Means of Semi-Specific PCR.* To distinguish the eight *G. jamesonii* cultivars, the utility of 12 semi-conservative primers was evaluated; those primers were partially specific for the coding fragment at the 15- and 18-nucleotide exon-intron junction (ET) designed by Dr. Andrzej Rafalski (Table 2) and prepared at the Institute of Biochemistry and Biophysics of the Polish Academy of Science (Instytut Biochemii i Biofizyki PAN).

Primer ET 6/18 (Figure 4) yielded the highest total number of bands and included the majority of differentiating bands. Next was the ET 2/18 primer which generated 11 bands in all, 90.9% of which were differentiating. The other 18-nucleotide primers would generate from 4 to 9 bands of which 49.9% were polymorphic. In the 15-nucleotide group, primer ET 32/15 (Figure 5) yielded the highest total number of bands (10). Except for ET 31/15 (0), the remaining primers generated from 3 to 9 bands with the average of 35.1% of polymorphic bands.

The evaluation of genetic similarity of the cultivars was based on the analysis of 59 DNA fragments obtained from the experiments using semi-specific PCR (Figure 2). The genetic distance coefficients between the tested lines were computed from the experiments using semi-specific PCR and ranged from 0.19 to 0.58 (average 0.40).

## 4. Discussion

Similar morphological features of the gerbera cultivars do permit to identify the varieties as the features are too similar between the varieties. Thus, methods based on molecular biology are being more frequently applied, for example, AFLP [10–12], SSR [13], RAPD [4, 14], and semi-specific PCR [8]. Methods such as AFLP and SSR require expensive equipment and reagents [10–13]. The here employed RAPD and semi-specific PCR methods presented in this study are relatively cheaper than the aforementioned alternatives, which may predispose them to being applied in cultivar identification, especially considering the cost and feasibility of analysis [8].

The conducted analyses of RAPD with the use of DNA from eight Polish cultivars of gerbera (*Amelia, Bartoszyce, Bolesławiec, Delfin, Kraków, Kreta, Safona, Samuraj*) stood out as generating few amplification products. In the present work 70 per cent of the used primers generated bands. Polymorphic bands were generated by 60 percent of the used-in-the-tests primers. Rezende et al. [4] carried out research on *G. jamesonii* testing 31 primers, of which 21 generated polymorphic bands (68%). Similar studies were conducted on gerbera by Chung et al. [7], and the elicited results showed that merely 36 of the 80 primers (45%) would generate polymorphic bands.

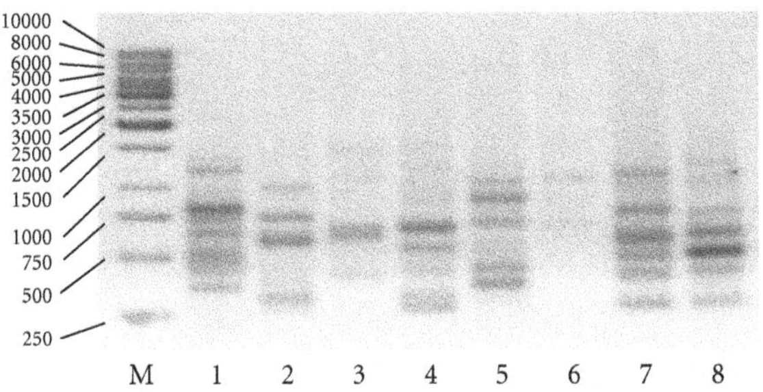


FIGURE 4: Amplification profile generated with the use of the primers ET 6/18, from the left: marker (M) of the size 1 kb, *G. jamesonii* cultivars: 1 *Amelia*, 2 *Bartoszyce*, 3 *Bolesławiec*, 4 *Delfin*, 5 *Kraków*, 6 *Kreta*, 7 *Safona*, *and* 8 *Samuraj*.

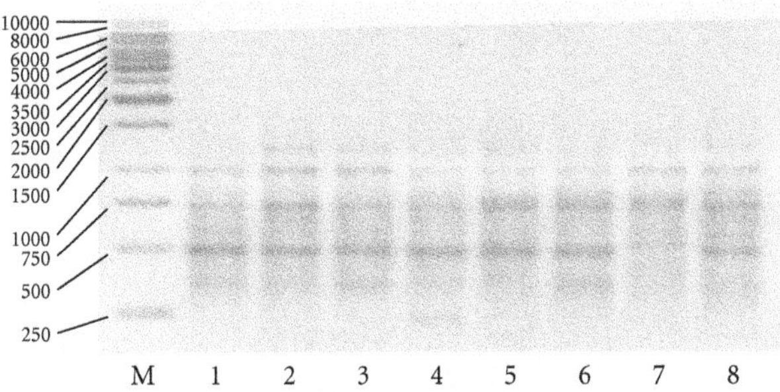


FIGURE 5: Amplification profile generated with the use of the primers ET 32/15, from the left: marker (M) of the size 1 kb, *G. jamesonii* cultivars: 1 *Amelia*, 2 *Bartoszyce*, 3 *Bolesławiec*, 4 *Delfin*, 5 *Kraków*, 6 *Kreta*, 7 *Safona*, *and* 8 *Samuraj*.

Although during the tests of genetic differentiation of gerbera cultivars different RAPD primers generated polymorphic bands, the use of a single primers to identify one lineage proved insufficient, as differentiating bands would show up for several cultivars simultaneously, and to distinguish them other primers had to be used. A similar result was yielded by Rafalski [8] when testing corn lineages.

The method utilizing semi-specific PCR has never been used in research on *Gerbera* sp. The use of partially specific primers is a relatively new method of investigating the DNA polymorphism. The suggested system is equally simple and quick as the RAPD method. Simultaneously the semi specific primers generate much more complex images of the amplification profiles in comparison to the RAPD method [8, 15].

When using the semi-specific PCR with the material from the eight *gerbera* cultivars (*Amelia, Bartoszyce, Bolesławiec, Delfin, Kraków, Kreta, Safona, Samuraj*) (Table 2). a polymorphism reaching 100% was observed, which means that one primer sufficed to distinguish all eight cultivars. The highest number of bands was generated by the 18-nucleotide primers of the ET group, similar results were yielded by Rafalski [8] who applied 15- and 18-nucoeotide primers of the ET group. Most of the used in the present work primers would generate 20 to 30 DNA fragments.

The yielded results served to compute the coefficients of genetic distance between the cultivars. Diagram of Euclidean distances is shown in Figures 1 and 2. The coefficients of genetic distance of the tested gerbera lines computed from the results obtained by means of semi-specific primers of the ET group ranged from 0.19 to 0.58, although the coefficients of genetic distance between the lines computed from experiments conducted with the use of the RAPD technique ranged from 0.09 to 0.62. Also in the Rafalski tests carried out on the corn lines [8], higher method sensitivity was achieved by means of semi-specific PCR in comparison to the RAPD technique. In the research on corn, the genetic distance computed from RAPD experiments was 0.19 on average, and 0.38 from multiplications by means of the semi-specific primers. The authors of this paper favor the thesis propounded by Rafalski, in which the PCR system using the semiconservative primers enable a precise measurement of the genetic distance between the lines manifesting significant consanguinity.

## Acknowledgment

One of the authors deeply grateful to Dr. A. Rafalski for his valuable assistance in the semi-specific PCR application and for donation of primers.

## References

[1] T. Orlikowska, "Metody biotechnologiczne w kwiaciarstwie," *Zeszyty Naukowe Instytutu Sadownictwa i Kwiaciarstwa*, vol. 7, pp. 69–78, 2000.

[2] B. Michalik, "Wykorzystanie biotechnologii w hodowli roślin. Hodowla roślin ogrodniczych u progu XX wieku," *Materiały VIII Ogólnopolskiego Zjazdu Naukowego Hodowców Roślin Ogrodniczych*, pp. 5–10, 1999.

[3] D. Grzebelus, "Zastosowanie metod biotechnologicznych w hodowli roślin," in *Drukrol*, B. Michalik, Ed., pp. 99–115, S.C. w Krakowie, 1996.

[4] R. K. S. Rezende, L. V. Paiva, R. Paiva, A. Chalfun Junior, P. P. Torga, and T. E. Masetto, "Genetic divergence among cultivars of gerbera using RAPD markers," *Ciencia Rural*, vol. 39, no. 8, pp. 2435–2440, 2009.

[5] T. L. D. Mata, M. I. Segeren, A. S. Fonseca, and C. A. Colombo, "Genetic divergence among gerbera accessions evaluated by RAPD," *Scientia Horticulturae*, vol. 121, no. 1, pp. 92–96, 2009.

[6] M. G. Murray and W. F. Thompson, "Rapid isolation of high molecular weight plant DNA," *Nucleic Acids Research*, vol. 8, no. 19, pp. 4321–4326, 1980.

[7] Y. Chung, H. Kim, K. Kim et al., "Morphological characteristics and genetic variation of gerbera (Gerbera hybrida hort.)," *Journal of Plant Biotechnology*, vol. 3, pp. 145–149, 2001.

[8] A. Rafalski,, "Semispecyficzny PCR w badaniach genetyczno-hodowlanych roślin," in *Monografie i rozprawy naukowe IHAR Radzików*, vol. 23, pp. 1–81, 2004.

[9] M. Nei and W. H. Li, "Mathematical model for studying genetic variation in terms of restriction endonucleases," *Proceedings of the National Academy of Sciences of the United States of America*, vol. 76, no. 10, pp. 5269–5273, 1979.

[10] S. Khan and W. Spoor, "Use of molecular and morphological markers as a quality control in plant tissue culture," *Pakistan Journal of Biological Sciences*, vol. 4, pp. 479–482, 2001.

[11] J. De Riek, "Are molecular markers strengthening plant variety registration and protection?" *Acta Horticulturae (ISHS)*, vol. 552, pp. 215–223, 2001.

[12] J. De Riek, M. Martens, J. Dendauw et al., "The use of fluorescent AFLP to assess genetic conformity of a breeder's collection of R. simsii hybrids," *Acta Horticulturae*, vol. 508, pp. 99–104, 2000.

[13] S. Rajapakse, D. H. Byrne, L. Zhang, N. Anderson, K. Arumuganathan, and R. E. Ballard, "Two genetic linkage maps of tetraploid roses," *Theoretical and Applied Genetics*, vol. 103, no. 4, pp. 575–583, 2001.

[14] K. Hamada and M. Hagimori, "RAPD-based method for cultivar-identification of calla lily (Zantedeschia spp.)," *Scientia Horticulturae*, vol. 65, no. 2-3, pp. 215–218, 1996.

[15] M. Gaweł, I. Wiśniewska, and A. Rafalski, "Semi-specific PCR for the evaluation of diversity among cultivars of wheat and triticale," *Cellular and Molecular Biology Letters*, vol. 7, no. 2, pp. 577–582, 2002.

# 12

# Genetics and Epigenetics of Parkinson's Disease

**Fabio Coppedè[1, 2]**

[1] Faculty of Medicine, University of Pisa, 56126 Pisa, Italy
[2] Genetics and Epigenetics of Complex Disease Program, Department of Neuroscience (DAI Neuroscience), Pisa University Hospital, Via S. Giuseppe 22, 56126 Pisa, Italy

Correspondence should be addressed to Fabio Coppedè, f.coppede@geog.unipi.it

Academic Editors: H. Cui and M. Hiltunen

In 1997 a mutation in the *a-synuclein* (*SNCA*) gene was associated with familial autosomal dominant Parkinson's disease (PD). Since then, several loci (PARK1-15) and genes have been linked to familial forms of the disease. There is now sufficient evidence that six of the so far identified genes at PARK loci (*a-synuclein, leucine-rich repeat kinase 2, parkin, PTEN-induced putative kinase 1, DJ-1,* and *ATP13A2*) cause inherited forms of typical PD or parkinsonian syndromes. Other genes at non-PARK loci (*MAPT, SCA1, SCA2, spatacsin, POLG1*) cause syndromes with parkinsonism as one of the symptoms. The majority of PD cases are however sporadic "idiopathic" forms, and the recent application of genome-wide screening revealed almost 20 genes that might contribute to disease risk. In addition, increasing evidence suggests that epigenetic mechanisms, such as DNA methylation, histone modifications, and small RNA-mediated mechanisms, could regulate the expression of PD-related genes.

## 1. Genetics of Parkinson's Disease (PD)

Parkinson's disease (PD) is a common neurodegenerative disorder affecting 1-2% of the population over the age of 65 years and reaching a prevalence of almost 4% in those aged above 85 years. Resting tremor, rigidity, bradykinesia, and postural instability are the main clinical symptoms of the disease often accompanied by nonmotor symptoms including autonomic insufficiency, cognitive impairment, and sleep disorders. The brain of PD individuals is pathologically characterized by a progressive and profound loss of neuromelanin containing dopaminergic neurons in the *substantia nigra* with the presence of eosinophilic, intracytoplasmic inclusions termed as Lewy bodies (LBs: containing aggregates of $\alpha$-synuclein as well as other substances), and Lewy neurites in surviving neurons. Unfortunately, only some improvements of the symptoms are offered by current treatments based on levodopa and dopaminergic therapy, but there is no currently available treatment to arrest the progression of the disease [1].

A familial history of PD is shown in approximately 20% of the cases, and in a minority of them the disease follows Mendelian inheritance patterns. Studies in PD families led to the identification of 15 PD loci (PARK1-15), and 11 genes for PARK loci have so far been described (Table 1). Although follow-up genetic studies have been inconsistent for some of them or conclusive data are still pending, there is evidence that five of those genes (*a-synuclein, parkin, PTEN-induced putative kinase 1, DJ-1,* and *leucine-rich repeat kinase 2*) cause typical PD [2], and mutations of *ATP13A2* (PARK9) cause Kufor-Rakeb disease, an autosomal recessive parkinsonism with many other features, including pyramidal tract dysfunction, supranuclear gaze paresis, and dementia [3]. The vast majority of PD cases are however sporadic (idiopathic) forms, likely resulting from a combination of polygenic inheritance, environmental exposures, and complex gene-environment interactions superimposed on slow and sustained neuronal dysfunction due to aging [4]. Therefore, linkage analyses in PD families have been paralleled by several hundreds of genetic association studies in order to identify genetic variants able to modify the individual risk to develop idiopathic forms of the disease. These studies have been conducted by either the candidate gene approach or, more recently, by means of genome-wide screenings, that is, genome-wide association studies (GWASs). The genetic screening has been successful with a common high-risk locus

TABLE 1: Loci and genes associated with familial PD.

| Designation | Locus | Gene | Inheritance* | Refs |
|---|---|---|---|---|
| *Validated loci* | | | | |
| PARK1/PARK4 | 4q21 | *SNCA* | AD | [5–16] |
| PARK8 | 12q12 | *LRRK2* | AD | |
| PARK2 | 6q25.2–q27 | *PARK* | AR | |
| PARK6 | 1p35-36 | *PINK1* | AR | |
| PARK7 | 1p36 | *DJ-1* | AR | |
| PARK9 | 1p36 | *ATP13A2* | AR | |
| *Other loci* | | | | |
| PARK3 | 2p13 | Unknown | AD | [17–23] |
| PARK5 | 4p14 | *UCH-L1* | AD | |
| PARK8 | 12q12 | *LRRK2* | AD | |
| PARK10 | 1p32 | Unknown | Not clear | |
| PARK11 | 2q36-37 | *GIGYF2* | AD | |
| PARK12 | Xq21-q25 | Unknown | X-linked | |
| PARK13 | 2p12 | *OMI/ HTRA2* | AD | |
| PARK 14 | 22q13.1 | *PLA2G6* | AR | |
| PARK 15 | 22q11.2-qter | *FBXO7* | AR | |

*AD: autosomal dominant; AR: autosomal recessive.

identified (*GBA*), and many common low-risk loci (*SNCA, MAPT, LRRK2*) were recently elucidated [24]. Moreover, the application of GWAS technology and the creation of international collaborative groups sharing samples and results are continuously revealing novel variants that could contribute to disease risk [25].

The last few years have seen the growing of evidence dealing with the possible contribution of epigenetic modifications to human diseases, including among others neurodegenerative ones [26–28]. The term "epigenetics" is used to describe those mechanisms able to modify the expression levels of selected genes without necessarily altering their DNA sequence, including DNA methylation, histone tail modifications, and chromatin remodeling, as well as mechanisms mediated by small RNA molecules. Epigenetic modifications are often environmentally induced, and tissue-specific phenomena that can have similar effects to those of pathogenic mutations or functional polymorphisms, since they are able to silence, increase or reduce the expression of a selected gene in a given tissue [29]. This paper aims at describing the current knowledge on genetic alterations and epigenetic modifications likely contributing to PD pathogenesis.

## 2. Familial Parkinson's Disease

There is sufficient evidence that six of the so far identified genes at PARK loci cause inherited forms of typical PD or parkinsonian syndromes. Only two of them (*a-synuclein* and *leucine-rich repeat kinase 2*) cause autosomal dominant forms, whilst the other four (*parkin, PTEN-induced putative kinase 1, DJ-1,* and *ATP13A2*) are inherited in an autosomal recessive fashion. Additional PARK loci have been so far described (Table 1), pending identification of a causative gene or confirmation/validation of the candidate one. Moreover, in addition to the genes mapped to the sequentially numbered PARK1-15 loci, other genes, such as *MAPT, ATXN2, ATXN3, spatacsin,* and *POLG1* (Table 2), are known to cause syndromes that can clinically manifest with parkinsonism as one of the symptoms [24].

TABLE 2: Additional genes causing parkinsonism.

| Gene | Disease | Refs |
|---|---|---|
| *MAPT* | Frontotemporal dementia with parkinsonism linked to chromosome 17 | [30–38] |
| *ATXN2* | Spinocerebellar ataxia type 2 and parkinsonism | |
| *ATXN3* | Spinocerebellar ataxia type 3 and parkinsonism | |
| *SPG11* | Hereditary spastic paraplegia and parkinsonism | |
| *POLG1* | Mitochondrial parkinsonism | |

### *2.1. Autosomal Dominant PD*

*2.1.1. α-Synuclein (SNCA): PARK1 and PARK4.* In 1997 Polymeropoulos et al. [5] described a mutation in the *a-synuclein* gene (*SNCA*) on 4q21 (PARK1), causing an A53T amino acidic substitution and segregating with PD in an Italian kindred and 3 unrelated families of Greek origin [5]. Two additional *SNCA* mutations (A30P and E46K) were subsequently described in other families with autosomal dominant PD [6, 7]. Moreover, a triplication of the *a-synuclein* gene (PARK4) was observed in a large family as causative of PD [8], and other PD families have been described with *a-synuclein* gene duplication and a disease course less severe of that observed in PARK4 carriers, suggesting the existence of a gene dosage effect [9]. Although *SNCA* has been the first PD gene identified, *SNCA* mutations and multiplications are both extremely rare causes of familial autosomal dominant PD [2]. α-Synuclein is expressed throughout the mammalian brain particularly in presynaptic nerve terminals, and mutated α-synuclein has an increased tendency to form aggregates critical to Lewy bodies (LBs) formation. These fibrillar aggregates are the major component of LBs in both familial and idiopathic PD, and aggregation of α-synuclein is thought to be a key event in dopaminergic neuronal cell death. The function of α-synuclein under normal physiological conditions is not yet completely clear, although there is evidence that implicates SNCA in neurotransmitter release and vesicle turnover at the presynaptic terminals [39, 40].

*2.1.2. Leucine-Rich Repeat Kinase 2 (LRRK2): PARK8.* The *LRRK2* gene was mapped on the PARK8 locus in 12q12 and was the second gene linked to autosomal dominant PD [10, 11]. Over 100 *LRRK2* mutations have been so far described in PD families and sporadic cases, but the pathogenic role of many of them has not yet been proven (a complete list can be found at: http://www.molgen.ua.ac.be/PDmutDB/). *LRRK2* encodes a protein named dardarin (from *dardara*, the Basque word for tremor) which contains several domains including the catalytic domain of a tyrosine kinase. The presence of a kinase domain suggests a role for dardarin in signaling cascades, likely relating to cytoskeletal dynamics [24]. The most prevalent *LRRK2* mutation is a G2019S missense mutation occurring in 1-2% of PD patients of European origin, 20% of Ashkenazi Jewish patients, and approximately 40% of Arab Berbers with PD. The Arg1441 codon is another frequent hotspot of *LRRK2* pathogenic mutations [2].

### 2.2. Autosomal Recessive PD

*2.2.1. Parkin: PARK2.* Autosomal recessive juvenile parkinsonism (AR-JP) is caused by mutations of the *parkin* gene on chromosome 6q25.2–q27 (PARK2) [12]. The disease is characterized by early onset and a marked response to levodopa treatment and differs from idiopathic PD in that there is usually no LBs formation. Over 100 mutations in *parkin*, including missense mutations and exonic deletions and insertions, have been observed in PD families [41]. Parkin is a ubiquitin E3 ligase preparing target proteins for their degradation mediated by the ubiquitin-proteosomal system [42]. Moreover, parkin is involved in mitochondrial maintenance, repair of mitochondrial DNA damage, might contribute to mitochondrial cytochrome c release, and induce subsequent autophagy of dysfunctional mitochondria [43–47].

*2.2.2. PTEN-Induced Putative Kinase 1 Gene (PINK-1): PARK6.* Mutations in the *PTEN-induced putative kinase 1* (*PINK-1*) gene on chromosome 1p35-36 (PARK6) have been linked to autosomal recessive early-onset PD [13]. Several different *PINK1* mutations, primarily missense and nonsense ones, have been identified in PD families worldwide and cause mitochondrial deficits contributing to PD pathogenesis (http://www.molgen.ua.ac.be/PDmutDB/). PINK1 is a kinase with an N-terminal mitochondrial targeting sequence, provides protection against mitochondrial dysfunction, and regulates mitochondrial morphology via fission/fusion machinery. PINK1 also acts upstream of parkin in a common pathway in the maintenance of mitochondrial quality via autophagy [48].

*2.2.3. DJ-1: PARK7.* Mutations in the *DJ-1* gene on 1p36 (PARK7), including exonic deletions and point mutations, have been associated with a monogenic early-onset autosomal recessive form of parkinsonism characterized by slow progression and response to levodopa [14, 15]. A complete list of DJ-1 mutations can be found at the PD mutation database (http://www.molgen.ua.ac.be/PDmutDB/). DJ-1 is a mitochondrial protein involved in the protection against oxidative stress and forms a complex with parkin and PINK1 to promote ubiquitination and degradation of parkin substrates, including parkin itself [49]. Recent evidence indicates that DJ-1 works in parallel to the PINK1/parkin pathway to maintain mitochondrial function in the presence of an oxidative environment [50].

*2.2.4. ATP13A2 Gene: PARK9.* Clinical features similar to those of idiopathic PD and pallydopyramidal syndrome, including parkinsonism, pyramidal tract dysfunction, supranuclear gaze paresis, and dementia, were observed in a Jordanian family. The pattern of transmission was autosomal recessive, and a region of linkage was identified on chromosome 1p36 (PARK9) [16]. The causative gene underlying PARK9 was then identified as the *ATP13A2* gene encoding a lysosomal 5 P-type ATPase [3] likely involved in the regulation of intracellular manganese homeostasis [51].

*2.3. Additional PARK Loci.* Additional putative PARK loci include: (1) a locus on 2p13, denoted as PARK3 but with no causative gene yet identified; (2) the *UCH-L1* gene on 4p14 (PARK5) coding for a protein that possesses both a hydrolase activity to generate the ubiquitin monomer and a ligase activity to link ubiquitin molecules to tag proteins for disposal; (3) a locus on 1p32, denoted as PARK10 but with no causative gene yet identified; (4) the *GYGYF2* gene on 2q36-37 (PARK11), encoding a protein that could participate in the regulation of signaling at endosomes; (5) a locus on Xq21-q25, denoted as PARK12 and showing X-linked inheritance, but still pending characterization of the causative gene; (6) the *OMI/HTRA2* gene on chromosome 2p12 (PARK 13) coding for a nuclear-encoded serine protease localized in the intermembrane space of the mitochondria and involved in mediating caspase-dependent and caspase-independent cellular death; (7) the *PLA2G6* gene on chromosome 22q13.1 (PARK14) encoding a calcium-independent group VI phospholipase A2; (8) the *FBXO7* gene on 22q11.2-qter (PARK15) encoding for a member of the F-box family of proteins, all of which may have a role in the ubiquitin-proteasome protein-degradation pathway [17–23].

*2.4. Other Loci.* Parkinsonism is often observed as one of the symptoms in other monogenic diseases (Table 2). For example, the *MAPT* gene encodes for the microtubule-associated protein tau, a protein that binds to microtubules and is primarily involved in the organization and integrity of the cytoskeleton. Hyperphosphorylated tau forms neurofibrillary tangles in Alzheimer's disease brains, and mutations of *MAPT* cause frontotemporal dementia with parkinsonism linked to chromosome 17 (FTDP-17) [30, 31]. Parkinsonism, dystonia, and postural tremor are particularly prevalent in spinocerebellar ataxias types 2 (SCA2) and 3 (SCA3). SCA2 can manifest either with a cerebellar syndrome or as Parkinson's syndrome, and SCA3, also known as Machado-Joseph disease (MJD), is the most common form of spinocerebellar ataxia worldwide. The diseases are caused by abnormal CAG trinucleotide repeat expansion of *ataxin-2* (*ATXN2*) and *ataxin-3* (*ATXN3*) genes, respectively [32, 33]. Ataxin-2 is an enzyme involved in RNA processing, whilst ataxin-3 is a deubiquitinating enzyme involved in the ubiquitin-proteasome system. The parkinsonian phenotype of both diseases is often observed in Asians [34, 35]. Autosomal recessive hereditary spastic paraplegia with thin corpus callosum (SPG11) is a rare neurodegenerative disorder often caused by mutations in the gene encoding for spatacsin at the SPG11 locus on chromosome 15q. Two patients, of Turkish descent, from the same consanguineous family, were affected with SPG11 in association with unusual early-onset parkinsonism that occurred during the very early stages of SPG11 in both patients [36]. Additional evidence indicates that mutation of *SPG11* is a rare cause of early-onset levodopa-responsive parkinsonism with pyramidal signs [37]. There is also indication that rearrangements of the gene coding for the mitochondrial DNA polymerase gamma (POLG1), involved in the repair of mitochondrial DNA, can directly cause parkinsonism [38].

## 3. Sporadic Parkinson's Disease

Several hundreds of genetic association studies have been performed in the last few decades by means of the candidate gene approach in order to identify genetic risk factors for non-Mendelian forms of PD.

The candidate gene was selected based on the knowledge of its function in a pathway related to PD pathogenesis, and only one single gene and one or a few polymorphisms were usually investigated. More recently, those studies have been replaced by GWAS where half a million or more polymorphisms are simultaneously investigated in large case-control cohorts. GWASs are currently considered as the gold standard to find loci at which common, normal genetic variability contributes to disease risk, and their introduction has revolutionized our knowledge in the genetics of sporadic PD [24]. The PDGene database [52] is a public and continuously updated database collecting data from PD genetic association studies and GWAS. Accessed on October 2011, the database contained information on 860 studies for a total of 909 candidate genes and 3434 polymorphisms within those genes. In addition, data from GWAS and other large-scale studies were available [52]. There is strong consensus from either GWAS or updated meta-analyses of the literature that variants at four loci (*SNCA, MAPT, GBA* and *LRRK2*) contribute to disease risk (Table 3). In addition, recent GWASs have revealed novel putative PD risk loci to be confirmed in future studies [25], and the meta-analysis of the literature suggests that several additional loci could contribute to disease risk [52] (Table 3).

Table 3: Genes or loci associated with idiopathic PD.

| Gene or locus | Methodologies employed | Refs |
|---|---|---|
| *SNCA* | Large-scale association studies | [25, 52–68] |
| *LRRK2* | | |
| *MAPT* | Meta-analyses of genetic association studies | |
| *GBA* | | |
| *HLA-DRB5* | | |
| *BST1* | GWAS | |
| *GAK* | | |
| *ACMSD* | Meta-analyses of GWAS | |
| *STK39* | | |
| *MCCC1/LAMP3* | | |
| *SYT11* | | |
| *CCDC62/HIP1R* | | |
| *PARK16/1q32* | | |
| *STX1B* | | |
| *GWA 8p22* | | |
| *STBD1* | | |
| *GPNMB* | | |
| *PM20D1* | | |
| *SETD1A* | | |
| *FAM47E* | | |
| *MED13* | | |

### 3.1. *α-Synuclein (SNCA).*

Common polymorphisms of causative PD genes have been frequently investigated as possible risk factors for the idiopathic forms of the disease. Particularly, genetic polymorphisms in the *SNCA* gene have been consistently associated with PD risk in genetic association studies and subsequently replicated in large-scale GWAS, including a dinucleotide repeat sequence (Rep1) within the promoter region and several single nucleotide polymorphisms (SNPs) at the 3′end of the gene, overall suggesting that *SNCA* alleles associated with increased disease risk correlate with higher α-synuclein expression, and pointing to a gene dosage effect [52–60]. Meta-analyses of those studies revealed that *SNCA* is a low-risk locus for idiopathic PD, with odds ratios (ORs) ranging from 1.2 to 1.4 [52].

### 3.2. *Leucine-Rich Repeat Kinase 2.*

Variants of *LRRK2* have been consistently associated with increased risk for sporadic PD in Asians, including a G2385R polymorphism [61] that represents one of the most frequent genetic risk factors for PD in Asian populations, with an estimated OR of 2.2 [52].

### 3.3. *Microtubule-Associated Protein Tau: MAPT.*

As described earlier in this paper, mutations of *MAPT* cause frontotemporal dementia with parkinsonism linked to chromosome 17 (FTDP-17). Moreover, large case-control studies, meta-analyses of the literature, and GWAS indicate a role for the *MAPT* haplotype H1 to disease risk [52, 62, 63].

### 3.4. *Glucocerebrosidase: GBA.*

Gaucher disease (GD) is an autosomal recessive lysosomal glycolipid storage disorder caused by mutations of the *GBA* gene encoding the enzyme glucocerebrosidase. A small subset of GD patients develop parkinsonism [64] and relatives of patients with GD have an increased incidence of parkinsonism [65]. Large cohort studies [66] and recent meta-analyses of the literature revealed that *GBA* loss of function variants are the most common genetic risk factor associated with parkinsonism, with an estimated OR of 3.4 for the common *GBA* N370S variant [52]. Although the mechanism for this association is unknown, several theories have been proposed, including protein aggregation, lipid accumulation, and impaired autophagy, mitophagy, or trafficking [67].

### 3.5. *Additional Loci.*

The application of GWAS to the understanding of the genetics of sporadic PD has significantly improved our knowledge in the field, and several loci have been associated with disease risk in recent years. A recent meta-analysis of published GWAS was performed by the International Parkinson Disease Genomics Consortium (IPDGC) for a total of 12,386 PD cases and 21,026 controls and suggested that, in addition to *SNCA*, *LRRK2*, and *MAPT* polymorphisms, variants at eight additional loci (*HLA-DRB5, BST1, GAK, ACMSD, STK39, MCCC1/LAMP3, SYT11*, and *CCDC62/HIP1R*) are significantly associated with disease risk [25]. A more recent two-stage meta-analysis performed by the IPDGC and the Wellcome Trust Case Control Consortium 2 (WTCCC2) revealed five additional loci associated with PD risk (*PARK16/1q32, STX1B, GWA 8p22, STBD1, GPNMB*) [68]. Up to date meta-analyses are available also at the PDGene database for each polymorphism that has been evaluated in at least four independent genetic association studies. Accessed on October 2011 the PDGene database contained 886 updated meta-analyses of the literature, overall suggesting that 18 loci could contribute to sporadic PD risk, including most of those already described (*SNCA, LRRK2, MAPT, GBA, HLA-DRB5, BST1, GAK, ACMSD, STK39, MCCC1/LAMP3, SYT11, CCDC62/HIP1R, GWA 8p22, GPNMB*) and additional ones (*PM20D1, SETD1A, FAM47E* and *MED13*) [52].

## 4. Epigenetics of Parkinson's Disease

*4.1. DNA Methylation.* DNA methylation represents one of the most important epigenetic processes, along with histone modifications and mechanisms involving small RNA molecules. Methylation of CpG sequences might induce chromatin conformational modifications and inhibit the access of the transcriptional machinery to gene promoter regions, thus altering gene expression levels. Therefore, promoter hypermethylation is commonly associated with gene silencing and promoter demethylation with gene expression, even if some exceptions to this rule are known [69]. Folate, other B vitamins (vitamins B6 and B12), and homocysteine (hcy) participate in one-carbon metabolism, a complex pathway required for the production of S-adenosylmethionine (SAM), the major intracellular methylating agent. DNA methylation is closely dependent on the DNA methylation potential, which is referred to as the ratio between SAM and S-adenosylhomocysteine (SAH) levels. DNA methyltransferases (DNMTs) are the key enzymes for DNA methylation and catalyze the transfer of a methyl group from SAM to cytosine, thus forming 5-methyl-cytosine. Impaired one-carbon metabolism, altered DNA methylation potential, reduced DNA methylation levels in the brain, and altered methylation and expression of several genes were observed in Alzheimer's disease (AD) patients, pointing to a role for epigenetic modifications in the neurodegenerative process [26]. Less studies have been so far performed in PD subjects; however, there is indication of impaired one-carbon metabolism in PD as well as altered DNA methylation potential [70, 71]. Epigenetic analyses of PD brains revealed that the *SNCA* gene could be subjected to epigenetic regulation [72, 73]. In addition, more recent large-scale studies suggest that other PD-related genes could be epigenetically modified in PD brains (Table 4) [68].

*4.1.1. α-Synuclein.* Several lines of evidence, including the identification of families with *SNCA* locus duplication and triplication and the association of both promoter and 3′UTR polymorphisms with sporadic forms, point to a gene-dosage effect for *SNCA* in PD pathogenesis [9]. Studies in individuals with alcoholism [74] and in anorexia patients [75] revealed hypermethylation of the *SNCA* promoter, suggesting that the gene could be epigenetically regulated. The analysis of *SNCA* alleles in a PD patient heterozygous for the A53T mutation, the first mutation to be implicated in PD pathogenesis, revealed that *SNCA* showed monoallelic expression in this patient, with epigenetic silencing of the mutated allele due to histone modifications but not DNA methylation, and upregulation of the wild-type allele resulting in higher mRNA levels than in matched control subjects [76]. Others observed that the methylation of human *SNCA* intron 1 decreases gene expression, while inhibition of DNA methylation activates *SNCA* expression. They also observed that DNA methylation of *SNCA* intron 1 was reduced in several brain regions of sporadic PD patients, including the *substantia nigra*, putamen, and cortex, pointing toward an epigenetic regulation of *SNCA* expression in PD [72]. Another research group identified an *SNCA* CpG island in which the methylation status altered along with increased *SNCA* expression. Postmortem brain analysis revealed regional nonspecific methylation differences in this CpG region in the anterior cingulate and putamen among controls and PD subjects; however, in the *substantia nigra* of PD individuals, the methylation of this region was significantly decreased [73]. Both findings are consistent with previous reports indicating increased *SNCA* mRNA levels in PD *substantia nigra* tissue [77, 78]. A recent paper suggested that α-synuclein sequesters DNA methyltransferase 1 (DNMT1) from the nucleus [79]. DNMT1 is the maintenance DNA methylation enzyme which is abundantly expressed in the adult brain and is mainly located in the nuclear compartment. The researchers observed a reduction of nuclear DNMT1 levels in human postmortem brain samples from PD and from patients with dementia with Lewy bodies (DLBs) as well as in the brains of α-synuclein transgenic mice models. Furthermore, sequestration of DNMT1 in the cytoplasm resulted in global DNA hypomethylation in human and mouse brains, involving CpG islands upstream of *SNCA* and other genes. The nuclear DNMT1 levels were partially rescued by overexpression of DNMT1 in neuronal cell cultures and in α-synuclein transgenic mice brains. Therefore, the authors suggested that the association of DNMT1 and α-synuclein might mediate aberrant subcellular localization of DNMT1, resulting in epigenetic modifications in the brain [79].

*4.1.2. Other Genes.* In addition to its association with ARJP, loss of heterozygosity of *parkin* has been found in several types of malignant tumors, including ovarian, breast, and hepatocellular tumors, and abnormal methylation of *parkin* promoter was observed in patients with cancer [80]. The levels of methylation of the *parkin* gene promoter were revealed in samples from PD patients with heterozygous parkin mutations, PD patients without parkin mutations, and normal controls. No difference was observed between the three

Table 4: Epigenetic changes of PD related genes or PD tissues.

| *Gene* | Observation | Refs |
|---|---|---|
| *α-synuclein (SNCA)* | Reduced *SNCA* methylation in the *substantia nigra* of PD patients | [72, 73] |
| | α-synuclein sequesters DNMT1 from the nucleus resulting in aberrant DNA methylation | [79] |
| | *SNCA* gene silencing mediated by histone modifications | [76] |
| | α-synuclein binds to histones and inhibits histone acetylation | [81–83] |
| | Histone deacetylase inhibitors are neuroprotective against α-synuclein mediated neurotoxicity in PD animal models | [82, 84–90] |
| | α-synuclein expression in *C. elegans* results in down-regulation of genes coding for histones | [91] |
| | miR-10a, -10b, -212, -132, -495 were impaired in presymptomatic *α-synuclein* transgenic mice | [92] |
| | miR-7 and mir-153 regulates α-synuclein levels | [93, 94] |
| | miR-64 and mir-65 and let-7 were co-under-expressed in *α-synuclein* transgenic *C. elegans* | [95] |
| *LRRK2* | Mutant LRRK2 antagonizes miR-184* and let7 in *Drosophila* PD models | [96] |
| *parkin* | let-7 family miRNAs let-7 were co-under-expressed in *parkin* transgenic *C. elegans* | [95] |
| *PARK16/1q32, GPNMB, STX1B* | Aberrant gene methylation observed in post-mortem PD brains | [68] |
| *Tissue* | Observation | Refs |
| PD brains | miR-133b was deficient in midbrain from PD patients | [97] |
| PD brains | miR-34b/c down-regulation was observed in pre-motor stages of PD and resulted in altered expression of DJ1 and parkin proteins | [98] |
| PD lymphocytes | Altered expression of miR-1, miR-16-2*, miR-22*, miR26a2*, miR29, miR30 | [99] |
| PD leukocytes | Altered methylation patterns of subtelomeric regions | [100] |

groups, suggesting that *parkin* promoter methylation is unlikely to play a role in the pathogenesis and development of PD [101]. The *UCHL-1* gene promoter was found to be hypermethylated in various types of cancer [102, 103]. However, the analysis of *UCHL-1* promoter methylation in the frontal cortex of PD patients and controls revealed no significant differences between the groups [104]. In the same study the authors evaluated the methylation profiles of *MAPT* promoter in the frontal cortex and hippocampus of controls, Alzheimer's disease patients, PD patients, and subjects with other tauopathies and synucleinopathies. No differences in the percentage of CpG methylation were found between control and disease samples or among the different pathological entities in any region analyzed [104]. Another study reported methylation differences in the gene coding for tumor necrosis alpha (*TNFA*) between the cortex and the *substantia nigra*, but these differences were present in PD cases and controls [105]. The analysis of the original Chilean family with Kufor-Rakeb syndrome that led to the discovery of the *ATP13A2* gene at the PARK9 locus also revealed that there was no significant correlation between DNA methylation of the *ATP13A2* promoter region and disease progression [106]. However, the recent collaborative study of the IPDGC and WTCCC2 on a dataset of post-mortem brain samples assayed for gene expression ($n = 399$) and methylation ($n = 292$), revealed methylation and expression changes associated with PD risk variants in *PARK16/1q32*, *GPNMB*, and *STX1B*, suggesting that the *SNCA* gene is unlikely to be the only one subjected to epigenetic regulation in PD brains [68]. In addition, an aging-associated alteration of subtelomeric methylation patterns was observed in peripheral leukocytes of Japanese PD patients that showed fewer short telomeres than healthy controls. Moreover, short telomeres in PD patients showed a constant methylation pattern, whilst an age-related demethylation of short telomeres was observed in controls [100].

*4.2. Histone Modifications.* Gene expression profiles are modulated not only by promoter methylation but also by the chromatin state. Indeed, chromatin can exist in a condensate inactive state (heterochromatin) or in a noncondensate and transcriptionally active state (euchromatin). Conformational changes in histone proteins or modifications of the way in which DNA wraps around the histone octamer in nucleosomes may either alter or facilitate the access of the transcriptional machinery to the promoter region of some genes, leading to gene silencing or activation, respectively. Histone tail modifications include acetylation, methylation, phosphorylation, ubiquitylation, sumoylation, and other post-translational modifications. Histone tail acetylation is associated with chromatin relaxation and transcriptional activation, while deacetylation is related to a more condensed chromatin state and transcriptional repression [107]. Histone acetyltransferases (HATs) catalyze the acetylation of lysine residues in histone tails, whereas histone deacetylation is mediated by histone deacetylases (HDACs). Another frequently studied modification of histone tails is methylation on either lysine or arginine residues mediated by histone methyltransferases. Methylation of histone tails can be associated with either condensation or relaxation of the chromatin structure, since several sites for methylation are present on each tail thus allowing several combinations [107, 108]. Little is still known concerning histone modifications in PD brains and most of the current knowledge is derived from studies in cell cultures and animal models of the disease, such as those induced by mitochondrial toxins, including 1-methyl-4-phenylpyridinium (MPP+), paraquat, and rotenone, or those overexpressing human α-synuclein (Table 4). However, those studies have revealed that α-synuclein interacts with histones and inhibits histone acetylation [81, 82] and that several histone deacetylase inhibitors (HDACIs) are neuroprotective against α-synuclein-mediated toxicity [82, 84–87]. Particularly, studies performed in nigral neurons of mice exposed to the herbicide paraquat revealed that α-synuclein translocates into the nucleus and binds with histones [81]. Studies in *Drosophila* showed that α-synuclein mediates neurotoxicity in the nucleus, binds directly to histone H3, and inhibits histone acetylation. The toxicity of α-synuclein was rescued by the administration of HDACIs [82]. The inhibition of the histone deacetylase Sirtuin 2 rescued α-synuclein-mediated toxicity in several models of PD [84]. In addition valproic acid (VPA) resulted in inhibition of histone deacetylase activity and in an increase of histone H3 acetylation in brain tissues of rats and resulted neuroprotective in a rat model of PD (obtained with the administration of the mitochondrial toxin rotenone), counteracting α-synuclein translocation into the nuclei [85]. Studies in rat and human neuronal cell cultures also revealed that HDACIs prevent MPP+ -mediated cytotoxicity [86]. Neurotoxic pesticides and paraquat were shown to increase histone acetylation in mice brains or cell culture models [87, 88], and additional studies reported protective effects of HDACIs on dopaminergic neurons following a neurotoxic-induced insult [89, 90]. A recent study in α-synuclein transgenic mice revealed that α-synuclein negatively regulates protein kinase Cδ expression to suppress apoptosis in dopaminergic neurons by reducing p300 histone acetyltransferase activity [83]. A genome-wide expression screen was performed in *C. elegans* overexpressing human α-synuclein, and nine genes that form histones H1, H2B, and H4 were downregulated [91]. Overall, these studies point to a role for histone modifications in α-synuclein-mediated as well as environmental-induced dopaminergic neuronal cell death. Moreover, the accumulation of misfolded proteins in proteinaceous inclusions termed aggresomes is a cytoprotective response serving to sequester potentially toxic misfolded proteins and facilitate their clearance by autophagy. Histone deacetylase 6 (HDAC6) plays a fundamental role in aggresome formation. HDAC6 is concentrated in Lewy bodies in PD and dementia with LBs, and the *Drosophila* histone deacetylase 6 (dHDAC6) was shown to play a critical role in the protection of dopaminergic neurons and promoted the formation of α-synuclein inclusions in a *Drosophila* PD model expressing human α-synuclein. On the contrary, mutation of dHDAC6 resulted in the accumulation of toxic α-synuclein oligomers [109]. It was suggested that the accumulation of HDAC6 might be specific to α-synucleinopathy and that LBs might represent cytoprotective responses to sequester toxic proteins [110]. Aggresome formation involves several regulators, including HDAC6, parkin, ataxin-3, and ubiquilin-1 [111]. It has been recently shown in neuronal cell cultures that HDAC6 participates in the degradation of MPP+ induced aggregates of α-synuclein by regulating the aggresome-autophagy pathway [112].

*4.3. RNA-Mediated Epigenetic Mechanisms.* MicroRNAs (miRNAs) are a group of small noncoding RNAs that bind to the 3′ untranslated region (3′UTR) of target mRNAs and mediate their posttranscriptional regulation leading to either degradation or translational inhibition, depending on the degree of sequence complementarity. In 2007 Kim et al. investigated the role of miRNAs in mammalian midbrain dopaminergic neurons and identified an miRNA, miR-133b, that is specifically expressed in midbrain dopaminergic neurons and is deficient in midbrain tissue from patients with PD. miR-133b regulates the maturation and function of midbrain dopaminergic neurons within a negative feedback circuit that includes the paired-like homeodomain transcription factor Pitx3 [97]. Subsequently, others observed that the *FGF20* rs12720208 polymorphism disrupts a binding site for miR-433, increasing translation of *FGF20 in vitro* and *in vivo*, and suggested that this could represent a risk factor for PD [113]. However, a recent study failed to confirm the association between rs12720208 and PD risk, or any effect of miR-433 variants to PD pathogenesis [114]. MicroRNA expression analysis in brains of early symptomatic α-synuclein(A30P)-transgenic mice showed that the levels of several miRNAs (miR-10a, -10b, -212, -132, -495) were significantly altered. MiR-132 was reported to be highly inducible by growth factors and to be a key regulator of neurite outgrowth [92]. Others observed that miR-7, which is expressed mainly in neurons, represses α-synuclein protein levels binding to α-synuclein mRNA. Further, miR-7 expression decreased in MPP+ -induced models of PD in cultured cells and in mice, thereby contributing to increased α-synuclein expression [93]. Another study confirmed that mir-7 regulates α-synuclein levels together with mir-153. Both miRNAs bind specifically to the 3′-untranslated region of α-synuclein and downregulate its mRNA and protein levels, with their effect being additive. They are expressed predominantly in the brain with a pattern that mirrors synuclein expression in different tissues as well as during neuronal development, likely playing a tuning role in the amount of α-synuclein produced [94]. The analysis of *C. elegans* models of PD revealed that several miRNAs were underexpressed in those animals; particularly the family of miR-64 and miR-65 were co-underexpressed in α-synuclein transgenic animals, and members of let-7 family co-underexpressed in both α-synuclein and parkin strains [95]. A recent study demonstrates that blood samples can be used as a source of miRNAs associated to PD. Six differentially expressed miRNAs were identified. While miR-1, miR-22*, and miR-29 expression levels allowed to distinguish nontreated PD patients from healthy subjects, miR-16-2*, miR-26a2*, and miR30a differentiated treated from untreated PD patients [99]. A recent miRNA profiling of PD brains identified early downregulation of miR-34b/c which modulate mitochondrial function. Particularly, misregulation of miR-34b/c was detected in premotor stages (stages 1–3) of the disease, and thus in cases that did not receive any PD-related treatment during life, and miR-34b/c downregulation was coupled to a decrease in the expression of DJ1 and Parkin proteins [98]. Moreover, studies in *Drosophila* bearing LRRK2 PD-associated mutations revealed that pathogenic LRRK2 antagonizes let-7 and miR-184*, leading to the overproduction of the E2F1/DP complex involved in cell cycle and survival control [96]. Overall, these studies indicate microRNA-mediated mechanisms in PD pathogenesis (Table 4).

## 5. Conclusions

Several advances have been gained in our understanding of the genetics of PD since the first report [5] of a *SNCA* gene mutation causing familial autosomal dominant PD in 1997 (Table 1). Studies on familial recessive forms of the disease have highlighted a central role for mitochondrial damage, repair, and turnover in the pathophysiology of the disease, with *parkin, DJ1, PINK1,* and *FBX07,* participating in the same or in similar/overlapping mitochondrial pathways. In addition, as suggested by Hardy [24], both glucocerebrosidase and ATP13A2 are lysosomal enzymes, indicating a second PD pathway involving lysosomes. In contrast, $\alpha$-synuclein and LRRK2 biology is still poorly understood, despite that both proteins are central to the disease etiology [24]. Association studies in sporadic cases and the recent application of genome-wide technology led to the identification of almost 20 genes that could modify the individual risk for the idiopathic forms of the disease (Table 3), and additional genes are expected to be unravelled within the next future. Public databases, such as the PD mutation database (http://www.molgen.ua.ac.be/PDmutDB/) and the PDGene database [52], have been created to continuously update the genetics of either familial or idiopathic forms, respectively. There is however a growing evidence that, in addition to genetic mutations, also epigenetic mechanisms could contribute to disease pathogenesis (Table 4). For example, studies in PD families as well as in sporadic PD cases suggest an *SNCA* gene-dosage effect critical to disease pathogenesis, but there is also indication of upregulated *SNCA* gene expression resulting from promoter demethylation in PD brains [72, 73]. $\alpha$-Synuclein itself was shown to exert epigenetic properties, such as histone tail modifications [81, 82, 84, 85] and DNMT1 sequestration [79]. In addition, microRNAs were found to modulate $\alpha$-synuclein expression [93, 94]. Other genes critical to PD pathogenesis have been found to be regulated by promoter methylation [68] or RNA-mediated mechanisms [98]. For example, a decrease in the expression of DJ1 and parkin proteins can result from microRNA-mediated mechanisms in PD brains, ultimately leading to mitochondrial impairments such as those caused by *parkin* or *DJ-1* gene mutations [98]. The increasing amount of papers aimed at understanding the epigenetics of PD has led to a better understanding of the molecular pathways involved in dopaminergic neuron degeneration, and several researchers are now working to understand the therapeutic potentials of epigenetic molecules to counteract age-related neurodegenerative diseases [115, 116]. Future research in PD should be aimed at understanding the complex interplay among genetic and epigenetic biomarkers, lifestyles, and environmental factors, to further characterize individuals at risk to develop the disease.

## References


[1] B. Thomas and M. F. Beal, "Molecular insights into Parkinson's disease," *F1000 Medicine Reports*, vol. 3, no. 7, 2011.

[2] K. Nuytemans, J. Theuns, M. Cruts, and C. Van Broeckhoven, "Genetic etiology of Parkinson disease associated with mutations in the *SNCA, PARK2, PINK1, PARK7,* and *LRRK2* genes: a mutation update," *Human Mutation*, vol. 31, no. 7, pp. 763–780, 2010.

[3] A. Ramirez, A. Heimbach, J. Gründemann et al., "Hereditary parkinsonism with dementia is caused by mutations in *ATP13A2*, encoding a lysosomal type 5 P-type ATPase," *Nature Genetics*, vol. 38, no. 10, pp. 1184–1191, 2006.

[4] L. Migliore and F. Coppedè, "Genetics, environmental factors and the emerging role of epigenetics in neurodegenerative diseases," *Mutation Research*, vol. 667, no. 1-2, pp. 82–97, 2009.

[5] M. H. Polymeropoulos, C. Lavedan, E. Leroy et al., "Mutation in the $\alpha$-*synuclein* gene identified in families with Parkinson's disease," *Science*, vol. 276, no. 5321, pp. 2045–2047, 1997.

[6] R. Krüger, W. Kuhn, T. Müller et al., "Ala30Pro mutation in the gene encoding $\alpha$-*synuclein* in Parkinson's disease," *Nature Genetics*, vol. 18, no. 2, pp. 106–108, 1998.

[7] J. J. Zarranz, J. Alegre, J. C. Gómez-Esteban et al., "The new mutation, E46K, of $\alpha$-*synuclein* causes Parkinson and Lewy body dementia," *Annals of Neurology*, vol. 55, no. 2, pp. 164–173, 2004.

[8] A. B. Singleton, M. Farrer, J. Johnson et al., "$\alpha$-*synuclein* locus triplication causes Parkinson's disease," *Science*, vol. 302, no. 5646, p. 841, 2003.

[9] M. C. Chartier-Harlin, J. Kachergus, C. Roumier et al., "$\alpha$-*synuclein* locus duplication as a cause of familial Parkinson's disease," *The Lancet*, vol. 364, no. 9440, pp. 1167–1169, 2004.

[10] C. Paisán-Ruíz, S. Jain, E. W. Evans et al., "Cloning of the gene containing mutations that cause *PARK8*-linked Parkinson's disease," *Neuron*, vol. 44, no. 4, pp. 595–600, 2004.

[11] A. Zimprich, S. Biskup, P. Leitner et al., "Mutations in *LRRK2* cause autosomal-dominant parkinsonism with pleomorphic pathology," *Neuron*, vol. 44, no. 4, pp. 601–607, 2004.

[12] T. Kitada, S. Asakawa, N. Hattori et al., "Mutations in the parkin gene cause autosomal recessive juvenile parkinsonism," *Nature*, vol. 392, no. 6676, pp. 605–608, 1998.

[13] E. M. Valente, P. M. Abou-Sleiman, V. Caputo et al., "Hereditary early-onset Parkinson's disease caused by mutations in *PINK1*," *Science*, vol. 304, no. 5674, pp. 1158–1160, 2004.

[14] C. M. Van Duijn, M. C. J. Dekker, V. Bonifati et al., "*PARK7*, a novel locus for autosomal recessive early-onset parkinsonism, on chromosome 1p36," *American Journal of Human Genetics*, vol. 69, no. 3, pp. 629–634, 2001.

[15] P. J. Lockhart, S. Lincoln, M. Hulihan et al., "*DJ-1* mutations are a rare cause of recessively inherited early onset parkinsonism mediated by loss of protein function," *Journal of Medical Genetics*, vol. 41, no. 3, p. e22, 2004.

[16] D. J. Hampshire, E. Roberts, Y. Crow et al., "Kufor-Rakeb syndrome, pallido-pyramidal degeneration with supranuclear upgaze paresis and dementia, maps to 1p36," *Journal of Medical Genetics*, vol. 38, no. 10, pp. 680–682, 2001.

[17] Y. Liu, L. Fallon, H. A. Lashuel, Z. Liu, and P. T. Lansbury, "The UCH-L1 gene encodes two opposing enzymatic activities that affect *α-synuclein* degradation and Parkinson's disease susceptibility," *Cell*, vol. 111, no. 2, pp. 209–218, 2002.

[18] C. Lautier, S. Goldwurm, A. Dürr et al., "Mutations in the *GIGYF2 (TNRC15)* gene at the PARK11 locus in familial Parkinson disease," *American Journal of Human Genetics*, vol. 82, no. 4, pp. 822–833, 2008.

[19] S. Higashi, E. Iseki, M. Minegishi, T. Togo, T. Kabuta, and K. Wada, "*GIGYF2* is present in endosomal compartments in the mammalian brains and enhances IGF-1-induced *ERK1/2* activation," *Journal of Neurochemistry*, vol. 115, no. 2, pp. 423–437, 2010.

[20] K. M. Strauss, L. M. Martins, H. Plun-Favreau et al., "Loss of function mutations in the gene encoding Omi/HtrA2 in Parkinson's disease," *Human Molecular Genetics*, vol. 14, no. 15, pp. 2099–2111, 2005.

[21] C. Paisán-Ruiz, R. Guevara, M. Federoff et al., "Early-onset L-dopa-responsive Parkinsonism with pyramidal signs due to *ATP13A2, PLA2G6, FBXO7* and *Spatacsin* mutations," *Movement Disorders*, vol. 25, no. 12, pp. 1791–1800, 2010.

[22] S. Shojaee, F. Sina, S. S. Banihosseini et al., "Genome-wide linkage analysis of a Parkinsonian-pyramidal syndrome pedigree by 500 K SNP arrays," *American Journal of Human Genetics*, vol. 82, no. 6, pp. 1375–1384, 2008.

[23] A. D. Fonzo, M. C. J. Dekker, P. Montagna et al., "FBXO7 mutations cause autosomal recessive, early-onset parkinsonian-pyramidal syndrome," *Neurology*, vol. 72, no. 3, pp. 240–245, 2009.

[24] J. Hardy, "Genetic analysis of pathways to parkinson disease," *Neuron*, vol. 68, no. 2, pp. 201–206, 2010.

[25] M. A. Nalls, V. Plagnol, D. G. Hernandez et al., "Imputation of sequence variants for identification of genetic risks for Parkinson's disease: a meta-analysis of genome-wide association studies," *The Lancet*, vol. 377, no. 9766, pp. 641–649, 2011.

[26] F. Coppedè, "One-carbon metabolism and Alzheimer's disease: focus on epigenetics," *Current Genomics*, vol. 11, no. 4, pp. 246–260, 2010.

[27] I. A. Qureshi and M. F. Mehler, "Advances in epigenetics and epigenomics for neurodegenerative diseases," *Current Neurology and Neuroscience Reports*, vol. 11, no. 5, pp. 464–473, 2011.

[28] S. C. F. Marques, C. R. Oliveira, C. M. F. Pereira, and T. F. Outeiro, "Epigenetics in neurodegeneration: a new layer of complexity," *Progress in Neuro-Psychopharmacology and Biological Psychiatry*, vol. 35, pp. 348–355, 2011.

[29] M. Chen and L. Zhang, "Epigenetic mechanisms in developmental programming of adult disease," *Drug Discovery Today*, vol. 16, no. 23-24, pp. 1007–1018, 2011.

[30] C. Dumanchin, A. Camuzat, D. Campion et al., "Segregation of a missense mutation in the microtubule-associated protein tau gene with familial frontotemporal dementia and parkinsonism," *Human Molecular Genetics*, vol. 7, no. 11, pp. 1825–1829, 1998.

[31] M. G. Spillantini and M. Goedert, "Tau mutations in familial frontotemporal dementia," *Brain*, vol. 123, no. 5, pp. 857–859, 2000.

[32] I. Lastres-Becker, U. Rüb, and G. Auburger, "Spinocerebellar ataxia 2 (SCA2)," *Cerebellum*, vol. 7, no. 2, pp. 115–124, 2008.

[33] C. A. Matos, S. de Macedo-Ribeiro, and A. L. Carvalho, "Polyglutamine diseases: the special case of ataxin-3 and Machado-Joseph disease," *Progress in Neurobiology*, vol. 95, no. 1, pp. 26–48, 2011.

[34] C. S. Lu, H. C. Chang, P. C. Kuo et al., "The parkinsonian phenotype of spinocerebellar ataxia type 3 in a Taiwanese family," *Parkinsonism and Related Disorders*, vol. 10, no. 6, pp. 369–373, 2004.

[35] J. M. Kim, S. Hong, P. K. Gyoung et al., "Importance of low-range CAG expansion and CAA interruption in SCA2 parkinsonism," *Archives of Neurology*, vol. 64, no. 10, pp. 1510–1518, 2007.

[36] M. Anheim, C. Lagier-Tourenne, G. Stevanin et al., "SPG11 spastic paraplegia: a new cause of juvenile parkinsonism," *Journal of Neurology*, vol. 256, no. 1, pp. 104–108, 2009.

[37] A. Guidubaldi, C. Piano, F. M. Santorelli et al., "Novel mutations in SPG11 cause hereditary spastic paraplegia associated with early-onset levodopa-responsive Parkinsonism," *Movement Disorders*, vol. 26, no. 3, pp. 553–556, 2011.

[38] D. Orsucci, E. Caldarazzo Ienco, M. Mancuso, and G. Siciliano, "POLG1-Related and other "Mitochondrial Parkinsonisms": an overview," *Journal of Molecular Neuroscience*, vol. 44, pp. 17–24, 2011.

[39] A. Abeliovich, Y. Schmitz, I. Fariñas et al., "Mice lacking *α-synuclein* display functional deficits in the nigrostriatal dopamine system," *Neuron*, vol. 25, no. 1, pp. 239–252, 2000.

[40] S. Liu, I. Ninan, I. Antonova et al., "*α-synuclein* produces a long-lasting increase in neurotransmitter release," *EMBO Journal*, vol. 23, no. 22, pp. 4506–4516, 2004.

[41] I. F. Mata, P. J. Lockhart, and M. J. Farrer, "Parkin genetics: one model for Parkinson's disease," *Human Molecular Genetics*, vol. 13, no. 1, pp. R127–R133, 2004.

[42] E. Leroy, R. Boyer, G. Auburger et al., "The ubiquitin pathway in Parkinson's disease," *Nature*, vol. 395, no. 6701, pp. 451–452, 1998.

[43] D. Narendra, A. Tanaka, D. F. Suen, and R. J. Youle, "Parkin is recruited selectively to impaired mitochondria and promotes their autophagy," *Journal of Cell Biology*, vol. 183, no. 5, pp. 795–803, 2008.

[44] H. Deng, M. W. Dodson, H. Huang, and M. Guo, "The Parkinson's disease genes *pink1* and parkin promote mitochondrial fission and/or inhibit fusion in *Drosophila*," *Proceedings of the National Academy of Sciences of the United States of America*, vol. 105, no. 38, pp. 14503–14508, 2008.

[45] A. C. Poole, R. E. Thomas, L. A. Andrews, H. M. McBride, A. J. Whitworth, and L. J. Pallanck, "The *PINK1/Parkin* pathway regulates mitochondrial morphology," *Proceedings of the National Academy of Sciences of the United States of America*, vol. 105, no. 5, pp. 1638–1643, 2008.

[46] A. K. Berger, G. P. Cortese, K. D. Amodeo, A. Weihofen, A. Letai, and M. J. LaVoie, "Parkin selectively alters the intrinsic threshold for mitochondrial cytochrome c release," *Human Molecular Genetics*, vol. 18, no. 22, pp. 4317–4328, 2009.

[47] O. Rothfuss, H. Fischer, T. Hasegawa et al., "Parkin protects mitochondrial genome integrity and supports mitochondrial DNA repair," *Human Molecular Genetics*, vol. 18, no. 20, pp. 3832–3850, 2009.

[48] S. Kawajiri, S. Saiki, S. Sato, and N. Hattori, "Genetic mutations and functions of *PINK1*," *Trends in Pharmacological Sciences*, vol. 32, no. 10, pp. 573–580, 2011.

[49] H. Xiong, D. Wang, L. Chen et al., "*Parkin, PINK1*, and *DJ-1* form a ubiquitin E3 ligase complex promoting unfolded protein degradation," *Journal of Clinical Investigation*, vol. 119, no. 3, pp. 650–660, 2009.

[50] K. J. Thomas, M. K. McCoy, J. Blackinton et al., "*DJ-1* acts in parallel to the *PINK1/parkin* pathway to control mitochondrial function and autophagy," *Human Molecular Genetics*, vol. 20, no. 1, pp. 40–50, 2011.

[51] J. Tan, T. Zhang, L. Jiang et al., "Regulation of intracellular manganese homeostasis by Kufor-Rakeb syndrome-associated *ATP13A2* protein," *Journal of Biological Chemistry*, vol. 286, no. 34, pp. 29654–29662, 2011.

[52] C. M. Lill, J. T. Roehr, M. B. McQueen et al., "The PDGene Database," Alzheimer Research Forum, http://www.pdgene.org/.

[53] D. M. Maraganore, M. De Andrade, A. Elbaz et al., "Collaborative analysis of *α-synuclein* gene promoter variability and Parkinson disease," *Journal of the American Medical Association*, vol. 296, no. 6, pp. 661–670, 2006.

[54] I. F. Mata, D. Yearout, V. Alvarez et al., "Replication of *MAPT* and *SNCA*, but not *PARK16-18*, as susceptibility genes for Parkinson's disease," *Movement Disorders*, vol. 26, no. 5, pp. 819–823, 2011.

[55] N. Pankratz, J. B. Wilk, J. C. Latourelle et al., "Genomewide association study for susceptibility genes contributing to familial Parkinson disease," *Human Genetics*, vol. 124, no. 6, pp. 593–605, 2009.

[56] J. Simón-Sánchez, C. Schulte, J. M. Bras et al., "Genome-wide association study reveals genetic risk underlying Parkinson's disease," *Nature Genetics*, vol. 41, no. 12, pp. 1308–1312, 2009.

[57] T. L. Edwards, W. K. Scott, C. Almonte et al., "Genome-Wide association study confirms *SNPs in SNCA* and the *MAPT* region as common risk factors for parkinson disease," *Annals of Human Genetics*, vol. 74, no. 2, pp. 97–109, 2010.

[58] W. Satake, Y. Nakabayashi, I. Mizuta et al., "Genome-wide association study identifies common variants at four loci as genetic risk factors for Parkinson's disease," *Nature Genetics*, vol. 41, no. 12, pp. 1303–1307, 2009.

[59] J. Fuchs, A. Tichopad, Y. Golub et al., "Genetic variability in the *SNCA gene* influences *α-synuclein* levels in the blood and brain," *FASEB Journal*, vol. 22, no. 5, pp. 1327–1334, 2008.

[60] A. Goris, C. H. Williams-Gray, G. R. Clark et al., "Tau and *α-synuclein* in susceptibility to, and dementia in, Parkinson's disease," *Annals of Neurology*, vol. 62, no. 2, pp. 145–153, 2007.

[61] M. J. Farrer, J. T. Stone, C. H. Lin et al., "Lrrk2 G2385R is an ancestral risk factor for Parkinson's disease in Asia," *Parkinsonism and Related Disorders*, vol. 13, no. 2, pp. 89–92, 2007.

[62] C. P. Zabetian, C. M. Hutter, S. A. Factor et al., "Association analysis of *MAPT* H1 haplotype and subhaplotypes in Parkinson's disease," *Annals of Neurology*, vol. 62, no. 2, pp. 137–144, 2007.

[63] A. Velayati, W. H. Yu, and E. Sidransky, "The role of glucocerebrosidase mutations in parkinson disease and lewy body disorders," *Current Neurology and Neuroscience Reports*, vol. 10, no. 3, pp. 190–198, 2010.

[64] N. Tayebi, J. Walker, B. Stubblefield et al., "Gaucher disease with parkinsonian manifestations: does glucocerebrosidase deficiency contribute to a vulnerability to parkinsonism?" *Molecular Genetics and Metabolism*, vol. 79, no. 2, pp. 104–109, 2003.

[65] A. Halperin, D. Elstein, and A. Zimran, "Increased incidence of Parkinson disease among relatives of patients with Gaucher disease," *Blood Cells, Molecules, and Diseases*, vol. 36, no. 3, pp. 426–428, 2006.

[66] E. Sidransky, M. A. Nalls, J. O. Aasly et al., "Multicenter analysis of glucocerebrosidase mutations in Parkinson's disease," *New England Journal of Medicine*, vol. 361, no. 17, pp. 1651–1661, 2009.

[67] W. Westbroek, A. M. Gustafson, and E. Sidransky, "Exploring the link between glucocerebrosidase mutations and parkinsonism," *Trends in Molecular Medicine*, vol. 17, no. 9, pp. 485–493, 2011.

[68] International Parkinson's Disease Genomics Consortium (IPDGC), "A two-stage meta-analysis identifies several new loci for Parkinson's disease," *PLoS Genetics*, vol. 7, no. 6, article e1002142, 2011.

[69] D. Gius, H. Cui, C. M. Bradbury et al., "Distinct effects on gene expression of chemical and genetic manipulation of the cancer epigenome revealed by a multimodality approach," *Cancer Cell*, vol. 6, no. 4, pp. 361–371, 2004.

[70] R. Obeid, A. Schadt, U. Dillmann, P. Kostopoulos, K. Fassbender, and W. Herrmann, "Methylation status and neurodegenerative markers in Parkinson disease," *Clinical Chemistry*, vol. 55, no. 10, pp. 1852–1860, 2009.

[71] W. Herrmann and R. Obeid, "Biomarkers of folate and vitamin B12 status in cerebrospinal fluid," *Clinical Chemistry and Laboratory Medicine*, vol. 45, no. 12, pp. 1614–1620, 2007.

[72] A. Jowaed, I. Schmitt, O. Kaut, and U. Wüllner, "Methylation regulates alpha-*synuclein* expression and is decreased in Parkinson's disease patients' brains," *Journal of Neuroscience*, vol. 30, no. 18, pp. 6355–6359, 2010.

[73] L. Matsumoto, H. Takuma, A. Tamaoka et al., "CpG demethylation enhances alpha-*synuclein* expression and affects the pathogenesis of Parkinson's disease," *PLoS ONE*, vol. 5, no. 11, Article ID e15522, 2010.

[74] D. Bönsch, B. Lenz, J. Kornhuber, and S. Bleich, "DNA hypermethylation of the alpha *synuclein* promoter in patients with alcoholism," *NeuroReport*, vol. 16, no. 2, pp. 167–170, 2005.

[75] H. Frieling, A. Gozner, K. D. Römer et al., "Global DNA hypomethylation and DNA hypermethylation of the alpha *synuclein* promoter in females with anorexia nervosa," *Molecular Psychiatry*, vol. 12, no. 3, pp. 229–230, 2007.

[76] G. E. Voutsinas, E. F. Stavrou, G. Karousos et al., "Allelic imbalance of expression and epigenetic regulation within the alpha-*synuclein* wild-type and p.Ala53Thr alleles in Parkinson disease," *Human Mutation*, vol. 31, no. 6, pp. 685–691, 2010.

[77] O. Chiba-Falek, G. J. Lopez, and R. L. Nussbaum, "Levels of alpha-*synuclein* mRNA in sporadic Parkinson disease patients," *Movement Disorders*, vol. 21, no. 10, pp. 1703–1708, 2006.

[78] J. Gründemann, F. Schlaudraff, O. Haeckel, and B. Liss, "Elevated *α-synuclein* mRNA levels in individual UV-laser-microdissected dopaminergic substantia nigra neurons in idiopathic Parkinson's disease," *Nucleic Acids Research*, vol. 36, no. 7, article e38, 2008.

[79] P. Desplats, B. Spencer, E. Coffee et al., "*α-synuclein* sequesters Dnmt1 from the nucleus: a novel mechanism for epigenetic alterations in Lewy body diseases," *Journal of Biological Chemistry*, vol. 286, no. 11, pp. 9031–9037, 2011.

[80] X. Agirre, J. Román-Gómez, I. Vázquez et al., "Abnormal methylation of the common *PARK2* and *PACRG* promoter is associated with downregulation of gene expression in acute lymphoblastic leukemia and chronic myeloid leukemia," *International Journal of Cancer*, vol. 118, no. 8, pp. 1945–1953, 2006.

[81] J. Goers, A. B. Manning-Bog, A. L. McCormack et al., "Nuclear localization of *α-synuclein* and its interaction with histones," *Biochemistry*, vol. 42, no. 28, pp. 8465–8471, 2003.

[82] E. Kontopoulos, J. D. Parvin, and M. B. Feany, "*α-synuclein* acts in the nucleus to inhibit histone acetylation and promote neurotoxicity," *Human Molecular Genetics*, vol. 15, no. 20, pp. 3012–3023, 2006.

[83] H. Jin, A. Kanthasamy, A. Ghosh, Y. Yang, V. Anantharam, and A. G. Kanthasamy, "*α-synuclein* negatively regulates protein kinase Cδ expression to suppress apoptosis in dopaminergic neurons by reducing p300 histone acetyltransferase activity," *Journal of Neuroscience*, vol. 31, no. 6, pp. 2035–2051, 2011.

[84] T. F. Outeiro, E. Kontopoulos, S. M. Altmann et al., "Sirtuin 2 inhibitors rescue *α-synuclein*-mediated toxicity in models of Parkinson's disease," *Science*, vol. 317, no. 5837, pp. 516–519, 2007.

[85] B. Monti, V. Gatta, F. Piretti, S. S. Raffaelli, M. Virgili, and A. Contestabile, "Valproic acid is neuroprotective in the rotenone rat model of Parkinson's disease: involvement of *α-synuclein*," *Neurotoxicity Research*, vol. 17, no. 2, pp. 130–141, 2010.

[86] S. K. Kidd and J. S. Schneider, "Protection of dopaminergic cells from $MPP^+$-mediated toxicity by histone deacetylase inhibition," *Brain Research*, vol. 1354, no. C, pp. 172–178, 2010.

[87] C. Song, A. Kanthasamy, H. Jin, V. Anantharam, and A. G. Kanthasamy, "Paraquat induces epigenetic changes by promoting histone acetylation in cell culture models of dopaminergic degeneration," *NeuroToxicology*, vol. 32, no. 5, pp. 586–595, 2011.

[88] C. Song, A. Kanthasamy, V. Anantharam, F. Sun, and A. G. Kanthasamy, "Environmental neurotoxic pesticide increases histone acetylation to promote apoptosis in dopaminergic neuronal cells: relevance to epigenetic mechanisms of neurodegeneration," *Molecular Pharmacology*, vol. 77, no. 4, pp. 621–632, 2010.

[89] S. K. Kidd and J. S. Schneider, "Protective effects of valproic acid on the nigrostriatal dopamine system in a 1-methyl-4-phenyl-1,2,3,6-tetrahydropyridine mouse model of Parkinson's disease," *Neuroscience*, vol. 194, pp. 189–194, 2011.

[90] S. H. Chen, H. M. Wu, B. Ossola et al., "Suberoylanilide hydroxamic acid, a histone deacetylase inhibitor, protects dopaminergic neurons from neurotoxin-induced damage," *British Journal of Pharmacology*, vol. 165, no. 2, pp. 494–505, 2012.

[91] S. Vartiainen, P. Pehkonen, M. Lakso, R. Nass, and G. Wong, "Identification of gene expression changes in transgenic C. elegans overexpressing human *α-synuclein*," *Neurobiology of Disease*, vol. 22, no. 3, pp. 477–486, 2006.

[92] F. Gillardon, M. Mack, W. Rist et al., "MicroRNA and proteome expression profiling in early-symptomatic *α-synuclein*(A30P)-transgenic mice," *Proteomics*, vol. 2, no. 5, pp. 697–705, 2008.

[93] E. Junn, K. W. Lee, S. J. Byeong, T. W. Chan, J. Y. Im, and M. M. Mouradian, "Repression of *α-synuclein* expression and toxicity by microRNA-7," *Proceedings of the National Academy of Sciences of the United States of America*, vol. 106, no. 31, pp. 13052–13057, 2009.

[94] E. Doxakis, "Post-transcriptional regulation of *α-synuclein* expression by mir-7 and mir-153," *Journal of Biological Chemistry*, vol. 285, no. 17, pp. 12726–12734, 2010.

[95] S. Asikainen, M. Rudgalvyte, L. Heikkinen et al., "Global microRNA expression profiling of caenorhabditis elegans Parkinson's disease models," *Journal of Molecular Neuroscience*, vol. 41, no. 1, pp. 210–218, 2010.

[96] S. Gehrke, Y. Imai, N. Sokol, and B. Lu, "Pathogenic LRRK2 negatively regulates microRNA-mediated translational repression," *Nature*, vol. 466, no. 7306, pp. 637–641, 2010.

[97] J. Kim, K. Inoue, J. Ishii et al., "A microRNA feedback circuit in midbrain dopamine neurons," *Science*, vol. 317, no. 5842, pp. 1220–1224, 2007.

[98] E. Miñones-Moyano, S. Porta, G. Escaramís et al., "MicroRNA profiling of Parkinson's disease brains identifies early downregulation of miR-34b/c which modulate mitochondrial function," *Human Molecular Genetics*, vol. 20, no. 15, pp. 3067–3078, 2011.

[99] R. Margis, R. Margis, and C. R.M. Rieder, "Identification of blood microRNAs associated to Parkinsonós disease," *Journal of Biotechnology*, vol. 152, no. 3, pp. 96–101, 2011.

[100] T. Maeda, J. Z. Guan, J. I. Oyama, Y. Higuchi, and N. Makino, "Aging-associated alteration of subtelomeric methylation in Parkinson's disease," *Journals of Gerontology A*, vol. 64, no. 9, pp. 949–955, 2009.

[101] M. Cai, J. Tian, G.-H. Zhao, W. Luo, and B.-R. Zhang, "Study of methylation levels of parkin gene promoter in Parkinson's disease patients," *International Journal of Neuroscience*, vol. 121, no. 9, pp. 497–502, 2011.

[102] I. Kagara, H. Enokida, K. Kawakami et al., "CpG hypermethylation of the *UCHL1* gene promoter is associated with pathogenesis and poor prognosis in renal cell carcinoma," *Journal of Urology*, vol. 180, no. 1, pp. 343–351, 2008.

[103] J. Yu, Q. Tao, K. F. Cheung et al., "Epigenetic identification of ubiquitin carboxyl-terminal hydrolase L1 as a functional tumor suppressor and biomarker for hepatocellular carcinoma and other digestive tumors," *Hepatology*, vol. 48, no. 2, pp. 508–518, 2008.

[104] M. Barrachina and I. Ferrer, "DNA methylation of Alzheimer disease and tauopathy-related genes in postmortem brain," *Journal of Neuropathology and Experimental Neurology*, vol. 68, no. 8, pp. 880–891, 2009.

[105] H. C. Pieper, B. O. Evert, O. Kaut, P. F. Riederer, A. Waha, and U. Wüllner, "Different methylation of the TNF-alpha promoter in cortex and substantia nigra: implications for selective neuronal vulnerability," *Neurobiology of Disease*, vol. 32, no. 3, pp. 521–527, 2008.

[106] M. I. Behrens, N. Brüggemann, P. Chana et al., "Clinical spectrum of Kufor-Rakeb syndrome in the Chilean kindred with *ATP13A2* mutations," *Movement Disorders*, vol. 25, no. 12, pp. 1929–1937, 2010.

[107] S. L. Berger, "The complex language of chromatin regulation during transcription," *Nature*, vol. 447, no. 7143, pp. 407–412, 2007.

[108] L. Chouliaras, B. P. F. Rutten, G. Kenis et al., "Epigenetic regulation in the pathophysiology of Alzheimer's disease," *Progress in Neurobiology*, vol. 90, no. 4, pp. 498–510, 2010.

[109] G. Du, X. Liu, X. Chen et al., "Drosophila histone deacetylase 6 protects dopaminergic neurons against *α-synuclein* toxicity by promoting inclusion formation," *Molecular Biology of the Cell*, vol. 21, no. 13, pp. 2128–2137, 2010.

[110] Y. Miki, F. Mori, K. Tanji, A. Kakita, H. Takahashi, and K. Wakabayashi, "Accumulation of histone deacetylase 6, an aggresome-related protein, is specific to Lewy bodies and glial cytoplasmic inclusions," *Neuropathology*, vol. 31, no. 6, pp. 561–568, 2011.

[111] J. A. Olzmann, L. Li, and L. S. Chin, "Aggresome formation and neurodegenerative diseases: therapeutic implications," *Current Medicinal Chemistry*, vol. 15, no. 1, pp. 47–60, 2008.

[112] M. Su, J.-J. Shi, Y.-P. Yang et al., "HDAC6 regulates aggresome-autophagy degradation pathway of *α-synuclein* in response to $MPP^+$-induced stress," *Journal of Neurochemistry*, vol. 117, no. 1, pp. 112–120, 2011.

[113] G. Wang, J. M. van der Walt, G. Mayhew et al., "Variation in the miRNA-433 binding site of FGF20 confers risk for Parkinson disease by overexpression of *α-synuclein*," *American Journal of Human Genetics*, vol. 82, no. 2, pp. 283–289, 2008.

[114] L. de Mena, L. F. Cardo, E. Coto et al., "*FGF20* rs*12720208* SNP and microRNA-433 variation: no association with Parkinson's disease in Spanish patients," *Neuroscience Letters*, vol. 479, no. 1, pp. 22–25, 2010.

[115] S. R. D'Mello, "Histone deacetylases as targets for the treatment of human neurodegenerative diseases," *Drug News and Perspectives*, vol. 22, no. 9, pp. 513–524, 2009.

[116] M. M. Harraz, T. M. Dawson, and V. L. Dawson, "MicroRNAs in Parkinson's disease," *Journal of Chemical Neuroanatomy*, vol. 42, no. 2, pp. 127–130, 2011.

# 13

# Genetic Dissection of Sympatric Populations of Brown Planthopper, *Nilaparvata lugens* (Stål), Using DALP-PCR Molecular Markers

**M. A. Latif,[1,2] M. Y. Rafii,[1,3] M. S. Mazid,[3] M. E. Ali,[4] F. Ahmed,[1] M. Y. Omar,[5] and S. G. Tan[6]**

[1] *Department of Crop Science, Faculty of Agriculture, Universiti Putra Malaysia, 43400, Serdang, Selangor, Malaysia*
[2] *Plant Pathology Division, Bangladesh Rice Research Institute (BRRI), Gazipur 1701, Bangladesh*
[3] *Institute of Tropical Agriculture, Universiti Putra Malaysia, 43400 UPM Serdang, Selangor, Malaysia*
[4] *Institute of Nano Electronic Engineering (INNE), Universiti Malaysia Perlis, 01000 Kangar, Malaysia*
[5] *Department of Biology, Faculty of Science, Universiti Putra Malaysia, 43400, Serdang, Selangor, Malaysia*
[6] *Department of Cell and Molecular Biology, Faculty of Biotechnology and Molecular Science, Universiti Putra Malaysia, 43400, Serdang, Selangor, Bangladesh*

Correspondence should be addressed to M. A. Latif, alatif1965@yahoo.com

Academic Editors: S. Mastana, T. Tanisaka, and Y. Yu

Direct amplified length polymorphism (DALP) combines the advantages of a high-resolution fingerprint method and also characterizing the genetic polymorphisms. This molecular method was also found to be useful in brown planthopper, *Nilaparvata lugens* species complex for the analysis of genetic polymorphisms. A total of 11 populations of *Nilaparvata* spp. were collected from 6 locations from Malaysia. Two sympatric populations of brown planthopper, *N. lugens*, one from rice and the other from a weed grass (*Leersia hexandra*), were collected from each of five locations. *N. bakeri* was used as an out group. Three oligonucleotide primer pairs, DALP231/DALPR′5, DALP234/DALPR′5, and DALP235/DALPR′5 were applied in this study. The unweighted pair group method with arithmetic mean (UPGMA) dendrogram based on genetic distances for the 11 populations of *Nilaparvata* spp. revealed that populations belonging to the same species and the same host type clustered together irrespective of their geographical localities of capture. The populations of *N. lugens* formed into two distinct clusters, one was insects with high esterase activities usually captured from rice and the other was with low esterase activities usually captured from *L. hexandra*. *N. bakeri*, an out group, was the most isolated group. Analyses of principal components, molecular variance, and robustness also supported greatly to the findings of cluster analysis.

## 1. Introduction

The brown plant hopper, *Nilaparvata lugens* (Stål) (Homoptera: Delphacidae), is a major pest of rice, which is widely distributed from tropical to temperate areas of Asia and Australia. The insect is a phloem-feeder and is restricted to cultivated and wild rice as host plants. It causes "hopper-burn" and complete wilting and drying of rice plants [1] and also transmits the grassy stunt and ragged stunt viral diseases [2]. Large-scale rice crop damage caused by the pest was reported in the 1970s in several South and Southeast Asian countries [1]. Another population of brown plant hopper was found to infest a weed grass, *Leersia hexandra*, which grows abundantly in canal near paddy fields in South East Asia [3, 4]. The weed infesting population of *N. lugens* fails to survive on rice plants. Conversely, rice infesting population of *N. lugens* does not thrive on grass [5]. Based on nymphal survival, virulence, ovipositional preference, mate choice, and hybridization experiments, Claridge et al. [5] suggested that the rice and *Leersia* infesting populations of brown planthopper (BPH) represented two distinct sympatric biological species. In recent studies, the analyses of isozymes and RAPD-PCR markers indicated that BPH with high esterase activity usually captured from rice plant, and those with esterase activity usually captured from *L. hexandra* in Malaysia represent two distinct closely related sibling species [6, 7].

Direct amplification of length polymorphisms (DALPs) is a technique which uses arbitrarily primed PCR (AP-PCR) to produce genomic finger prints and to enable sequencing of DNA polymorphisms in any species. Oligonucleotide pairs were designed to produce a specific multibanded pattern for each individual of a population and between populations. This strategy combines the advantages of a high-resolution fingerprint technique and also characterizing the polymorphisms [8]. Higher number of polymorphic loci could be detected and isolated for sequencing in only one step. Therefore, this method is not simply another supplementary molecular fingerprinting technique but was designed from the very beginning to obtain nucleotide sequence information on DNA fragments from any genome with no need for a genomic library.

Genetic polymorphic markers, such as isozymes, RAPD and SSR, and nuclear or organelle DNA polymorphism, have been developed for a variety of studies on genetic diversity, population structures, and subdivisions [6, 7, 9–11]. The present study was undertaken to analyze genetic diversity as well as to detect genetic structures between two sympatric populations of *N. lugens*, one from rice and the other from weed grass, *L. hexandra*. We hypothesized that the molecular method newly applied in rice brown planthopper could be able to detect structures among the populations of brown planthopper species complex.

## 2. Experimental Section

*2.1. Collection of Insect Populations.* A total of 11 populations were collected from 6 locations. Two sympatric populations of *N. lugens*, one from rice and other from *L. hexandra*, were collected from each of five locations. The locations were Universiti Putra Malaysia (UPM), Tanjung Karang (TK), Melaka (MK), Perak (PK), and Sabah (SB). An out group, *N. bakeri* was also collected from Cameron Highlands (CH), Malaysia. Locations, host type, date of collection, population code are shown in Table 1. Each population consisted of twenty insects. All collected insects were frozen at −70°C for further use.

*2.2. Esterase Activity Test.* The individual insect used for DALP analysis was tested for esterase activity on a simple filter paper using the method reported by Pasteur and Georghiou [13].

*2.3. DNA Extraction.* DNA from individual insect was extracted by grinding single frozen adult insect with a glass rod in a 1.5 mL tube containing 20 $\mu$L extraction buffer (0.1 M Nacl, 0.2 M Sucrose, 0.1 M Tris-HCL (pH 9.0) 0.05 M EDTA, 0.5% SDS). The glass rod was washed with an additional 40 $\mu$L of extraction buffer and the homogenate was incubated at 65°C for 40 min. An amount of 10 $\mu$L of 8 M potassium acetate was added and the tube was placed on ice for 40 min. The tube was spun at 14000 rpm for 20 min. The supernatant was transferred into a fresh 1.5 mL tube. One hundred microliters of chilled (−20°C) 100% ethanol was added and the DNA was allowed to precipitate at room temperature for 10 min. The tube was spun for 20 min and the ethanol was carefully removed with a pipette. The DNA pellet was washed with 100 $\mu$L of chilled 70% ethanol and spun for 10 min. The DNA pellets were dried by pouring off the ethanol. The tubes were kept for 10 min at room temperature. The dried DNA pellet was suspended in 50 $\mu$L TE (Tris EDTA, pH 8.0) and gently mixed for 10 min. The DNA concentration was measured using LKB-Ultrastep III UV/visible spectrophotometer at the absorbance of 260 nm and 280 nm. The DNA was considered pure if the ratio of $OD_{260}/OD_{280}$ was within the range of 1.6–1.9 [12].

*2.4. Primers Used in This Study.* A total of three forward sequencing primer denoted as DALP231, DALP234, DALP235 and a universal reverse primer, DALPR (5′TTTCACACAGGAAACAGCTATGAC-3′), were used for the PCR amplification (Table 2). Primer DALPR was end labeled with $Y^{33}$ PATP (10 m Cie/mL) [14].

*2.5. PCR Protocols.* The PCR reaction mixture contained 60 ng of insect DNA, 1.8 mM of $Mg^{+}$, 0.15 $\mu$M of oligonucletide primer, 200 $\mu$M of each dNTP, 1 unit Taq DNA polymerase (Promega), and 1x PCR buffer in a total volume of 25 $\mu$L. Amplification reactions were carried out in a programmable thermal cycler (GeneAmp, PCR system 2400, Perkin Elmer) programmed as follows: predenaturation at 95°C for 2 min; followed by 30 cycles of denaturation at 91°C for 30 sec, annealing temperature at 55°C for 45 sec, and extension at 70°C for 30 sec. After the last cycle, final extension was at 70°C for 5 min. The protocols were modified from [8].

*2.6. Electrophoresis of the Multilocus Amplification Products.* Electrophoresis was performed on 6% denaturing polyacrylamide gels and run on a 50 cm long gel apparatus. The samples were mixed with 5 $\mu$L 100% formamide loading dye and then heated for 10 min at 96°C before loading. The gel was run at 55 W for 3 hours.

*2.7. Autoradiography.* After electrophoresis, the gel was transferred to a Whatman paper and dried and developed after 5 days of exposures to X-ray film.

*2.8. Statistical Analysis*

*Band Scoring.* DALP-PCR band profiles were scored visually for each DNA sample for each primer pair. The data was recorded according to the presence/absence criterion (1 = presence; 0 = absence of band).

*Cluster Analysis.* The Dice algorithm was used for similarity index. The similarity index was calculated between two samples from within or between populations according to [15]:

$$S_{xy} = \frac{2m_{xy}}{\left(m_x + m_y\right)}, \tag{1}$$

Table 1: Host types, sites of collection, and coding for 11 populations of *Nilaparvata* spp.

| Species | Locations | Host plants | Population code |
|---|---|---|---|
| *N. lugens* | Universiti Putra Malaysia (UPM), Selangor, Malaysia | Rice | UPM1 |
| *N. lugens* | UPM, Selangor, Malaysia | *L. hexandra* | UPM2 |
| *N. lugens* | Tanjung Karang (Tk), Selangor, Malaysia | Rice | TK1 |
| *N. lugens* | Tanjung Karang (Tk), Selangor, Malaysia | *L. hexandra* | TK2 |
| *N. lugens* | Malim, Melaka (Mk), Malaysia | Rice | MK1 |
| *N. lugens* | Malim, Melaka(Mk), Malaysia | *L. hexandra* | MK2 |
| *N. lugens* | Bander Seberang, Perak (Pk), Malaysia | Rice | PK1 |
| *N. lugens* | Bander, Seberang, Perak (Pk), Malaysia | *L. hexandra* | PK2 |
| *N. lugens* | Tuaran, Sabah (SB), Malaysia | Rice | SB1 |
| *N. lugens* | Tuaran, Sabah (SB), Malaysia | *L. hexandra* | SB2 |
| *N. bakeri* | Cameron Highlands (CH), Pahang, Malaysia | *L. hexandra* | CH |

Table 2: Optimisation of DALP primers used for the PCR protocols.

| Primers | Sequence | Mg+2 conc. | Taq polymerase used | Annealing temperature (°C) |
|---|---|---|---|---|
| DALP231F | 5′-GTTTTCCCAGTCACGACAGC-3′ | 1.8 mM | Promega | 55 |
| DALP234F | 5′-GTTTTCCCAGTCACGACCAG-3′ | 1.8 mM | Promega | 55 |
| DALP235F | 5′-GTTTTCCCAGTCACGACCAC-3′ | 1.8 mM | Promega | 55 |
| Universal DALPR | 5′-TTTCACACAGGAAACAGCTATGAC-3′ | | | |

where $m_{xy}$ is the number of bands showed by sample $x$ and sample $y$ and $m_x$ and $m_y$ are the number of bands in sample $x$ and sample $y$, respectively. The value produced by this index ranges from 0 (representing no band sharing) to 1 (representing complete identity). The within or between population values are based on pairwise comparisons between individuals for a particular primer. The values obtained are then averaged over primers.

The between population similarity indices were also converted to distance values using the relationship $D = 1 - S$ [16, 17]. These distance matrices were used as the input matrix for the unweighted pair group method with arithmetic mean (UPGMA) tree [18] to find population relationships graphically using NTSYS-PC software (version 1.8; [19]).

*Test of Robustness.* The test of robustness or bootstrapping was performed using the Phylogeny Inference Package (PHYLIP; version 3.5p) developed by Felsenstein [20]. The bootstrap values were obtained using gene frequencies option within the program PHYLIP. A consensus tree was produced based on the 1000 bootstrapped replicates as reported by Haymer et al. [21].

*Principle Component Analysis.* A principal component analysis was performed based on the distance matrix among the populations using the NTSYS-PC software. The relationship among the populations is expressed in a three-dimensional graph based on the first three components.

*Analysis of Molecular Variance (AMOVA).* The distance between two samples was calculated according to the formula of Excoffier et al. [22]:

$$D = N\left\{1 - \left(\frac{N_{xy}}{N}\right)\right\}, \qquad (2)$$

where $N$ is the total number of bands and $N_{xy}$ is the number of bands shared by two samples. The resulting distance matrix was used in an AMOVA [22]. In the AMOVA, the sources of variation were divided into three nested levels: among the host types, among the populations within host types, and among individuals within populations. Mean square deviation was calculated by dividing sum of squared deviation by the degrees of freedom. The variance component was expressed as percentage. The significance of components of variance was tested by the random permutation.

## 3. Results

Band or marker frequency was calculated for each marker pair for each population for DALP primers. Figure 1 shows the banding patterns obtained from rice and *Leersia* infesting populations of *N. lugens* using primer DALP235/DALPR. A hundred percent marker frequency represented monomorphism while 0% showed complete absence of the particular marker. The data showed a range of 28.3–42.9% polymorphic markers for rice infesting populations of *N. lugens* while *Leersia* infesting populations and an out group, *N. bakeri* showed 31.9–45.5% and 17.1% polymorphic markers, respectively. The overall data for 10 populations of *N. lugens* based on three primers showed 29 (42.6%) polymorphic markers. Frequency of DALP markers, total number of markers, number of polymorphic markers, % polymorphic markers for each population are shown in Table 3.

*3.1. Cluster Analysis.* All data from three pairs of DALP primers were incorporated for cluster analysis. In this analysis, pairwise genetic distances were calculated between all individuals in order to make comparison. The distances within rice infesting populations of *N. lugens* ranged from

Table 3: Frequency of presence of bands (%) markers in *Nilaparvata* spp. obtained from three DALP primers.

| Primer | Marker (M. Wt.) | UPM | | Tanjung Karang | | Melaka | | Perak | | Sabah | | Cameron High lands |
|---|---|---|---|---|---|---|---|---|---|---|---|---|
| | | *N. lugens* (Rice) | *N. lugens* (*Leersia*) | *N. lugens* (Rice) | *N. lugens* (*Leersia*) | *N. lugens* (Rice) | *N.lugens* (*Leersia*) | *N. lugens* (Rice) | *N.lugens* (*Leersia*) | N. lugens (Rice) | *N.lugens* (*Leersia*) | *N.bakeri* (*Leersia*) |
| DALP231 | DALP231.1 (2.15 kb) | 10 | 10 | 0 | 0 | 10 | 20 | 10 | 0 | 45 | 10 | 0 |
| | DALP231.2 (2.0 kb) | 0 | 25 | 0 | 0 | 40 | 30 | 15 | 0 | 50 | 10 | 30 |
| | DALP231.3 (1.9 kb) | 25 | 0 | 0 | 0 | 40 | 35 | 15 | 10 | 25 | 10 | 0 |
| | DALP231.4 (1.8 kb) | 100 | 65 | 100 | 90 | 100 | 90 | 100 | 80 | 70 | 100 | 80 |
| | DALP231.5 (1.7 kb) | 100 | 100 | 100 | 100 | 100 | 100 | 100 | 100 | 100 | 100 | 90 |
| | DALP231.6 (1.6 kb) | 100 | 100 | 100 | 100 | 100 | 100 | 100 | 100 | 100 | 100 | 0 |
| | DALP231.7 (1.4 kb) | 60 | 100 | 100 | 80 | 50 | 100 | 100 | 60 | 70 | 100 | 90 |
| | DALP231.8 (1.3 kb) | 100 | 100 | 100 | 100 | 100 | 100 | 100 | 100 | 100 | 100 | 100 |
| | DALP231.9 (1.2 kb) | 100 | 100 | 100 | 100 | 100 | 100 | 100 | 100 | 100 | 100 | 0 |
| | DALP231.10 (1.0 kb) | 0 | 60 | 0 | 50 | 0 | 40 | 0 | 60 | 0 | 40 | 100 |
| | DALP231.11 (0.9 kb) | 0 | 10 | 10 | 0 | 0 | 20 | 0 | 0 | 0 | 30 | 0 |
| | DALP231.12 (0.85 kb) | 30 | 40 | 25 | 50 | 30 | 30 | 20 | 30 | 40 | 35 | 80 |
| | DALP231.13 (0.75 kb) | 0 | 0 | 0 | 0 | 0 | 0 | 0 | 0 | 0 | 0 | 100 |
| | DALP231.14 (0.70 kb) | 0 | 0 | 0 | 0 | 0 | 0 | 0 | 0 | 0 | 0 | 100 |
| | DALP231.15 (0.60 kb) | 0 | 0 | 0 | 0 | 0 | 0 | 0 | 0 | 30 | 30 | 0 |
| | DALP231.16 (0.50 kb) | 0 | 0 | 0 | 0 | 0 | 0 | 0 | 0 | 0 | 0 | 100 |
| | DALP231.17 (0.40 kb) | 100 | 100 | 100 | 100 | 100 | 100 | 100 | 100 | 100 | 100 | 100 |
| | DALP231.18 (0.30 kb) | 100 | 100 | 100 | 100 | 100 | 100 | 100 | 100 | 100 | 100 | 0 |
| | DALP231.19 (0.25 kb) | 100 | 100 | 100 | 100 | 100 | 100 | 100 | 100 | 100 | 100 | 100 |
| | DALP231.20 (0.20 kb) | 80 | 60 | 100 | 100 | 100 | 100 | 100 | 100 | 100 | 100 | 0 |

Table 3: Continued.

| Primer | Marker (M. Wt.) | UPM | | Tanjung Karang | | Melaka | | Perak | | Sabah | | Cameron High lands |
|---|---|---|---|---|---|---|---|---|---|---|---|---|
| | | *N. lugens* (Rice) | *N. lugens* (*Leersia*) | *N. lugens* (Rice) | *N. lugens* (*Leersia*) | *N. lugens* (Rice) | *N.lugens* (*Leersia*) | *N. lugens* (Rice) | *N.lugens* (*Leersia*) | N. lugens (Rice) | *N.lugens* (*Leersia*) | *N.bakeri* (*Leersia*) |
| | DALP231.21 (0.10 kb) | 100 | 100 | 100 | 100 | 100 | 100 | 100 | 100 | 100 | 100 | 100 |
| | DALP231.22 (0.09 kb) | 0 | 0 | 0 | 0 | 0 | 0 | 0 | 0 | 0 | 0 | 100 |
| | DALP234.1 (2.0 kb) | 100 | 100 | 100 | 100 | 100 | 100 | 100 | 100 | 100 | 100 | 0 |
| | DALP234.2 (1.9 kb) | 100 | 100 | 100 | 100 | 100 | 100 | 100 | 100 | 100 | 100 | 0 |
| | DALP234.3 (1.8 kb) | 20 | 50 | 25 | 50 | 20 | 50 | 20 | 30 | 30 | 50 | 0 |
| | DALP234.4 (1.75 kb) | 0 | 0 | 0 | 0 | 0 | 0 | 0 | 0 | 0 | 0 | 100 |
| | DALP234.5 (1.70 kb) | 100 | 100 | 100 | 100 | 100 | 100 | 100 | 100 | 100 | 100 | 0 |
| | DALP234.6 (1.65 kb) | 0 | 70 | 0 | 40 | 0 | 40 | 0 | 60 | 0 | 40 | 100 |
| DALP234 | DALP234.7 (1.6 kb) | 0 | 10 | 30 | 0 | 0 | 20 | 0 | 0 | 0 | 30 | 0 |
| | DALP234.8 (1.5 kb) | 40 | 40 | 60 | 50 | 30 | 30 | 20 | 30 | 40 | 35 | 100 |
| | DALP234.9 (1.45 kb) | 0 | 0 | 0 | 0 | 0 | 0 | 0 | 0 | 0 | 0 | 100 |
| | DALP234.10 (1.3 kb) | 100 | 100 | 100 | 100 | 100 | 100 | 100 | 100 | 100 | 100 | 100 |
| | DALP234.11 (1.25 kb) | 0 | 0 | 0 | 0 | 0 | 0 | 0 | 0 | 30 | 30 | 0 |
| | DALP234.12 (1.10 kb) | 0 | 0 | 0 | 0 | 0 | 0 | 0 | 0 | 0 | 0 | 100 |
| | DALP234.13 (1.0 kb) | 100 | 100 | 100 | 100 | 100 | 100 | 100 | 100 | 100 | 100 | 100 |
| | DALP234.14 (0.9 kb) | 100 | 100 | 100 | 100 | 100 | 100 | 100 | 100 | 100 | 100 | 0 |
| | DALP234.15 (0.85 kb) | 100 | 100 | 100 | 100 | 100 | 100 | 100 | 100 | 100 | 100 | 100 |

Table 3: Continued.

| Primer | Marker (M. Wt.) | UPM | | Tanjung Karang | | Melaka | | Perak | | Sabah | | Cameron High lands |
|---|---|---|---|---|---|---|---|---|---|---|---|---|
| | | *N. lugens* (Rice) | *N. lugens* (*Leersia*) | *N. lugens* (Rice) | *N. lugens* (*Leersia*) | *N. lugens* (Rice) | *N.lugens* (*Leersia*) | *N. lugens* (Rice) | *N.lugens* (*Leersia*) | N. lugens (Rice) | *N.lugens* (*Leersia*) | *N.bakeri* (*Leersia*) |
| | DALP234.16 (0.7 kb) | 100 | 100 | 100 | 100 | 100 | 100 | 100 | 100 | 100 | 100 | 0 |
| | DALP234.17 (0.65 kb) | 90 | 90 | 100 | 100 | 100 | 100 | 90 | 100 | 100 | 90 | 100 |
| | DALP234.18 (0.5 kb) | 0 | 0 | 0 | 0 | 0 | 0 | 0 | 0 | 0 | 0 | 100 |
| | DALP234.19 (0.35 kb) | 40 | 50 | 100 | 100 | 100 | 100 | 100 | 100 | 80 | 90 | 100 |
| | DALP234.20 (0.30 kb) | 80 | 20 | 80 | 70 | 60 | 40 | 50 | 90 | 70 | 80 | 100 |
| | DALP234.21 (0.15 kb) | 40 | 50 | 90 | 100 | 80 | 100 | 100 | 100 | 80 | 90 | 100 |
| | DALP235.1 (2.2 kb) | 5 | 10 | 10 | 8 | 0 | 10 | 9 | 20 | 40 | 35 | 0 |
| | DALP235.2 (2.1 kb) | 100 | 100 | 100 | 100 | 100 | 100 | 100 | 100 | 100 | 100 | 100 |
| | DALP235.3 (2.0 kb) | 100 | 100 | 100 | 100 | 100 | 100 | 100 | 100 | 100 | 100 | 0 |
| | DALP235.4 (1.9 kb) | 100 | 100 | 100 | 100 | 100 | 100 | 100 | 100 | 100 | 100 | 100 |
| | DALP235.5 (1.8 kb) | 10 | 40 | 15 | 50 | 20 | 35 | 16 | 30 | 10 | 50 | 0 |
| | DALP235.6 (1.75 kb) | 0 | 0 | 0 | 0 | 0 | 0 | 0 | 0 | 0 | 0 | 100 |
| | DALP235.7 (1.70 kb) | 100 | 100 | 100 | 100 | 100 | 100 | 100 | 100 | 100 | 100 | 0 |
| | DALP235.8 (1.65 kb) | 0 | 60 | 0 | 50 | 0 | 40 | 0 | 60 | 0 | 40 | 100 |
| | DALP235.9 (1.6 kb) | 0 | 10 | 10 | 0 | 0 | 20 | 0 | 0 | 0 | 30 | 0 |
| | DALP235.10 (1.5 kb) | 30 | 40 | 25 | 50 | 30 | 30 | 20 | 30 | 40 | 35 | 100 |

Table 3: Continued.

| Primer | Marker (M. Wt.) | UPM | | Tanjung Karang | | Melaka | | Perak | | Sabah | | Cameron High lands |
|---|---|---|---|---|---|---|---|---|---|---|---|---|
| | | *N. lugens* (Rice) | *N. lugens* (*Leersia*) | *N. lugens* (Rice) | *N. lugens* (*Leersia*) | *N. lugens* (Rice) | *N.lugens* (*Leersia*) | *N. lugens* (Rice) | *N.lugens* (*Leersia*) | N. lugens (Rice) | *N.lugens* (*Leersia*) | *N.bakeri* (*Leersia*) |
| DALP235 | DALP235.11 (1.4 kb) | 0 | 0 | 0 | 0 | 0 | 0 | 0 | 0 | 0 | 0 | 100 |
| | DALP235.12 (1.3 kb) | 0 | 0 | 0 | 0 | 0 | 0 | 0 | 0 | 0 | 0 | 100 |
| | DALP235.13 (1.2 kb) | 0 | 0 | 0 | 0 | 0 | 0 | 0 | 0 | 30 | 30 | 0 |
| | DALP235.14 (1.15 kb) | 0 | 0 | 0 | 0 | 0 | 0 | 0 | 0 | 0 | 0 | 100 |
| | DALP235.15 (1.0 kb) | 100 | 100 | 100 | 100 | 100 | 100 | 100 | 100 | 100 | 100 | 100 |
| | DALP235.16 (0.9 kb) | 100 | 100 | 100 | 100 | 100 | 100 | 100 | 100 | 100 | 100 | 100 |
| | DALP235.17 (0.8 kb) | 100 | 100 | 100 | 100 | 100 | 100 | 100 | 100 | 100 | 100 | 100 |
| | DALP235.18 (0.7 kb) | 100 | 100 | 100 | 100 | 100 | 100 | 100 | 100 | 100 | 100 | 0 |
| | DALP235.19 (0.6 kb) | 100 | 100 | 100 | 100 | 100 | 100 | 100 | 100 | 100 | 100 | 100 |
| | DALP235.20 (0.5 kb) | 0 | 0 | 0 | 0 | 0 | 0 | 0 | 0 | 0 | 0 | 100 |
| | DALP235.21 (0.35 kb) | 30 | 50 | 100 | 100 | 100 | 100 | 100 | 100 | 80 | 90 | 70 |
| | DALP235.22 (0.30 kb) | 90 | 20 | 80 | 70 | 60 | 40 | 50 | 90 | 70 | 80 | 100 |
| | DALP235.23 (0.20 kb) | 60 | 90 | 50 | 30 | 80 | 100 | 90 | 60 | 40 | 75 | 100 |
| | DALP235.24 (0.15 kb) | 100 | 100 | 80 | 100 | 100 | 100 | 100 | 100 | 90 | 100 | 90 |
| | DALP235.25 (0.10 kb) | 100 | 100 | 100 | 100 | 100 | 100 | 100 | 100 | 100 | 100 | 100 |
| Total Number of markers | | 45 | 51 | 46 | 46 | 45 | 52 | 46 | 47 | 49 | 55 | 41 |
| Number of polymorphic markers | | 17 | 23 | 14 | 14 | 13 | 18 | 13 | 15 | 21 | 25 | 7 |
| % Polymorphic markers | | 37.77 | 45.09 | 30.43 | 30.43 | 28.88 | 34.61 | 28.26 | 31.91 | 42.85 | 45.45 | 17.08 |
| Overall number of markers based on the populations of *N. lugens* | | | | | | | | | | | | 68 |
| Overall number of polymorphic markers based on the populations of *N. lugens* | | | | | | | | | | | | 29 |
| Overall % poymorphism based on the populations of *N. lugens* | | | | | | | | | | | | 42.64 |

Frequency based upon 20 individuals per population.

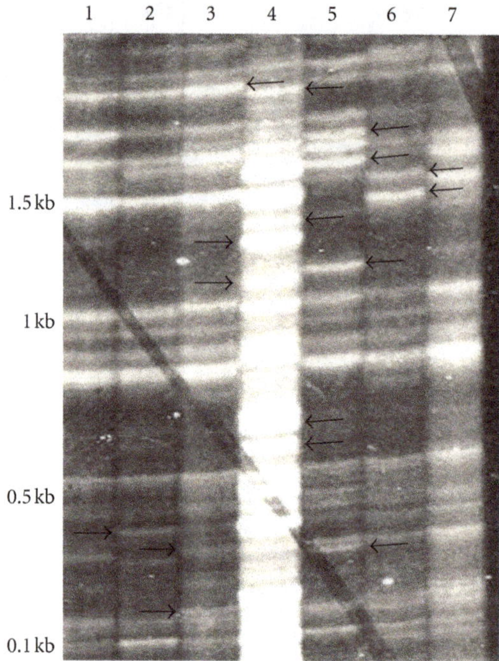


FIGURE 1: DALP-PCR amplicons obtained from rice and *Leersia* infesting populations of *N. lugens* using primer DALP235/DALPR (lanes 1–3 = rice infesting populations, lanes 5–7 = *Leersia* infesting populations; lane 4 = An out group, *N. bakeri*). Polymorphic markers showed in arrow sign.

0.112487 to 0.285200 (average 0.2245843) while distances within *Leersia* infesting populations ranged from 0.152379 to 0.235396 (average 0.2078274). The genetic distances between two sympatric populations of *N. lugens*, one from rice and the other from grass, ranged from 0.24019 to 0.390182 (average 0.31672).

In addition to that genetic distances between rice infesting population of *N. lugens* and *Leersia* infesting populations of *N. bakeri* (out group) ranged from 0.555389 to 0.564963 (average 0.540276) but it was ranged from 0.499403 to 0.578171 (average 0.539168) between the populations of *Leersia* infesting *N. lugens* and *N. bakeri* (Table 4).

UPGMA dendrogram revealed the genetic relationships among the 11 populations of *Nilaparvata* species. The cluster analysis divided the individuals into three main clusters. Among the three clusters, one was the most distinct and distant and the other two were closely related. All rice infesting populations like UPM1, MK1, TK1, PK1, and SB1 were included in one cluster, likewise the *Leersia* infesting populations such as UPM2, TK2, PK2, MK2, and SB2 were separated into another group. A common branch was shared by both groups. The isolated population CH (*N. bakeri*) was far away from either rice or *Leersia* infesting populations of *N. lugens* (Figure 2).

*3.2. Test of Robustness.* The UPGMA tree was subjected to numerical resampling by bootstrapping [23] and the resultant bootstrap values were shown at the tree branch points. Each value represents the number of times that the represented groupings occurred in the resamplings. The consensus tree showed 100% confidence levels between rice (MK1, UPM1, TK1, PK1, and SB1) and *Leersia* infesting (MK2, TK2, UPM2, SB2, and PK2) population. Within rice and *Leersia* infesting populations, confidence level ranged from 37–56% to 35–51%, respectively (Figure 2). The confidence level between *N. lugens* and *N. bakeri* was also 100%.

*3.3. Principal Component Analysis (PCA).* A principal component analysis was performed based on the distance matrix among the populations using the NTSYS-PC software. The relationship among the populations was expressed in a three-dimensional graph. In PCA graph, 11 populations were clustered into 3 groups. The cluster I consisted of rice infesting population of *N. lugens* while *Leersia* infesting populations showed another group. The population of *N. bakari* showed an out group. The first three principal components accounted for 78.31% of the total variation among the 11 populations of *Nilaparvata* spp. and these 3 components, PC1, PC2, and PC3 showed 41.19, 27.42, 9.70% variation, respectively (Figure 3).

*3.4. Analysis of Molecular Variance (AMOVA).* Three level nested structures for each pair of primer of DALP are shown in Table 5. All primers showed variance among the host types, among the populations, and among the individuals in a population. Out of three primers, DALP 235 determined the highest variance among the groups (rice versus *Leersia*) (26.90%), followed by DALP231 (10.10%), and DALP234 (9.68%). The percentage of the variance component among groups (rice versus *Leersia*) was greater than the percentage of the variance component among the populations detected by the three primers DALP235, DALP234 and DALP231. The results of AMOVA as well as dendrogram confirmed that genetic variation exists between the brown plant hopper of rice and *Leersia*.

## 4. Discussion

Cluster and principal component analyses revealed the genetic relationships among the different populations of *Nilaparvata* species. Three major clusters were observed in the dendrogram as well as in the graph. The results showed that population of *N. bakeri* formed the most isolated cluster from populations of either rice or *Leersia* infesting populations of *N. lugens*. The rice infesting populations of UPM, Tanjung Karang, Melaka, Perak, and Sabah, Malaysia clustered together as a group. On the other hand, *Leersia* infesting populations of the same localities formed another distinct cluster. *Leersia* infesting populations with low esterase activities seem to be formed, a different structure from rice infesting populations of brown planthopper, *N. lugens*. These results were also confirmed by bootstrapping analysis as described by Felsenstein [23] and Latif et al. [6]. Bootstrapping was initially used to evaluate the accuracy of a tree obtained by the parsimony method and could increase the confidence level of the results obtained from the DALP assay. The results showed 100% confidence level for the separate clusterings

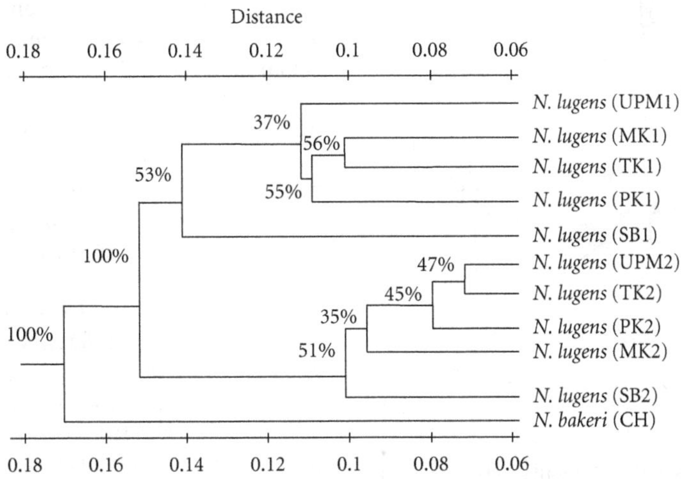


FIGURE 2: UPGMA dendrogram of the 11 populations of *Nilaparvata* spp. based on genetic distance from Dice's index for DALP markers (Rice infesting population of *N. lugens* = UPM1, TK1, MK1, PK1, and SB1; *Leersia*-infesting population of *N. lugens* = UPM2, TK2, MK2, PK2, and SB2; an out group, *N. bakeri* = CH); bootstrap values from 1000 bootstraps are given at each fork.

TABLE 4: Genetic distance matrix of the 11 populations of *Nilaparvata* spp. based on Nei and Li's similarity index.

| | UPM 1 | TK1 | MK1 | PK1 | SB1 | UPM2 | TK2 | MK2 | PK2 | SB2 | CH |
|---|---|---|---|---|---|---|---|---|---|---|---|
| UPM1 | 0 | | | | | | | | | | |
| TK1 | 0.112487 | 0 | | | | | | | | | |
| MK1 | 0.169136 | 0.158045 | 0 | | | | | | | | |
| PK1 | 0.281682 | 0.285200 | 0.244057 | 0 | | | | | | | |
| SB1 | 0.242807 | 0.250772 | 0.228062 | 0.273595 | 0 | | | | | | |
| UPM2 | 0.290265 | 0.292717 | 0.299515 | 0.372207 | 0.327829 | 0 | | | | | |
| TK2 | 0.300214 | 0.285250 | 0.313183 | 0.378135 | 0.343346 | 0.152379 | 0 | | | | |
| MK2 | 0.240193 | 0.253426 | 0.273773 | 0.331707 | 0.284036 | 0.167285 | 0.183129 | 0 | | | |
| PK2 | 0.306063 | 0.316361 | 0.311206 | 0.338466 | 0.326583 | 0.208954 | 0.222898 | 0.197168 | 0 | | |
| SB2 | 0.328876 | 0.332774 | 0.34628 | 0.390182 | 0.335617 | 0.235396 | 0.24859 | 0.228597 | 0.233878 | 0 | |
| CH | 0.555389 | 0.509043 | 0.497337 | 0.574649 | 0.564963 | 0.499403 | 0.525974 | 0.554652 | 0.537638 | 0.578171 | 0 |

UPM1, TK1, MK1, PK1, and SB1 are rice-infesting populations; UPM2, TK2, MK2, PK2, and SB2 are *Leersia*-infesting populations of *N. lugens*. CH is *N. bakeri*.

TABLE 5: Analysis of molecular variance (AMOVA) of 10 populations of *N. lugens* based on three DALP primers.

| Primer | Source of variation | Degree of freedom | Sum of squared deviation | Mean squared deviation | Variance component | % Total | $P^*$ |
|---|---|---|---|---|---|---|---|
| DALP235/DALPR | Rice versus *Leersia* (Among groups) | 1 | 19.50 | 16.81 | 0.159 | 26.90 | <0.001 |
| | Populations within group | 8 | 22.34 | 2.87 | 0.172 | 26.29 | <0.001 |
| | Individuals within population | 190 | 49.50 | 0.29 | 0.280 | 52.03 | <0.001 |
| DALP231/DALPR | Rice versus *Leersia* (Among groups) | 1 | 9.58 | 7.58 | 0.075 | 10.10 | <0.001 |
| | Populations within group | 8 | 9.99 | 2.00 | 0.041 | 6.98 | <0.001 |
| | Individuals within population | 190 | 75.43 | 0.48 | 0.474 | 85.06 | <0.001 |
| DALP234/DALPR | Rice versus *Leersia* (Among groups) | 1 | 3.04 | 3.09 | 0.034 | 9.68 | 0.031 |
| | Populations within group | 8 | 7.07 | 0.82 | 0.03 | 5.78 | 0.006 |
| | Individuals within population | 190 | 66.90 | 0.35 | 0.38 | 95.02 | <0.001 |

*After 1000 random permutations; $P$ = Probability level.

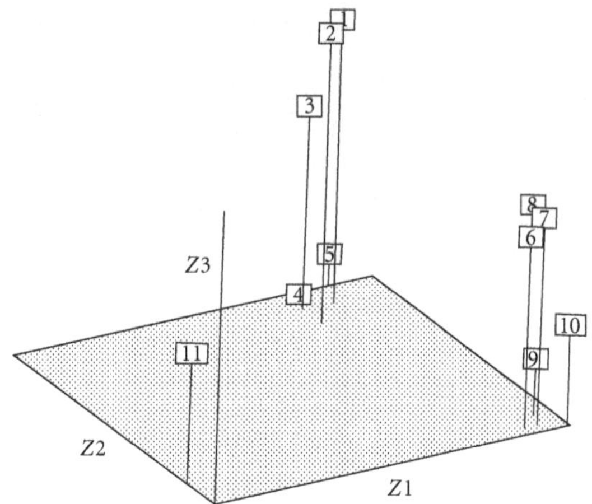


FIGURE 3: Patterns of relationships of the 11 populations revealed by the principal component analysis based on short primer DALP data. Proportion of the total variance explained by the first three principal components (PCs) is 78.31%: PC1 = 41.19%; PC2 = 27.42%; PC3 = 9.70% (Rice infesting populations of *N. lugens*: 1 = UPM1, 2 = TK1, 3 = MK1, 4 = PK1 and 5 = SB1; Grass infesting populations of *N. lugens*: 6 = UPM2, 7 = TK2, 8 = MK2, 9 = PK2, and 10 = SB2; An out group, *N. bakeri*, 11 = CH).

between the rice and *Leersia* infesting populations of *N. lugens* and also for the genetically isolated group, *N. bakeri.*

In DALP fingerprinting method, we did not get any diagnostic markers between two sympatric populations of *N. lugens.* Saxena and Barrion [24] reported that karyoytpe, idiogram, nuclear organelles, chromosomes with nucleolus organizing region (site of RNA synthesis) showed clear differences between rice and *Leersia* infesting populations. Despite their morphological similarities, a distinct cytological incongruity and a certain degree of genetic isolation between the two populations were inferred. Species differentiation in early stages of a species formation may not be associated with substantial genetic change [25–27]. Many ecologists have accepted that the evolutionary processes are common in animals with specialized food habits [28, 29]. There was no distinct electrophoretic differentiation between *Lethe eurydice* L. and *Lethe appalacia* L., although the two species were found to be good species [27, 30]. Latif et al. [6] reported that the closely related sibling species in the *N. lugens* complex might have developed through insecticide exposures that were heavier in rice-infesting populations than in grass populations, through RAPD-PCR analysis.

The genetic distance indicates the magnitude of genetic variation between populations. Genetic distance commonly ranged from nearly 0.01 for populations within species, 0.1 for different subspecies, and 1.0 for different species [31]. So, the genetic distances (average 0.31672) between rice infesting populations (high esterase activities) and *Leersia* infesting populations (low esterase activities) of brown planthoppers indicated that these sympatric populations represented two distinct but closely related biological species.

The results of AMOVA in single primer yielded highly significant variance among group (rice versus *Leersia)* and among population components. The total genetic variation, an average 15.56% was attributable to group divergence (Rice versus *Leersia*), 13.01% to population differences and 77.37% to individual differences within a population. The percentage of variance component among groups (rice versus *Leersia*) was larger than the percentage of variance component among populations for bands detected by three DALP primers and these were tested by random permutation. These results revealed that there was genetic differentiation between the brown planthopper of rice versus *Leersia* (two sympatric populations of *N. lugens*). AMOVA was performed and was confirmed the differentiation into two groups of *Aphid gossypii* [32], five groups of *Acorus gramineus* [33], and two groups of natural populations of the wild rice, *Oryza rufipogon* [34]. Therefore, our molecular data of DALP-PCR indicated that brown plant hopper (BPH) with high esterase activity usually captured from rice plant and those with low esterase activity, usually captured from *L. hexandra* in Malaysia, represent two distinct closely related species and supported previous results as reported by Latif et al. [6, 7, 35]. Although DALP molecular method is not new, but so far to our knowledge this study is the first to detect genetic polymorphism in rice brown planthopper complex using this method.

## Acknowledgment

This work was supported by RUGS fund, University Putra Malaysia, Ministry of Science, Technology and Environment, Malaysia.

## References

[1] V. A. Dyck and B. Thomas, "The brown planthopper problem," in *Brown planthopper. Treat to Rice Production in Asia*, pp. 3–17, IRRI, Philippines, 1979.

[2] D. S. Park, S. K. Lee, J. H. Lee et al., "The identification of candidate rice genes that confer resistance to the brown planthopper (*Nilaparvata lugens*) through representational difference analysis," *Theoretical and Applied Genetics*, vol. 115, no. 4, pp. 537–547, 2007.

[3] I. T. Domingo, E. A. Heinrichs, and R. S. Saxena, "Occurrence of brown planthopper on *Leersia hexandra* in the Phillipines," *International Rice Research Newsletter*, vol. 8, p. 17, 1983.

[4] K. Sogawa, D. Kilin, and A. Kusmayadi, "A *Leersia* feeding brown planthopper (BPH) biotype in North Sumatra, Indonesia," *International Rice Research Newsletter*, vol. 9, p. 20, 1984.

[5] M. F. Claridge, J. Den Hollander, and J. C. Morgan, "The status of weed-associated populations of the brown planthopper, *Nilaparvata lugens* (Stal)—host race or biological species?" *Zoological Journal of the Linnean Society*, vol. 84, no. 1, pp. 77–90, 1985.

[6] M. A. Latif, T. Soon Guan, Y. O. Mohd, and S. S. Siraj, "Evidence of sibling species in the brown planthopper complex (*Nilaparvata lugens*) detected from short and long primer random amplified polymorphic DNA fingerprints," *Biochemical Genetics*, vol. 46, no. 7-8, pp. 520–537, 2008.

[7] M. A. Latif, M. Y. Omar, S. G. Tan, S. S. Siraj, and A. R. Ismail, "Interpopulation crosses, inheritance study, and genetic variability in the brown planthopper complex, *Nilaparvata lugens*

(Homoptera: Delphacidae)," *Biochemical Genetics*, vol. 48, no. 3-4, pp. 266–286, 2010.

[8] E. Desmarais, I. Lanneluc, and J. Lagnel, "Direct amplification of length polymorphisms (DALP), or how to get and characterize new genetic markers in many species," *Nucleic Acids Research*, vol. 26, no. 6, pp. 1458–1465, 1998.

[9] N. Abdullah, M. Rafii Yusop, I. Maizura, S. Ghizan, and M. A. Latif, "Genetic variability of oil palm parental genotypes and performance of its' progenies as revealed by molecular markers and quantitative traits," *Comptes Rendus*, vol. 334, no. 4, pp. 290–299, 2011.

[10] J. C. Avise, *Molecular Markers, Natural History and Evolution*, Chapman and Hall, London, UK, 1993.

[11] M. A. Latif, M. Rafii Yusop, M. Motiur Rahman, and M. R. Bashar Talukdar, "Microsatellite and minisatellite markers based DNA fingerprinting and genetic diversity of blast and ufra resistant genotypes," *Comptes Rendus*, vol. 334, no. 4, pp. 282–289, 2011.

[12] J. Sambrook, E. F. Frilsch, and T. Maniatis, *Molecular Cloning: A Laboratory Manual*, Cold Spring Harbor Laboratory Press, Cold Spring Harbor, NY, USA, 2nd edition, 1989.

[13] N. Pasteur and G. P. Georghiou, "Analysis of esterase as a means of determining organophosphate resistance in field populations of *Culex pipiend* mosquito," in *Proceedings of the 48th Annual Conference California Mosquito and Vector Control Association*, pp. 74–77, 1980.

[14] S. Usmany, S. G. Tan, and K. M. Yousoff, "Development of direct amplification of length polymorphisms in Mystus nemurus," in *Proceedings of the 3rd National Congress on Genetics*, pp. 93–95, UKM, Bangi, Malaysia, 1998.

[15] M. Nei and W. H. Li, "Mathematical model for studying genetic variation in terms of restriction endonucleases," *Proceedings of the National Academy of Sciences of the United States of America*, vol. 76, no. 10, pp. 5269–5273, 1979.

[16] D. L. Swofford and G. J. Olsen, "Phylogeny construction," in *Molecular Systematics*, D. Hillis and C. Moritz, Eds., pp. 411–500, Sinauer, Sunderland, NY, USA, 1990.

[17] S. Kambhampati, W. C. Black, and K. S. Rai, "Random amplified polymorphic DNA of mosquito species and populations (Diptera: Culicidae): techniques, statistical analysis, and applications," *Journal of Medical Entomology*, vol. 29, no. 6, pp. 939–945, 1992.

[18] P. H. A. Sneath and R. R Sokal, *Numerical Taxonomy*, W. H. Freeman, San Fransico, Calif, USA, 1973.

[19] F. J. Rohlf, *NTSYS-PC Numerical Taxonomy and Multivariate Analysis System*, Exeter, New York, NY, USA, 1993.

[20] J. Felsenstein, "PHYLIP (Phylogeny Inference Package) version 3.5," Department of Genetics, University of Washington, Seattle, Wash, USA, 1993.

[21] D. S. Haymer, M. He, and D. O. McInnis, "Genetic marker analysis of spatial and temporal relationships among existing populations and new infestations of the Mediterranean fruit fly (*Ceratitis capitata*)," *Heredity*, vol. 79, no. 3, pp. 302–309, 1997.

[22] L. Excoffier, P. E. Smouse, and J. M. Quattro, "Analysis of molecular variance inferred from metric distances among DNA haplotypes: application to human mitochondrial DNA restriction data," *Genetics*, vol. 131, no. 2, pp. 479–491, 1992.

[23] J. Felsenstein, "Confidence limits on phylogenies: an approach using the bootstrap," *Evolution*, vol. 39, pp. 783–791, 1985.

[24] R. C. Saxena and A. A. Barrion, "Biotypes of the brown planthoppers, *Nilaparvata lugens* Stal and strategies in development of host plant resistance," *International Science Application*, vol. 6, pp. 271–289, 1985.

[25] R. Lewontin, *The Genetic Basis of Evolutionary Change*, Columbia University Press, New York, NY, USA, 1974.

[26] G. L. Bush, "Sympatric speciation in phytophagus parasitic insects," in *Evolutionary Strategies of Parasitoids*, P. W. Price, Ed., pp. 187–207, Plenum, New York, NY, USA, 1975.

[27] A. Harrison and A. Vawter, "Allozyme differentiation between pheromone strains of the corn borer, *Ostrinia nubilalis*," *Annals of the Entomological Society of America*, vol. 70, pp. 717–720, 1977.

[28] P. W. Price, *Evolutionary Biology of Parasites*, Princeton University, NJ, USA, 1980.

[29] T. R. E. Southwood, "The components of diversity," in *The Diversity of Insect Faunas (Symposium of the Royal Entomolgical Society of London)*, L. A. Mound and N. Waloff, Eds., vol. 9, pp. 19–40, London, UK, 1978.

[30] R. T. Cardé, W. L. Roelofs, R. G. Harrison et al., "European corn borer: pheromone polymorphism or sibling species?" *Science*, vol. 199, no. 4328, pp. 555–556, 1978.

[31] M. Nei, "Genetic distance and molecular phylogeny," in *Population Genetics and Fisheries Management*, N. Ryman and F. M. Utter, Eds., pp. 193–224, University of Washington Press, 1988.

[32] F. Vanlerberghe-Masutti and P. Chavigny, "Host-based genetic differentiation in the aphid Aphis gossypii Glover, evidenced from RAPD fingerprints," *Molecular Ecology*, vol. 7, no. 7, pp. 905–914, 1998.

[33] L. C. Liao and J. Y. Hsiao, "Relationship between population genetic structure and riparian habitat as revealed by RAPD analysis of the rheophyte *Acorus gramineus* Soland. (Araceae) in Taiwan," *Molecular Ecology*, vol. 7, no. 10, pp. 1275–1281, 1998.

[34] G. E. Song, G. C. X. Oliveira, B. A. Schaal, L. Z. Gao, and D. Y. Hong, "RAPD variation within and between natural populations of the wild rice Oryza rufipogon from China and Brazil," *Heredity*, vol. 82, no. 6, pp. 638–644, 1999.

[35] M. A. Latif, M. Y. Omar, S. G. Tan, S. S. Siraj, and A. R. Ismail, "Biochemical studies on malathion resistance, inheritance and association of carboxylesterase activity in brown planthopper, *Nilaparvata lugens* complex in Peninsular Malaysia," *Insect Science*, vol. 17, no. 6, pp. 517–526, 2010.

# 14

# Random Amplified Polymorphic Markers as Indicator for Genetic Conservation Program in Iranian Pheasant (*Phasianus colchicus*)

**Ghorban Elyasi Zarringhabaie,[1] Arash Javanmard,[2] and Ommolbanin Pirahary[3]**

[1] *Department of Animal Science, East Azarbaijan Research Center for Agriculture and Natural Resources, Tabriz, Iran*
[2] *Department of Genomics, Agricultural Biotechnology Research Institute for Northwest and West of Iran, Tabriz, Iran*
[3] *Animal Science Section, East Azerbaijan Jahad-e-Keshavarzi Organization, Tabriz, Iran*

Correspondence should be addressed to Ghorban Elyasi Zarringhabaie, gh.elyasi@gmail.com

Academic Editor: Maria Fiammetta Romano

The objective of present study was identification of genetic similarity between wild Iran and captive Azerbaijan Pheasant using PCR-RAPD markers. For this purpose, in overall, 28 birds were taken for DNA extraction and subsequently 15 arbitrary primers were applied for PCR-RAPD technique. After electrophoresis, five primers exhibited sufficient variability which yielded overall 65 distinct bands, 59 polymorphic bands, for detalis, range of number of bands per primer was 10 to 14, and produced size varied between 200 to 1500 bp. Highest and lowest polymorphic primers were OPC5, OPC16 (100%) and OPC15 (81%), respectively. Result of genetic variation between two groups was accounted as nonsignificant (8.12%) of the overall variation. According to our expectation the wild Iranian birds showed higher genetic diversity value than the Azerbaijan captive birds. As general conclusion, two pheasant populations have almost same genetic origin and probably are subpopulations of one population. The data reported herein could open the opportunity to search for suitable conservation strategy to improve richness of Iran biodiversity and present study here was the first report that might have significant impact on the breeding and conservation program of Iranian pheasant gene pool. Analyses using more regions, more birds, and more DNA markers will be useful to confirm or to reject these findings.

## 1. Introduction

Pheasants are one of the most endangered species of birds in the world [1] which appeared on the most recent list of endangered species list [2]. Generally, pheasants species refer to any member of the subfamily of Phasianidae in the order Galliformes. There are 35 species of pheasant in 11 genera [3]. The pheasants are Asian bird in their native distributions, with the single exception of the peafowl, which is endemic to the central Africa [4]. The best known is the common pheasant, which is widespread throughout the world in introduced feral populations and in farm operations. It is native to Russia and has been widely introduced elsewhere as a game bird.

Iranian pheasant is one of the unique and unmixed species of world which is on the edge of extinction now. These species are on the list of the world's threatened species. Therefore, conservation program aims to preserve the genetic distinctiveness of the species in the north of Iran [5]. The belief is that dwindling wild populations of any such species can be supplemented (i.e., restocked) or reintroduced after extinction from the species' native range by releasing individuals from ex situ populations back into the wild. Pheasant propagation programs and subsequence releasing of young pheasant to their habitat could be possible practical way to increase number of Iranian pheasant conservation [5].

Azerbaijan country as neighbor country of Iran has many captive pheasant farms and could be as good potential for importing eggs or live birds for pheasant propagation programs. There is several seasonal immigrations between borders of these counties reported by environmental protection organization of Iran. There is limited scientific information about genetic structure of pheasant in Iran. Identification of genetic similarity between two countries could provide scientific evidence of genetic relatedness between two counties pheasant population.

Conservation attempts for wild birds have been limited; management practices are based almost solely on controlled burning, leaving areas of varying postfire age to maintain optimal habitat availability. The analysis of genetic variability is an essential ingredient for conservation programs, and the approach must be based on a combination of phenotypic and genetic data [6].

These indigenous birds such as pheasant are now subjected to fast genetic degradation and dilution because of unplanned conservation and introduction of exotic germplasm, so evaluation of DNA marker is prerequired before designing of any strategy for conservation. The use of molecular markers can aid in the choice of breeds and populations to be conserved, when there is a shortage of resources, as well as the estimation of genetic variability of species breeds and populations [7].

Random amplified polymorphic DNA (RAPD) is a useful approach to assessing genetic variation for conservation of wild populations; it is based on PCR amplification of genomic DNA with arbitrary nucleotide sequence primers. The RAPD marker can detect high levels of DNA polymorphism and can produce fine genetic markers [8, 9]. This method is simple and quick to perform when there is no prior knowledge about the genetic make-up of the organism [10]. Nevertheless, RAPD analysis has some limitations. It shows dominant inheritance, and marker/marker homozygotes cannot be distinguished from marker/null heterozygote [8].

The objective of present study was identification of genetic similarity between Iran and Azerbaijan Pheasant birds using PCR-RAPD Markers.

## 2. Material and Methods

*2.1. Birds.* Birds were taken from Arasbaran region (Jolfa, Karanlu) one of the most important border areas of Iran with a varying altitude from 256 m in the vicinity of Aras River to 2896 m and covering an area of 78560 hectares and also commercial farm of Bardeh city near Baku in Azerbaijan country. In overall, 28 birds (Iranian wild pheasant $n = 16$ (coded: 1–16) captives Azerbaijan Pheasant $n = 12$ (coded: 17–28)) were used for present study.

*2.2. DNA Extraction.* Blood samples for DNA genotyping were collected from hunted and live birds using feathertrap and stored at −20°C for few weeks or −70°C up to several months. DNA in tissues or feathers from museum skins was used for DNA extraction as well. DNA extraction was done using commercial kit [11]. Relative purity of the DNA was determined using a spectrophotometer based on absorbance at 260 and 280 nm. The sequences of the primers and the annealing temperatures used were as previously reported studies [12]. Table 1 shows sequence, GC content, and melting temperature of primers.

*2.3. PCR Protocol.* PCR was carried out in 25 $\mu$L volumes comprising of 1.5 mM $MgCl_2$, 0.2 mM dNTP, 0.01 mM of primer (Fermentas), 50 ng of genomic DNA, and 1 U Taq DNA polymerase (Promega). The PCR-RAPD protocol included initial denaturation for 3 min at 94°C, followed by 34 cycles of denaturation for 45 s at 94°C, annealing for 45 s at 37°C, extension for 1 min at 72°C, and a final extension at 72°C for 10 min. PCR products were electrophoresed at 85 V for 45 min in 2.5% agarose gels and viewed under UV light. The sizes of alleles were determined in relation to a 100 bp DNA size standard (Fermentas) using a computer software BIO 1D++.

Table 1: Sequence, GC content, and melting temperature of primers.

| Primer ID | Sequence (5′→3′) | GC | Tm |
|---|---|---|---|
| OPA04 | GGCACGCGTT | 70 | 34 |
| OPA07 | GAAACGGGTG | 60 | 32 |
| OPC02 | GTGAGGCGTC | 70 | 34 |
| OPC05 | GATGACCGCC | 70 | 34 |
| OPC06 | ATGCCCCTGT | 60 | 32 |
| OPC07 | AAAGCTGCGG | 60 | 32 |
| OPC08 | TGGACCGGTG | 70 | 34 |
| OPC10 | CTGCTCGAGT | 60 | 32 |
| OPC11 | TGGACCGGTG | 70 | 34 |
| OPC15 | TGAGCGGACA | 60 | 32 |
| OPC16 | CACACTCCAG | 60 | 32 |
| OPE02 | AGGCCCCTGT | 70 | 34 |
| OPE05 | ATGCCCCTGT | 70 | 34 |
| OPM10 | TCTGGCGCAC | 70 | 34 |
| OPP11 | AACGCGTCGG | 70 | 34 |

*2.4. Statistical Analysis.* Produced loci (band), number of polymorphic band, gene diversity Shannon's information, and polymorphic percentage (%) were calculated with POPGENE 3.1 [13]. The analyses of molecular variance (AMOVA), which estimated population differentiation, were carried out by Arlequin ver. 3.1 software [14]. Principle component analyses were graphed using PAST and NTSYS softwares.

## 3. Results

Five primers among total 15 random primers exhibited sufficient variability for studied populations. To score the band pattern, we assumed that one band corresponded to one locus. After the duplication tests, it was concluded that the RAPD bands acquired in this study are reproducible. The number of polymorphic bands varied from 10 to 14, with a range of 200–1500 bp (Table 2).

The five primers yielded 65 distinct bands, 59 of which were polymorphic (Figures 1, 2, 3, 4 and 5). Highest and lowest polymorphism were OPC5, OPC16 (100%) and OPC15 (81%), respectively (Table 2). AMOVA also revealed that genetic variation between the two countries accounted for nonsignificant 8.12% of the total variation (Table 3).

Table 2: Statistics of polymorphic primers for investigated populations.

| Primer | Amplified bands | Polymorphic bands | Band size (bp) | Polymorphism |
|---|---|---|---|---|
| OPC02 | 16 | 14 | 100–1450 | 87.5 |
| OPC05 | 13 | 13 | 100–1500 | 100 |
| OPC08 | 12 | 10 | 100–1380 | 83 |
| OPC15 | 16 | 13 | 200–1500 | 81 |
| OPC16 | 11 | 11 | 100–1300 | 100 |
| Total | 68 | 59 | | 86.7 |

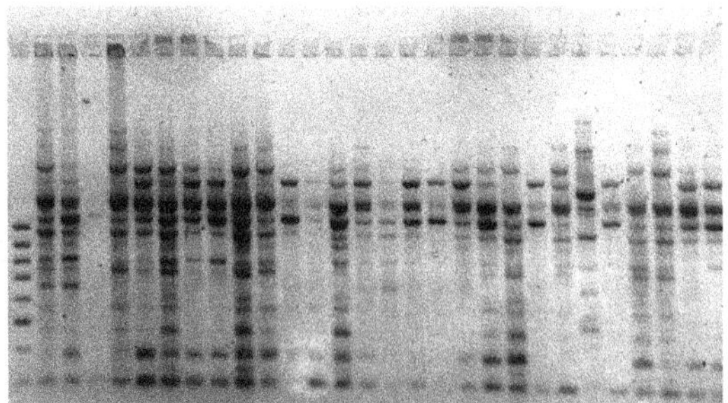

Figure 1: Amplified PCR product using OPC-02 primer.

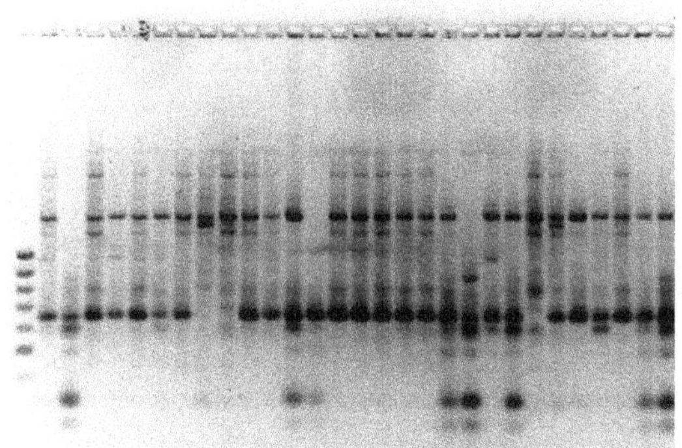

Figure 2: Amplified PCR product using OPC-05 primer.

Table 3: Result of AMOVA for two groups of pheasants.

| Source of variation | Variance Component | % Variance | $P$ value |
|---|---|---|---|
| Between populations | 0.84 $V_a$ | 8.32 | ns |
| Within populations | 9.2 $V_b$ | 91.88 | <0.01 |

Estimation of polymorphic loci, intrapopulation similarity indices suggested 96.5 percent similarity between two populations. It concluded that two investigated pheasant populations have same genetic origin and probably we can assume them as subpopulations of one population (Figures 6 and 7).

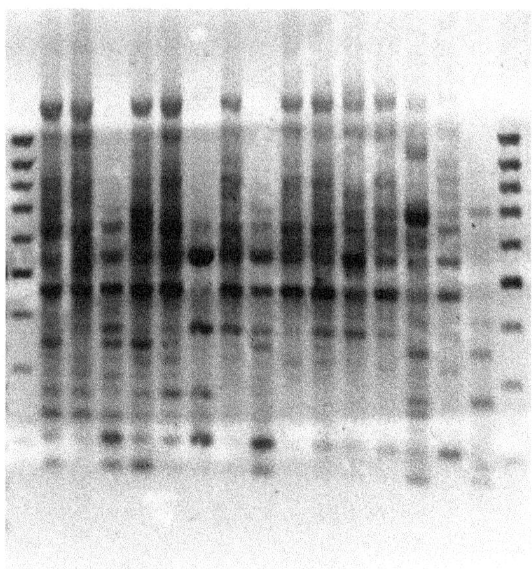

Figure 3: Amplified PCR product using OPC-08 primer.

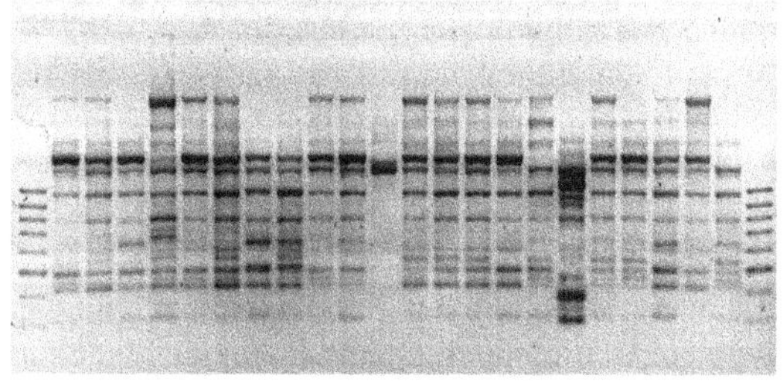

Figure 4: Amplified PCR product using OPC-15 primer.

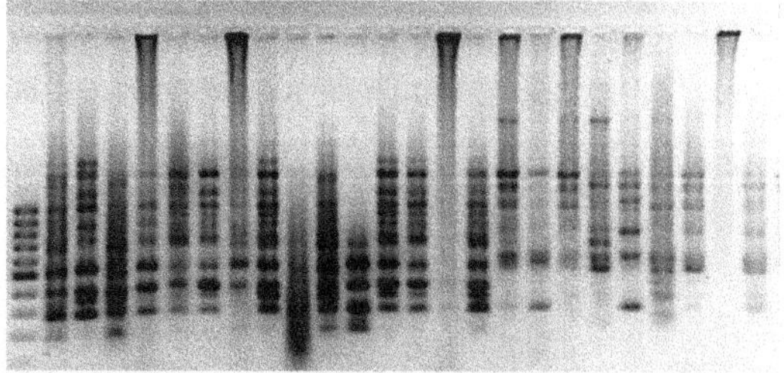

Figure 5: Amplified PCR product using OPC-16 primer.

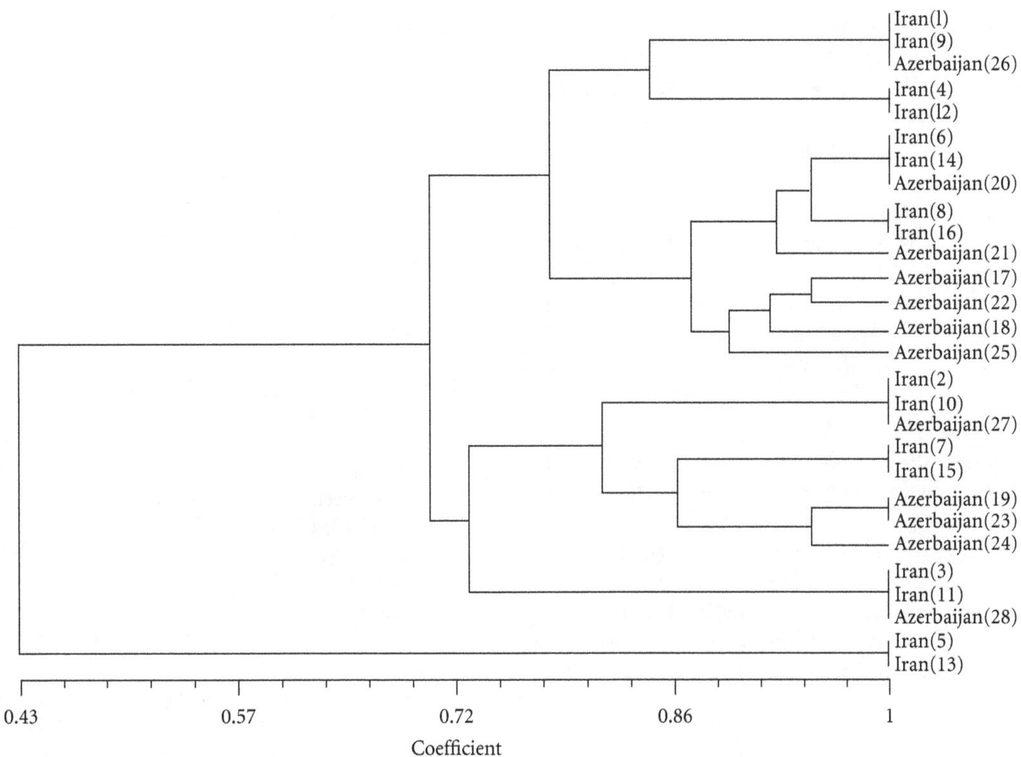


Figure 6: Cluster analysis of PCR-RAPD result using NTSYS software.

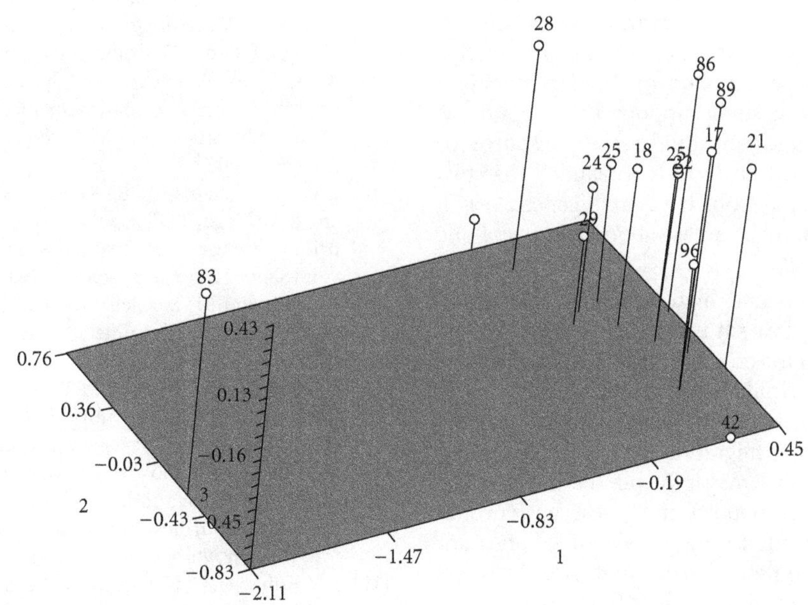


Figure 7: PCA analysis for classification of birds using NTSYS software.

## 4. Discussion

Genetic diversity is a major issue of conservation biology recognized by the IUCN [15]. Within a population it reflects the evolutionary potential to adapt to novel environmental changes.

Therefore, during the past few years, the genetic diversity of many threatened mammals, birds, fish, insects, and plants has been investigated [15]. Recently, concerns have become focused on the issue of the genetic integrity and conservation status of free ranging populations of pheasant.

The common pheasant *(Phasianus colchicus)* occurs widely in forested regions in northern Iran. Management of populations for sport hunting has centred on the south Caspian region, where special management areas have been set aside and stocks manipulated. Hunting of this bird is illegal in all seasons and some of its important residences such as the national park of Golestan, the preserved area of central Alborz, Miankaleh, Samkandeh, and Arasbaran are among the preserved areas. Unfortunately, the illegal hunting and destruction of its environment have exposed this species to a serious danger.

From the analysis conducted in present study, it is clear that the wild birds have higher genetic diversity than the captive individuals. Therefore genetic diversity of wild Iranian pheasant was higher than Azerbaijan captive birds. There is no doubt that the reduction of genetic diversity has the tendency to compromise the ability of the populations to evolve to cope with novel environmental changes and reduces their chances of long-term existence.

A good understanding of population genetic structure is critical to the design of an effective conservation programme for this species. Once genetic resources have been identified and characterized, two basic conservation activities follow, which may be defined as in situ and ex situ. Molecular markers provide important measures for population genetic structures and geographic differentiations, which are especially widely used in analyses of intraspecific phylogeographical patterns.

The belief is that dwindling wild populations of any such species can be supplemented (i.e., re-stocked) or re-introduced after extinction from the species' native range by releasing individuals from ex situ populations back into the wild. Pheasant propagation programs and subsequence releasing of young pheasant to their habitat could be possible practical way to increase number of Iranian pheasant conservation. Result of this study supported the hypothesis that captives Azerbaijan pheasant could be good resource for richness of genetic resource of Iran. It concluded that two investigated pheasant populations have same genetic origin and probably we can assume them as subpopulations of one population.

These results demonstrate high genetic similarity of Iranian and Azerbaijan pheasant using RAPD markers. The reason for a difference in the polymorphism level for present populations and other similar species seems particularly due to different genetic makeup, species adaptation, natural selection, history of birds, migration, and mutation, even more by technical staff, which may influence the analysis. The data reported here open the opportunity to search for proper conservation strategy for increasing population of pheasant in Iran. Analyses using more birds and more DNA markers will be useful to confirm or to reject these findings.

## Abbreviation

RAPD: Randomly amplified polymorphic DNA
PCR: Polymerase chain reaction
bp: Base pair
dNTP: Dinucleotide triphosphate
$MgCl_2$: Magnesium chloride
ng: Nano-gram
1X: One time
°C: Centigrade Celsius.

## Acknowledgment

This work was completely funded by the Environment Protection Organization for Scientific Research and Technology Application. The authors thank molecular genetics laboratory staff of Agriculture Biotechnology Institute for providing good atmosphere of research and facility of conducting this project and necessary travels during of these experiments.

## References

[1] W. Beebe, *A Monograph of the Ppheasants*, vol. 1-2, Dover, New York, NY, USA, 1990.

[2] CITES, Convention on International Trade in Endangered Species of Wild Fauna and Flora (CITES). Wild Animal and Plant Protection and Regulation of International and Interprovincial Trade Act, 2001.

[3] P. A. Johnsgard, *The Pheasants of the World*, Oxford University Press, Oxford, UK, 5th edition, 1999.

[4] R. Fuller, Pheasants: status survey and conservation action plan 2000–2004. WPA/BirdLife/SSC Pheasant Specialist Group, 2004.

[5] J. Tavakkolian, *An Introduction to Genetic Resources of Native Farm Animals in Iran*, Animal Science Research Institute, Karaj, Iran, 2000.

[6] D. J. S. Hetzel and R. D. Drinkwater, "The use of DNA technologies for the conservation and improvement of animal genetic resources," in *The Management of Global Animal Genetic Resources. Proceedings of an FAO Expert Consultation*, J. Hodges, Ed., p. 309, Rome, Italy, 1992.

[7] A. A. Egito, B. Fuck, A. L. Spritze et al., "RAPD markers utilization on the formation or maintenance of conservation nuclei of livestock species," *Archivos de Zootecnia*, vol. 54, pp. 277–281, 2005.

[8] J. G. K. Williams, A. R. Kubelik, K. J. Livak, J. A. Rafalski, and S. V. Tingey, "DNA polymorphisms amplified by arbitrary primers are useful as genetic markers," *Nucleic Acids Research*, vol. 18, no. 22, pp. 6531–6535, 1990.

[9] J. Welsh and M. McClelland, "Fingerprinting genomes using PCR with arbitrary primers," *Nucleic Acids Research*, vol. 18, no. 24, pp. 7213–7218, 1990.

[10] H. Hadrys, M. Balick, and B. Schierwater, "Applications of random amplified polymorphic DNA (RAPD) in molecular ecology," *Molecular Ecology*, vol. 1, no. 1, pp. 55–63, 1992.

[11] R. Boom, C. J. A. Sol, M. M. M. Salimans, C. L. Jansen, P. M. E. W. van Dillen, and J. van der Noordaa, "Rapid and simple method for purification of nucleic acids," *Journal of Clinical Microbiology*, vol. 28, no. 3, pp. 495–503, 1990.

[12] I. V. Kulikova, G. N. Chelomina, and Y. N. Zhuravlev, "RAPD-PCR analysis of genetic diversity in the Manchurian pheasant," *Genetika*, vol. 38, no. 6, pp. 836–841, 2002.

[13] F. C. Yeh, T. Boyle, and R. Yang, *POPGENE Version 1.31. Microsoft Window Based Freeware for Population Genetic Analysis*, University of Alberta, Alberta, Canada, 1999.

[14] E. Laurent, G. Laval, and S. Schneider, *Arlequin ver 3.1 An Integrated Software Package for Population Genetics Data Analysis.Computational and Molecular Population Genetics Lab (CMPG)*, Institute of Zoology, University of Berne, Baltzerstrasse 6, Bern, Switzerland, 2001.

[15] R. Frankham, J. D. Ballou, and D. A. Briscoe, *Introduction to Conservation Genetics*, Cambridge University Press, Cambridge, UK, 2002.

# Mutations in *MC1R* Gene Determine Black Coat Color Phenotype in Chinese Sheep

**Guang-Li Yang,[1] Dong-Li Fu,[2] Xia Lang,[3] Yu-Tao Wang,[2] Shu-Ru Cheng,[2] Su-Li Fang,[2] and Yu-Zhu Luo[2]**

[1] *Department of Life Sciences, Shangqiu Normal University, Shangqiu 476000, China*
[2] *Gansu Province Key Laboratory of Herbivorous Animal Biotechnology, Gansu Agricultural University, Lanzhou 730070, China*
[3] *Lanzhou Institute of Animal and Veterinary Pharmaceutics Sciences, Chinese Academy of Agricultural Sciences, Lanzhou 730050, China*

Correspondence should be addressed to Yu-Zhu Luo; luoyz@gsau.edu.cn

Academic Editors: N. L. Tang and A. Walley

The melanocortin receptor 1 (MC1R) plays a central role in regulation of animal coat color formation. In this study, we sequenced the complete coding region and parts of the 5′- and 3′-untranslated regions of the *MC1R* gene in Chinese sheep with completely white (Large-tailed Han sheep), black (Minxian Black-fur sheep), and brown coat colors (Kazakh Fat-Rumped sheep). The results showed five single nucleotide polymorphisms (SNPs): two non-synonymous mutations previously associated with coat color (c.218 T>A, p.73 Met>Lys. c.361 G>A, p.121 Asp>Asn) and three synonymous mutations (c.429 C>T, p.143 Tyr>Tyr; c.600 T>G, p.200 Leu>Leu. c.735 C>T, p.245 Ile>Ile). Meanwhile, all mutations were detected in Minxian Black-fur sheep. However, the two nonsynonymous mutation sites were not in all studied breeds (Large-tailed Han, Small-tailed Han, Gansu Alpine Merino, and China Merino breeds), all of which are in white coat. A single haplotype AATGT (haplotype3) was uniquely associated with black coat color in Minxian Black-fur breed ($P = 9.72E - 72$, chi-square test). The first and second A alleles in this haplotype 3 represent location at 218 and 361 positions, respectively. Our results suggest that the mutations of *MC1R* gene are associated with black coat color phenotype in Chinese sheep.

## 1. Introduction

Animal coloration is an ideal model for studying the genetic mechanisms that determine phenotype [1]. Coat color in domestic animals is one of the most strikingly variable and visible traits and has been widely used as a unique phenotype in the morphological selection for breed identification and attribution. In a large number of mammalian species, the coat color diversity is mainly determined by the relative amount of two basic melanins, eumelanin (black/brown), and pheomelanin (yellow/red), which are genetically controlled by the *Extension* (*E*) and *Agouti* (A) loci, respectively [2]. The *Agouti* locus encodes for the agouti signalling protein (ASIP) [3], a small paracrine signaling molecule that interacts with the product of the *Extension* locus. The *E* locus encodes the melanocortin receptor 1 (MC1R), which is a seven-transmembrane domains protein belonging to the G-protein coupled receptor present on the surface of the melanocyte membrane [4].

Functional mutations of the *MC1R* gene causing variation in coat colors have been described in domestic animals, such as cattle [5], pigs [6, 7], horses [8], goats [9], and sheep [10–13]. A notable example is the conserved role of the *MC1R* in mammalian pigmentation [14]. Studies of *MC1R* have provided valuable insights not only into the biology of pigmentation but also the evolution of domesticated animals [15, 16].

China has more than 40 native sheep breeds [17]. During the long-term selective breeding, it has resulted in diverse coat color phenotypes in Chinese indigenous breeds, including black, white, and brown pigment types. There are three ecosystem sheep groups (Kazak, Tibetan, and Mongolian) in China as well as other local populations or breeds (Minxian Black-fur, Small-tailed Han, Large-tailed Han, Tan, Gansu

Figure 1: Illustration of sheep coat colors. (a) Minxian Black-fur sheep: black; (b) Mongolian sheep: white coat with black or brown face; (c) Tibetan sheep: white coat with black or brown face; (d) Tan sheep: white coat with black or brown face; (e) Small-tailed Han sheep: white; (f) Large-tailed Han: white; (g) China Merino: white; (h) Kazakh Fat-Rumped: Brown; (i) Gansu Alpine Merino: white; (j) Duolang sheep: white or gray coat with black or brown face.

Alpine Merino, China Merino, and Duolang) [17]. In most breeds, all animals share the same color pattern as breed character, such as Minxian Black-fur, Kazakh Fat-Rumped, and Mongolian sheep breeds (Ujimqin, Bayinbuluk, Wuranke, Sunite, and Hulun Buir) [17] (Figure 1). The Minxian Black-fur sheep is the predominant among Chinese indigenous breed that has uniform black coat [17].

In sheep, at least *ASIP*, *MC1R*, and *TYRP1* genes have been implicated in coat color [10–13, 18–21]. There are lots of papers describing the effect of *MC1R* gene in coat color trait [10–13]. *MC1R* gene is located on chromosome 14 (OAR14) in sheep [11] and has three main alleles ($E^+$, $E^D$, and $e$), which are defined by three mutations in the coding region and associated with variation in coat color [10–12]. However, so far there is no report regarding the *MC1R* gene and the potential association of its mutations with coat colors in Chinese indigenous sheep. Therefore, we characterized the *MC1R* gene by sequencing DNA pools comprising 30 sheep

TABLE 1: Sample collection: breed name, sample size, coat color phenotype, and sampling location.

| Breed | Number | Coat color phenotype | Sampling location |
|---|---|---|---|
| Minxian Black-fur | 46 | Black | Gansu, Min county |
| Tibetan | 42 | White, with black or brown face | Qinghai, Hainan county |
| Large-tailed Han | 48 | White | Henan, Jia county |
| Small-tailed Han | 34 | White | Shandong, Heze county |
| Mongolian | 51 | White, with black or brown face | Mongolian |
| Tan | 45 | White, with black or brown face | Gansu, Jintai county |
| Kazakh Fat-Rumped | 18 | Brown | Xinjiang, Akesu |
| Gansu Alpine Merino | 34 | White | Gansu, Sunan county |
| China Merino | 27 | White | Xinjiang, Yili state |
| Duolang | 28 | White or gray, with black or brown face | Xinjiang, Maigaiti county |

individuals belonging to three native breeds with different coat colors: Minxian Black-fur sheep (Black), Large-tailed Han (White), and Kazakh Fat-Rumped (Brown) (Figure 1) and subsequent analysis of mutations in 10 different Chinese sheep breeds. The purpose of this study was to investigate the variability in *MC1R* and their possible association with the coat color in Chinese sheep breeds.

## 2. Material and Methods

*2.1. Animals.* A total of 373 blood samples were collected from 10 Chinese sheep breeds representing a range of distinct coat colors (Figure 1). Breed name, sample size, coat color phenotype, and sampling location for each breed were shown in Table 1. Coat colors were determined by direct visual inspection. Genomic DNA was extracted from blood specimens by using the TIANamp blood DNA kit (Tianjin, Beijing, China).

*2.2. SNPs Identification and Genotyping.* SNPs were identified by sequencing amplicons of the whole coding domain sequences (CDS, 954 bp) and parts of the 5′- and 3′-untranslated regions (35 and 125 bp, resp.) of *MC1R* in both directions. Three DNA pools comprise thirty individuals with 10 individuals DNA (100 ng/$\mu$L, 5 $\mu$L for each individual) from each breed of Large-tailed Han sheep (White), Minxian Black-fur sheep (Black), and Kazakh Fat-Rumped sheep (Brown) and were used for identification mutation sites. Primers (MF: GAGAGCAAGCACCCTTTCCT, MR: GAGAGTCCTGTGATTCCCCT) for *MC1R* amplification and sequencing were designed with the program Primer 3 (http://fokker.wi.mit.edu/) based on the published coding region sequences in sheep (GenBank accession number: Y13965) and the complete sequences in bovine and goat which include 5′- and 3′-untranslated flanking regions (GenBank accession numbers: AF445641 and FM212940).

All amplifications were performed on Eppendorf Mastercycler (Hamburg, Germany). The reaction was performed in a total of 25 $\mu$L containing 50 ng DNA template (DNA pools), 100 $\mu$M dNTPs, 10 pM of *MC1R* specific primers (MF and MR), and 2.5 U Taq polymerase (Bocai, Shanghai, China). After denaturation at 94°C for 3 min, 35 amplification cycles were performed comprising a denaturation step at 94°C for 30 s and an annealing step at 62°C for 30 s, an extension at 72°C for 45 s, followed by a last extension at 72°C for 10 min. The PCR products were separated and visualized by electrophoresis on 1.5% agarose gels ethidium bromide staining. PCR products were purified with the QIAquick PCR Purification Kit (Qiagen, Hilden, Germany). Sequences were analyzed using DNAStar software (DNAStar Inc., Madison, WI, USA) to identify polymorphisms. Identified highly informative SNPs were (minor allele frequencies >0.3) chosen for genotyping by sequencing in a larger sample of animals belonging to the 10 breeds. PCR amplification and SNPs genotyping were performed as described above.

*2.3. Data Analysis.* Deviations from Hardy-Weinberg equilibrium (HWE) between SNPs were tested by POPGENE 3.1 [22]. Haplotypes of the SNPs within *MC1R* gene were determined using the PHASE program v. 2.1 [23]. The association analyses between haplotypes and coat colors were performed using crosstabs with fisher exact test implemented in the procedure descriptive statistics with the SPSS version 16.0 software (SPSS Inc. Chicago, IL, USA).

## 3. Results

*3.1. SNPs Identification and Genotyping.* By analysing and comparing the obtained sequence electropherograms from DNA pools of 30 sheep individuals. The results showed that five single nucleotide polymorphisms (SNPs), two nonsynonymous mutations previously associated with coat color (c.218 T>A, p.73 Met>Lys. c.361 G>A, p.121 Asp>Asn) and three synonymous mutations (c.429 C>T, p.143 Tyr>Tyr; c.600 T>G, p.200 Leu>Leu. c.735 C>T, p.245 Ile>Ile), were identified in the CDS of *MC1R* gene (Figure 2) (GenBank accession number: KF198511). These polymorphisms were reported by Våge et al. [10] and Fontanesi et al. [12]. However, we did not find recessive allele *e* (c.199 C>T), which was reported by Fontanesi et al. [12].

These SNPs were further screened in a larger number of animals of 10 Chinese sheep breeds. Genotypes and allele frequencies were shown in Table 2. A chi-square test showed that 10 breeds were in Hardy-Weinberg equilibrium, while Kazakh Fat Rumped and Minxian Black-fur breed showed significant ($P < 0.05$) and very significant ($P < 0.01$)

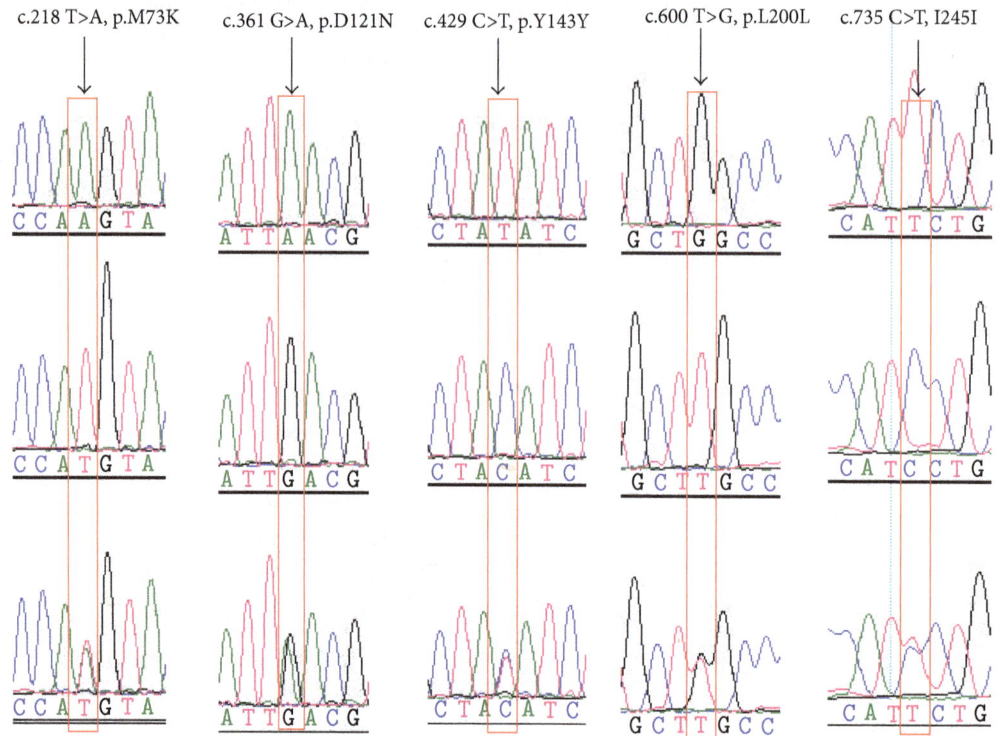


Figure 2: Identified SNPs and alignment of the MC1R protein regions around the deduced amino acid substitutions.

departures from Hardy-Weinberg equilibrium at *MC1R* c.218 T>A and *MC1R* c.361 G>A.

All mutation alleles (c.218A, c.361A, c.429T, c.600G, and c.735T) were detected in Minxian Black-fur sheep breed, and each mutation site has two genotypes. In particular, two nonsynonymous mutations (c.218 T>A, p.73 Met>Lys. c.361 G>A, p.121 Asp>Asn) determining the dominant black ($E^D$) allele [10], were not at all identified in Large-tailed Han, Small-tailed Han, Gansu Alpine Merino, and China Merino, all of which are in white coat color. The Kazakh Fat-Rumped and Mongolian populations have three genotypes for each nonsynonymous mutation loci. But two nonsynonymous mutations frequencies were very low or rare in Kazakh Fat-Rumped, Mongolian, and other three Chinese native sheep breeds (Tibetan, Tan, and Duolang). Three silent mutations in other sheep breeds, except for Minxian Black-fur and Kazakh Fat-Rumped breeds, have three genotypes. Interestingly, in this study, we found almost all mutation alleles in the Minxian Black-fur sheep breed at a rather high frequency (0.6630 and 0.8913). Three silent mutation alleles have also higher frequency (0.667) than two nonsynonymous mutation sites (0.333) in Kazakh Fat-Rumped breed.

*3.2. Haplotype.* Table 3 reports individual diplotype types and haplotype frequencies among the investigated breeds. Three haplotypes (haplotype1 [TGCTC], haplotype2 [TGTGT], and haplotype3 [AATGT]) and six individual diplotype types (haplotype1/haplotype1, haplotype1/haplotype2, haplotype2/haplotype2, haplotype1/haplotype3, haplotype2/haplotype3, and haplotype3/haplotype3) were identified. The haplotype3 of the all individual mutations was observed only in the Minxian Black-fur sheep breed. 17 Minxian Black-fur sheep were homozygous for haplotype3/haplotype3, 29 Minxian Black-fur sheep were heterozygous for haplotype1/haplotype3 (10), and for haplotype2/haplotype3 (19). The white coat color breeds (Large-tailed Han, Small-tailed Han, Gansu Alpine Merino, and China Merino) were only found in three diplotype types (haplotype1/haplotype1, haplotype1/haplotype2, and haplotype2/haplotype2). The Mongolian has six diplotype types. Tan and Kazakh Fat-Rumped breeds have five diplotype types. Tibetan and Duolang have four similar diplotype types. Interestingly, we also observed that the haplotype3 frequency was the highest in Minxian Black-fur sheep population (0.6848). The haplotype3 was absent in four Chinese sheep breeds with white coat (Large-tailed Han, Small-tailed Han, Gansu Alpine Merino, and China Merino) and was very low (0.0119–0.333) in other five Chinese indigenous breeds (Tibetan, Mongolian, Tan, Kazakh Fat-Rumped, and Duolang).

*3.3. Association Analysis.* Among a total of 373 Chinese sheep individuals, 46 were black coat color phenotype (Minxian Black-fur) and 18 were classified as brown (Kazakh Fat-Rumped). The genotyping and haplotyping data (Tables 2 and 3) clearly indicated that polymorphisms in the *MC1R* gene affect coat color in Minxian Black-fur. First of all, all animals with a mutation sites haplotype3 (46) showed uniform apparent black coat color, and almost all animals without a mutation sites haplotype3 (143) were completely white coat color (Large-tailed Han, Small-tailed Han, Gansu

ABLE 2: Genotype and allele frequencies of the 5 SNPs in *MC1R* in Chinese sheep breeds.

| Breed | *MC1R1* c.218 T>A | | | | | *MC1R1* c.361 G>A | | | | | *MC1R1* c.429 C>T | | | | | *MC1R1* c.600 T>G | | | | | *MC1R1* c.735 C>T | | | | |
|---|---|---|---|---|---|---|---|---|---|---|---|---|---|---|---|---|---|---|---|---|---|---|---|---|---|
| | Genotype | | | Allele frequency | | Genotype | | | Allele frequency | | Genotype | | | Allele frequency | | Genotype | | | Allele frequency | | Genotype | | | Allele frequency | |
| | AA | AT | TT | A | T | AA | AG | GG | A | G | TT | CT | CC | T | C | GG | GT | TT | G | T | TT | TC | CC | T | C |
| Minxian Black-fur | 17 | 29 | 0 | 0.6630 | 0.3370 | 17 | 29 | 0 | 0.6630 | 0.3370 | 36 | 10 | 0 | 0.8913 | 0.1087 | 36 | 10 | 0 | 0.8913 | 0.1087 | 36 | 10 | 0 | 0.8913 | 0.1087 |
| Tibetan | 0 | 1 | 41 | 0.0119 | 0.9881 | 0 | 1 | 41 | 0.0119 | 0.9881 | 17 | 20 | 5 | 0.6429 | 0.3571 | 17 | 20 | 5 | 0.6429 | 0.3571 | 17 | 20 | 5 | 0.6429 | 0.3571 |
| Large-tailed Han | 0 | 0 | 48 | 0 | 1 | 0 | 0 | 48 | 0 | 1 | 2 | 14 | 32 | 0.1875 | 0.8125 | 2 | 14 | 32 | 0.1875 | 0.8125 | 2 | 14 | 32 | 0.1875 | 0.8125 |
| Small-tailed Han | 0 | 0 | 34 | 0 | 1 | 0 | 0 | 34 | 0 | 1 | 10 | 16 | 8 | 0.5294 | 0.4706 | 10 | 16 | 8 | 0.5294 | 0.4706 | 10 | 16 | 8 | 0.5294 | 0.4706 |
| Mongolian | 1 | 8 | 42 | 0.0980 | 0.9020 | 1 | 8 | 42 | 0.0980 | 0.9020 | 9 | 28 | 14 | 0.4510 | 0.5490 | 9 | 28 | 14 | 0.4510 | 0.5490 | 9 | 28 | 14 | 0.4510 | 0.5490 |
| Tan | 0 | 2 | 43 | 0.0222 | 0.9778 | 0 | 2 | 43 | 0.0222 | 0.9778 | 8 | 20 | 17 | 0.4000 | 0.6000 | 8 | 20 | 17 | 0.4000 | 0.6000 | 8 | 20 | 17 | 0.4000 | 0.6000 |
| Kazakh Fat-Rumped | 2 | 8 | 8 | 0.3333 | 0.6667 | 2 | 8 | 8 | 0.3333 | 0.6667 | 6 | 12 | 0 | 0.6667 | 0.3333 | 6 | 12 | 0 | 0.6667 | 0.3333 | 6 | 12 | 0 | 0.6667 | 0.3333 |
| Gansu Alpine Merino | 0 | 0 | 34 | 0 | 1 | 0 | 0 | 34 | 0 | 1 | 6 | 17 | 11 | 0.4265 | 0.5735 | 6 | 17 | 11 | 0.4265 | 0.5735 | 6 | 17 | 11 | 0.4265 | 0.5735 |
| China Merino | 0 | 0 | 27 | 0 | 1 | 0 | 0 | 27 | 0 | 1 | 1 | 16 | 10 | 0.3333 | 0.6667 | 1 | 16 | 10 | 0.3333 | 0.6667 | 1 | 16 | 10 | 0.3333 | 0.6667 |
| Duolang | 0 | 1 | 27 | 0.0179 | 0.9821 | 0 | 1 | 27 | 0.0179 | 0.9821 | 6 | 14 | 8 | 0.4643 | 0.5357 | 6 | 14 | 8 | 0.4643 | 0.5357 | 6 | 14 | 8 | 0.4643 | 0.5357 |

TABLE 3: Haplotype and haplotype frequencies at *MC1R* in 10 Chinese sheep breeds.

| Breed | Haplotype | | | | | | Haplotype frequency | | |
|---|---|---|---|---|---|---|---|---|---|
| | 1/1 | 1/2 | 2/2 | 1/3 | 2/3 | 3/3 | Haplotype1 [TGCTC] | Haplotype2 [TGTGT] | Haplotype3 [AATGT] |
| Minxian Black-fur | 0 | 0 | 0 | **10** | **19** | **17** | 0.1087 | 0.2065 | 0.6848 |
| Tibetan | 5 | 18 | 18 | 0 | 1 | 0 | 0.3571 | 0.6310 | 0.0119 |
| Large-tailed Han | **32** | **14** | **2** | 0 | 0 | 0 | 0.8125 | 0.1875 | 0.0000 |
| Small-tailed Han | **8** | **16** | **10** | 0 | 0 | 0 | 0.4706 | 0.5294 | 0.0000 |
| Mongolian | 14 | 22 | 6 | 6 | 2 | 1 | 0.5490 | 0.3529 | 0.0981 |
| Tan | 17 | 19 | 7 | 1 | 1 | 0 | 0.6000 | 0.3778 | 0.2222 |
| Kazakh Fat-Rumped | 0 | 7 | 1 | 5 | 3 | 2 | 0.3333 | 0.3333 | 0.3333 |
| Gansu Alpine Merino | **11** | **18** | **5** | 0 | 0 | 0 | 0.5735 | 0.4265 | 0.0000 |
| China Merino | **9** | **16** | **2** | 0 | 0 | 0 | 0.6667 | 0.3333 | 0.0000 |
| Duolang | 8 | 14 | 5 | 0 | 1 | 0 | 0.5357 | 0.4464 | 0.0179 |

Notes: Haplotypes are indicated following the SNP positions in the *MC1R* gene: c.218 T>A, c.361 G>A, c.429 C>T, c.600 T>G, and c.735 C>T.

Alpine Merino, and China Merino). Secondly, the association analyses between haplotypes and coat colors are also showing that all the mutation alleles of haplotype3 were highly significantly associated with Minxian Black-fur coat color ($P = 9.72E - 72$). But only a few animals did not follow the above rules: twenty-one of 184 (13/166 white coat color with black or brown patches in the head and 8/18 brown animals) carried out haplotype3. The alleles of haplotype3 have not been associated with white coat color with black or brown patches in the head (Tibetan, Mongolian, Tan, and Duolang) and brown coat color (Kazakh Fat-Rumped) animals.

## 4. Discussion

Classical genetic studies had proved two alleles ($E^D$ and $E^+$) at the *Extension* locus affecting sheep coat color phenotypes [2, 24]. Subsequently, Våge et al. [10] characterized two missense mutations (p.M73 K and p.D121N) determining the dominant black ($E^D$) allele in the Norwegian Dala breed. The presence of two mutations was also observed in other sheep breeds: Corriedale, Damara, Black Merino, Black Castellana, and Karakul [11, 19]. The allele $E^D$ was directly involved in affecting sheep pigmentation at the molecular level and causes the dominant black coat color.

The recessive *e* allele of the *Extension* locus has also been clearly documented in sheep. One SNP (c.199 C>T) caused a predicted amino acid substitution (p.R67C) in a highly conserved position of the first intracellular loop of the MC1R protein [12]. The same substitution causes recessive pheomelanism in other species [7, 25]. Therefore, they propose that the p.67C allele represents the recessive *e* allele at the sheep *Extension* series that was not completely recognized in sheep by classical genetic studies. This polymorphism was analysed in Italian sheep breeds or populations. Confirming the effect of this novel allele on coat color will lead to new perspectives.

Chinese sheep breeds have more variations on coat color among and, in some cases, within breeds. Therefore, five SNPs were also identified in Chinese sheep breeds in the *MC1R* gene by direct sequencing (Figure 2). The recessive allele *e* (c.199 C>T), which has been linked to the control coat color in sheep, was not detected in the Chinese sheep. Two of five polymorphisms (c.218 T>A and c.361 G>A) were deduced as nonsynonymous substitutions causing a p.M73K and the p.D121N amino acid change, respectively. In the sheep, two amino acid (p.M73K and p.D121N) changes resided in the extracellular second transmembrane region (p.M73K) and in the third transmembrane domain (p.D121N) [10]. Both mutations in sheep have been associated with coat color variation. Additionally, both mutations could explain the dominant black coat color in sheep [10, 11, 19].

Five SNPs were genotyped in 10 Chinese sheep breeds with different coat color phenotypes. All mutations were detected in Minxian Black-fur sheep breed, and nonsynonymous mutation sites were not at all identified in white coat coloration breeds (Table 2). This finding demonstrated that five mutations were completely associated with the black coat color in Minxian Black-fur sheep population. Meanwhile, three haplotypes (haplotype1, haplotype2, and haplotype3) were defined by the mutations SNPs in the *MC1R* gene. It was interesting that haplotype3 was almost fixed in the Minxian Black-fur sheep breed (two missense mutations causing the $E^D$ allele were inserted in a haplotype3). Other four completely white sheep breeds had not carried the haplotype3 (Table 3). Furthermore, association analysis also indicated that the alleles of haplotype3 were significantly associated with the black coat color ($P = 9.72E - 72$, Chi-square test). Therefore, the alleles of haplotype3 might be a possible result that can interpret black coat color mechanisms in the Minxian Black-fur sheep breed that shaped the genetic pool of this sheep breed.

In Kazakh Fat-Rumped, two nonsynonymous mutation sites have three genotypes, and three silent mutations results were in accordance with Minxian Black-fur sheep breed (Table 2). Moreover, this breed has five diplotype types (Table 3). Thus, it is worthwhile to caution that the brown phenotype in Kazakh Fat-Rumped breed seems not to be caused by the identified *MC1R* mutations. However, Gratten et al. [20] report the identification of the *TYRP1* gene and causal mutation underlying coat color variation in a free-living population of Soay sheep. They identified a nonsynonymous substitution in exon IV that was perfectly associated with coat color. This polymorphism is predicted to cause the loss of a cysteine residue that is highly evolutionarily conserved and likely to be of functional significance. They eliminated the possibility that this association is due to the presence of strong linkage disequilibrium with an unknown regulatory mutation by demonstrating that there is no difference in relative TYRP1 expression between color morphs. Analysis of this putative causal mutation in a complex pedigree of more than 500 sheep revealed almost perfect cosegregation with coat color and very tight linkage between coat color and TYRP1.

In addition, according to the phenotype observed in Chinese-Tibetan having the same brown coat color [26], Ren et al. [27] performed a genome-wide association study (GWAS) on Tibetan and Kele pigs and found that brown colors in Chinese breeds are controlled by a single locus on pig chromosome 1. Then, by using a haplotype-sharing analysis, they refined the critical region to a 1.5 Mb interval that encompasses only one pigmentation gene: *TYRP1*. Lastly, mutation screens of sequence variants in the coding region of *TYRP1* revealed a strong candidate causative mutation (c.1484-1489del). The protein-altering deletion showed complete association with the brown coloration across Chinese-Tibetan, Kele, and Dahe breeds by occurring exclusively in brown pigs and lacking in all nonbrown-coated pigs from 27 different breeds. The findings provide the compelling evidence that brown colors in Chinese indigenous pigs are caused by the same ancestral mutation in TYRP1. Moreover, Beraldi et al. [28] have shown an effect of dilution of pigmentation in Soay sheep that maps to chromosome 2, in a region where the candidate gene for brown coat color, *TYRP1*, is located. Therefore, we can rule out the possibility of *MC1R* mutations determining the brown coat color phenotype. The brown coat color phenotype in Kazakh Fat-Rumped sheep may be caused by *TYRP1* gene mutations that need to be further investigated.

Tibetan, Duolang, Tan, and Mongolian breeds usually include completely white coat animals together with black or brown patches in the head (around the eyes and/or in the ears or cheeks) (Figure 1). According to results from genotype and haplotype, the same substitution and haplotype (haplotype frequencies) were present in Tibetan, Duolang, Tan, and Mongolian breeds. However, there was no complete association between the presence of black or brown spots in the face and the presence of the AATGT alleles or haplotype3. Fontanesi et al. [9] reported missense and nonsense mutations in *MC1R* gene of different goat breeds. According to the results obtained that *MC1R* mutations may determine eumelanic and pheomelanic phenotypes, however, they are probably not the only factors. In particular, the surprising not complete association of the nonsense mutation (p.Q225X) with red coat colour raises a few hypotheses on the determination of pheomelanic phenotypes in goats that should be further investigated. Sponenberg et al. [24] showed that the wild allele at the Spotting locus allows full extension of pigmentation with no white spotting. The spotting of the recessive allele usually involves the distal legs and top of head before other areas and tends to result in reasonably recognizable patterns of spotting. Adalsteinsson [29] also suggested that the variation in the spotted ($S^S$) effect can be explained by the action of modifiers, and white head spot occurs in animal heterozygous for white markings by incomplete dominance of the dominant allele ($S^+$) for full pigmentation. Hence, these (Tibetan, Duolang, Tan, and Mongolian) breeds were probably due to incomplete fixation of different alleles at the spotting locus. The spotting locus or other loci with similar phenotypic effects might act through inhibition or disregulation of melanocyte migration from the neural crest at the embryonic level. This complicates the interpretation of the results as a complete characterization of the spotting locus in sheep is lacking. Therefore, when spots are present it could be possible to evaluate if different mutations are associated with the presence of eumelanic or pheomelanic colors.

Norris and Whan [18] characterized the sheep *ASIP* gene showing that a 190 kb tandem duplication encompassing this gene, the AHCY coding region (CDS), and the ITCH promoter region should be the cause of the white coat colour of dominant white and tan ($A^{Wt}$) Agouti sheep. In addition, a not yet characterized regulatory mutation as well as a deletion of 5 bp in exon 2 and a missense mutation in exon 4 was identified as the causes of the black recessive nonagouti ($A^a$) allele [18, 19, 21]. Analysis of the *ASIP* gene was also performed in the same Chinese sheep breeds by Fu et al. [30] (in press). The results showed that two deletion mutations and three SNPs were identified: a 9 bp deletion (c.10-18del) and 5-bp deletion (c.100-105del), both of which were located in exon 2, and three SNPs (g.672 G>A, g.1580 G>A and g.1617 G>A) were located in intron 2. Two deletion mutations were presented in 10 Chinese sheep breeds. Moreover, only two sheep have the $D_5D_5$ genotype, one in Minxian Black-fur sheep and one in Duolang sheep, and no homozygosis $D_9D_9$ was found in all sheep that we detected. The genotype results suggested that these mutations are not associated or not completely associated with coat color in the investigated sheep breeds. The above results indicated that the variation in the protein coding region of *ASIP* did not explain the coat colour phenotypes variation of Chinese indigenous sheep breeds. These investigated results are also proved evidence that the black coat color phenotype in Chinese sheep was caused by the *MC1R* gene mutations.

## 5. Conclusion

The present study results further confirm that the *MC1R* gene is an important candidate gene because its mutations are associated with black color phenotype in Chinese indigenous sheep breed. In addition, we can rule out the mutations of *MC1R* determining the brown coat color phenotype.

## Authors' Contribution

Guang-Li Yang and Dong-Li Fu contributed equally to this work and should be considered as cofirst authors.

## Acknowledgments

This study was supported by He'nan Research Program of Foundation and Advanced Technology of China (102300410143, 132300410398) and Foundation of He'nan Educational Committee of China (12B230011).

## References

[1] J. K. Hubbard, J. A. C. Uy, M. E. Hauber, H. E. Hoekstra, and R. J. Safran, "Vertebrate pigmentation: from underlying genes to adaptive function," *Trends in Genetics*, vol. 26, no. 5, pp. 231–239, 2010.

[2] A. G. Searle, *Comparative Genetics of Coat Colour in Mammals*, Logos Press, London, UK, 1968.

[3] S. J. Bultman, E. J. Michaud, and R. P. Woychik, "Molecular characterization of the mouse agouti locus," *Cell*, vol. 71, no. 7, pp. 1195–1204, 1992.

[4] L. S. Robbins, J. H. Nadeau, K. R. Johnson et al., "Pigmentation phenotypes of variant extension locus alleles result from point mutations that alter MSH receptor function," *Cell*, vol. 72, no. 6, pp. 827–834, 1993.

[5] F. Rouzaud, J. Martin, P. F. Gallet et al., "A first genotyping assay of french cattle breeds based on a new allele of the extension gene encoding the *melanocortin-1 receptor* (*MC1R*)," *Genetics Selection Evolution*, vol. 32, no. 5, pp. 511–520, 2000.

[6] J. M. H. Kijas, M. Moller, G. Plastow, and L. Andersson, "A frameshift mutation in *MC1R* and a high frequency of somatic reversions cause black spotting in pigs," *Genetics*, vol. 158, no. 2, pp. 779–785, 2001.

[7] J. M. H. Kijas, R. Wales, A. Törnsten, P. Chardon, M. Moller, and L. Andersson, "Melanocortin receptor 1 (*MC1R*) mutations and coat color in pigs," *Genetics*, vol. 150, no. 3, pp. 1177–1185, 1998.

[8] L. Marklund, M. J. Moller, K. Sandberg, and L. Andersson, "A missense mutation in the gene for melanocyte-stimulating hormone receptor (*MC1R*) is associated with the chestnut coat color in horses," *Mammalian Genome*, vol. 7, no. 12, pp. 895–899, 1996.

[9] L. Fontanesi, F. Beretti, V. Riggio et al., "Missense and nonsense mutations in *melanocortin 1 receptor* (*MC1R*) gene of different goat breeds: association with red and black coat colour phenotypes but with unexpected evidences," *BMC Genetics*, vol. 10, Article ID 47, 2009.

[10] D. I. Våge, H. Klungland, L. Dongsi, and R. D. Cone, "Molecular and pharmacological characterization of dominant black coat color in sheep," *Mammalian Genome*, vol. 10, no. 1, pp. 39–43, 1999.

[11] D. I. Våge, M. R. Fleet, R. Ponz et al., "Mapping and characterization of the dominant black colour locus in sheep," *Pigment Cell Research*, vol. 16, no. 6, pp. 693–697, 2003.

[12] L. Fontanesi, F. Beretti, V. Riggio et al., "Sequence characterization of the *melanocortin 1 receptor* (*MC1R*) gene in sheep with different coat colours and identification of the putative *e* allele at the ovine Extension locus," *Small Ruminant Research*, vol. 91, no. 2-3, pp. 200–207, 2010.

[13] L. Fontanesi, S. Dall'Olio, F. Beretti, B. Portolano, and V. Russo, "Coat colours in the Massese sheep breed are associated with mutations in the *agouti signalling protein (ASIP)* and *melanocortin 1 receptor* (*MC1R*) genes," *Animal*, vol. 5, no. 1, pp. 8–17, 2011.

[14] L. Andersson, "Melanocortin receptor variants with phenotypic effects in horse, pig, and chicken," *Annals of the New York Academy of Sciences*, vol. 994, pp. 313–318, 2003.

[15] M. Y. Fang, G. Larson, H. S. Ribeiro, N. Li, and L. Andersson, "Contrasting mode of evolution at a coat color locus in wild and domestic pigs," *PLoS Genetics*, vol. 5, no. 1, Article ID e1000341, 2009.

[16] J. Li, H. Yang, J.-R. Li et al., "Artificial selection of the melanocortin receptor 1 gene in Chinese domestic pigs during domestication," *Heredity*, vol. 105, no. 3, pp. 274–281, 2010.

[17] P. Q. Zheng, *The Editorial Section of Sheep and Goat Breeds in China. Sheep and Goat Breeds in China, Shanghai*, Scientific and Technical Publishers, Shanghai, China, 1989.

[18] B. J. Norris and V. A. Whan, "A gene duplication affecting expression of the ovine *ASIP* gene is responsible for white and black sheep," *Genome Research*, vol. 18, no. 8, pp. 1282–1293, 2008.

[19] L. J. Royo, I. Álvarez, J. J. Arranz et al., "Differences in the expression of the *ASIP* gene are involved in the recessive black coat colour pattern in sheep: evidence from the rare Xalda sheep breed," *Animal Genetics*, vol. 39, no. 3, pp. 290–293, 2008.

[20] J. Gratten, D. Beraldi, B. V. Lowder et al., "Compelling evidence that a single nucleotide substitution in *TYRP1* is responsible for coat-colour polymorphism in a free-living population of Soay sheep," *Proceedings of the Royal Society B*, vol. 274, no. 1610, pp. 619–626, 2007.

[21] J. Gratten, J. G. Pilkington, E. A. Brown, D. Beraldi, J. M. Pemberton, and J. Slate, "The genetic basis of recessive self-colour pattern in a wild sheep population," *Heredity*, vol. 104, no. 2, pp. 206–214, 2010.

[22] F. C. Yeh, R. C. Yang, and T. Boyle, "Popgene version 1.31 Microsoft Widows-based freeware for population genetic analysis," Quick User Guide, 1999.

[23] M. Stephens, N. J. Smith, and P. Donnelly, "A new statistical method for haplotype reconstruction from population data," *American Journal of Human Genetics*, vol. 68, no. 4, pp. 978–989, 2001.

[24] D. P. Sponenberg, C. H. S. Dolling, R. S. Lundie, A. L. Rae, C. Renieri, and J. J. Lauvergne, "Coat colour loci (category 1)," in *Mendelian Inheritance in Sheep (MIS 96)*, J. J. Lauvergne, C. H. S. Dolling, and C. Renieri, Eds., pp. 13–57, University of Camerino, Camerino, Italy, 1998.

[25] H. Klungland, D. I. Vage, L. Gomez Raya, S. Adalsteinsson, and S. Lien, "The role of melanocyte-stimulating hormone (MSH) receptor in bovine coat color determination," *Mammalian Genome*, vol. 6, no. 9, pp. 636–639, 1995.

[26] H. R. Mao, J. Ren, N. S. Ding, S. J. Xiao, and L. S. Huang, "Genetic variation within coat color genes of *MC1R* and *ASIP* in Chinese brownish red Tibetan pigs," *Animal Science Journal*, vol. 81, no. 6, pp. 630–634, 2010.

[27] J. Ren, H. R. Mao, Z. Y. Zhang, S. J. Xiao, N. S. Ding, and L. S. Huang, "A 6-bp deletion in the *TYRP1* gene causes the brown colouration phenotype in Chinese indigenous pigs," *Heredity*, vol. 106, no. 5, pp. 862–868, 2011.

[28] D. Beraldi, A. F. McRae, J. Gratten, J. Slate, P. M. Visscher, and J. M. Pemberton, "Development of a linkage map and mapping of phenotypic polymorphisms in a free-living population of soay sheep (*Ovis aries*)," *Genetics*, vol. 173, no. 3, pp. 1521–1537, 2006.

[29] S. Adalsteinsson, "Color inheritance in Icelandic sheep and relation between colour fertility and fertilization," *Journal of Agricultural Research*, vol. 2, no. 1, pp. 3–135, 1970.

[30] D. L. Fu, G. L. Yang, X. Lang et al., "Research on *ASIP* gene mutations in Chinese sheep breeds," *Journal of Gansu Agriculture University*, 2014 (Chinese).

# 16

# Wild *Termitomyces* Species Collected from Ondo and Ekiti States Are More Related to African Species as Revealed by ITS Region of rDNA

**Victor Olusegun Oyetayo**

*Department of Microbiology, Federal University of Technology, P.M.B. 704, Akure, Nigeria*

Correspondence should be addressed to Victor Olusegun Oyetayo, ovofuta@yahoo.com

Academic Editor: Bernard Paul

Molecular identification of eighteen *Termitomyces* species collected from two states, Ondo and Ekiti in Nigeria was carried out using the internal transcribed spacer (ITS) region. The amplicons obtained from rDNA of *Termitomyces* species were compared with existing sequences in the NCBI GenBank. The results of the ITS sequence analysis discriminated between all the *Termitomyces* species (obtained from Ondo and Ekiti States) and *Termitomyces* sp. sequences obtained from NCBI GenBank. The degree of similarity of T1 to T18 to gene of *Termitomyces* sp. obtained from NCBI ranges between 82 and 99 percent. *Termitomyces* species from Garbon with ascension number AF321374 was the closest relative of T1 to T18 except T12 that has T. eurhizus and T. striatus as the closet relative. Phylogenetic tree generated with ITS sequences obtained from NCBI GenBank data revealed that T1 to T18 are more related to *Termitomyces* species indigenous to African countries such as Senegal, Congo, and Gabon.

## 1. Introduction

*Termitomyces* species belongs to a group called "termitophilic Agaricales" This group was created for these fungi by Heim [1]. There is symbiosis association that exists between the termite and the fungus, *Termitomyces*, since neither of the two partners can exist without the other. Hence, artificial cultivation had been difficult. *Termitomyces species* is a well known edible mushroom in Nigeria.

These mushrooms make their appearance after heavy rains [2] and grow in contact with termite nests in forest soil. They usually appear between the months of April through October. *Termitomyces* species is an important source of enzymes of industrial importance such as xylanase, amylase, and cellulase [3], antioxidant compounds such as polyphenol and vitamin C [4]; protein (31.4–36.4%) [5] and immunostimulatory agent [6]. There is evidence that the extract can activate splenocytes [6].

For a long time, most researchers in Nigeria examine mushrooms with the naked eye based on phenotypic characters. It has been impossible to distinguish between genetically related species by this method. Morphologically, mushrooms belonging to different genera may look similar. The present study mainly focuses on ascertaining the phylogenetic relationship between *Termitomyces* species found in Ondo and Ekiti States Nigeria by sequencing of their ITS zone. Moreover, comparing the gene sequence of *Termitomyces* species from Ondo and Ekiti States Nigeria with sequences obtained from the NCBI GeneBank.

## 2. Materials and Methods

*2.1. Fungal Material.* Fungal material Fresh fruiting body of Termitomyces species were collected from Ekiti and Ondo States (Figure 1), Nigeria (Table 1). The fruitbodies were kept dry by wrapping in tissue paper and keeping in a polythene paper containing silica gel. The polythene bags containing the samples were well labeled for easy identification.

*2.2. Extraction of DNA.* Standard DNA isolation methods employing CTAB lysis buffer [7] was used. Briefly, dried *Termitomyces* fruitbodies were ground in mortal. The grinded

Table 1: Information on *Termitomyces* species collected from Ondo and Ekiti States.

| *Termitomyces* sp. | Location where collection was made | State | Year of collection | Name of collector |
|---|---|---|---|---|
| T1 | Ado Ekiti | Ekiti | October, 2006 | Oyetayo, V. O. |
| T2 | Ado Ekiti | Ekiti | October, 2006 | Oyetayo, V. O. |
| T3 | FUT, Akure | Ondo | September, 2006 | Oyetayo, V. O. |
| T4 | FUT, Akure | Ondo | September, 2006 | Oyetayo, V. O. |
| T5 | Ado Ekiti | Ekti | September, 2006 | Oyetayo, V. O. |
| T6 | Akure | Ondo | September, 2006 | Oyetayo, V. O. |
| T7 | Aule | Ondo | October, 2007 | Fakoya, S. |
| T8 | Aule | Ondo | October, 2007 | Fakoya, S. |
| T9 | Igbatoro | Ondo | July, 2009 | Oyetayo, V. O. |
| T10 | Igbatoro | Ondo | July, 2009 | Oyetayo, V. O. |
| T11 | UNAD Road | Ekiti | October, 2009 | Oyetayo, V. O. |
| T12 | Orita Obele, Akure | Ondo | July, 2009 | Oyetayo, V. O. |
| T13 | Obanla, FUTA | Ondo | September, 2008 | Oyetayo, V. O. |
| T14 | Orita Obele, Akure | Ondo | September, 2009 | Fakoya, S. |
| T15 | Ilara Mokin | Ondo | August, 2009 | Fakoya, S. |
| T16 | Ogbese | Ondo | August, 2009 | Fakoya, S. |
| T17 | Owena | Ondo | August, 2009 | Fakoya, S. |
| T18 | Obanla, FUTA | Ondo | September, 2009 | Fakoya, S. |

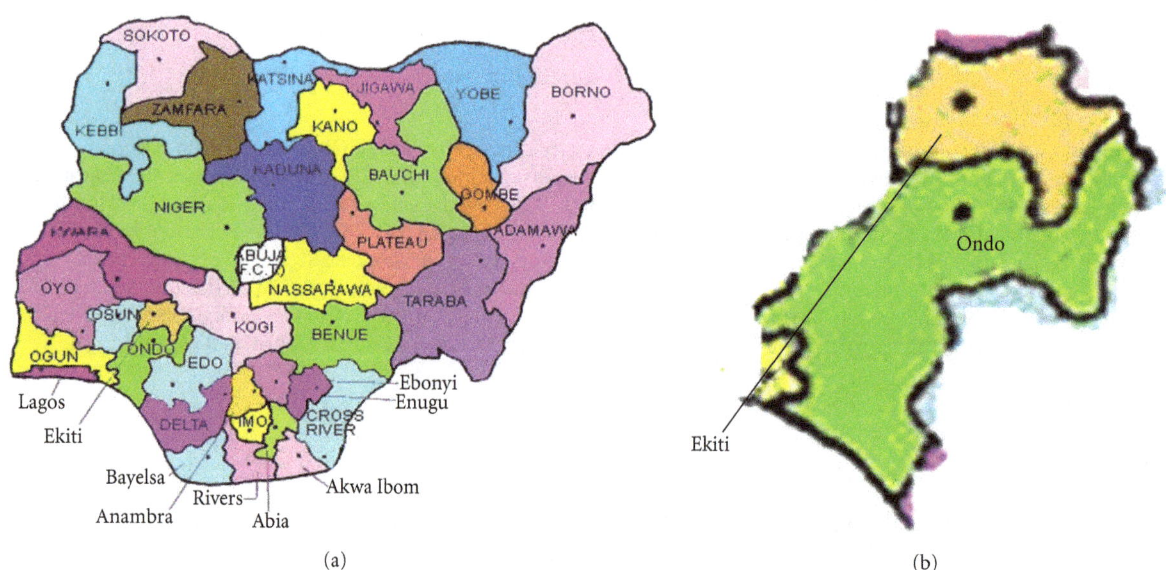


Figure 1: Map of Nigeria (a) and map of the states Ondo and Ekiti States (b) where *Termitomyces* samples were collected.

materials were transferred into well-labeled tube. Prewarmed extraction buffer (CTAB) was added, and the tubes were incubated at 65°C for 30 to 60 minutes. Equal volume of chloroform and alcohol (24 : 1) was added and mixed by inverting tubes for 15 minutes. The tubes were centrifuged for 10 minutes at 10,000 g (13000 rpm). The process was repeated, but the time of mixing was 3 minutes and time of centrifugation was 5 minutes at the same speed as above. Upper aqueous layers were removed into clean tubes, and 40 $\mu$L NaAc was added followed by 260 $\mu$L of cold isopropanol. This was gently mixed by inverting tubes. The tubes were incubated at −20°C overnight. On the second day, the mixture was centrifuged at 10,000 g (13000 rpm) for 10 minutes. The supernatant was discarded and pellets rinsed with 70% alcohol and mixed for sometimes. This procedure was repeated three times. After discarding the supernatant, the sample was dried in a dryer for 20 minutes at room temperature. Pellets were resuspended in 30 $\mu$L TE. DNA concentration and quality was checked on an ethidium-stained agarose gel (0.7%) using 0.2 $\mu$L of each sample.

*2.3. PCR Amplification of the ITS Region.* The entire region of ITS4 and ITS5 was amplified by PCR. The reaction mix was made up to a total volume of 25 $\mu$L, composed of 23 $\mu$L of *Taq* polymerase "Ready to Go" (Pharmacia) with 0.2 $\mu$L of each primer (100 pM) and 2 $\mu$L of DNA solution. The

TABLE 2: Genomic identification based on the ITS gene sequences of *Termitomyces* species collected from Ondo and Ekiti States, Nigeria.

| Termitomyces | Phenotypic identity | Closest relative in NCBI GenBank | Ascension number of closest relative | % Identity with sequence from NCBI GenBank |
|---|---|---|---|---|
| T1 | *T. clypeatus* | *T. striatus* | AB073519 | 89 |
| T2 | *T. clypeatus* | *T. striatus* | AF321367 | 91 |
| T3 | *T. robustus* | *T. eurhizus* | AF321366 | 91 |
| T4 | *T. robustus* | *T. striatus* | AF321367 | 93 |
| T5 | *T. rubustus* | *T. striatus* | AB073519 | 89 |
| T6 | *T. robustus* | *T. striatus* | AF321374 | 93 |
| T7 | *T. clypeatus* | *T. striatus* | AF321367 | 93 |
| T8 | *T. robustus* | *T. striatus* | AB073519 | 93 |
| T9 | *T. clypeatus* | *T. striatus* | AF321367 | 91 |
| T10 | *T. clypeatus* | *T. striatus* | AF321367 | 93 |
| T11 | *T. clypeatus* | *T. striatus* | AF321367 | 93 |
| T12 | *T. clypeatus* | *T. eurhizus* | AB073529 | 88 |
| T13 | *T. clypeatus* | *T. striatus* | AF321374 | 98 |
| T14 | *Termitomyces sp.* | *T. striatus* | AF321374 | 99 |
| T15* | *T. microcarpus* | *T. microcarpus* | AB073529 | 82 |
| T16 | *Termitomyces sp.* | *T. striatus* | AF321374 | 99 |
| T17 | *Termitomyces sp.* | *T. striatus* | AF321374 | 99 |
| T18 | *Termitomyces sp.* | *T. striatus* | AF321374 | 99 |

* Phenotypic identification confirmed with genomic data.

TABLE 3: Information on gene sequence of *Termitomyces* species from NCBI GenBank with close identity with $T_1$ to $T_{18}$.

| Ascension number | Name | Location |
|---|---|---|
| AB073519 | *Termitomyces* sp. group3 | Thailand: Saraburi |
| AF321367 | *Termitomyces striatus* | Republic of Congo |
| AF321366 | *Termitomyces eurhizus* | Republic of Congo |
| AF321374** | *Termitomyces* sp. AGI | Gabon |
| AB073529 | *Termitomyces* sp. group 8 | Thailand: Khao Kitchagoot |
| AF321364 | *Termitomyces* sp. OSI | Senegal |
| AF321365 | *Termitomyces* sp. ASI | Senegal |

** Gene sequence of *Termitomyces* sp. from NCBI GenBank with the closest identity with most *Termitomyces* sp. from Ondo and Ekiti states, Nigeria.

tubes were placed in a thermal cycler (GenAmp PCR System 2400; Perkin-Elmer) for amplification under the following conditions: 30 cycles of (1) denaturation at 95°C for 30 s, (2) annealing at 50°C for 1 min, (3) extension at 72°C for 1 min. The amplification products were purified using a PCR Purification Kit and electrophoresed on agarose gel. The amplified products were purified using a PCR Purification Kit and electrophoresed on ethidium-stained agarose gel (0.7%) to check the purity. DNA sequencing was performed using the primers (ITS 4 and ITS 5) in an Applied Biosystem DNA Analyser.

*2.4. Sequencing of DNA and Alignment of Sequence.* Alignments were performed with the Clustal W package [8]. The aligned sequences were corrected manually, focusing on gap positions. DNA sequence data were analyzed to provide pairwise percentage sequence divergence. The data obtained from the sequence alignment were used to plot a tree diagram (Tree View, Win 32).

## 3. Results and Discussion

The results of the ITS sequence analysis discriminated between all the 18 *Termitomyces* species obtained from Ondo and Ekiti States, Nigeria (T1 to T18) and *Termitomyces* sp. sequences obtained from NCBI GenBank. The ITS region of the rDNA is the most used genomic region for molecular characterization of fungi [9, 10] (Gardes and Bruns, 1993). The degree of similarity of T1 to T18 to gene of *Termitomyces* sp. Obtained from NCBI ranges between 82 and 99 percent (Table 2). *Termitomyces* species from Garbon with ascension number AF321374 was the closest relative of T1 to T18 (Table 3).

Phylogenetic tree generated with ITS sequences obtained from NCBI GenBank data base revealed that T1 to T18 are more related to *Termitomyces* species indigenous to African countries such as Senegal, Congo, and Gabon (Figure 2). Five clades were observed in the final phylogenetic tree; Clade 1 was made up of *Termitomyces* species (Con 1 and Con 2)

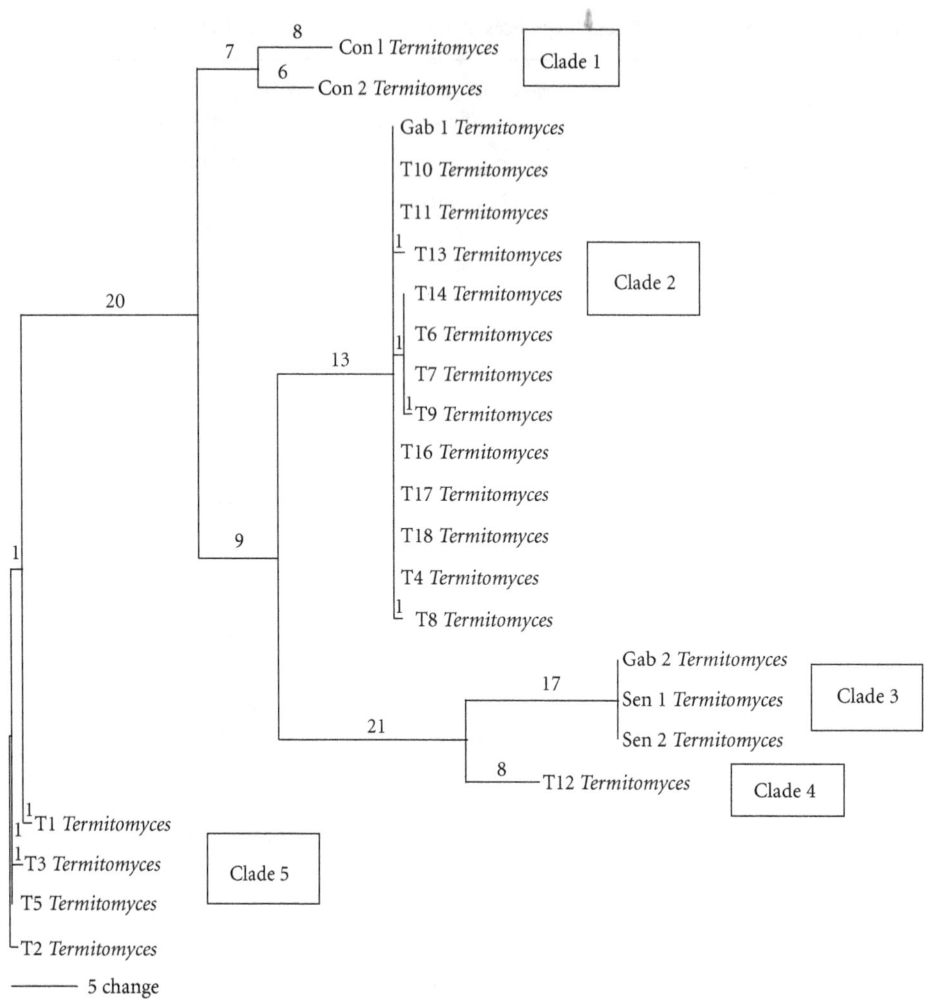


Figure 2: Phylogenetic tree showing positions of *Termitomyces* species collected from Akure and Ado Ekiti (T1 to T18) relative to existing sequences obtained from NCBI Genbank ITS sequence data.

from Congo DR. Clade 2 was made up of *Termitomyces* species (Gab 1) from Gabon and *Termitomyces* species (T4, T7, T8, T9, T10, T11, T12, T13, T14, T16, T17, and T18) from Nigeria. This implies that the *Termitomyces* species from Nigeria and Gab 1 may be from the same ancestral stock. Clade 3 was made up of *Termitomyces* species from Gabon (Gab2) and Senegal (Sen 1 and 2). Clade 4 was made up of only *Termitomyces* species T12 while clade 5 was made up of *Termitomyces* species (T1, T2, T3, and T5) from Nigeria. This suggests that they may be new species.

The closest relatives of T1 to T18 which were phenotypically identified as *T. clypeatus* and *T. robustus* were *T. striatus* and *T. eurhizus* except T15 which was *T. microcarpus* as revealed by BLAST search (Table 2). Earlier report by Frøslev et al. [11] showed that *Sinotermitomyces carnosus*, *S. griseus* and *S. rugosiceps* are synonyms of *T. mammiformis*. Moreover, Frøslev et al. [11] also found that *S. cavus* and *S. taiwanensis* are, respectively, conspecific with *T. heimii* and *T. clypeatus*. Another study by Oyetayo [12] revealed that phylogenetic tree generated from the ITS sequence obtained from *Termitomyces* species earlier identified phenotypically as *T. clypeatus* was found to be 100% homologous to *T. eurhizus* found in NCBI GenBank. This shows that *T. clypeatus* from Nigeria may be conspecific of *T. eurhizus*.

This study showed that not all the gene sequence of *Termitomyces* species indigenous to Nigeria are 100% homologous with existing gene sequences in NCBI GenBank. *Termitomyces* species from some countries in Africa such as Congo, Gabon, and Senegal are more closely related to *Termitomyces* species indigenous to Nigeria. This may suggest common origin. An earlier phylogenetic study of some African *Termitomyces* revealed that they are from monophyletic origin [13]. Clades 4 and 5 shows that *Termitomyces* species (T1, T2, T3, T5, and T12) are totally different from others species whose gene sequences are already in NCBI GenBank.

## Acknowledgments

The author wishes to acknowledge the financial support of CAS-TWAS. V. O. Oyetayo is a recipient of the CAS-TWAS Postdoctoral fellowship to China. Yao is also gratefully acknowledged for hosting V. O. Oyetayo in his laboratory (Key Laboratory of Systematic Mycology and Lichenology,

Institute of Microbiology, Chinese Academy of Sciences, Beijing, China).

## References


[1] R. Heim, "Etudes descriptives et expérimentales sur les agarics termitophiles d'Afrique tropicale," *Mémoire de l'Académie des Sciences*, vol. 64, pp. 25–29, 1941.

[2] S. O. Alasoadura, "Studies in the higher fungi of Nigeria. 1-The Genus *Termitomyces* Heim," *Journal of West African Science Association*, vol. 12, no. 2, pp. 136–146, 1967.

[3] S. Khowala and S. Sengupta, "Secretion of $\beta$-glucosidase by *Termitomyces* clypeatus: regulation by carbon catabolite products," *Enzyme and Microbial Technology*, vol. 14, no. 2, pp. 144–149, 1992.

[4] J. L. Mau, C. N. Chang, S. J. Huang, and C. C. Chen, "Antioxidant properties of methanolic extracts from *Grifola frondosa*, *Morchella esculenta* and *Termitomyces albuminosus* mycelia," *Food Chemistry*, vol. 87, no. 1, pp. 111–118, 2004.

[5] S. K. Ogundana and O. E. Fagade, "Nutritive value of some Nigerian edible mushrooms," *Food Chemistry*, vol. 8, no. 4, pp. 263–268, 1982.

[6] S. Mondal, I. Chakraborty, D. Rout, and S. S. Islam, "Isolation and structural elucidation of a water-soluble polysaccharide (PS-I) of a wild edible mushroom, *Termitomyces striatus*," *Carbohydrate Research*, vol. 341, no. 7, pp. 878–886, 2006.

[7] M. E. Zolan and P. J. Pukkila, "Inheritance of DNA methylation in Coprinus cinereus," *Molecular and Cellular Biology*, vol. 6, no. 1, pp. 195–200, 1986.

[8] J. D. Thompson, T. J. Gibson, F. Plewniak, F. Jeanmougin, and D. G. Higgins, "The CLUSTAL X windows interface: flexible strategies for multiple sequence alignment aided by quality analysis tools," *Nucleic Acids Research*, vol. 25, no. 24, pp. 4876–4882, 1997.

[9] M. Gardes and T. D. Bruns, "Rapid characterization of ectomycorrhizae using RFLP pattern of their PCR amplified-ITS," *Mycological Society Newsletter*, vol. 41, pp. 44–45, 1991.

[10] J. B. Anderson and E. Stasovski, "Molecular phylogeny of Northern Hemisphere species of *Armillaria*," *Mycologia*, vol. 84, pp. 505–516, 1992.

[11] T. G. Frøslev, D. K. Aanen, T. Læssøe, and S. Rosendahl, "Phylogenetic relationships of *Termitomyces* and related taxa," *Mycological Research*, vol. 107, no. 11, pp. 1277–1286, 2003.

[12] V. O. Oyetayo, "Molecular characterisation of *Termitomyces* species collected from Ado Ekiti and Akure, Nigeria," *Nigerian Journal of Microbiology*, vol. 23, no. 1, pp. 1933–1938, 2009.

[13] C. Rouland-Lefevre, M. N. Diouf, A. Brauman, and M. Neyra, "Phylogenetic relationships in *Termitomyces* (family agaricaceae) based on the nucleotide sequence of ITS: a first approach to elucidate the evolutionary history of the symbiosis between fungus-growing termites and their fungi," *Molecular Phylogenetics and Evolution*, vol. 22, no. 3, pp. 423–429, 2002.

17

# New Insights in the Sugarcane Transcriptome Responding to Drought Stress as Revealed by Supersage

**Éderson Akio Kido,[1] José Ribamar Costa Ferreira Neto,[1] Roberta Lane de Oliveira Silva,[1] Valesca Pandolfi,[1] Ana Carolina Ribeiro Guimarães,[2] Daniela Truffi Veiga,[2] Sabrina Moutinho Chabregas,[2] Sérgio Crovella,[1] and Ana Maria Benko-Iseppon[1]**

[1] *Department of Genetics, Federal University of Pernambuco (UFPE), 50670-901 Recife, PE, Brazil*
[2] *Biotechnology Division, Sugarcane Technology Center (CTC), 13400-970 Piracicaba, SP, Brazil*

Correspondence should be addressed to Éderson Akio Kido, kido.ufpe@gmail.com and Ana Maria Benko-Iseppon, ana.iseppon@gmail.com

Academic Editor: Luigi Cattivelli

In the scope of the present work, four SuperSAGE libraries have been generated, using bulked root tissues from four drought-tolerant accessions as compared with four bulked sensitive genotypes, aiming to generate a panel of differentially expressed stress-responsive genes. Both groups were submitted to 24 hours of water deficit stress. The SuperSAGE libraries produced 8,787,315 tags (26 bp) that, after exclusion of singlets, allowed the identification of 205,975 unitags. Most relevant BlastN matches comprised 567,420 tags, regarding 75,404 unitags with 164,860 different ESTs. To optimize the annotation efficiency, the Gene Ontology (GO) categorization was carried out for 186,191 ESTs (BlastN against Uniprot-SwissProt), permitting the categorization of 118,208 ESTs (63.5%). In an attempt to elect a group of the best tags to be validated by RTqPCR, the GO categorization of the tag-related ESTs allowed the *in silico* identification of 213 upregulated unitags responding basically to abiotic stresses, from which 145 presented no hits after BlastN analysis, probably concerning new genes still uncovered in previous studies. The present report analyzes the sugarcane transcriptome under drought stress, using a combination of high-throughput transcriptome profiling by SuperSAGE with the Solexa sequencing technology, allowing the identification of potential target genes during the stress response.

## 1. Introduction

Sugarcane (*Saccharum* spp.) is an outstanding crop throughout the tropical regions of the world [1]. It represents an important food and bioenergy source, being cultivated in many tropical and subtropical countries [2], and covering more than 23 million hectares worldwide, with a production of 1.6 billion metric tons of crushable stems [3]. This crop is responsible for almost two thirds of the global sugar production [1]. Brazil, the world's largest sugarcane producer, processed and generated in 2008 about 31 million tons of sugar [4]. In contrast to most plants, sugarcane stores sucrose—rather than polymeric compounds such as starch, proteins, or lipids—as the primary carbon and energy reserve [1]. Hence, sugarcane byproducts have received greater attention, due to their multiple uses, with the ethanol generation being highlighted, as an important renewable biofuel source [5]. Moreover, the bagasse of sugarcane has been largely used for energy cogeneration at distilleries, production of animal feed and also for paper production [6]. Nevertheless, similarly to other meaningful agronomical crops, sugarcane cultivation faces considerable losses due to inappropriate or unfavorable edaphoclimatic conditions.

Abiotic stresses are among the main causes of major crops worldwide productivity losses [7], causing negative impacts on crop adaptation and productivity. In this scenario, drought figures as the most significant stress and is considered an extremely important factor when it comes of losses in the productivity of sugarcane [8]. Several plant biotechnology programs have been initiated aiming to increase drought stress tolerance in crop plants using genetic engineering and traditional breeding [9]. Although breeding activities have provided significant progress for the understanding of the physiological and molecular responses of plants to water deficit, there is still a large gap between yields in optimal and stress conditions [10]. For this purpose, case-sensitive methods are demanded, not only to discover new genes associated to those stress conditions, but also to effectively detect differentially expressed genes on a drought tolerant variety. The identification and expression profile of such responsive genes may be helpful to unravel the basic mechanism of stress tolerance [11]. In this sense, previous works uncovered genes associated to important roles in stress perception, signal transduction, and transcriptional regulatory networks in cellular responses, useful for the improvement of stress tolerance in plants by gene transfer [12, 13].

Molecular approaches concerning drought and salinity performance in sugarcane were carried out using techniques based on molecular hybridization such as *Suppression Subtractive Hybridization* (SSH) [11] and micro-/macroarrays [14]. In general, the main limitations of these methods are their low sensibility and specificity [15]. Among the methodologies for transcriptomic analysis, the SuperSAGE [16] approach represents one of the most recent and informative methods [17], especially with its association to the high-performance sequencing platforms [pyrosequencer (454 Roche), Solexa (Illumina), and SOLiD (Applied Biosystems)]. SuperSAGE regards an evolution of the traditional *Serial Analysis of Gene Expression* [18] generating longer (26 bp) tags and thus allowing most reliable annotation analysis. Since, it is an open architecture method (i.e., allowing the discovery of new genes), it presents the potential to provide a global and quantitative gene expression analysis, based on the study of the entire transcriptome produced in a given time and tissue, under a given stimulus. Additionally, SuperSAGE permits a simultaneous analysis of two interacting eukaryotic organisms, full-length cDNAs amplification using tags as primers, potential use of tags via RNA interference (RNAi) in gene function studies, identification of antisense and rare transcripts, and identification of transcripts with alternative splicing [19]. Besides, this method has been recently associated to the next generation sequencing technologies, allowing a less expensive and faster covering of the analyzed transcriptomes, permitting a deep insight of the modulated responses under different physiological conditions. The association of SuperSAGE with the rapid advances in high throughput sequencing opened the possibility of performing genome-wide transcriptome studies in non model organisms. Additionally, this technique has been successfully applied in plant species such as rice [16], banana [20], chickpea [21, 22], chili pepper [23], tobacco [24], and tropical crops (cowpea, soybean, sugarcane; [25]). In the present work, we profit from the high resolution power of SuperSAGE coupled to the Illumina sequencing to characterize the transcriptome of drought-stressed sugarcane roots after 24 hours of submission to this stress, aiming to elect a best group of tags to be validated by RTqPCR.

## 2. Methodology

*2.1. Identification of Drought-Tolerant and Sensitive Sugarcane Accessions.* For the selection of the drought-tolerant and sensitive accessions used in the present evaluation, a previous assay was carried out in order to identify contrasting genotypes for these features. For this purpose, 20 commercial sugarcane varieties (CTC 1 to 15, SP83-2847, SP83-5073, CT94-3116, SP90-1638, and SP90-3414) from CTC (Sugarcane Technology Center, Piracicaba, Brazil) were evaluated. Among these, the four above-mentioned varieties were used as a standard for the interpretation of results, including two varieties (SP83 and SP83-2847-5073) identified as drought-tolerant and other two (SP90 and SP90-1638-3414) indicated as drought-sensitive based on field empirical observations performed by specialized technicians during several years in sugarcane commercial fields.

For this assay, mini-cuttings from the 20 varieties above were planted in 50 L pods containing inert substrate (Plantmax) in order to allow the slow increase of water deficit by removing irrigation. Tests were performed with six-months-old plants under greenhouse conditions and the treatments included plant permanently irrigated (without stress), suppression of irrigation for three days (72 hours stress), suppression of irrigation for 10 days (240 hours stress), and suppression of irrigation for 20 days (480 hours stress). Physiological measurements applied in all treatments included chlorophyll content using an SPAD-507.B Chlorophyll Meter; analysis of chlorophyll fluorescence ratio between variable and maximum chlorophyll-*a* (Fv/Fm); estimation of chlorophyll content with a fluorometer; determining the relative water content. For the parameters of chlorophyll-*a* fluorescence and chlorophyll content, three measurements were taken from three plants from each treatment. Data analysis was performed by comparing the percentage change considering the parameters mentioned above. After this assay, four drought-tolerant and four sensitive accessions could be selected according to the parameters used, revealing a gradient of water stress tolerance among the varieties analyzed. Considering the classification of the standard varieties identified previously as drought-tolerant (SP83-2847 and SP83-5073) and drought-sensitive (SP90-1638 e SP90-3414) and also considering the measurements taken after stress under glasshouse conditions (these results will be presented in a separated manuscript) four varieties were considered as drought-tolerant (CTC15, CTC6, SP83-2847 and SP83-5073) and other four as drought-sensitive (CTC9, CTC13, SP90-3414, and SP90-1638).

*2.2. Drought Stress Application and the SuperSAGE Libraries.* Plants of each selected accession were grown under glasshouse conditions in 40 L pods, in randomized experimental design (comprising six repetitions) under daily irrigation until the age of three months. After that, part of the material was submitted to drought by interruption of irrigation during 24 hours. Roots of both, stressed and nonstressed plants, were collected and frozen in liquid $N_2$, being maintained in a deep freezer until total RNA extraction using

Trizol (Invitrogen). The extracted samples were quantified by spectrophotometry, digested with DNAse and purified with the aid of the RNeasy Mini kit (Qiagen). The samples were quantified again by spectrophotometry, allowing the composition of the bulks using equimolar amounts of poli-$A^+$ messenger RNA, for all treatments. Four libraries have been generated: TD (bulk of four tolerant accessions under stress); TC (bulk of four tolerant genotypes without stress, as tolerant negative control); SD (bulk of four sensitive materials after stress); SC (bulk of nonstressed sensitive accessions, as sensitive negative control). The procedures for SuperSAGE library generation followed Matsumura et al. [26], including the attachment of library-specific adaptors carried out by GenXPro GmbH (Frankfurt am Main, Germany) allowing the identification of library-specific reads after SOLEXA sequencing.

*2.3. Statistical Analysis and Tag-Gene Annotation.* The 26-bp tags were extracted from each library. Singlets (reads appearing only once) were excluded from the present evaluation. Statistical tests were applied to the remaining tags (Audic Test, Claverie; $P \leq 0.05$) with aid of the DiscoverySpace 4.1 software [27] regarding the four contrasting treatments [T (TD *versus* TC); S (SD *versus* SC); D (TD *versus* SD); C (TC *versus* SC)]. The tests allowed the identification of the total number of expressed unitags (or tag species) for each situation and contrast, as well as the differentially expressed tags, including up- (UR) and downregulated (DR) tags. The tag-gene annotation was performed by independent evaluations via BlastN [28] against different EST databases: NCBI: (i) dbEST including only *Saccharum* ESTs; (ii) Gene Index (including *Arabidopsis thaliana*, AtGI 15.0, and Poaceae species: *S. officinarum*, SOGI 3.0; *Sorghum bicolor*, SBGI 9.0; *Zea mays*, ZMGI 19.0; *Panicum virgatum*, PAVIGI 1.0; *Oryza sativa*, OsGI 18.0; *Triticum aestivum*, TAGI 12.0; *Hordeum vulgare*, HVGI 11.0; *Festuca arendinaceae*, FAGI 3.0; *Secare cereale*, RYEGI 4.0); and (iii) KEGG (including *A. thaliana* and Fabaceae ESTs)]. Valid BlastN alignments were considered when the following parameters were observed: score from 42 to 52; integrity of the CATG sequence at the 5′ end; plus/plus alignments. Inferences about the modulation of a specific tag (Fold Change; FC) were carried out considering the ratio of the observed frequencies of a given library in relation to the other.

*2.4. Gene Ontology of SuperSAGE Hits.* Matching ESTs to the analyzed tags were categorized via GO using the software Blast2GO [29] after BlastX alignment against the Uniprot-SwissProt protein database (e-value $\leq e^{-10}$). ESTs related to the GO subcategories concerning abiotic stress response (to water deprivation, GO: 0009414; to heat/cold, GO: 0009408/GO: 0009409; to osmotic stress, GO: 0006970, to oxidative stress, GO: 0006979, to abscisic acid stimulus, GO: 0009737; to jasmonic acid stimulus, GO: 0009753) were identified, as well as UR tags related to these classes. Sets of UR tags considering the different contrasting situations (T, S, D, and C) were annotated, generating Venn diagrams, aiming the visualization of specific or shared tags considering the different treatments.

## 3. Results and Discussion

*3.1. Qualitative and Quantitative Analysis of the SuperSAGE Libraries.* The four SuperSAGE libraries produced 8,787,315 tags, from which 1,862,064 (21.2%) regarded singlets (tags sequenced only once), and were excluded from this evaluation. The most representative libraries considering the number of tags were TC (drought-tolerant control; 2,516,454 tags) and SD (drought-sensitive under stress; 2,133,587 tags), while the less representative were TD (drought-tolerant under stress; 750,226 tags) and SC (drought-sensitive control; 762,492 tags). The coverage of the transcriptome by the tags was estimated considering the total number of tags per genotype (3,266,680 for the tolerant bulk and 2,896,079 for the sensitive bulk) in relation to the number of expected transcripts per cell. The total number of average-sized transcripts was estimated to range from 100,000 [30] to 500,000 [31] per cell in higher plants. Considering the high value (500,000), the coverage provided by the tags in relation to the sugarcane transcriptome was 6.5 times higher for the tolerant bulk and 5.8 for the sensitive bulk, that is, the number of expected single copy transcripts per cell should be represented by their tags in the absolute frequencies of around six in each library. Taking the less represented libraries (TD and SC) in account, the coverage of the transcriptome regarded 1.5 times higher for both, tolerant and sensitive bulks. Considering this value, we established the $n < 2$ frequency as cutoff threshold, allowing the exclusion of singlet tags. Coverage of this magnitude allowed a comprehensive evaluation of a given transcriptome, also including rare transcripts expressed during the response to the evaluated stress.

Taking all valid tags ($n \geq 2$) into account, a total of 205,975 unitags remained for evaluation. In a recent approach, Yamaguchi et al. [32] observed similar amounts ($\approx$190,000 unitags) in the roots of *Solanum torvum* under heavy metal stress ($CdCl_2$ 0.1 $\mu$M). The high number of unitags, here observed, shows the diversity of transcripts (and expressed genes), possibly also reflecting the allopolyploid nature of sugarcane, since tags diverging in a single nucleotide were considered to be distinct unitags. It has been speculated that, in some cases, unitags could be the result of artifacts generated by the amplification process during library construction [33] or incomplete digestion of the synthesized cDNA by the *Nla*III enzyme [34], and also by PCR amplifications associated to innate features of the sequencing technology [32]. In order to minimize error sources, some precautions were taken during library development in this study, including double digestion of the total RNA extracted with DNAse, double digestion with the *Nla*III enzyme, and exclusion of singlet tags. An additional way to minimize potential errors would be the exclusion of unitags related to other similar sister-tags, grouping them to other most frequent, so called mother-tags. On the other hand, this procedure would eliminate transcripts bearing important single nucleotide polymorphisms (SNPs). Still, another possibility would be to establish a minimum frequency ($n$) of a given tag to be considered valid. In the present work, only canonical tags were accepted, with

Table 1: Total number of differentially expressed (DE; $P \leq 0.05$) up- or downregulated tags observed in different contrasting SuperSAGE root libraries from sugarcane under drought stress (24 hours without irrigation) as compared with negative control (irrigated materials).

| Contrasting | Upregulated | Downregulated | DE | Total |
|---|---|---|---|---|
| T (TD × TC) | 12,179 | 12,482 | 24,661 | 152,049 |
| S (SD × SC) | 12,085 | 16,339 | 28,424 | 141,946 |
| D (TD × SD) | 15,591 | 12,269 | 27,860 | 148,657 |
| C (TC × SC) | 12,961 | 16,342 | 29,303 | 148,631 |

TD: bulk of tolerant genotypes under stress; TC: bulk of tolerant genotypes without stress (control); SD: bulk of sensitive genotypes under stress; SC: bulk of sensitive genotypes without stress (control).

complete adapter sequences (removed by *in silico* procedures) bearing the full CATG restriction site and with $n > 2$. A more stringent value ($n > 10$) was adopted by Yamaguchi et al. [32], to reduce the number of unitags per library (from 300,000 to 450,000) for each 33 thousand tags, in an attempt to reach the number of expected genes for model species as rice (32,000 genes) and *A. thaliana* (26,000 genes). However, this procedure impairs the identification of rare and alternative transcripts that possibly play important roles in the cell metabolism.

Statistical analysis considering $P \leq 0.05$ (Audic-Claverie test) among libraries permitted the identification of differentially expressed tags including up- (UR) or downregulated (DR) tags for the four contrasting situations (**T**; **S**; **D**; **C**), as shown in Table 1.

*3.2. Primary Annotation of SuperSAGE Tags.* Relevant BlastN alignments comprised 567,420 tags (75,404 unitags with 164,860 different ESTs). Details about the results obtained after alignment to different databases are not itemized here, since this is not the aim of the present evaluation. Despite that the choice of the databases and the adopted criteria allowed the following: (a) the identification of ESTs related to most tags, preferentially concerning sequences from sugarcane or taxonomic the related species sequences; (b) annotation of a considerable number of tags considering a minimum alignment of 21 bp (similar to a LongSAGE tag); (c) identification of tags with perfect alignments (100% identity) or with a maximum of a single mismatch among tag and EST, important for future development of primers; (d) avoidance of plus/minus alignments, minimizing false NATs (natural antisense transcripts).

The strategy of considering the alignments without the election of a best hit allowed the maximization of annotation chances, since no alignment was disregarded in the acceptable score range. Thus, alignments with annotated ESTs could be more informative than similar alignments with a slightly superior score in relation to nonannotated ESTs. Moreover, tags aligned with distinct ESTs could be analyzed, minimizing the chance of a wrong choice that could compromise the validation of the expression results, especially considering that they are used as targets for RTqPCR primer design. In this context, seeking the maximization of the annotation procedures, the use of the Gene Index database for tag identity annotation was carried out trying to circumvent at least two limitations, when compared with the partial dbEST bank additionally used: (a) no need of clusterization concerning ESTs deposited at dbEST, since the Gene Index project provides tentative clusters (TC); (b) best functional annotation, with the Uniref100 (Uniprot) bank as reference. Thus, in view of the posterior need of primer design for RTqPCR and data validation of SuperSAGE tags, alignments with tolerance of a maximum of a single mismatch (TSM) tag-hit represented up to one third (186,191 or 32.8%) of the data, indicating high identity among 26 bp tags and similar ESTs, since a minimum of 21 bp alignment (size of a LongSAGE tag) was considered relevant. Almost all valid alignments (471,672 or 83.12%) regarded *Saccharum* spp. (partial dbEST) and *S. officinarum* (Gene Index SOGI), as expected. TSM alignments restricted to these databases comprised 163,742 tags. Considering TSM alignments with sequences of the SOGI only, from 26,884 ESTs, 73.0% presented informative gene descriptions and/or their functions, allowing the identification of molecular targets and gene-feature association. Despite of the higher number of TSM matches concerning alignments with dbEST sequences (136,858), the EST annotation was not informative for most contemplated ESTs (97.0%). To overcome this deficiency, the Gene Ontology categorization proved to bring a valuable contribution.

*3.3. Functional Categorization of SuperSAGE Tags.* BlastX evaluations (e-value $\leq e^{-10}$) of the 186,191 ESTs (diverse databases and TSM alignments) against the peptide Uniprot-SwissProt bank allowed the characterization of 118,208 ESTs (63.5%) that presented at least one GO reference. From this categorization, the Biological Process (BP) subcategories in response to abiotic stress were considered more informative to evaluate the sugarcane response to drought conditions. The first interesting indicators were UR tags associated to EST in the BP subcategories responding to water deprivation (GO: 0009414), heat (GO: 0009408), cold (GO: 0009409), osmotic stress (GO: 0006970), oxidative stress (GO: 0006979), abscisic acid stimulus (GO: 0009737), and jasmonic acid stimulus (GO: 0009753). By the analysis of the UR tags observed in the above-mentioned subcategories (Table 2), it was possible to generate the Venn diagrams presented in Figure 1, where Figure 1(a) represents the UR tags evaluated in the contrasting situations T (TD *versus* TC; 20 tags) and D (TD *versus* SD; 25 tags), both important for future gene validation. The first case (T) related to tags from the tolerant bulk induced after water deficit when

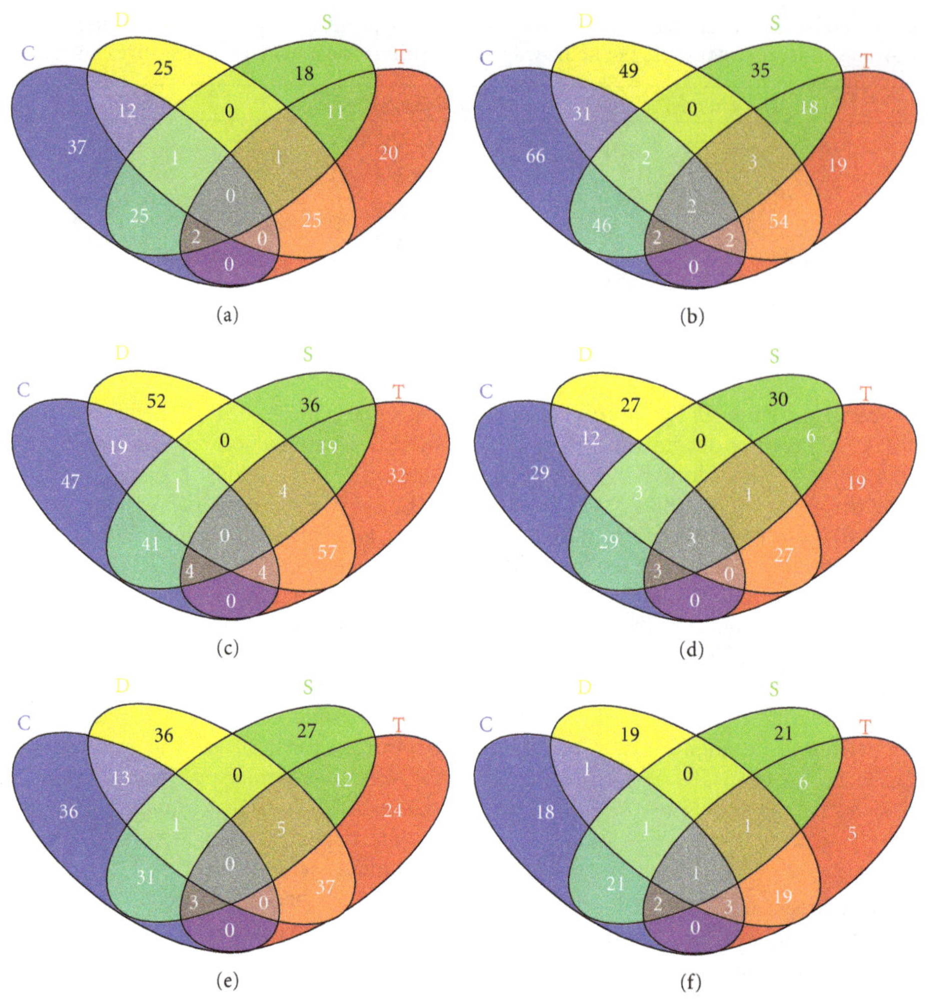


FIGURE 1: Venn diagrams with numbers of differentially upregulated (UR) tags from sugarcane roots ($P \leq 0.05$) under drought stress, considering different comparisons between SuperSAGE libraries [contrasts: T (TD *versus* TC); S (SD *versus* SC); D (TD *versus* SD); C (TC *versus* SC)]. UR tags associated with gene ontology (GO) response to (a) water deprivation; (b) heat/cold; (c) osmotic stress; (d) oxidative stress; (e) abscisic acid stimulus; (f) jasmonic acid stimulus. Libraries: TD (drought tolerant bulk under stress); TC (tolerant bulk control); SD (drought sensitive bulk under stress); SC (sensitive bulk control).

compared with the bulk control; the second group refers to induced tags from both bulks submitted to drought stress (tolerant *versus* sensitive), with higher expression (UR) in the tolerant bulk. The first group exhibited 17 non annotated tags and only three identified genes (encoding 18S ribosomal RNA, membrane integral protein, and viviparous-14). The second group included 17 tags without annotation and other eight bearing descriptions (18S ribosomal RNA gene (two tags); ABA responsive element binding factor 2; Auxin-induced protein; DRF-like transcription factor DRFL2a; ERF/AP2 domain containing transcription factor; GST; RAPB protein) are discussed latter in this manuscript. Additionally, 11 tags are worth mentioning, since they were UR in both tolerant (T) and sensitive (S) comparisons after stress (Figure 1(a)), when compared to the expression of the respective controls. Despite of being not genotype-dependent, these tags may influence positively in the plant adaptation process under drought stress. Such results and other for similar subcategories are presented in Table 2. This table comprises the total number of UR tags induced in the tolerant bulk under stress, highlighting the exclusive (T comparison) or differentially expressed tags in comparison to the sensitive bulk (D comparison), bringing interesting candidates for validation via RTqPCR. Since the same tags may be involved in different stresses, the identified tags (exclusive in T and common in the comparisons T and D; Table 2) may not be exclusive of a given condition or response. Thus, the total number of UR tags (alone or in combination) in response to water deprivation (W), heat/cold (H), osmotic stress (Os), and oxidative stress (Ox) is presented in Figure 2(a). Likewise, the number of tags induced in response to osmotic and oxidative stress is presented in Figure 2(b), while the tags responsive to hormonal stimuli (abscisic and jasmonic acids) is shown in the Figure 2(c), and a Venn diagram showing all the categories is presented in Figure 2(d).

Table 2: Total number of sugarcane upregulated (UR) root tags observed on contrasting SuperSAGE libraries when associated with ESTs classified by Gene Ontology (GO) in the subcategories related to abiotic stress response.

| Response against | GO categories | Exclusive UR tags | Common UR tags after comparison | |
|---|---|---|---|---|
| | | T | T and D | T and S |
| Water deprivation | 0009414 | 20 | 25 | 11 |
| Heat and cold | 0009408; 0009409 | 19 | 54 | 18 |
| Osmotic stress | 0006970 | 32 | 57 | 19 |
| Oxidative stress | 0006979 | 19 | 27 | 6 |
| Abscisic acid stimulus | 0009737 | 24 | 37 | 12 |
| Jasmonic acid stimulus | 0009753 | 5 | 19 | 6 |

EST: expressed sequence tag; contrast of libraries [T (TD *versus* TC); D (TD *versus* SD); S (SD *versus* SC)]; Libraries [TD: drought-tolerant bulk under stress; TC: tolerant bulk control; SD: drought-sensitive bulk under stress; SC: sensitive bulk control].

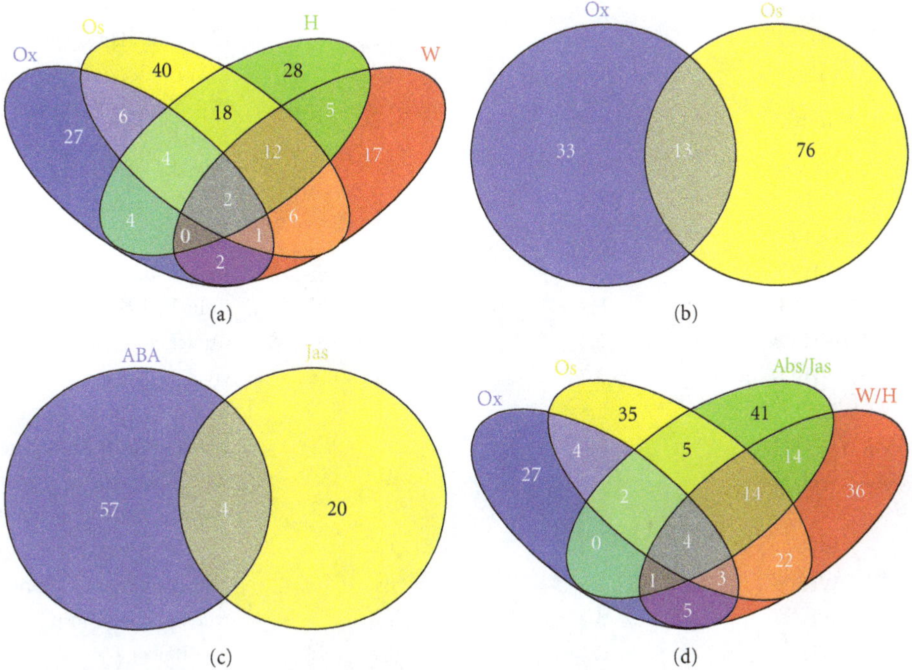


Figure 2: Venn diagrams with numbers of SuperSAGE tags overexpressed ($P \leq 0.05$) in sugarcane roots under drought stress, considering different tag sets related to gene ontology (GO) subcategories associated in response to: W (water deprivation), H (heat/cold), Os (osmotic stress), Ox (oxidative stress), ABA (abscisic acid stimulus), Jas (jasmonic acid stimulus).

In relation to the 213 UR tags, including the exclusive ones from the T contrast and those presented in both T and D contrasts (Table 2), the gene-function annotation together with the GO descriptions were available for 68 of them, while 145 tags remained unknown candidates. The annotations of these 68 UR tags and respective GO subcategories, as well as the fold change (FC) for both most relevant contrasts (T and D), are listed in Table 3. Some of them will be further addressed.

*3.3.1. Response to Hormone Stimulus.* Response to hormonal stimulus, such as jasmonic (JA) and abscisic acid (ABA), together with other plant hormones, as salicylic acid (SA) and ethylene (ET), form a complex network that plays major roles in disease resistance and response to abiotic stresses, including drought [35, 36]. In our study, 21 potential hormone-responsive tags were identified (Table 3) and some of them are thereafter discussed.

*(a) ZIM Motif Family Protein.* According to the database of *Arabidopsis* transcription factors (DATF; http://datf.cbi.pku.edu.cn/index.php), this short motif is associated to a panel of plant transcription factors and JA signaling, which is among the most important defense-related signals in plants, acting under environmental stresses, such as UV radiation, osmotic shock, heat, and drought [37]. Examining a jasmonate-insensitive 3 (jai3-1) mutant gene, Chini et al. [38] identified a novel family of jasmonate-regulated nuclear targets of SCFCOI1, named jasmonate ZIM-domain (JAZ) proteins repressing JA signaling and targeted by the E3-ubiquitin ligase SCFCOI1 for proteasome degradation. The overexpression of this hormone activated a damping

Table 3: Upregulated SuperSAGE tags associated via Gene Ontology (GO) to abiotic stress, with fold change for T ($FC_T$ = TD/TC) and D ($FC_D$ = TD/SD) comparisons for tag frequencies in the sugarcane roots libraries, as well as the annotation of the best aligned EST.

| Tag | GO | $FC_T$ | $FC_D$ | Annotation |
|---|---|---|---|---|
| SD159390 | AJ | 5.6 | ns | 50S ribosomal prot. L5, chloroplast |
| SD191288 | AJ | 2.9 | 2.3 | AP2 domain transcription factor EREBP |
| SD122727 | AJ | 2.2 | 2.7 | Bet v I allergen-like |
| SD75453 | AJ | 2.8 | 2.8 | Chromatin-remodeling factor CHD3 |
| SD123546 | AJ | 1.6 | ns | Chromatin-remodeling factor CHD3 |
| SD9608 | AJ | 2.4 | 3.3 | Initiator-binding prot.; ibp |
| SD15756 | AJ | 1.9 | 1.6 | OSK1; SNF1-related prot. Kinase |
| SD108270 | AJ | 6.4 | 2.6 | ZIM motif family prot. |
| SD258836 | AJ | 3.1 | ns | ZIM motif family prot. |
| SD108269 | AJ | 2.9 | 2.1 | ZIM motif family prot. |
| SD133809 | AJ | 2.8 | 2.6 | ZIM motif family prot. |
| SD196399 | AJ/Os | 3.2 | 3.2 | P18; Nucleoside diphosphate kinase I; NDK1 |
| SD169158 | AJ/Os/Ox | 3.2 | ns | Peptidyl-prolyl cis-trans isomerase |
| SD252082 | AJ/WH | 3.6 | 3.6 | Auxin-induced prot. |
| SD237930 | AJ/WH/Os | 2.4 | ns | 18S ribosomal RNA gene |
| SD282917 | AJ/WH/Os | 1.5 | 1.6 | ABA responsive element binding factor 2 |
| SD237939 | AJ/WH/Os | 6.8 | 6.8 | Branched-chain-amino-acid aminotransf. |
| SD238059 | AJ/WH/Os | 2.4 | 2.4 | Branched-chain-amino-acid aminotransf. |
| SD140270 | AJ/WH/Os | 3.2 | ns | viviparous-14 |
| SD237936 | AJ/WH/Os/Ox | 2.8 | 2.8 | Ribosomal prot. L28e domain cont. prot. |
| SD178862 | AJ/WH/Os/Ox | 2.8 | 2.8 | 18S ribosomal RNA gene |
| SD203616 | WH | 1.3 | ns | RAP2-like prot. |
| SD246714 | WH | 2.8 | 2.8 | CoA-thioester hydrolase CHY1 |
| SD286424 | WH | 1.4 | 2.6 | ERF/AP2 domain cont. transcription factor |
| SD279457 | WH | 2.8 | ns | Mitochondrial uncoupling prot. 2 |
| SD107875 | WH | 3.7 | 5.2 | Nucleic acid binding |
| SD191687 | WH | 2.1 | 6.9 | RAPB prot. |
| SD147607 | WH | 5.6 | ns | Salt tolerance prot. |
| SD109060 | WH | 1.4 | 1.3 | Transposable element Mu1 sequence |
| SD199146 | Os | 1.1 | 1.1 | Alpha tubulin-4[a] |
| SD54073 | Os | 2.3 | 6.8 | Calreticulin-like prot. |
| SD102228 | Os | 6.8 | 6.8 | Endo-1,4-beta-glucanase Cel1 |
| SD80163 | Os | 4.7 | ns | Endo-1,4-beta-glucanase Cel1 |
| SD13344 | Os | 1.9 | ns | Eukaryotic translation if 2 alpha sub family |
| SD182876 | Os | 4.4 | ns | Phosphopantetheine adenylyl transf. dephospho CoA kinase |
| SD129463 | Os | 3.0 | ns | Serine/threonine-prot. kinase SAPK1 |
| SD87319 | Os | 2.4 | 1.6 | Serine/threonine-prot. kinase SAPK6 |
| SD270381 | Ox | 6.5 | 4.5 | Allene oxide synthase |
| SD272257 | Ox | 2.4 | ns | Allene oxide synthase |
| SD63148 | Ox | 2.0 | 2.5 | Allene oxide synthase |
| SD113907 | Ox | 2.4 | 2.4 | Brassinosteroid biosynthesis-like prot. |
| SD219102 | Ox | 3.2 | ns | Glutathione peroxidase |
| SD213044 | Ox | 2.2 | 2.4 | Na+/H+ antiporter |
| SD54454 | Ox | 2.0 | 3.1 | Nicotianamine aminotransferase A |

TABLE 3: Continued.

| Tag | GO | $FC_T$ | $FC_D$ | Annotation |
|---|---|---|---|---|
| SD125582 | Ox | 3.8 | 1.9 | Nicotinate phosphoribosyltransferase-like |
| SD122742 | Ox | 3.2 | ns | Nucleotide repair prot. |
| SD102844 | Ox | 1.6 | ns | Peroxidase precursor |
| SD17103 | Ox | 6.4 | ns | Tyrosine/nicotianamine aminotransf. family |
| SD17107 | Ox | 2.5 | ns | Tyrosine/nicotianamine aminotransf. family |
| SD17108 | Ox | 1.8 | ns | Tyrosine/nicotianamine aminotransf. family |
| SD151691 | WH/Os | 1.8 | 1.6 | DRF-like transcription factor DRFL2a |
| SD9805 | WH/Os | 25.0 | 25.0 | Glycine-rich RNA binding prot. |
| SD9802 | WH/Os | 14.7 | 14.7 | Glycine-rich RNA binding prot. |
| SD9806 | WH/Os | 13.1 | 13.1 | Glycine-rich RNA binding prot. |
| SD9767 | WH/Os | 2.8 | 2.8 | Glycine-rich RNA binding prot. |
| SD9803 | WH/Os | 2.4 | 2.4 | Glycine-rich RNA binding prot. |
| SD9800 | WH/Os | 2.4 | 2.4 | Glycine-rich RNA binding prot. |
| SD9801 | WH/Os | 1.1 | ns | Glycine-rich RNA binding prot. |
| SD108120 | WH/Os | 6.0 | 6.0 | Glycine-rich RNA-binding prot. |
| SD108115 | WH/Os | 1.3 | 1.3 | Glycine-rich RNA-binding prot. 2; GRP2 |
| SD264077 | WH/Os | 3.6 | ns | Membrane integral prot. |
| SD92627 | WH/Ox | 5.7 | 2.4 | Glutathione transferase III |
| SD243418 | WH/Ox | 4.0 | ns | Serine hydroxymethyltransferase |
| SD179937 | WH/Ox | 3.8 | 2.5 | Serine hydroxymethyltransferase |
| SD21923 | WH/Ox | 1.3 | 1.7 | Whitefly-induced gp91-phox |
| SD184083 | Os/Ox | 3.2 | 3.2 | Delta-1-pyrroline-5-carboxylate dehydrog. |
| SD8088 | Os/Ox | 3.2 | ns | MutT domain prot.-like |
| SD251703 | Os/Ox | 7.1 | ns | P5cs; delta 1-pyrroline-5-carboxylate synth. |

Libraries [TD: drought-tolerant bulk under stress; TC: tolerant bulk control; SD: drought-sensitive bulk under stress]; ns: fold change of tag not significant ($P \leq 0.05$). WH: response to water deprivation and to heat/cold; Os: response to osmotic stress, Ox: response to oxidative stress; AJ: response to abscisic acid stimulus and to jasmonic acid stimulus.

mechanism concerning the JA signaling cascades after stress initiation [39]. In our evaluation, five UR tags were hormone related, with one candidate (SD108270) presenting expressive fold change in both contrasts ($FC_T = 6.4$ and $FC_D = 2.6$; Table 3).

*(b) Chromatin-Remodeling Factor.* CHD3 has been implicated in the repression of transcription [40]. Association of these proteins to drought-responsive genes was related during *Arabidopsis* seed germination process by regulating the ABA-dependent and gibberellic acid (GA) dependent responses, modulating the plant reaction to mild osmotic stresses and limiting the expression levels of transcription factors, preventing a maladapted growth arrest. In other words, it refines the pace of seed germination in response to ABA and maintains embryonic characters silent in response to GA [41]. Our results indicate a differential expression of CHD3 also in roots of adult sugarcane plants undergoing water deficit, with two UR tags (SD75453 and SD123546) with FC values of 1.6 (SD123546; $FC_T$) and 2.8 (SD75453; $FC_T = FC_D$; Table 3).

*(c) AP2/EREBP.* It is a large family of plant transcriptional regulators that plays key roles in the development and environmental stress response pathways. Transcription factors encoded by AP2/EREBP genes contain the highly conserved AP2/ERF DNA binding domain [42] constituting a plant supergene family [43] subdivided into five subfamilies according to the number of AP2/ERF motifs [44]. The AP2/EREBP subgroup induced by biotic and abiotic stresses was identified by Sharoni et al. [45]. Among the upregulated genes, 52 were induced in response to diverse abiotic stress, such as cold, drought, and salt. Lin et al. [46] working with a full-length cDNA *OsEBP2* (ethylene-responsive-element binding protein2) in japonica rice leaves infected by blast fungus *Magnaporthe grisea* observed that *OsEBP2* responded transiently to the treatments with methyl jasmonate (MeJA), ABA, and ethophen (ethylene generator). In our analysis, a UR tag was annotated as *APETALA 2/ethylene response element binding protein* (AP2/EREBP) showing expressive modulation ($FC_T = 2.9$ and $FC_D = 2.3$; Table 3). Additionally, one UR tag (SD286424; $FC_T = 1.4$ and $FC_D = 2.6$; Table 3) annotated as *AP2/ERF domain containing transcription factor* was associated to our WH group (response to water deprivation + response to heat/cold; Table 3), indicating an important candidate for validation, since the overexpression of an ERF transcription factor GmERF3 from soybean in tobacco plants raised the tolerance to salinity (up to 400 mM, NaCl) and drought [47] in transgenic plants.

*3.3.2. Response to Water Deprivation, Oxidative and Osmotic Stress.* In our analysis, 47 potential stress-responsive UR tags with acceptable annotation were identified (Table 3) and some of them deserve special mentioning when considering their GO categorization and the fold change data.

*(a) Glycine-Rich RNA Binding Protein (GRP) Superfamily.* This superfamily, characterized by the presence of a glycine-rich domain arranged in (Gly)n-X repeats, was recently reviewed by Mangeon et al. [48] that highlighted the diversity in structure, expression pattern, and subcellular localization, suggesting that these proteins perform different functions in plants, such as processing, transport, localization, stability, and translation of mRNA molecules. This supposition is consistent with literature data regarding GRPs and biotic and abiotic stresses [49, 50]. Wang et al. [50] analyzing the transcriptome of *Malus prunifolia* (an apple relative with strong drought tolerance) identified a GRP (*MpGR-RBP1*) expressed in roots and leaves, which plays a role in the response to plant dehydration. Among the most representative tags found to be water-deprivation responsive in our analysis, nine tags with FC ranging near 1.1 up to 25.0 (both $FC_T$ and $FC_D$; Table 3) in roots showed to be upregulated in the drought-tolerant bulk under stress when compared to nonstressed control (TD *versus* TC) or in relation to the drought-sensitive bulk also under stress (TD *versus* SD).

*(b) CoA-Thioester Hydrolase (CHY1; Synonym: β-Hydroxy-isobutyryl-CoA Hydrolase).* In our analysis, one UR tag of this class showing FC of 2.8 ($FC_T$ = $FC_D$; Table 3) was identified. This peroxissomal metabolic enzyme is needed for valine catabolism and fatty acid b-oxidation. Analyzing freezing sensitive *Arabidopsis* mutants (*chy1-10*) after cold acclimation, Dong et al. [51] observed that the disruption of CHY1 function leads to an excess of methylacrylyl-CoA, causing accumulation of *Reactive Oxygen Species* (ROS), electrolyte leakage, impairing cold-induced gene expression. Additionally, methylacrylyl-CoA may be sequestered in the peroxisome leading to localized changes in this sub cellular region and influencing peroxisome-derived signals after cold-induction. Potential alterations in auxin response or homeostasis in the *chy1* mutant may contribute to the impaired cold stress tolerance of the mutant, since peroxisome-defective mutants showed resistance to the inhibitory effects of exogenous IBA, analogous to the IAA molecule (a hormone that inhibits the root elongation and promotes lateral root formation).

*(c) Glutathione Transferase (GST; EC 2.5.1.18).* GSTs encode an ancient, heterogeneous, and widely distributed protein group in living organisms catalyzing a variety of reactions [52], including hormonal metabolism, vacuolar sequestration of anthocyanin, tyrosine metabolism, hydroxyperoxide detoxification, and regulation of apoptosis [52, 53]. In our study, one UR tag (SD92627) associated to GST showed a significant expression modulation ($FC_T$ = 5.7 and $FC_D$ = 2.4; Table 3). GST expression is induced by a wide variety of stresses, as oxidative stress [54], xenobiotic-type of stresses [55], and dehydration [56]. Expression of *TaGSTU1B (Triticum aestivum)* was induced by drought stress in four genotypes investigated, but high transcript amounts were detected only in drought-tolerant genotypes [57]. George et al. [58] reported the subcellular localization and the ability of GST from *Prosopis juliflora* (*PjGSTU1*), a drought-tolerant woody Fabaceae species, to confer drought tolerance in transgenic tobacco. Ji et al. [59] working with tobacco plants overexpressing a GST gene from *Glycine soja* showed six-fold higher GST activity enhanced dehydration tolerance than wild-type plants.

*(d) Serine Hydroxymethyltransferase (SHMT; EC 2.1.2.1).* The SHMT genic family comprises five genes in *A. thaliana* [60] bearing both cytosolic and mitochondrial isoforms in eukaryotes [61] with activity associated to the Serine and Glycine metabolism (EMBL, 2010). In our evaluation two SHMT candidates [SD243418 ($FC_T$ = 4.0; $FC_D$ = ns); SD179937 ($FC_T$ = 3.8; $FC_D$ = 2.5); Table 3] were identified. According to Moreno et al. [62], *Arabidopsis* SHMT1 functions in the photorespiratory pathway and influences resistance to biotic and abiotic stresses. The *Arabidopsis* SHMT1 mutant (*shmt1-1*) showed enhanced susceptibility to pathogens, as well as to abiotic stresses (50 mM NaCl and high light intensity). The reduced activity in *shmt1-1* mutant appears to hinder the ability of the plant to cope with any kind of additional stress, compromising the cellular mechanisms during oxidative stress. In proteome analysis [63], ten out of twelve drought responsive proteins identified from rice leaf sheaths were upregulated including an SHMT. The authors suggested that SHMT was induced for protection from oxidative degradation under drought stress.

*(e) Peptidyl-Prolyl Cis-Trans Isomerase (PPIase).* It is also known as rotamases or immunophilins (cyclophilins included), which is an enzyme superfamily with catalytic function, facilitating metabolism regulation through a chaperone or a cis-trans isomerization of proline residues during protein folding [64, 65]. A UR tag (SD169158) showing an $FC_T$ = 3.2 (Table 3) concerns a potential PPIase. In plants PPIases have been associated with the response to adverse environmental conditions. Using contrasting genotypes of *Sorghum bicolor* under water deficit, Sharma and Singh [64] observed a significant increase in leaf- and root-PPIase activity in the drought-tolerant cultivar. Similarly, various rice PPIases were differentially expressed under water deficit and salinity (200 mM NaCl) stresses [65]. Also, a correlation with plant hormones was pointed out by Godoy et al. [66] working with cyclophilins (CyPs) of *Solanum tuberosum*. CyPs are ubiquitous proteins with an intrinsic enzymatic activity of PPIase that catalyzes the rotation of X-Pro peptide bonds. *StCyP* mRNA accumulation was stimulated by the application of abscisic acid (ABA) and methyl jasmonate (MeJA) in potato tubers. The accumulation of *StCyP* transcripts was also detected when the potato tubers were exposed to heat-shock treatment.

*(f) Viviparous14.* It is a key enzyme involved in the biosynthesis of the phytohormone abscisic acid [67], represented

in our analysis by the SD140270 tag with $FC_T$ of 3.2 (Table 3). Viviparous genes are encoded in the process of plant vivipary, also reported as early germination. Of the 15 genes described so far for maize, 12 control specific steps in ABA biosynthesis [68, 69], with *vip14* (viviparous-14), associated to the control of final steps of ABA synthesis, encoding a 9-cis-epoxycarotenoid dioxygenase 1 (NCED1) enzyme that catalyzes the cleavage of the C40 neoxanthin chain into the C15 ABA skeleton xanthoxin [70]. Maize mutants for the *nced1* gene have strongly reduced kernel ABA content [71] while in *Arabidopsis*, NCED1 overexpression conferred a significant increase in ABA accumulation in the plant and also in drought tolerance [72]. Wan and Li [73] demonstrated that the expression of *AhNCED1* gene in peanut plants was significantly upregulated by dehydration and high salinity (250 $mmol\cdot L^{-1}$ NaCl).

*(g) Branched-Chain Amino Acid Transaminase.* BCATs are enzymes that play a crucial role in the metabolic pathway of BCAAs (branched-chain amino acids that include leucine, isoleucine, and valine) by catalyzing the last step of synthesis and the initial step of degradation of these amino acids [74]. Plants contain a small family of *bcat* genes, which have been characterized in *Solanum tuberosum* (potato), *Hordeum vulgaris,* and *A. thaliana* [75, 76]. Malatrasi et al. [77] evaluated the role of these enzymes in the drought tolerance process. In this study, the transcriptional levels of *Hvbcat-1*, in *H. vulgaris*, increased seven folds (results obtained by double checking with RTqPCR) after progressive drought stress (up to 14 days of water deprivation). Physiologically, the authors associated the overregulation to the activation of the BCAAs catabolism, since this is the first enzyme in the branched-chain amino acid (BCAA) catabolic pathway. In high concentrations, these amino acids are toxic to the cells; therefore, activation of their catabolism may play an important role as detoxification mechanism. In our analysis, two UR tags annotated as BACTs were identified exhibiting an expressive modulation of the FC, mainly for the SD237939 tag (FC of 6.8 for both $FC_T$ and $FC_D$; Table 3), while the other tag (SD238059) showed an FC of 2.4 ($FC_T = FC_D$).

*(h) Allene Oxide Synthase.* AOS is the first enzyme in the pathway leading to the biosynthesis of Jasmonic acid (JA), catalyzing the production of unstable allene epoxides that cyclize to form cyclopentenone acids, the precursors for JA [78]. Three tags of this category were identified (SD270381, SD272257, SD63148) being upregulated in most comparisons (Table 3). For example, the SD270381 tag presented high FC values in both T and D comparisons. The overexpression of AOSs has been observed also in other drought assays, as reported by Ozturk et al. [79] and Talamè et al. [80] with barley (*H. vulgare*) and peanut (*Arachis hypogaea*) [81].

*(i) Na+/H+ Antiporter.* Membrane proteins involved in the Na+ and H+ transport of both eukaryotes and prokaryotes act in the homeostasis maintenance of such ions [82]. In our analysis, the SD213044 tag, annotated as potential Na+/H+ antiporter, was overexpressed in both analyzed contrasting situations (Table 3). Assays evaluating those proteins under salinity stress showed that these salt-responsive genes may be able to activate the expression of drought-related genes in the tolerance acquisition [83]. Thus, ions are stored in vacuoles, acting as osmolytes, decreasing the hydric potential of the cell. Evaluations with transgenic plants overexpressing those genes, including *Petunia hybrida* [83], *A. thaliana* [84], and *A. hypogaea* [85], conferred higher tolerance to dehydration under drought and salinity.

*(j) Glutathione Peroxidase (EC 1.11.1.9).* In the present approach, a UR GPX candidate (SD219102) was overexpressed $FC_T$ of 3.2 (Table 3). These enzymes are known as cell protectors against oxidative damage generated by reactive oxygen species [86]. They present a very broad distribution in the cell, occurring in several subcellular compartments [87]. Miao et al. [88] suggested that ATGPX3 might play dual and distinctive roles in $H_2O_2$ homeostasis, acting as a general scavenger and relaying the $H_2O_2$ signal, and also as an oxidative ABA signal transducer during drought stress signaling. Their differential regulation during biotic and abiotic stresses was reported by Navrot et al. [87], indicating their importance for plant breeding.

*(l) Serine-Threonine Kinase SAPK1 (Also Known as JNK).* It belongs to the MAPK family [89], including important proteins active in the osmosensory signal transduction pathways in cells exposed to osmotic stress [90]. A wheat candidate (*W55a*) with about 90% homology to rice SAPK1 was evaluated by Xu et al. [91]. Transgenic *Arabidopsis* plants overexpressing W55a exhibited higher tolerance to drought, being also upregulated by salt, exogenous abscisic acid, salicylic acid, ethylene, and methyl jasmonate. In addition, W55a transcripts were abundant in leaves, but not in roots or stems, under environmental stresses. Expression of *SAPK* members analyzed by RNA gel blot hybridization with samples of leaves (blades and sheath), roots, and treatments with ABA (50 $\mu$M), NaCl (150 mM), or mannitol (600 mM) showed that *SAPK1* was upregulated by all three treatments in both roots and leaves, although the effect of ABA was weaker than those of the other two treatments. *SAPK6* was weakly upregulated by all treatments in the blades and the sheaths, and weakly by ABA or NaCl but strongly by mannitol treatment in the roots [92]. Overexpressed candidates analyzed here (SD129463; $FC_T$ = 3.0; Table 3) included an SAPK1 as well as a second tag matching SAPK6 (SD87319; $FC_T$ = 2.4 and $FC_D$ = 1.6; Table 3), both in roots, indicating their activation also in this tissue.

*(m) Delta-1-Pyrroline-5-Carboxylate Synthetase.* P5CS is an enzyme that catalyzes the initiation of the proline biosynthesis in plants [93]. The excessive production of this amino acid would increase the osmotolerance in plants [94]. Rice plants transformed with the P5CS gene underwent 10 days of irrigation withdrawal with higher growth rates, when compared to the control group [94]. Effects of salt in transgenic tobacco transformed with *P5CS* gene revealed the overexpression of P5CS after 24–48 h exposure to

NaCl (300 mM), when compared with non-transgenic plants under the same stress [95]. Transgenic lines of petunia [96] and tobacco [97] with enhanced accumulation of proline showed also high drought tolerance. Transcripts involved in amino acid metabolism, such as *P5CS*, *OAT* and *AS*, were also induced more than 10 folds during the identification of drought-responsive genes during sucrose accumulation and water deficit in sugarcane [98]. In our study, a UR tag (SD251703) showing an expressive induction ($FC_T$ 7.1) was annotated as a potential *P5CS* candidate (Table 3).

## 4. Concluding Remarks

The present report is the first to analyze contrasting sugarcane accessions under drought stress with a combination of the high-throughput transcriptome profiling SuperSAGE technology coupled with a next-generation sequencing platform. This approach allowed the identification of many potential target candidates in the drought stress response. The adopted methodology of annotation and GO categorization revealed the success of the work in accessing genes from very different pathways, ranging from those controlling the perception and first reaction against the stress (as transcription factors) to those known as classic genes of the osmotic stress (as *P5CS*). The number of induced tags (213) with GO categorization and high modulation is surprising, especially considering the short time (24 h) after drought stress application. Besides, a high number of important gene candidates with no hits (145)—probably completely new to the research community—will demand additional efforts for the recognition of their function. Validation procedures as well as transient expression assays are planned for future works, aiming to collaborate with breeding and biotechnological approaches for the benefit of the sugarcane culture, especially facing the scenario of future climate changes.

## Conflict of Interests

The authors declare that they have no conflict of interests.

## Acknowledgments

The authors thank Professor Günter Kahl (Frankfurt University, Germany). This work has been funded by Brazilian institutions: Financiadora de Estudos e Projetos (FINEP), Fundação de Amparo à Ciência e Tecnologia do Estado de Pernambuco (FACEPE), and Conselho Nacional de Desenvolvimento Científico e Tecnológico (CNPq).

## References

[1] A. D'Hont, G. M. Souza, M. Menossi et al., "Sugarcane: a major source of sweetness, alcohol, and bio-energy," in *Genomics of Tropical Crop Plants*, P. H. Moore and R. Moore, Eds., pp. 483–513, Springer, New York, NY, USA, 2008.

[2] A. J. Waclawovsky, P. M. Sato, C. G. Lembke, P. H. Moore, and G. M. Souza, "Sugarcane for bioenergy production: an assessment of yield and regulation of sucrose content," *Plant Biotechnology Journal*, vol. 8, no. 3, pp. 263–276, 2010.

[3] FAOSTAT, "Food and Agriculture Organization of the Unite Nations. In: FAO Statistical Databases," 2010, http://faostat.fao.org/.

[4] UNICA, "União da indústria de Cana-de-açúcar. In: Statistics of sugarcane sector—2009," 2009, http://www.unica.com.br/dadosCotacao/estatistica/.

[5] J. Goldemberg, "Ethanol for a sustainable energy future," *Science*, vol. 315, no. 5813, pp. 808–810, 2007.

[6] M. Menossi, M. C. Silva-Filho, M. Vincentz, M. A. Van-Sluys, and G. M. Souza, "Sugarcane functional genomics: gene discovery for agronomic trait development," *International Journal of Plant Genomics*, vol. 2008, Article ID 458732, 11 pages, 2008.

[7] E. A. Bray, J. Bailey-Serres, and E. Weretilnyk, "Responses to abiotic stresses," in *Biochemistry and Molecular Biology of Plants*, W. Gruissem, B. Buchannan, and R. Jones, Eds., pp. 1158–1249, American Society of Plant Physiologists, Rockville, Md, USA, 2000.

[8] L. Taiz and E. Zeiger, *Fisiologia Vegetal*, S. A. Artmed Editora, Porto Alegre, Brazil, 2004.

[9] W. Wang, B. Vinocur, and A. Altman, "Plant responses to drought, salinity and extreme temperatures: towards genetic engineering for stress tolerance," *Planta*, vol. 218, no. 1, pp. 1–14, 2003.

[10] L. Cattivelli, F. Rizza, F. W. Badeck et al., "Drought tolerance improvement in crop plants: an integrated view from breeding to genomics," *Field Crops Research*, vol. 105, no. 1-2, pp. 1–14, 2008.

[11] V. Y. Patade, A. N. Rai, and P. Suprasanna, "Expression analysis of sugarcane shaggy-like kinase (SuSK) gene identified through cDNA subtractive hybridization in sugarcane (*Saccharum officinarum* L.)," *Protoplasma*, vol. 248, no. 3, pp. 613–621, 2010.

[12] K. Nakashima and K. Shinozaki, "Regulons involved in osmotic stress-responsive and cold stress-responsive gene expression in plants," *Physiologia Plantarum*, vol. 126, no. 1, pp. 62–71, 2006.

[13] T. Umezawa, M. Fujita, Y. Fujita, K. Yamaguchi-Shinozaki, and K. Shinozaki, "Engineering drought tolerance in plants: discovering and tailoring genes to unlock the future," *Current Opinion in Biotechnology*, vol. 17, no. 2, pp. 113–122, 2006.

[14] F. A. Rodrigues, M. L. de Laia, and S. M. Zingaretti, "Analysis of gene expression profiles under water stress in tolerant and sensitive sugarcane plants," *Plant Science*, vol. 176, no. 2, pp. 286–302, 2009.

[15] R. A. Shimkets, "Gene expression quantitation technology summary," in *Gene Expression Profile: Methods and Protocol*, R. A. Shimkets, Ed., pp. 1–12, Humana Press, New Haven, Conn, USA, 2004.

[16] H. Matsumura, S. Reich, A. Ito et al., "Gene expression analysis of plant host-pathogen interactions by SuperSAGE," *Proceedings of the National Academy of Sciences of the United States of America*, vol. 100, no. 26, pp. 15718–15723, 2003.

[17] R. Terauchi, H. Matsumura, D. H. Krüger, and G. Kahl, "SuperSAGE: the most advanced through comparative genomics," in *The Handbook of Plant Functional Genomics: Concepts and Protocols*, G. Kahl and K. Meksem, Eds., pp. 37–54, Wiley-VCH, Weinheim, Germany, 2008.

[18] V. E. Velculescu, L. Zhang, B. Vogelstein, and K. W. Kinzler, "Serial analysis of gene expression," *Science*, vol. 270, no. 5235, pp. 484–487, 1995.

[19] H. Matsumura, K. H. Bin Nasir, K. Yoshida et al., "SuperSAGE array: the direct use of 26-base-pair transcript tags in oligonucleotide arrays," *Nature Methods*, vol. 3, no. 6, pp. 469–474, 2006.

[20] B. Coemans, H. Matsumura, R. Terauchi, S. Remy, R. Swennen, and L. Sági, "SuperSAGE combined with PCR walking allows global gene expression profiling of banana (*Musa acuminata*), a non-model organism," *Theoretical and Applied Genetics*, vol. 111, no. 6, pp. 1118–1126, 2005.

[21] C. Molina, B. Rotter, R. Horres et al., "SuperSAGE: the drought stress-responsive transcriptome of chickpea roots," *BMC Genomics*, vol. 9, article 553, 2008.

[22] C. Molina, M. Zaman-Allah, and F. Khan, "The salt-responsive transcriptome of chickpea roots and nodules via deepSuperSAGE," *BMC Plant Biology*, vol. 11, article 31, 2011.

[23] H. Hamada, H. Matsumura, R. Tomita, R. Terauchi, K. Suzuki, and K. Kobayashi, "SuperSAGE revealed different classes of early resistance response genes in *Capsicum chinense* plants harboring L3-resistance gene infected with Pepper mild mottle virus," *Journal of General Plant Pathology*, vol. 74, no. 4, pp. 313–321, 2008.

[24] P. A. Gilardoni, S. Schuck, R. Jungling, B. Rotter, I. T. Baldwin, and G. Bonaventure, "SuperSAGE analysis of the Nicotiana attenuata transcriptome after fatty acid-amino acid elicitation (FAC): identification of early mediators of insect responses," *BMC Plant Biology*, vol. 10, article 66, 2010.

[25] E. A. Kido, V. Pandolfi, L. M. Houllou-Kido et al., "Plant antimicrobial peptides: an overview of superSAGE transcriptional profile and a functional review," *Current Protein and Peptide Science*, vol. 11, no. 3, pp. 220–230, 2010.

[26] H. Matsumura, D. H. Krüger, G. Kahl, and R. Terauchi, "SuperSAGE: a modern platform for genome-wide quantitative transcript profiling," *Current Pharmaceutical Biotechnology*, vol. 9, no. 5, pp. 368–374, 2008.

[27] N. Robertson, M. Oveisi-Fordorei, S. D. Zuyderduyn et al., "DiscoverySpace: an interactive data analysis application," *Genome Biology*, vol. 8, no. 1, article R6, 2007.

[28] S. F. Altschul, W. Gish, W. Miller, E. W. Myers, and D. J. Lipman, "Basic local alignment search tool," *Journal of Molecular Biology*, vol. 215, no. 3, pp. 403–410, 1990.

[29] A. Conesa, S. Götz, J. M. García-Gómez, J. Terol, M. Talón, and M. Robles, "Blast2GO: a universal tool for annotation, visualization and analysis in functional genomics research," *Bioinformatics*, vol. 21, no. 18, pp. 3674–3676, 2005.

[30] M. Kiper, D. Bartels, F. Herzfeld, and G. Richter, "The expresion of a plant genome in hnRNA and mRNA," *Nucleic Acids Research*, vol. 6, no. 5, pp. 1961–1978, 1979.

[31] J. C. Kamalay and R. B. Goldberg, "Regulation of structural gene expression in tobacco," *Cell*, vol. 19, no. 4, pp. 935–946, 1980.

[32] H. Yamaguchi, H. Fukuoka, T. Arao et al., "Gene expression analysis in cadmium-stressed roots of a low cadmium-accumulating solanaceous plant, *Solanum torvum*," *Journal of Experimental Botany*, vol. 61, no. 2, pp. 423–437, 2010.

[33] P. A. F. Galante, D. O. Vidal, J. E. de Souza, A. A. Camargo, and S. J. de Souza, "Sense-antisense pairs in mammals: functional and evolutionary considerations," *Genome Biology*, vol. 8, no. 3, pp. R40.1–R40.14, 2007.

[34] M. Gowda, C. Jantasuriyarat, R. A. Dean, and G. L. Wang, "Robust-LongSAGE (RL-SAGE): a substantially improved LongSAGE method for gene discovery and transcriptome analysis," *Plant Physiology*, vol. 134, no. 3, pp. 890–897, 2004.

[35] R. A. Hassanein, A. A. Hassanein, A. B. El-din, M. Salama, and H. A. Hashem, "Role of jasmonic acid and abscisic acid treatments in alleviating the adverse effects of drought stress and regulating trypsin inhibitor production in soybean plant," *Australian Journal of Basic and Applied Sciences*, vol. 3, no. 2, pp. 904–919, 2009.

[36] C. D. Rock, Y. Sakata, and R. S. Quatrano, "Stress signaling I: the role of abscisic acid (ABA)," in *Abiotic Stress Adaptation in Plants: Physiological, Molecular and Genomic Foundation*, A. Pareek, S. K. Sopory, and H. J. Bohnert, Eds., pp. 33–73, Springer, Dordrecht, The Netherlands, 2010.

[37] S. A. Anjum, L. Wang, M. Farooq, I. Khan, and L. Xue, "Methyl jasmonate-induced alteration in lipid peroxidation, antioxidative defence system and yield in soybean under drought," *Journal of Agronomy and Crop Science*, vol. 197, no. 4, pp. 296–301, 2011.

[38] A. Chini, S. Fonseca, G. Fernández et al., "The JAZ family of repressors is the missing link in jasmonate signalling," *Nature*, vol. 448, no. 7154, pp. 666–671, 2007.

[39] B. Thines, L. Katsir, M. Melotto et al., "JAZ repressor proteins are targets of the $SCF^{COI1}$ complex during jasmonate signalling," *Nature*, vol. 448, no. 7154, pp. 661–665, 2007.

[40] J. Ogas, S. Kaufmann, J. Henderson, and C. Somerville, "PICKLE is a CHD3 chromatin-remodeling factor that regulates the transition from embryonic to vegetative development in Arabidopsis," *Proceedings of the National Academy of Sciences of the United States of America*, vol. 96, no. 24, pp. 13839–13844, 1999.

[41] C. Belin and L. Lopez-Molina, "Arabidopsis seed germination responses to osmotic stress involve the chromatin modifier PICKLE," *Plant Signaling & Behavior*, vol. 3, no. 7, pp. 478–479, 2008.

[42] J. L. Riechmann and E. M. Meyerowitz, "The AP2/EREBP family of plant transcription factors," *Biological Chemistry*, vol. 379, no. 6, pp. 633–646, 1998.

[43] J. L. Riechmann, J. Heard, G. Martin et al., "Arabidopsis transcription factors: genome-wide comparative analysis among eukaryotes," *Science*, vol. 290, no. 5499, pp. 2105–2110, 2000.

[44] Y. Sakuma, Q. Liu, J. G. Dubouzet, H. Abe, K. Shinozaki, and K. Yamaguchi-Shinozaki, "DNA-binding specificity of the ERF/AP2 domain of Arabidopsis DREBs, transcription factors involved in dehydration- and cold-inducible gene expression," *Biochemical and Biophysical Research Communications*, vol. 290, no. 3, pp. 998–1009, 2002.

[45] A. M. Sharoni, M. Nuruzzaman, K. Satoh et al., "Gene structures, classification and expression models of the AP2/EREBP transcription factor family in rice," *Plant & Cell Physiology*, vol. 52, no. 2, pp. 344–360, 2011.

[46] R. Lin, W. Zhao, X. Meng, and Y. L. Peng, "Molecular cloning and characterization of a rice gene encoding AP2/EREBP-type transcription factor and its expression in response to infection with blast fungus and abiotic stresses," *Physiological and Molecular Plant Pathology*, vol. 70, no. 1–3, pp. 60–68, 2007.

[47] G. Zhang, M. Chen, L. Li et al., "Overexpression of the soybean *GmERF3* gene, an AP2/ERF type transcription factor for increased tolerances to salt, drought, and diseases in transgenic tobacco," *Journal of Experimental Botany*, vol. 60, no. 13, pp. 3781–3796, 2009.

[48] A. Mangeon, R. M. Junqueira, and G. Sachetto-Martins, "Functional diversity of the plant glycine-rich proteins superfamily," *Plant Signaling and Behavior*, vol. 5, no. 2, pp. 99–104, 2010.

[49] G. Sachetto-Martins, L. O. Franco, and D. E. de Oliveira, "Plant glycine-rich proteins: a family or just proteins with

a common motif?" *Biochimica et Biophysica Acta*, vol. 1492, no. 1, pp. 1–14, 2000.

[50] S. Wang, D. Liang, S. Shi, F. Ma, H. Shu, and R. Wang, "Isolation and characterization of a novel drought responsive gene encoding a glycine-rich RNA-binding protein in *Malus prunifolia* (Willd.) borkh," *Plant Molecular Biology Reporter*, vol. 29, no. 1, pp. 125–134, 2011.

[51] C. H. Dong, B. K. Zolman, B. Bartel et al., "Disruption of *Arabidopsis CHY1* reveals an important role of metabolic status in plant cold stress signaling," *Molecular Plant*, vol. 2, no. 1, pp. 59–72, 2008.

[52] P. G. Sappl, A. J. Carroll, R. Clifton et al., "The Arabidopsis glutathione transferase gene family displays complex stress regulation and co-silencing multiple genes results in altered metabolic sensitivity to oxidative stress," *The Plant Journal*, vol. 58, no. 1, pp. 53–68, 2009.

[53] D. P. Dixon, M. Skipsey, and R. Edwards, "Roles for glutathione transferases in plant secondary metabolism," *Phytochemistry*, vol. 71, no. 4, pp. 338–350, 2010.

[54] S. S. Gill and N. Tuteja, "Reactive oxygen species and antioxidant machinery in abiotic stress tolerance in crop plants," *Plant Physiology and Biochemistry*, vol. 48, no. 12, pp. 909–930, 2010.

[55] R. Edwards and D. P. Dixon, "The role of glutathione transferases in herbicide metabolism," in *Herbicides and Their Mechanisms of Action*, A. H. Cobb and R. C. Kirkwood, Eds., pp. 38–71, Sheffield Academic Press, Sheffield, UK, 2000.

[56] M. W. Bianchi, C. Roux, and N. Vartanian, "Drought regulation of GST8, encoding the Arabidopsis homologue of ParC/Nt107 glutathione transferase/peroxidase," *Physiologia Plantarum*, vol. 116, no. 1, pp. 96–105, 2002.

[57] Á. Gallé, J. Csiszár, M. Secenji et al., "Glutathione transferase activity and expression patterns during grain filling in flag leaves of wheat genotypes differing in drought tolerance: response to water deficit," *Journal of Plant Physiology*, vol. 166, no. 17, pp. 1878–1891, 2009.

[58] S. George, G. Venkataraman, and A. Parida, "A chloroplast-localized and auxin-induced glutathione S-transferase from phreatophyte Prosopis juliflora confer drought tolerance on tobacco," *Journal of Plant Physiology*, vol. 167, no. 4, pp. 311–318, 2010.

[59] W. Ji, Y. Zhu, Y. Li et al., "Over-expression of a glutathione S-transferase gene, GsGST, from wild soybean (*Glycine soja*) enhances drought and salt tolerance in transgenic tobacco," *Biotechnology Letters*, vol. 32, no. 8, pp. 1173–1179, 2010.

[60] C. R. McClung, M. Hsu, J. E. Painter, J. M. Gagne, S. D. Karlsberg, and P. A. Salomé, "Integrated temporal regulation of the photorespiratory pathway. Circadian regulation of two Arabidopsis genes encoding serine hydroxymethyltransferase," *Plant Physiology*, vol. 123, no. 1, pp. 381–391, 2000.

[61] E. A. Cossins and L. Chen, "Folates and one-carbon metabolism in plants and fungi," *Phytochemistry*, vol. 45, no. 3, pp. 437–452, 1997.

[62] J. I. Moreno, R. Martín, and C. Castresana, "Arabidopsis SHMT1, a serine hydroxymethyltransferase that functions in the photorespiratory pathway influences resistance to biotic and abiotic stress," *The Plant Journal*, vol. 41, no. 3, pp. 451–463, 2005.

[63] G. M. Ali and S. Komatsu, "Proteomic analysis of rice leaf sheath during drought stress," *Journal of Proteome Research*, vol. 5, no. 2, pp. 396–403, 2006.

[64] A. D. Sharma and P. Singh, "Comparative studies on drought-induced changes in peptidyl prolyl cis-trans isomerase activity in drought-tolerant and susceptible cultivars of Sorghum bicolor," *Current Science*, vol. 84, no. 7, pp. 911–918, 2003.

[65] J. C. Ahn, D. W. Kim, Y. N. You et al., "Classification of rice (*Oryza sativa* L. *Japonica nipponbare*) immunophilins (FKBPs, CYPs) and expression patterns under water stress," *BMC Plant Biology*, vol. 10, article 253, 2010.

[66] A. V. Godoy, A. S. Lazzaro, C. A. Casalongué, and B. San Segundo, "Expression of a Solanum tuberosum cyclophilin gene is regulated by fungal infection and abiotic stress conditions," *Plant Science*, vol. 152, no. 2, pp. 123–134, 2000.

[67] S. A. J. Messing, S. B. Gabelli, I. Echeverria et al., "Structural insights into maize viviparous14, a key enzyme in the biosynthesis of the phytohormone abscisic acid," *The Plant Cell*, vol. 22, no. 9, pp. 2970–2980, 2010.

[68] D. R. McCarty, "Genetic control and integration of maturation and germination pathways in seed development," *Annual Review of Plant Physiology and Plant Molecular Biology*, vol. 46, pp. 71–93, 1995.

[69] D. Durantini, A. Giulini, A. Malgioglio et al., "Vivipary as a tool to analyze late embryogenic events in maize," *Heredity*, vol. 101, no. 5, pp. 465–470, 2008.

[70] S. H. Schwartz, B. C. Tan, D. A. Gage, J. A. D. Zeevaart, and D. R. McCarty, "Specific oxidative cleavage of carotenoids by VP14 of maize," *Science*, vol. 276, no. 5320, pp. 1872–1874, 1997.

[71] B. C. Tan, S. H. Schwartz, J. A. D. Zeevaart, and D. R. Mccarty, "Genetic control of abscisic acid biosynthesis in maize," *Proceedings of the National Academy of Sciences of the United States of America*, vol. 94, no. 22, pp. 12235–12240, 1997.

[72] S. Iuchi, M. Kobayashi, T. Taji et al., "Regulation of drought tolerance by gene manipulation of 9-cis-epoxycarotenoid dioxygenase, a key enzyme in abscisic acid biosynthesis in Arabidopsis," *The Plant Journal*, vol. 27, no. 4, pp. 325–333, 2001.

[73] X. R. Wan and L. Li, "Regulation of ABA level and water-stress tolerance of Arabidopsis by ectopic expression of a peanut 9-cis-epoxycarotenoid dioxygenase gene," *Biochemical and Biophysical Research Communications*, vol. 347, no. 4, pp. 1030–1038, 2006.

[74] F. Gao, C. Wang, C. Wei, and Y. Li, "A branched-chain aminotransferase may regulate hormone levels by affecting KNOX genes in plants," *Planta*, vol. 230, no. 4, pp. 611–623, 2009.

[75] M. A. Campbell, J. K. Patel, J. L. Meyers, L. C. Myrick, and J. L. Gustin, "Genes encoding for branched-chain amino acid aminotransferase are differentially expressed in plants," *Plant Physiology and Biochemistry*, vol. 39, no. 10, pp. 855–860, 2001.

[76] R. Diebold, J. Schuster, K. Däschner, and S. Binder, "The branched-chain amino acid transaminase gene family in *Arabidopsis* encodes plastid and mitochondrial proteins," *Plant Physiology*, vol. 129, no. 2, pp. 540–550, 2002.

[77] M. Malatrasi, M. Corradi, J. T. Svensson, T. J. Close, M. Gulli, and N. Marmiroli, "A branched-chain amino acid aminotransferase gene isolated from *Hordeum vulgare* is differentially regulated by drought stress," *Theoretical and Applied Genetics*, vol. 113, no. 6, pp. 965–976, 2006.

[78] M. J. Mueller, "Enzymes involved in jasmonic acid biosynthesis," *Physiologia Plantarum*, vol. 100, no. 3, pp. 653–663, 1997.

[79] Z. N. Ozturk, V. Talamé, M. Deyholos et al., "Monitoring large-scale changes in transcript abundance in drought- and salt-stressed barley," *Plant Molecular Biology*, vol. 48, no. 5-6, pp. 551–573, 2002.

[80] V. Talamè, N. Z. Ozturk, H. J. Bohnert, and R. Tuberosa, "Barley transcript profiles under dehydration shock and drought stress treatments: a comparative analysis," *Journal of Experimental Botany*, vol. 58, no. 2, pp. 229–240, 2007.

[81] K. M. Devaiah, G. Bali, T. N. Athmaram, and M. S. Basha, "Identification of two new genes from drought tolerant peanut up-regulated in response to drought," *Plant Growth Regulation*, vol. 52, no. 3, pp. 249–258, 2007.

[82] E. Padan and S. Schuldiner, "$Na^+/H^+$ antiporters, molecular devices that couple the $Na^+$ and $H^+$ circulation in cells," *Journal of Bioenergetics and Biomembranes*, vol. 25, no. 6, pp. 647–669, 1993.

[83] K. Xu, P. Hong, L. Luo, and T. Xia, "Overexpression of *AtNHX1*, a vacuolar $Na^+/H^+$ antiporter from *Arabidopsis thalina*, in *Petunia hybrida* enhances salt and drought tolerance," *Journal of Plant Biology*, vol. 52, no. 5, pp. 453–461, 2009.

[84] F. Brini, M. Hanin, I. Mezghani, G. A. Berkowitz, and K. Masmoudi, "Overexpression of wheat $Na^+/H^+$ antiporter *TNHX1* and $H^+$-pyrophosphatase TVP1 improve salt- and drought-stress tolerance in *Arabidopsis thaliana* plants," *Journal of Experimental Botany*, vol. 58, no. 2, pp. 301–308, 2007.

[85] M. A. Asif, Y. Zafar, J. Iqbal et al., "Enhanced expression of *AtNHX1*, in transgenic groundnut (*Arachis hypogaea* L.) improves salt and drought tolerance," *Molecular Biotechnology*, vol. 49, no. 3, pp. 250–256, 2011.

[86] M. A. R. Milla, A. Maurer, H. A. Rodríguez, and J. P. Gustafson, "Glutathione peroxidase genes in *Arabidopsis* are ubiquitous and regulated by abiotic stresses through diverse signaling pathways," *The Plant Journal*, vol. 36, no. 5, pp. 602–615, 2003.

[87] N. Navrot, V. Collin, J. Gualberto et al., "Plant glutathione peroxidases are functional peroxiredoxins distributed in several subcellular compartments and regulated during biotic and abiotic stresses," *Plant Physiology*, vol. 142, no. 4, pp. 1364–1379, 2006.

[88] Y. Miao, D. Lv, P. Wang et al., "An Arabidopsis glutathione peroxidase functions as both a redox transducer and a scavenger in abscisic acid and drought stress responses," *The Plant Cell*, vol. 18, no. 10, pp. 2749–2766, 2006.

[89] Y. Fleming, C. G. Armstrong, N. Morrice, A. Paterson, M. Goedert, and P. Cohen, "Synergistic activation of stress-activated protein kinase 1/c-Jun N-terminal kinase (SAPK1/JNK) isoforms by mitogen-activated protein kinase kinase 4 (MKK4) and MKK7," *Biochemical Journal*, vol. 352, no. 1, pp. 145–154, 2000.

[90] S. Kültz, "Evolution of osmosensory MAP kinase signaling pathways," *Integrative and Comparative Biology*, vol. 41, no. 4, pp. 743–757, 2001.

[91] Z. S. Xu, L. Liu, Z. Y. Ni et al., "W55a encodes a novel protein kinase that Is involved in multiple stress responses," *Journal of Integrative Plant Biology*, vol. 51, no. 1, pp. 58–66, 2009.

[92] Y. Kobayashi, S. Yamamoto, H. Minami, Y. Kagaya, and T. Hattori, "Differential activation of the rice sucrose nonfermenting1-related protein kinase2 family by hyperosmotic stress and abscisic acid," *The Plant Cell*, vol. 16, no. 5, pp. 1163–1177, 2004.

[93] P. B. K. Kishor, S. Sangam, R. N. Amrutha et al., "Regulation of proline biosynthesis, degradation, uptake and transport in higher plants: its implications in plant growth and abiotic stress tolerance," *Current Science*, vol. 88, no. 3, pp. 424–438, 2005.

[94] B. Zhu, J. Su, M. Chang, D. P. S. Verma, Y. L. Fan, and R. Wu, "Overexpression of a $\Delta^1$-pyrroline-5-carboxylate synthetase gene and analysis of tolerance to water- and salt-stress in transgenic rice," *Plant Science*, vol. 139, no. 1, pp. 41–48, 1998.

[95] R. Razavizadeh and A. A. Ehsanpour, "Effects of salt stress on proline content, expression of delta-1-pyrroline-5-carboxylate synthetase, and activities of catalase and ascorbate peroxidase in transgenic tobacco plants," *Biological Letters*, vol. 46, no. 2, pp. 63–75, 2009.

[96] M. Yamada, H. Morishita, K. Urano et al., "Effects of free proline accumulation in petunias under drought stress," *Journal of Experimental Botany*, vol. 56, no. 417, pp. 1975–1981, 2005.

[97] J. Gubis, R. Vaňková, V. Červená et al., "Transformed tobacco plants with increased tolerance to drought," *South African Journal Botany*, vol. 73, no. 4, pp. 505–511, 2007.

[98] H. M. Iskandar, R. Casu, A. Fletcher et al., "Identification of drought-response genes and a study of their expression during sucrose accumulation and water deficit in sugarcane culms," *BMC Plant Biology*, vol. 11, article12, 2011.

18

# Biomarker Identification for Prostate Cancer and Lymph Node Metastasis from Microarray Data and Protein Interaction Network Using Gene Prioritization Method

**Carlos Roberto Arias,[1,2] Hsiang-Yuan Yeh,[3] and Von-Wun Soo[1,3]**

[1] *Institute of Information Systems and Applications, National Tsing Hua University, Hsinchu 30013, Taiwan*
[2] *Facultad de Ingeniería, Universidad Tecnológica Centroamericana, Tegucigalpa 11101, Honduras*
[3] *Computer Science Department, National Tsing Hua University, Hsinchu 30013, Taiwan*

Correspondence should be addressed to Carlos Roberto Arias, carlos.r.arias@gmail.com

Academic Editor: Pierre-Olivier Angrand

Finding a genetic disease-related gene is not a trivial task. Therefore, computational methods are needed to present clues to the biomedical community to explore genes that are more likely to be related to a specific disease as biomarker. We present biomarker identification problem using gene prioritization method called gene prioritization from microarray data based on shortest paths, extended with structural and biological properties and edge flux using voting scheme (GP-MIDAS-VXEF). The method is based on finding relevant interactions on protein interaction networks, then scoring the genes using shortest paths and topological analysis, integrating the results using a voting scheme and a biological boosting. We applied two experiments, one is prostate primary and normal samples and the other is prostate primary tumor with and without lymph nodes metastasis. We used 137 truly prostate cancer genes as benchmark. In the first experiment, GP-MIDAS-VXEF outperforms all the other state-of-the-art methods in the benchmark by retrieving the truest related genes from the candidate set in the top 50 scores found. We applied the same technique to infer the significant biomarkers in prostate cancer with lymph nodes metastasis which is not established well.

## 1. Introduction

Genetic diseases have been around for a long time. In the past they were just not understood or known. Nowadays we do have a better knowledge of the underlying mechanisms behind these diseases, for instance now it is understood that cancer is a mutated genetic disease [1] and many researchers in molecular genetics have identified a number of key genes and potential drug targets for various types of cancer [2]. Cancer is extremely complex and heterogeneous and it has been suggested that 5% to 10% of the human genes probably contribute to oncogenesis [3]. However, our current understanding is still limited, this is due to the very nature of the genetic mechanisms of life. It is not a trivial task to discover new genes involved with genetic diseases like cancer, as they usually do not work alone, but as a part of a mechanism inside the machinery of the cells. Current research in the discovery of new cancer related genes consists of several approaches, one direct approach called in vitro, and another one called "in-silico". The in vitro approach is done by the biomedical community, they perform wet-lab experiments where they experiment with live tissue, comparing control and case cells. This approach is very accurate, but it is time consuming and extremely expensive, and sometimes it is not successful, since they might be investigating a gene that is not related with the disease. Here it is where bioinformatics provides tools to perform these studies in the so-called in silico environment. The bioinformatics studies are less accurate than in vitro ones, due to makeup of its source data that is usually noisy and incomplete [4]. On the other hand bioinformatics studies offer clues and hints to the biomedical researchers that help narrow the search for key genes and key mechanisms involved with a given disease, and it does it in a much cheaper fashion. Advances in this direction

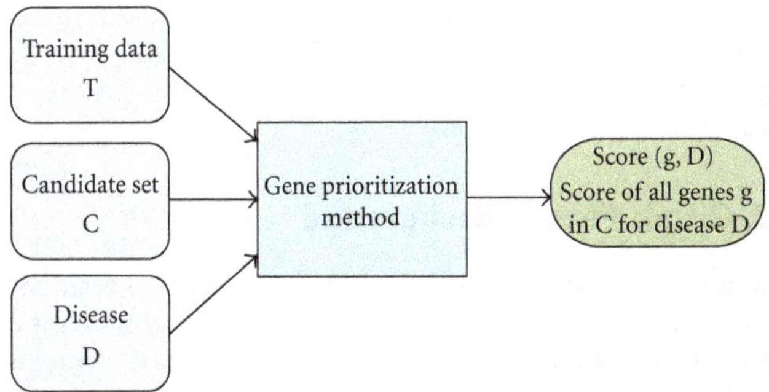


FIGURE 1: General gene prioritization overview.

are essential for identifying new disease genes as biomarker in complex diseases. To achieve this goal our research is oriented to a line of bioinformatics investigation called whole genome "disease gene prioritization" (DGP). This line of research objective is to find disease-related genes, and to assign more relevant genes to the disease a higher score in such a way that higher scores have higher probability of being related to the disease in question. In general DGP is described in Figure 1, were it can be seen that DGP methods take as input training data, that is, information to indicate previous knowledge about the disease. Along with this data comes the candidate set, that represents the whole set of genes being studied, and that are going to be ranked or prioritized by the method. Finally the disease information is also input to the DGP method. Once the method has finished processing the input it will output the set of genes with a score, high scoring genes are believed to be relevant to the input disease.

In general there are two types of DGP, one of them is data and text-mining based, and the other is network based. Data and text-mining-based DGP methods rely on data mining techniques to mine disease relevant genes from literature or different bioinformatics sources like sequence information. Among these methods there are: the eVOC method that performs candidate gene selection based on the coocurrence of the disease name in PubMed abstracts using data mining methods [5], GeneSeeker that is a web-based tool that selects candidate genes of the disease under study based in gene expression and phenotypic data of human and mouse [6], the method proposed by Piro et al. [7] that uses spatial gene expression profiles and linkage analysis, disease gene prediction [8], and prospects [9] that use basic sequence information to classify genes as likely or unlikely to be involved with the given disease, SUSPECTS [10] an extended version of PROSPECTS that integrates annotation data from gene ontology (GO) [11], InterPro [12] and expression data. Along this line of research is also MedSim [13] that uses GO enrichment and their own similarities measures [14].

Additionally there are network based methods, these ones are based on network analysis tools applied on biological networks. Network methods have the advantage that there is an increasing availability of human protein interaction data, along with the maturity of network analysis. In the case of these kind of methods the training set is usually a set of genes that are called "seed genes," these are genes that have been validated by wet lab experiments. Furthermore methods in this category can be classified in local and global methods, local methods use local information to the seeds, basically classifying by network proximity through the inspection of the seed genes or higher order neighbors in other words nodes in the network that are not directly adjacent to the seed nodes but are easily accessible by them. Global methods model the flow through the whole network to provide a score of the connectivity and impact of the seed genes. Either type, local or global, usually relies on the assumption that genes that are associated with diseases have a heavy interaction between each other [22]. The general idea behind network-based DGP is to assess how much genes interact together and how close they are to known diseases, integration of expression data from microarray into the network would improve its the accuracy, for more relevant biological information would be used. Among network-based methods: a method proposed by Chen et al. applies link-based strategies widely used in social and web network analysis such as Hits with Priors, PageRank and K-Step Markov to prioritize disease candidate genes based on protein interaction networks (PINs) [18]. Some various network-based approaches that predicted disease genes based on the protein network have obtained much better performance than traditional disease gene prediction approaches only based on the genome sequence alignment [23]. These kinds of the researches are associated with the long-held assumption that genes likely to interact directly or indirectly with each other are more likely to cause the same or similar diseases [24]. Wu et al. proposed a novel method, CIPHER (correlating protein interaction network and phenotype network to predict disease genes) [25] based on the phenotypic similarity and protein networks and they supposed that the phenotypic similarity among diseases can extract the disease-related genes on the network by measuring the direct neighbor (CIPHER-DN), shortest path (CIPHER-SP), or diffusion kernels. However, the direct neighbor strategy has some limitations to extract those indirect interaction genes and is more likely to be true for cases where two genes function in the same protein complex than in a pathway [26]. Shortest path analysis may yield a higher coverage and more novel predictions that are not so obvious to observe directly from the protein interaction data. The advantages of CIPHER perform genome-wide candidate gene prioritization for almost all human diseases but it does not work well for specific cancer due to not taking the relevant experimental data. They do not further select the active interaction relationships among protein while only a part of the interactions among a set of proteins may be active. These kinds of methods are inconsistent with previous studies which found that not all protein interactions occur at a specific condition [27]. More recently, Vavien is a system that uses the notion of topological profile to characterize a protein with respect to others [28]. Most of the aforementioned methods belong to the global or the local type of methods, integrating various sources of information to enrich their scoring. Another interesting fact is that as time passes the line the divides network based methods and data mining methods become less clear, this is due to the integration of data and text mining sources to network-based methods. This is the case of ENDEAVOR that takes a machine learning approach that builds a model with seed genes, and then that model is used to rank the candidate set according to a similarity score using multiple genomic data sources [17].

The accumulation of high-throughput data greatly promotes computational investigation of the expressions of thousands of genes and uses to manifest the expressions of genes under particular conditions. However, based on differential expression of the genes in the microarray data is likely to be incomplete, because there may be genes that are not differentially expressed but may be subtly involved in a pathway. The differential expression analysis only focuses on the selection of genes and does not pay attention to analyzing the interactions among them. Ma et al. proposes another method that performs the gene prioritization by Combining Gene expression and PIN (CGI) using Markov random field theory [19], CANDID that uses information from publications, protein domain descriptions, cross-species conservation measures, gene expression profiles and PIN to do a prioritization on the candidate genes that influence complex human traits [20], GeneRanks uses Google's PageRank algorithm and expression data to do gene prioritization [29], Mani et al. proposed a method called interactome dysregulation enrichment analysis (IDEA) to predict cancer related genes using interactome and microarray data [21], Karni et al. attempted to predict the causal gene from expression profile data and they identified a set of disease-related genes that could best explain the expression changes of the disease-related genes in terms of probable pathways leading from the causal to the affected genes in the network [30]. A summary table with the aforementioned methods can be found in Appendix D.

Scientific understanding of the biological mechanisms of cancer will help with the development of improved treatments for this disease, researchers around the world are attacking this issue with different approaches with a common goal: find a protocol to increase the probability of recovery from cancer. For this reason, our research is oriented towards this goal, and we have selected as our test domain prostate cancer. Motivated by the availability of rich information about this disease, and the fact that it is the third most common cause of death from cancer in male subjects according to the United States of America Library of Medicine.

In this paper we propose a method called gene prioritization based on microarray data with shortest paths, using voting extra scoring and edge flux strategy: (GP-MIDAS-VXEF), that is, a network-based DGP and a hybrid local and global method at it. It is based on the premise that disease relevant genes are on shortest paths involving seed genes like a local method. However uses the core of NetWalk to obtain disease relevant interactions making it also a global methods. Additionally boosting the scores using topological properties of the nodes that are considered to be "broker genes" as proposed by [31]. Finally is integrated differential expression data in the final score to increase biological meaning to the results.

This paper is organized as follows, following the introduction a brief background on graph theory is introduced, after which the Materials and Methods will follow where the source data and details on GP-MIDAS-VXEF will be presented. Then our results will be presented, finalizing this paper with the conclusions.

## 2. Graph Theory Background

A graph is a data structure that represents a set of relationships between elements or objects. Formally a graph $G$ is a pair defined by $G = (V, E)$, where $V$ is a set of elements that represent the nodes or vertices of the graph, the vertices in most applications hold the name of the attribute being represented. $E$ is the set of edges, where each edge represents a relation between two vertices, an edge is defined by $E = \{(u, v) \mid u, v \in V\}$, which may hold additional information as weight, confidence or distance between nodes, therefore having $E = \{(u, v, w) \mid u, v \in V \text{ and } w \in Real\}$. The edges may have directions, where $(u, v) \neq (v, u)$, in which case the graph is called directed graph, and when a direction is not important, the graph is called undirected graph.

*2.1. Graph Properties.* Among the intrinsic properties of a graph there are: *Nodes*, the number of nodes in the network, formally $n = |V|$. *Edges*, the number of edges in the graph, formally $e = |E|$. *Graph Path*, is a sequence of vertices of the form $\{v_1, v_2, v_3, \ldots, v_k\}$ where $v_1$ is the starting node and $v_k$ is the destination node, and $(v_i, v_{i+1}) \in E$; the length of the path is defined by $l = \sum_{i=1}^{k-1} w_i$ where $w_i \in (v_i, v_{i+1}, w_i)$, when all weights are equal to 1 then the length of the path is $k - 1$. A shortest path from vertex $v$ to $u$ is one of the paths that have the least accumulated weight from $u$ to $v$, note that there can be multiple shortest paths from one node to another.

*2.2. Nodes Properties.* The most basic node property is the *degree* $(v)$ that denotes the number of connections a node $v$ has; in directed graphs there can be a distinction between incoming and outgoing connections, called *in-degree* and *out-degree*, respectively. Another property of the nodes is the *Clustering Coefficient*, this property measures which nodes in the network tend to cluster together and shows how close a node and its immediate neighbors are to become a full connected graph. Clustering coefficient is formally defined by (1) where $\mathrm{ec}(v)$ is the number of edges in the subgraph made only of node $v$ and its neighbors, and $\mathrm{nc}(v)$ is the number of nodes of that graph:

$$\mathrm{cc}(v) = \frac{2\mathrm{ec}(v)}{\mathrm{nc}(v)(\mathrm{nc}(v) - 1)}. \quad (1)$$

Many other graph and node properties have been defined, a good reference for these properties can be found at [32].

## 3. Materials and Methods

*3.1. Materials.* Current public PIN databases provide rich information and they mostly differ on the way they acquire or validate their data. For example, HPRD, BIND, MINT, and MIPS are manually curated. On the other hand, DIP and IntAct are based on literature mining and they achieve these using computational methods that retrieve the interaction knowledge automatically from published papers. Prieto and De Las Rivas have shown a limited intersection and overlap between the six major databases (BioGRID, BIND, MINT, HPRD, IntAct, DIP) [33]. The information contained in these databases is partly complementary and the knowledge of the protein interactions can be increased and improved by combining multiple databases.

To build more complete protein-protein interaction networks, we integrated PIN data warehouse included HPRD, DIP, BIND, IntAct, MIPS, MINT, and BioGrid databases (see Appendix C for details on these databases) which has successfully gathered 54,283 available and nonredundant human PIN pairs among 10,710 human related proteins into BioIR database [34]. This integration is a result of the availability of public protein interaction databases. Prostate cancer is a worldwide leading cancer and it is characterized by its aggressive metastasis. It is considered by the American Cancer Society as the second leading cause of cancer death among men, thus making this disease an important issue to study. However, up to date there are no reliable biomarkers can reliably associated with them. Understanding the differences in the biology of metastatic prostate cancer and non-metastatic primary tumors is essential for developing new prognostic markers and therapeutic targets. We used microarray data taken from [35] that consists of 62 primary tumors, 9 lymph nodes metastasis and 41 normal control samples. We applied those data into two groups, one is prostate primary tumor and normal samples and the other is prostate primary tumor with and without lymph nodes metastasis, respectively. We assign the weights to the protein networks for the edge flux step of our method and to boost the score in the final stage of the scoring phase. According to the genes from the microarray data, we extracted 8,123 genes and 43,468 protein-protein interactions to identify the prostate-related genes and subnetworks. These extracted genes are used as our training set for the prioritization process. The initial seed genes known to be related to the prostate cancer are extracted from public OMIM [36] database which stores gene-disease associations provided by summaries of publications and the list of the 15 seed genes are shown in Table 1, these genes are selected from [37] so that we can use a common ground for comparison. We took the KEGG pathway database [38] and PGDB database [39] that are manually curated database for prostate cancer and obtained 137 genes (Target Genes) as the truly disease-related genes for prostate cancer to compare the performance with the previous methods: CIPHER, Endeavour, HITS with priors, PageRank with priors, K-step Markov, and plain Random Walk with Restarts.

*3.2. Methods.* GP-MIDAS-VXEF is a hybrid local and global network based method for disease gene prioritization. The backbone of the method is based on the PIN, hence it relies on network analysis tools. The network analysis tools used along the method rest on the following well-documented assumptions:

(i) genes that have strong relationship between each other in the network tend to be closer together [22, 40, 41];

(ii) important genes in the network show high degree and low clustering coefficient, since these genes are significant they are called in published literature "broker genes" [31].

Although it is network based, it integrates expression data to find relevant interactions using a random walk with restarts (RWR) strategy, thus its global nature, after all, the RWR processes the network in its full extent (EF stage). Subsequently it does a shortest paths analysis along the networks that were generated by the EF stage. In each of those analysis it uses an extension of the basic scoring by incorporating a score boosting by means of considering the clustering coefficient of each of the genes in the network, covering the local nature. The previous two steps combined are called GP-MIDAS-XEF, GP-MIDAS for the basic shortest paths analysis, $X$ for the extension using topological features, and EF for the incorporation of edge flux networks. Once the prioritization is done over all the EF networks, a set of scores is available for the voting phase, where the scores are integrated to produce a single-score base (voting stage). Finally, each gene score is boosted again using the average differential expression of each gene. Figure 2 presents the general overview of GP-MIDAS-VXEF and how the input data is used on each of the stages. Following this introduction to the method, each of the stages are going to be presented.

*3.2.1. Edge Flux Filter.* This is the first formal stage of the prioritization process, where the input PIN is analyzed using RWR with the purpose of finding relevant interactions to the specific domain under study. This model is applied using $\gamma = 0.3$ as probability of restart, as suggested by [22]. One disadvantage about the network based DGP is the use of noisy source data, therefore some steps are needed to filter out the source PINs in such a way that more relevant interactions are used in the core of the method. This is an open issue of research, nevertheless good results have been achieved like those of Komurov et al. that proposed a method called NetWalk [42]. This method is based on the execution of random walks on the network to obtain disease relevant interaction in the network. The steps that are executed during this stage are described in Algorithm 1.

Using the available microarray data the first step is to calculate the pearson correlation Coefficient (PCC) to determine the coexpressed relationship of the interactions in the PIN [43]. Two sets of PCC are calculated one for control sample ($\mathrm{PCC}^N$) and one for the case sample ($\mathrm{PCC}^D$). These Pearson correlation coefficients will be used as weights of the PIN that is input to the computation of the stochastic matrix

$$\text{Stochastic Matrix}_{ij} = \begin{cases} 0, & \text{if there is no edge } (i,j), \\ \dfrac{\text{Weight}(i,j)}{\sum_{kadji} \text{Weight}(i,k)}, & \text{otherwise,} \end{cases} \quad (2)$$

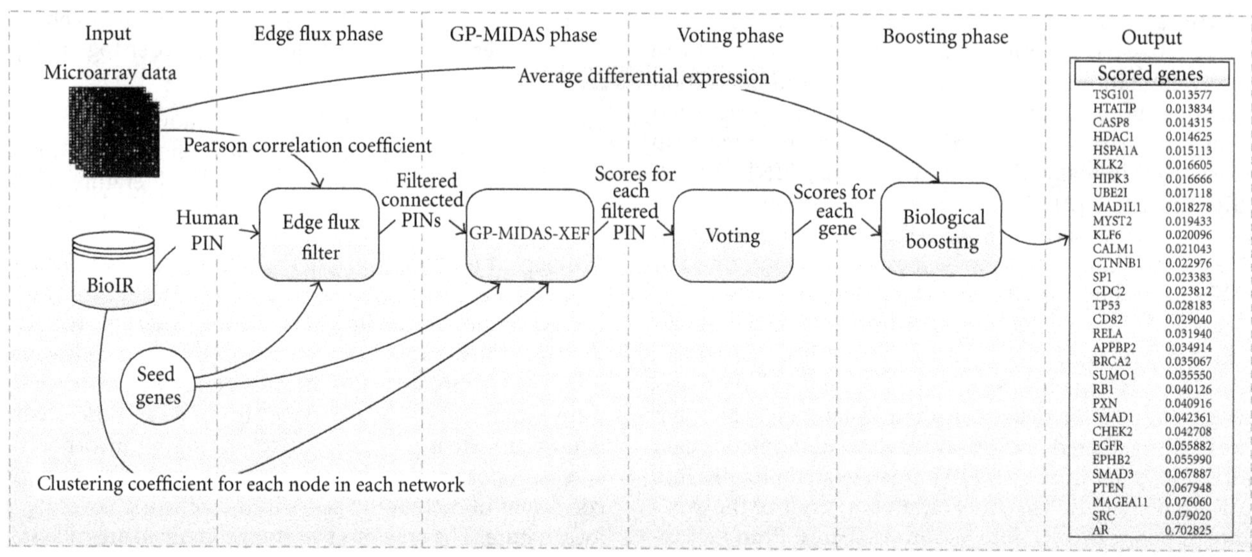


FIGURE 2: GP-MIDAS-VXEF workflow.

**Input:** Differential Expression PPI `DE_PIN`, Pearson Correlation Coefficiente Networks (Control and Case) `D_PCC_PIN` and `N_PCC_PIN`, Seed Set as Training set of the RWR `ρSeedSet`, Boundaries and Step of Filtering `ThresholdStart`, `ThresholdEnd`, `ThresholdStep`
**output:** Set of filtered networks `DE_FilteredPPI`
(1) Create Stochastic Matrix `StochasticMatrix` from Case Samples Pearson Correlation Coefficient $PCC^D$ PIN according to (2)
(2) Create Reference Stochastic Matrix `StochasticMatrixRef` from Control Samples Pearson Correlation Coefficient $PCC^N$ PIN according to (2)
(3) Run `RWR` with $\gamma = 0.3$ using `StochasticMatrix` save the results in `RWR_Scores`
(4) Run `RWR` with $\gamma = 0.3$ using `StochasticMatrixRef` save the results in `RWR_RefScores`
(5) Compute raw edge flux value for case sample $ef_{ij}$ according to (3)
(6) Compute raw edge flux value for control (reference) sample: $efref_{ij}$ according to (3)
(7) Compute Normalized Edge Flux according to (4)
(8) **for** $i \leftarrow$`ThresholdStart` to `ThresholdEnd` **step** `ThresholdStep` **do**
(9) Create Edge List `EdgeList` using threshold $+i$ ($i \to +\infty$)
(10) Change Weights in `EdgeList` to weights in `DE_PIN` using only the edges in `EdgeList` store results in $\texttt{DE\_FilteredPPl}_{(+i)}$
(11) Create Edge List using `EdgeList` threshold $-i$ ($-\infty \to -i$)
(12) Change Weights in `EdgeList` to weights in `DE_PIN` using only the edges in `EdgeList` store results in $\texttt{DE\_FilteredPPl}_{(-i)}$
(13) Create Edge List `EdgeList` using threshold $Ci$ ($-\infty \to i \cup i \to \infty$)
(14) Change Weights in `EdgeList` to weights in `DE_PIN` using only the edges in `EdgeList` store results in $\texttt{DE\_FilteredPPI}_{(Ci)}$
(15) **return** *Set of Filtered PPI Networks* `DE_FilteredPPl`

ALGORITHM 1: Overview of NetWalk Phase.

$$ef_{ij} = \text{RWR}_{\text{Scores}_i} * \text{StochasticMatrix}_{ij}, \quad (3)$$

$$EF_{ij} = \log_2\left(\frac{ef_{ij}}{\text{ref\_}ef_{ij}}\right). \quad (4)$$

The main difference between Algorithm 1 and the one presented by [42] is that this method normalizes using weights from control expression data, and Komurov et al. use an unweighted network. The purpose of this normalization serves two purposes to unbias the results from structural bias that is natural in the RWR method and to unbias interactions that are similar between control and case samples. Notice that the resulting networks will no longer possess the *EF* values as weights but the weights of differential expression weights (diff_expr$(u, v)$) that are explained next, this sets the networks ready for shortest path analysis. Once this stage is over, the result is a set of networks that will be processed by the next phase.

To assign the expression weights to the EF-filtered PINA, the microarray data must be transformed in such a way

TABLE 1: Seed genes of prostate cancer from omim database.

| Gene ID | Gene symbol | Gene name |
|---|---|---|
| 367 | AR | Androgen receptor |
| 675 | BRCA2 | Breast cancer type 2 susceptibility protein |
| 3732 | CD82 | CD82 antigen |
| 11200 | CHEK2 | Serine/threonine-protein kinase Chk2 |
| 60528 | ELAC2 | Zinc phosphodiesterase ELAC protein 2 |
| 2048 | EPHB2 | Ephrin type-B receptor 2 precursor |
| 3092 | HIP1 | Huntingtin-interacting protein 1 |
| 1316 | KLF6 | Kruppel-like factor 6 |
| 8379 | MAD1L1 | Mitotic spindle assembly checkpoint protein MAD |
| 4481 | MSR1 | Macrophage scavenger receptor types I and II |
| 4601 | MXI1 | MAX-interacting protein 1 |
| 7834 | PCAP | Predisposing for prostate cancer |
| 5728 | PTEN/PTENP1 | Phosphatidylinositol-3,4,5-trisphosphate 3-phosphatase, and dual-specificity protein phosphatase PTEN |
| 6041 | RNASEL | 2–5A-dependent ribonuclease |
| 5513 | HPC1 | Hereditary prostate cancer 1 |

TABLE 2: Target genes found across methods.

| Method | Target genes |
|---|---|
| CIPHER | ATM, BRCA1, CAV1, CCND1, CDKN1A, CDKN1B, EGFR, EGR1, ESR1, ESR2, HIF1A, HRAS, MME, MSH2, MYC, NCOA3, NCOA4, PGR, RB1, RNF14, SMARCA4, TP53 |
| ENDEAVOUR | ACPP, ANXA7, APC, ARMET, ATM, BCL2, BMP6, BRCA1, BTRC, CAV1, CCND1, CD44, CDH1, CDH13, CDKN1A, CDKN1B, CDKN2A, CTCF, CTNNA1, CTNNB1, CYP1B1, DAPK1, EDNRB, EGFR, EGR1, ERBB2, ERCC5, ESR1, ESR2, FAF1, FHIT, GGT1, GSTP1, HIF1A, HOXA13, HRAS, IGFBP3, IL12A, IL8, KLK10, KLK2, KLK3, MAP2K4, MME, MSH2, MYC, NAT1, NCOA3, NCOA4, NEFL, PGK1, PGR, PLAU, POLB, PTPN13, RARB, RASSF1, RB1, RNF14, SLC2A2, SMARCA4, SOX2, STMN1, TCEB1, TMEPAI, TNF, TP53, TYR, VDR |
| ToppNet (K-Step Markov, HITS with Priors, PageRank with Priors | ACPP, ANXA7, APC, ATM, BCL2, BMP6, BRCA1, BTRC, CAV1, CCND1, CD44, CDH1, CDH13, CDKN1A, CDKN1B, CDKN2A, CTCF, CTNNA1, CTNNB1, CYP1B1, DAPK1, EDNRB, EGFR, EGR1, ERBB2, ERCC5, ESR1, ESR2, FAF1, FHIT, GGT1, GSTP1, HIF1A, HOXA13, HRAS, IGFBP3, IL12A, IL8, KLK10, KLK2, KLK3, MAP2K4, MC1R, MME, MSH2, MYC, NAT1, NCOA3, NCOA4, NEFL, NME1, PGK1, PGR, PLAU, POLB, PTPN13, RARB, RASSF1, RB1, RNF14, SLC2A2, SMARCA4, SOX2, STMN1, TCEB1, TNF, TP53, TYR, VDR |
| GP-MIDAS-VXEF | ACPP, ANXA7, APC, ARMET, ATM, BCL2, BMP6, BRCA1, BTRC, CAV1, CCND1, CD44, CDH1, CDH13, CDKN1A, CDKN1B, CDKN2A, CTCF, CTNNA1, CTNNB1, CYP1B1, DAPK1, EDNRB, EGFR, EGR1, **EIF3S3**, ERBB2, ERCC5, ESR1, ESR2, FAF1, FHIT, GGT1, GSTP1, HIF1A, HOXA13, HRAS, IGFBP3, IL12A, IL8, KLK10, KLK2, KLK3, MAP2K4, MC1R, MME, MSH2, MYC, NAT1, NCOA3, NCOA4, NEFL, NME1, PGK1, PGR, PLAU, POLB, PTPN13, RARB, RASSF1, RB1, RNF14, SLC2A2, SMARCA4, SOX2, STMN1, TCEB1, TMEPAI, TNF, TP53, TYR, VDR, **VEGF** |

that can be used to represent weights in the PIN, and that large weights indicate less interaction than small weights. This transformation has two steps, initially the values are updated using a sample of control expression microarray data, the effect of this operation is that values that are very similar between normal and cancer samples should have less impact on our analysis. To accomplish this we subtract the value from the cancer microarray data to the value of the control expression data as shown in (5), where there are $N$ samples of control tissue and $M$ samples of case tissue. The next step is to transform the values, the rationale behind this transformation is that expression values may be negative for underexpressed genes, and if these values are used as they are, our network may have negative weights, thus making shortest paths analysis more difficult. Equation (6) shows how the expression values are transformed:

$$\text{Expression Value}_i = \left| \frac{\sum_{n=1}^{N}\left(\text{control\_expr}_{n,i}\right)}{N} - \frac{\sum_{m=1}^{M}\left(\text{case\_expr}_{m,i}\right)}{M} \right|. \tag{5}$$

Transformed Expression $\text{Value}_i$

$$= -\ln\left(\frac{|\text{Expression Value}_i| - \min}{\max - \min}\right). \quad (6)$$

Considering that the sign of the value in the microarray data represents over- or under-expression, and the fact that we want to make a representation of distance, for this is what we want in our quantitative analysis, we use the absolute value of the microarray data, then these results are normalized, using the max and min values found, by doing these two steps we get values in the range $[0, 1]$, where values closer to 1 mean that they are more expressed (either over expressed or under expressed). Finally we compute the negative of the natural logarithm on the previous results, this is to make smaller numbers (less expression level) become large distances, and bigger numbers (higher expression level) become short distances. The result of this step is a transformation of the gene expression, where more expressed genes have smaller value, and less expressed genes have higher values, in the next step we convert this values into distances between genes, thus more expressed genes relationships will become shorter distances than less expressed genes relationships. In the case the $|\text{Expression Value}_i| = \min$ we just set the whole result to be a big value, since $\text{in}(0)$ is not defined. The result of this process is diff_expr$(u, v)$, that represents the differential expression as a distance between nodes $u$ and $v$ in the PIN. Once the microarray expression data is transformed, it is ready to be integrated as weights into the PIN. Since we need the network to become a weighted one, where these weights are related to the specific interactions in disease-related network, we use the transformed values of the microarray data. However the microarray data provides transformed expression values for the genes, not for the relationship between genes. To overcome this issue, we combine the values of the two interacting genes together. For instance if we have microarray values {(SEPW1, 4.097), (BRCA1, 1.395), (AKT1, 2.006), (BACH1, 2.823), (AHNAK, 3.597)} and we have the following edges in our graph {(AKT1, AHNAK), (BACH1, BRCA1), (BRCA1, AKT1)}, then the first edge weight would be the addition of the transformed expression values of each of the vertices 2.006 + 3.597 = 5.603 providing the weight of the first edge. The resulting weighted edges of this instance would be {(AKT1, AHNAK, 5.603), (BACH1, BRCA1, 4.218), (BRCA1, AKT1, 3.401)}, this process results in all the relations in diff_expr$(u, v)$ where $(u, v) \in$ Interactions of *PIN*.

*3.2.2. Shortest Paths and Structural Prioritization: GP-MIDAS-XEF.* At this phase each of the networks created by the NetWalk stage is going to be used as input of the GP-MIDAS-XEF. GP-MIDAS will do its prioritization based on the shortest paths, and then by boosting each gene score using the clustering coefficient of the gene in the specific network.

*Scoring of Genes with Shortest Paths.* For this analysis all the shortest paths are computed, that is, for each pair of genes in the network the shortest paths are computed. As each of

Table 3: Top 50 Genes.

| Rank | Using prostate and normal tissue | Using prostate and metastatis tissue |
|---|---|---|
| 1 | CAV1 | CAV1 |
| 2 | TP53 | MAGEA11 |
| 3 | MAGEA11 | CALM1 |
| 4 | CALM1 | CALR |
| 5 | EGFR | TP53 |
| 6 | UBE2I | FHL2 |
| 7 | CALR | EGFR |
| 8 | SMAD3 | APP |
| 9 | FHL2 | JUN |
| 10 | HDAC1 | SMAD3 |
| 11 | APP | SMAD2 |
| 12 | MYC | ESR1 |
| 13 | JUN | RB1 |
| 14 | ESR1 | HIPK3 |
| 15 | GNB2L1 | BRCA1 |
| 16 | HIPK3 | SMAD1 |
| 17 | SMAD2 | GNB2L1 |
| 18 | APPBP2 | XRCC6 |
| 19 | CDC2 | UBE2I |
| 20 | BRCA1 | HDAC1 |
| 21 | RB1 | CDC2 |
| 22 | SMAD1 | AES |
| 23 | PXN | STAT3 |
| 24 | XRCC6 | IL6ST |
| 25 | IL6ST | APPBP2 |
| 26 | STAT3 | PCAF |
| 27 | DLG1 | REPS2 |
| 28 | AES | FLNA |
| 29 | TRAF6 | RAF1 |
| 30 | FLNA | MYC |
| 31 | TRIM29 | MAPK1 |
| 32 | PCAF | TRAF6 |
| 33 | REPS2 | CCND1 |
| 34 | AKT1 | SMARCA4 |
| 35 | PRKCA | HLA-B |
| 36 | RAF1 | TRAF2 |
| 37 | HLA-B | RANBP9 |
| 38 | TRAF2 | PIAS4 |
| 39 | SMARCA4 | GSK3B |
| 40 | MAPK1 | TRIM29 |
| 41 | CHGB | FOS |
| 42 | RANBP9 | IDE |
| 43 | CCND1 | SRC |
| 44 | GSK3B | PXN |
| 45 | HSPA1A | SLC25A4 |
| 46 | BCL2 | SP1 |

TABLE 3: Continued.

| Rank | Using prostate and normal tissue | Using prostate and metastatis tissue |
|---|---|---|
| 47 | VCL | NR5A1 |
| 48 | RAI17 | YWHAG |
| 49 | TGFBR1 | AKT1 |
| 50 | SELENBP1 | CCNE1 |

them is computed, the path is verified to check if any of the seed genes is on the resulting path, if so, these paths are added to the list of paths PathList to be considered in the scoring. Finally a score is computed for each gene.

*Compute the Score Function.* Having all the paths stored in PathList we can compute the denominator denom using (7):

$$\text{denom} = \sum_{i=1}^{n} \frac{1}{l_i}, \tag{7}$$

where $l_i$ is the total length of the $i$th shortest path. Once the denominator is ready, we proceed to compute the score. For each gene $g$ on the network we compute the score according to (8):

$$\text{Score}(\text{Gene}_i) = \sum_{\text{Gene}_i \in \text{Path}_j}^{\text{PathList}} \frac{1/l_j}{\text{denom}}. \tag{8}$$

The motivation behind (8) is that a gene that appears in more shortest paths or more times in the paths list is going to achieve higher score, the highest being 1 if the gene appears in all the found paths.

*Extending the Score of Genes.* Cai et al. have demonstrated that disease genes in the network show particularly high degree and low clustering coefficient, defined in (1), they called this special genes *broker genes* [31]. Based on this idea, each of the previously computed scores are updated using (9). The boosting is computed from locally computed clustering coefficient of the node, and it affects that node alone:

$$\begin{aligned} &\text{Score}(\text{Gene}_i) \\ &\quad = \text{Score}(\text{Gene}_i) * \left(2 - \text{Clustering\_Coefficient}_{(\text{Gene}_i)}\right). \end{aligned} \tag{9}$$

By doing this boosting, genes with low clustering coefficient will have higher boosting, and high clustering coefficient will have lower boosting. The consequence is that disease-related genes are expected to have increased scores, a result that was achieved as will be demonstrated in the results section.

*3.2.3. Voting Phase.* Since we are getting a set of thresholds $T$ in the edge flux filtering phase to produce $|T|$ different coexpressed networks. Those networks are built using the edges that have the values that result from the steps 8 to 14 in Algorithm 1, in other words the values on the ends of the two tails of the edge flux distribution. Second, we compute a score on each gene $g$ of those networks and have a matrix of ranked genes where each row represents the position of the gene, as expressed in (10), where $S_i$ is the score achieved by GP-MIDAS-XEF with threshold $i$. For all the tested ranked lists, we used rank aggregation to re-rank the genes. Borda count has been extensively studied which is originally a voting method based on positional-scoring rankings [44, 45]. We generate a weight vector w as follows: The top 1 ranked receive weight 1, top 2 ranked receive weight 1/2, by the same way to the weight $1/k$ for each last ranked (where $k$ denotes the number of the genes in the network). This ranking is denoted in (11). Finally ScoreMatrix is summarized to provide a single score, this is done using (12) where $\text{pos}_i(g)$ denotes the position of gene $g$ in the $i$th network:

$$\begin{aligned} \text{Score Matrix} &= \text{EF} - \text{Phase\_Results\_Matrix} \\ &= \langle S_1, S_2, \ldots, S_{|T|} \rangle, \end{aligned} \tag{10}$$

$$\text{Score Matrix}_{i,j} = \frac{1}{j}, \tag{11}$$

$$\text{Voting Score}\left(\text{Gene}_g\right) = \sum_{i=1}^{|T|} \left(\text{Score Matrix}_{i,\text{pos}_i(g)}\right). \tag{12}$$

The rationale behind this equation is that a gene that appears more times in a higher position would get higher weight those genes that appear in lower positions. Finally, the final score of the genes is sum up the weights of the position of the genes from different coexpressed networks. The newly rank denotes the largest score wins higher positions from different network topology.

*3.2.4. Biological Boosting.* Before the voting phase we focus more on the gene prioritization based on the edge score in the network, however each gene in the candidate set will have a corresponding average differential expression (ADE) as defined by (5), where the average of the control expression and case expression samples are calculated and then substracted from each other. The rationale behind this computation is that larger values will indicate larger difference between disease tissue and normal tissue, and on the contrary values closer to zero will represent genes that their expression level does not change much between control and case samples. These values will serve in the last stage of the prioritization where the biological score boosting takes place, thus more boosting for higher differential expression and less boosting for lower differential expression.

Once the ADE is ready, the final stage of the prioritization takes in the single-score list from the Voting phase, this scores are boosted one last time using a normalized absolute differential expression (ADE) as was computed previously. The ADE values are normalized to ensure that the range of values are $[0, 1]$, numbers closer to 1 will represent numbers with higher difference between control and case samples, as opposed to values closer to zero. Equation (13) shows how the boosting is done. Notice that ADE in the boosting process uses only the gene specific expression data as a value to

Table 4: Top 50 genes overlap.

| Set | Total genes | Genes in set |
|---|---|---|
| Overlapped genes | 41 | HDAC1 IL6ST SMAD1 RB1 TRAF2 RAF1 BRCA1 APP CDC2 EGFR AKT1 FLNA AES SMAD2 REPS2 GSK3B SMARCA4 GNB2L1 STAT3 UBE2I TRAF6 MAPK1 MYC CAV1 JUN CCND1 RANBP9 HLA-B PCAF FHL2 TP53 TRIM29 CALR APPBP2 SMAD3 CALM1 MAGEA11 HIPK3 ESR1 PXN XRCC6 |
| Genes in metastasis analysis | 9 | PIAS4 FOS YWHAG SLC25A4 SP1 SRC IDE CCNE1 NR5A1 |
| Genes in nonmetastasis analysis | 9 | DLG1 VCL TGFBR1 CHGB SELENBP1 BCL2 PRKCA RAI17 HSPA1A |

Table 5: Available biological networks sites.

| Name | Acronym | URL |
|---|---|---|
| Human Protein Reference Database | HPRD | http://www.hprd.org/ |
| Biomolecular Interaction Network Database | BIND | http://bond.unleashedinformatics.com/ |
| Biological General Repository for Interaction Datasets | BioGRID | http://thebiogrid.org/ |
| Database of Interacting Proteins | DIP | http://dip.doe-mbi.ucla.edu/ |
| IntAct Molecular Interaction Database | IntAct | http://www.ebi.ac.uk/intact/ |
| The MIPS Mammalian Protein-Protein Interaction Database | MIPS | http://mips.helmholtz-muenchen.de/proj/ppi/ |
| Molecular Interaction Database | MINT | http://mint.bio.uniroma2.it/mint/ |
| Kyoto Encyclopedia of Genes and Genomes | KEGG | http://www.genome.jp/kegg/ |
| National Center for Biotechnology Information | NCBI | http://www.ncbi.nlm.nih.gov/ |

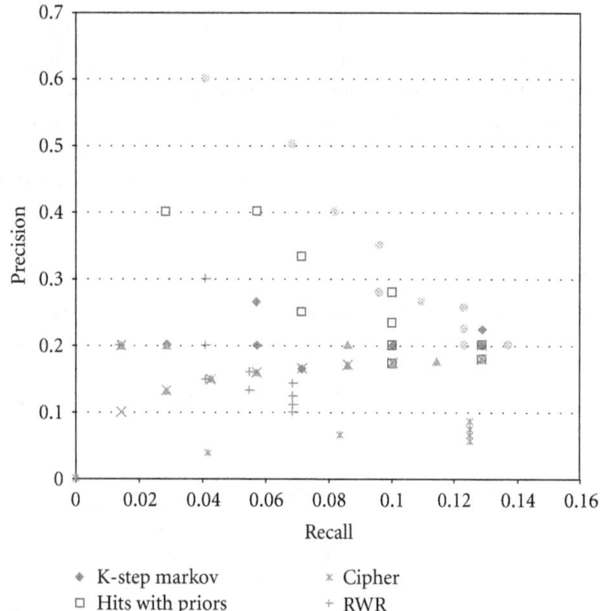


Figure 3: ROC curves comparing the performance of GP-MIDAS-VXEF with existent state-of-the-art network-based prioritization methods.

express how much important a node is in this analysis, unlike the diff_expr$(u, v)$ that represented weights in the network:

$$\text{Final Score}(\text{Gene}_i) = \text{Voting Score}(\text{Gene}_i) * (1 + \text{Normalized\_ADE}_i). \tag{13}$$

## 4. Results and Discussion

*4.1. Performance.* The method was tested using Prostate Cancer as the domain for the experiments, as it is explained in the Materials section previously. The method is compared to HITS with Priors, K-step Markov and PageRank with Priors all from the ToppNet suite [18]; other methods in the benchmark are ENDEAVOR [17], CIPHER [25] and plain Random Walk with Restarts [46] using Pearson Correlation Coefficient for weights of the network. These methods were selected because they belong to the Network Based DGP methods class, and they do not integrate data and text mining capabilities in their prioritization. Seed Genes (Training Set) is not considered in the benchmark in any of the methods, therefore our method does include them in our benchmark. It is worth to mention that all 13 seed genes are recovered in the top 20 rank of the method. Another reason for the selection of these methods is their public availability, Vavien [28] is not considered because it can only handle 50 candidate genes, and the methods in the benchmark handle any amount of candidate genes.

Figure 3 shows a precision-recall diagram were it is evident that our method has the best results among the rest of the methods in the benchmark. In Figure 4 absolute count of found genes per rank is presented. Additionally there is a number "Average Position" that represent the average position of the known cancer-related genes in the rank on the top of each bar. The figure clearly shows that GP-MIDAS-VXEF outperforms the other methods in the benchmark, and to resolve ties average position in the rank is also shown. For instance in the Top 40 ranks there is a tie between K-Step Markov and GP-MIDAS-VXEF, where both methods find 9 known cancer related gens. However GP-MIDAS-VXEF has

Table 6: Data and Text Mining Gene Prioritization Methods.

| Method | Brief description | Reported results |
|---|---|---|
| Gene seeker | Gathers gene expression and phenotypic data from human and mouse from nine databases. Relies on the assumption that disease genes are likely to be expressed in tissues affected by that disease [6] | Offers a web-service to find disease-related genes to the input genetic localisation and phenotypic/expression terms |
| eVOC | Co-occurrence of disease name on PubMed Abstracts. It selects the disease genes according to expression profiles [5] | It was tested on 417 candidate genes, using 17 known disease genes. It successfully retrieved 15 of the 17 known disease genes and shrunk the candidate set by 63.3% |
| DPG | Basic Sequence Information [8]. | They concluded that disease proteins tend to be long, conserved, phylogenetically extended, and without close paralogues. |
| Prospectr | Basic Sequence Information [10]. | It achieved an enrichment of list of disease genes twofold 77% of the time, fivefold 37% of the time and twentyfold 11% of the time |
| Suspects | Extension of prospectr, incorporates GO [9, 15]. | On average the target gene was on the top 31.23% of the resulting ranking list. |
| MedSim | GO enrichment and functional comparison [13]. | It accomplished a performance of up to 0.90 in their ROC curve. |
| *Limitations* | Generally imposed by the source data which carries little knowledge about the disease. For instance GO terms include brief description of the corresponding biological function of the genes but only 60% of all human genes have associated GO terms, and they may be inconsistent due to differences in curators' judgement [16] | |

Table 7: Network based gene prioritization methods.

| Method | Brief description |
|---|---|
| Endeavor | Machine learning: using initial known disease genes; then multiple genomic data sources to rank [17] |
| HITS with priors<br>Page rank<br>K-Step markov | 310 cm prioritization based on networks using social and web networks analysis [18] |
| CGI | Combination of protein interaction network and gene expression using markov random field theory [19] |
| CANDID | Uses publications, protein domain descriptions, cross species conservation measures, gene expression profiles and Protein Interaction Networks [20] |
| IDEA | Uses the interactome and microarray data [21] |
| *Limitations* | Most of these approaches include additional interactions predicted from coexpression, pathway, functional or literature data, but still fail to incorporate weights expressing the confidence on the evidence of the interactions. Another issue is that previous methods start with the given PIN without filtering its edges, to keep more relevant interactions to the disease |
| GP-MIDAS-VXEF | Our proposed method, integrates protein interaction network with normal and disease microarray data, using this integration we apply all-pairs shortest paths to find the significant networks and calculate the score for the genes. Additionally our method filters interactions, in such way the most relevant interactions are left for analysis |

an AP score of 13 which is less than the value of K-Step Markov with 22.11.

Additionally a Venn diagram is presented in Figure 5 where it is shown that most of the genes are found using ToppGene and GP-MIDAS-VXEF methods, where GP-MIDAS-VXEF outperforms all by finding two target genes that no other method finds. Furthermore, there are 22 overlapping genes showing that our method is consistent with previously found results. Appendix A shows a list of target genes found by major methods.

*4.2. Results Comparing Prostate Cancer and Normal Samples.* We used the mean and variance to calculate the top 5% area as lower limits of the 95th percentile confidence interval with two tails in the distribution of the edge flux score which is shown in Figure 6. We present the two networks induced by top 50 genes from two kind of experiments which is shown in Figures 7 and 8. In Figure 7, we discovered that overexpressed gene androgen receptor (AR) being annotated in KEGG database as oncogene in the prostate cancer pathways also support the disease-related proteins in prostate cancer

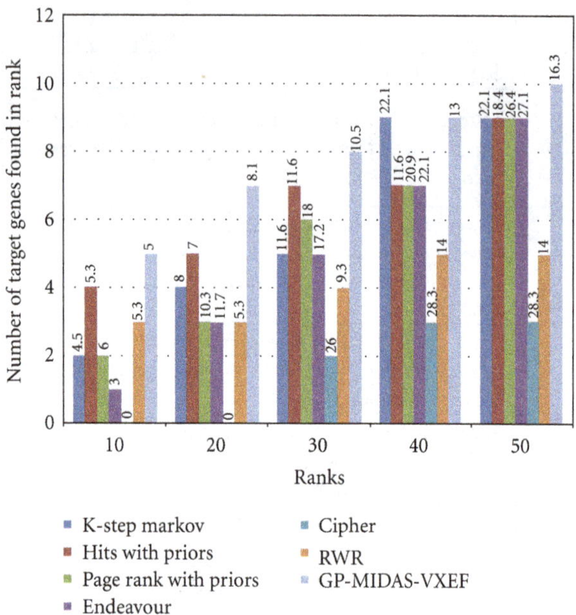


FIGURE 4: Target genes retrieved. Showing the amount of target genes retrieve on different ranks, on top of each bar the average position of the found genes is shown.

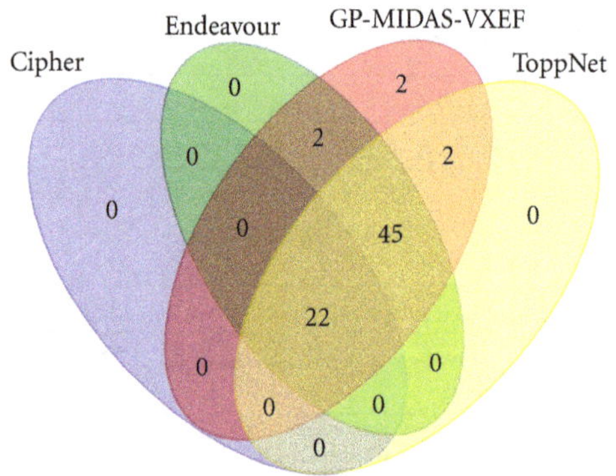


FIGURE 5: Venn diagram shows how the set of target genes is found amongst the different methods tested.

growth [23, 47]. BRCA1 and BRCA2 proteins play important role in DNA repair in both S and G2 checkpoint phase of the cell cycle and the results denote prostate cancer are strongly related to the tumor suppress genes (TP53, BRCA1, MYC, and PTEN) which have effect on the regulation of the cell cycle or promote apoptosis. Epidermal growth factor receptor (EGFR) family is also expressed in prostate cancer cells and their stimulation by EGF activates the mitogen-activated protein kinase (MAPK) and phosphatidylinositol-3 kinase (PI3K)/AKT pathways [48]. Those signal pathways stimulate cell cycle progression or survival which associated with cyclin D1 (CCND1) transcription and translation and the level of the BCL2. We found gene CALM1 that are associated with androgen receptor processes and interleukin 6 (IL 6) type of cytokine signaling pathways and their interactions with p38

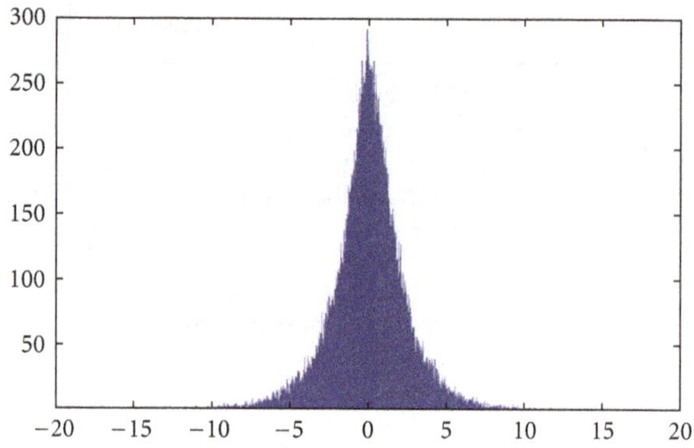


FIGURE 6: Edge flux values distribution.

MAPK may be the important factors related to the prostate cancer. Overexpression of MYC occurred frequently in most human prostate tumor databases revealed modules of human genes [49]. MXI1 protein associated with MYC which has also been suggested to play a role in prostate cancer [50–52]. Additionally we compare to previous results found in [53] that is focused on the construction of the regulatory network with emphasis in transcription factors. Transcription factors STAT3, MYC, and JUN are overlapped in both studies, providing some evidence to support the relationship of this genes to the prostate cancer. However our list does not overlap more since this study is not focussed in transcription factors only as in [53].

*4.3. Results Comparing Prostate Cancer and Lymph Node Metastasis.* In Figure 8, Raf-1 kinase inhibitor protein (RKIP) was identified as the first physiologic inhibitor of the Raf/mitogen-activated protein kinase kinase/extracellular signal-regulated kinase (ERK) pathway [54]. Recently, RKIP has been recognized as a strong candidate for a metastasis suppressor gene in our experiments and we investigated RKIP expression is altered in clinical human lymph node metastases. Studies in cell cultures and animal models have suggested RKIP were found to be reduced or absent in metastatic variants of established cell lines derived from prostate cancer [54]. Androgen receptor coregulator, Filamin A (FlnA) is corresponded to hormone-dependence in prostate cancer and may be related to increased metastatic capacity [55]. We sought to determine FlnA expression across prostate cancer progression in human prostate cancer corresponded with metastatic potential. Histone deacetylase-1 (HDAC1) is association with SP1 was much weaker in lymph node metastatic than in nonmetastatic prostate cancer [56]. Our experimental data suggests induction of signalling activity via EGFR in prostate tumor cells and may provide a rationale for the use of EGFR inhibition in systemic prevention or treatment of lymph node metastatic [57]. In particular our experiments observed a properly designed inhibitor of nuclear receptor subfamily 5 (NR5A1) may be predicted to have therapeutic utility in the treatment of metastatic lymph node through suppression of androgen receptor. Previous studies have been studies that cyclin d1 (CCND1) is a

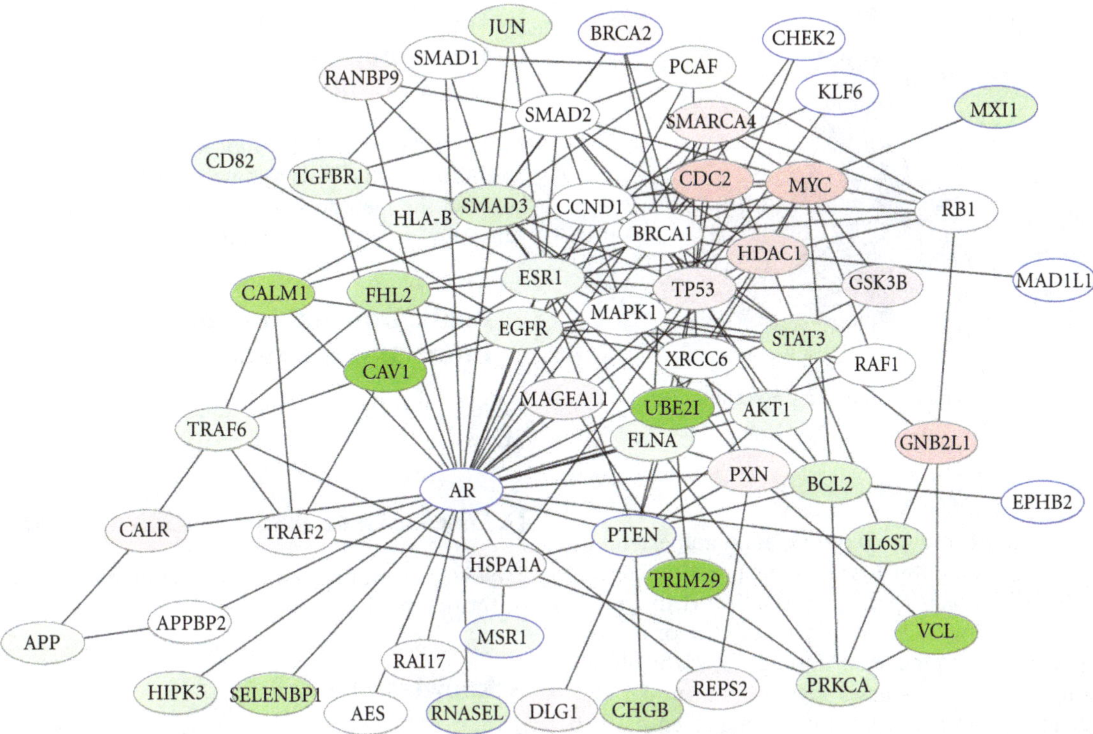


FIGURE 7: Prostate-normal experiment top 50 genes induced result network. Red color shows the higher difference in expression between prostate cancer and normal tissue sample, on the other hand the green color shows the smaller difference in expression between the samples. Nodes with node circle denote seed genes.

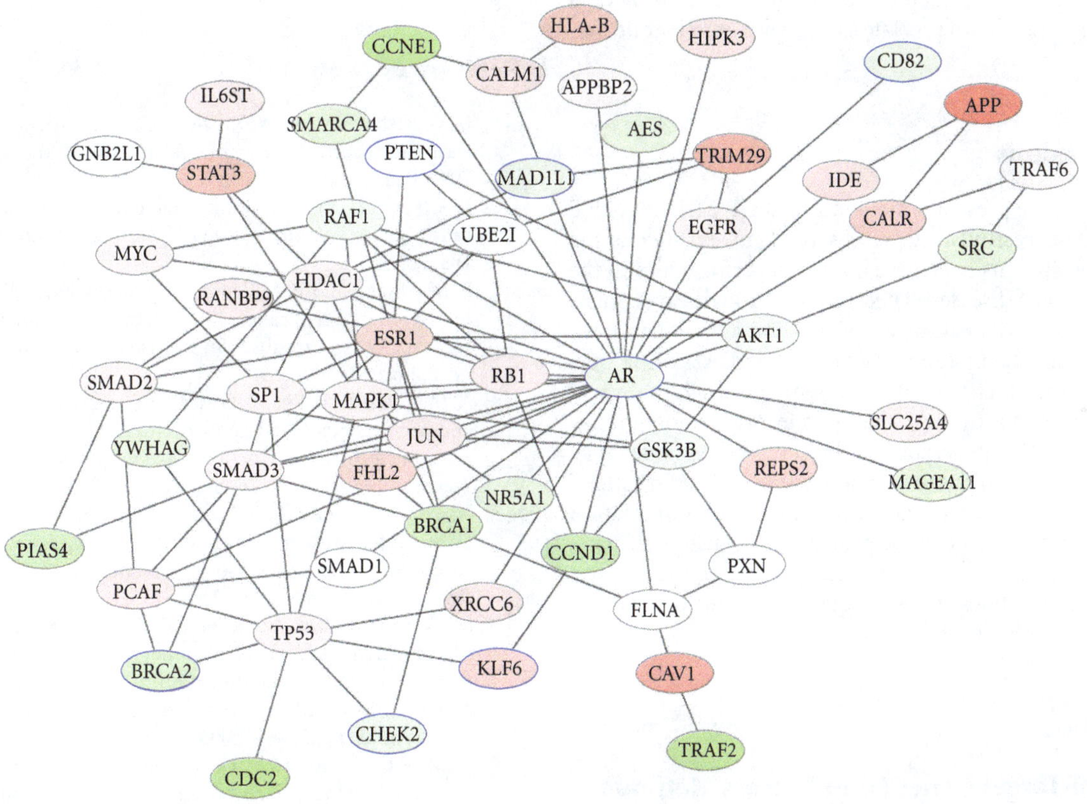


FIGURE 8: Prostate-metastatis experiment top 50 genes induced result network. Red color shows the higher difference in expression between prostate cancer and lymph node metastasis tissue sample, on the other hand the green color shows the smaller difference in expression between the samples. Nodes with node circle denote Seed Genes.

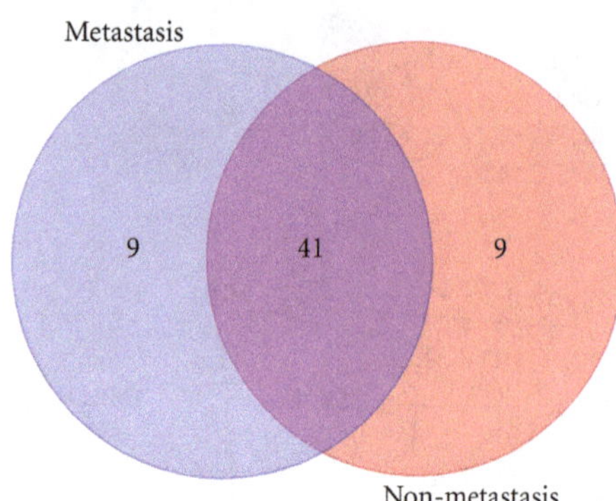


FIGURE 9: Venn diagram of top 50 genes.

mediator of prostate tumour cell proliferation and extend to lymph node metastasis [58]. V-src sarcoma viral oncogene homolog (SRC) has been specifically implicated in tumor growth and progression and resulting in both tumor growth and development of lymph node metastases [59]. This shows that targeting SRC family kinases may inhibit growth and lymph node metastases of prostate cancer. Not all biomarker genes found in lymph node metastasis (See Appendix B for details) can be explained at this moment. However our investigation shows that the molecular effects of lymph node metastasis related to AKT/GSK-3/AR signaling network along with the data presented above, that it may provide a biomarker indicative of prostate cancer with lymph node metastasis.

## 5. Conclusions

In this paper we present a method called GP-MIDAS-VXEF in which is successfully integrates several current state of the art acomplishments to achieve improved performance in the identification of disease-related genes. Through experimentation using Prostate Cancer as the domain, it has been shown that for the first top 50 genes GP-MIDAS-VXEF outperforms other methods, thus presenting an alternative in the gene prioritization field, that is, in terms of finding ranking known disease genes among the candidate gene set. The reason for our results are attributed to: the filtering phase where we obtain more relevant interactions, the combination of global and local network prioritization, using all-pairs shortest paths to find relevant routes for the seed genes, and for the particular boosting techniques that add structural and biological meaning to the results.

## Appendices

### A. List of Target Genes Found across Methods

Table 2 presents the list of target genes that were found in major gene prioritization methods. Notice that *EIF3S3* and *VEGF* are found only in GP-MIDAS-VXEF.

### B. Top 50 Genes from Prostate Cancer and Lymph Node Metastasis Extracted by Our Methods

Table 3 presents the list of biomarkers found using our methods. Additionally Table 4 shows the genes that are overlapped between the metastasis and nonmetastasis prioritization results, this is also shown graphically in Figure 9.

### C. Public Domain Protein Interaction Databases

Table 5 presents a list of publicly available biological networks databases.

### D. Disease Gene Prioritization Methods

Tables 6 and 7 show a summary on different disease gene prioritization methods.

## Acknowledgments

This paper is partially supported by the Bioresources Collection and Research Center of Linko Chang Gung Memorial Hospital and National Tsing Hua University of Taiwan R. O. C. under the Grant no. 98N2424E1, and the Universidad Tecnológica Centroamericana in Honduras.

## References

[1] B. Vogelstein and K. W. Kinzler, "Cancer genes and the pathways they control," *Nature Medicine*, vol. 10, no. 8, pp. 789–799, 2004.

[2] P. A. Futreal, L. Coin, M. Marshall et al., "A census of human cancer genes," *Nature Reviews Cancer*, vol. 4, no. 3, pp. 177–183, 2004.

[3] S. M. Huang and P. M. Harari, "Epidermal growth factor receptor inhibition in cancer therapy: biology, rationale and preliminary clinical results," *Investigational New Drugs*, vol. 17, no. 3, pp. 259–269, 1999.

[4] A. M. Edwards, B. Kus, R. Jansen, D. Greenbaum, J. Greenblatt, and M. Gerstein, "Bridging structural biology and genomics: assessing protein interaction data with known complexes," *Trends in Genetics*, vol. 18, no. 10, pp. 529–536, 2002.

[5] N. Tiffin, J. F. Kelso, A. R. Powell, H. Pan, V. B. Bajic, and W. A. Hide, "Integration of text- and data-mining using ontologies successfully selects disease gene candidates," *Nucleic Acids Research*, vol. 33, no. 5, pp. 1544–1552, 2005.

[6] M. A. van Driel, K. Cuelenaere, P. P. C. W. Kemmeren, J. A. M. Leunissen, and H. G. Brunner, "A new web-based data mining tool for the identification of candidate genes for human genetic disorders," *European Journal of Human Genetics*, vol. 11, no. 1, pp. 57–63, 2003.

[7] R. M. Piro, I. Molineris, U. Ala, P. Provero, and F. di Cunto, "Candidate gene prioritization based on spatially mapped gene expression: an application to XLMR," *Bioinformatics*, vol. 26, no. 18, pp. i618–i624, 2010, in *Proceedings of the 9th European Conference on Computational Biology*, Ghent, Belgium, SEP 26-29, 2010.

[8] N. López-Bigas and C. A. Ouzounis, "Genome-wide identification of genes likely to be involved in human genetic disease," *Nucleic Acids Research*, vol. 32, no. 10, pp. 3108–3114, 2004.

[9] E. A. Adie, R. R. Adams, K. L. Evans, D. J. Porteous, and B. S. Pickard, "Speeding disease gene discovery by sequence based candidate prioritization," *BMC Bioinformatics*, vol. 6, article 55, 2005.

[10] E. A. Adie, R. R. Adams, K. L. Evans, D. J. Porteous, and B. S. Pickard, "SUSPECTS: enabling fast and effective prioritization of positional candidates," *Bioinformatics*, vol. 22, no. 6, pp. 773–774, 2006.

[11] The Gene Ontology Consortium, "The gene ontology project in 2008," *Nucleic Acids Research*, vol. 36, pp. 440–444, 2008.

[12] S. Hunter, R. Apweiler, T. K. Attwood et al., "InterPro: the integrative protein signature database," *Nucleic Acids Research*, vol. 37, no. 1, pp. D211–D215, 2009.

[13] A. Schlicker, T. Lengauer, and M. Albrecht, "Improving disease gene prioritization using the semantic similarity of gene ontology terms," *Bioinformatics*, vol. 26, no. 18, pp. i561–i567, 2010, in *Proceedings of the 9th European Conference on Computational Biology*, Ghent, Belgium, SEP 26-29, 2010.

[14] A. Schlicker, J. Rahnenführer, M. Albrecht, T. Lengauer, and F. S. Domingues, "GOTax: investigating biological processes and biochemical activities along the taxonomic tree," *Genome Biology*, vol. 8, no. 3, article R33, 2007.

[15] M. Ashburner, C. A. Ball, J. A. Blake et al., "Gene ontology: tool for the unification of biology," *Nature Genetics*, vol. 25, no. 1, pp. 25–29, 2000.

[16] M. E. Dolan, L. Ni, E. Camon, and J. A. Blake, "A procedure for assessing GO annotation consistency," *Bioinformatics*, vol. 21, no. 1, pp. i136–i143, 2005.

[17] S. Aerts, D. Lambrechts, S. Maity et al., "Gene prioritization through genomic data fusion," *Nature Biotechnology*, vol. 24, no. 5, pp. 537–544, 2006.

[18] J. Chen, E. E. Bardes, B. J. Aronow, and A. G. Jegga, "Toppgene Suite for gene list enrichment analysis and candidate gene prioritization," *Nucleic Acids Research*, vol. 37, no. 2, pp. W305–W311, 2009.

[19] X. Ma, H. Lee, L. Wang, and F. Sun, "CGI: a new approach for prioritizing genes by combining gene expression and protein-protein interaction data," *Bioinformatics*, vol. 23, no. 2, pp. 215–221, 2007.

[20] J. E. Hutz, A. T. Kraja, H. L. McLeod, and M. A. Province, "CANDID: a flexible method for prioritizing candidate genes for complex human traits," *Genetic Epidemiology*, vol. 32, no. 8, pp. 779–790, 2008.

[21] K. M. Mani, C. Lefebvre, K. Wang et al., "A systems biology approach to prediction of oncogenes and molecular perturbation targets in B-cell lymphomas," *Molecular Systems Biology*, vol. 4, article 169, 2008.

[22] S. Erten and M. Koyuturk, "Role of centrality in network-based prioritization of disease genes," in *Evolutionary Computation, Machine Learning and Data Mining in Bioinformatics*, C. Pizzuti, M. Ritchie, and M. Giacobini, Eds., vol. 6023 of *Lecture Notes in Computer Science*, pp. 13–25, Springer, Berlin, Germany, 2010.

[23] M. Benson and R. Breitling, "Network theory to understand microarray studies of complex diseases," *Current Molecular Medicine*, vol. 6, no. 6, pp. 695–701, 2006.

[24] C. Perez-Iratxeta, P. Bork, and M. A. Andrade-Navarro, "Update of the G2D tool for prioritization of gene candidates to inherited diseases," *Nucleic Acids Research*, vol. 35, pp. W212–216, 2007.

[25] X. Wu, R. Jiang, M. Q. Zhang, and S. Li, "Network-based global inference of human disease genes," *Molecular Systems Biology*, vol. 4, article 189, 2008.

[26] K. Lage, E. O. Karlberg, Z. M. Størling et al., "A human phenome-interactome network of protein complexes implicated in genetic disorders," *Nature Biotechnology*, vol. 25, no. 3, pp. 309–316, 2007.

[27] J. D. J. Han, N. Berlin, T. Hao et al., "Evidence for dynamically organized modularity in the yeast protein-protein interaction network," *Nature*, vol. 430, no. 6995, pp. 88–93, 2004.

[28] S. Erten, G. Bebek, and M. Koyutuerk, "Disease gene prioritization based on topological similarity inprotein-protein interaction networks," in *Research in Computational Biology*, V. Bafna and S. Sahinalp, Eds., vol. 6577 of *Lecture Notes in Bioinformatics*, pp. 54–68, Springer, Berlin, Germany, 2011, in *Proceedings of the 15th Annual International Conference on Research in Computational Molecular Biology*, Simon Fraser University, Lab for Computational Biology, Vancouver, Canada, 2011.

[29] A. Özgür, T. Vu, G. Erkan, and D. R. Radev, "Identifying gene-disease associations using centrality on a literature mined gene-interaction network," *Bioinformatics*, vol. 24, no. 13, pp. i277–i285, 2008.

[30] S. Karni, H. Soreq, and R. Sharan, "A network-based method for predicting disease-causing genes," *Journal of Computational Biology*, vol. 16, no. 2, pp. 181–189, 2009.

[31] J. J. Cai, E. Borenstein, and D. A. Petrov, "Broker genes in human disease," *Genome Biology and Evolution*, vol. 2, pp. 815–825, 2010.

[32] B. H. Junker and F. Schreiber, Eds., *Analysis of Biological Networks*, Wiley Series on Bioinformatics: Computational Techniques and Engineering, John Wiley & Sons, Hoboken, NJ, USA, 2008.

[33] C. Prieto and J. D. L. Rivas, "APID: agile protein interaction DataAnalyzer," *Nucleic Acids Research*, vol. 34, pp. W298–W302, 2006.

[34] H.-C. Liu, C. R. Arias, and V.-W. Soo, "Bioir: an approach to public domain resource integration of humanprotein-protein interaction," in *Proceeding of the 7th Asia Pacific Bioinformatics Conference*, 2009.

[35] J. Lapointe, C. Li, J. P. Higgins et al., "Gene expression profiling identifies clinically relevant subtypes of prostate cancer," *Proceedings of the National Academy of Sciences of the United States of America*, vol. 101, no. 3, pp. 811–816, 2004.

[36] A. Hamosh, A. F. Scott, J. S. Amberger, C. A. Bocchini, and V. A. McKusick, "Online Mendelian Inheritance in Man (OMIM), a knowledgebase of human genes and genetic disorders," *Nucleic Acids Research*, vol. 33, pp. D514–D517, 2005.

[37] A. Özgür, T. Vu, G. Erkan, and D. R. Radev, "Identifying gene-disease associations using centrality on a literature mined gene-interaction network," *Bioinformatics*, vol. 24, no. 13, pp. i277–i285, 2008.

[38] M. Kanehisa, S. Goto, S. Kawashima, Y. Okuno, and M. Hattori, "The KEGG resource for deciphering the genome," *Nucleic Acids Research*, vol. 32, pp. D277–D280, 2004.

[39] L. C. Li, H. Zhao, H. Shiina, C. J. Kane, and R. Dahiya, "PGDB: a curated and integrated database of genes related to the prostate," *Nucleic Acids Research*, vol. 31, no. 1, pp. 291–293, 2003.

[40] K. I. Goh, M. E. Cusick, D. Valle, B. Childs, M. Vidal, and A. L. Barabási, "The human disease network," *Proceedings of the National Academy of Sciences of the United States of America*, vol. 104, no. 21, pp. 8685–8690, 2007.

[41] D. R. Rhodes and A. M. Chinnaiyan, "Integrative analysis of the cancer transcriptome," *Nature Genetics*, vol. 37, no. 6, pp. S31–S37, 2005.

[42] K. Komurov, M. A. White, and P. T. Ram, "Use of data-biased random walks on graphs for the retrieval of context-specific networks from genomic data," *PLoS Computational Biology*, vol. 6, no. 8, Article ID e1000889, 2010.

[43] A. Grigoriev, "A relationship between gene expression and protein interactions on the proteome scale: analysis of the bacteriophage T7 and the yeast Saccharomyces cerevisiae," *Nucleic Acids Research*, vol. 29, no. 17, pp. 3513–3519, 2001.

[44] J. C. de Borda, *Memoire sur les Elections au Scrutin*, Histoire de l'Academie Royale des Sciences, Paris, France, 1781.

[45] M. van Erp and L. Schomaker, "Variants of the borda count method for combining ranked classifier hypotheses," in *Proceedings of the 7th International Workshop of Frontiers in Handwriting Recognition*, B. Zhang, D. Ding, and L. Zhang, Eds., The Learning Methodology Inspired by Human's Intelligence, pp. 443–452, 2000.

[46] X. Chen, G. Yan, W. Ren, and J.-B. Qu, "Modularized random walk with restart for candidate diseasegenes prioritization," *Systems Biology*, pp. 353–360, 2009.

[47] J. Edwards and J. M. S. Bartlett, "The androgen receptor and signal-transduction pathways in hormone-refractory prostate cancer. Part 1: modifications to the androgen receptor," *BJU International*, vol. 95, no. 9, pp. 1320–1326, 2005.

[48] C. Festuccia, G. L. Gravina, L. Biordi et al., "Effects of EGFR tyrosine kinase inhibitor erlotinib in prostate cancer cells in vitro," *Prostate*, vol. 69, no. 14, pp. 1529–1537, 2009.

[49] B. Gurel, T. Iwata, C. M. Koh et al., "Nuclear MYC protein overexpression is an early alteration in human prostate carcinogenesis," *Modern Pathology*, vol. 21, no. 9, pp. 1156–1167, 2008.

[50] C. Vandenberg, X. Guan, D. Vonhoff et al., "DNA-sequence amplification in human prostate-cancer identified by chromosome microdissection—potential prognostic implications," *Clinical Cancer Research*, vol. 1, no. 1, pp. 11–18, 1995.

[51] L. Bubendorf, J. Kononen, P. Koivisto et al., "Survey of gene amplifications during prostate cancer progression by high-throughput fluorescence in situ hybridization on tissue microarrays," *Cancer Research*, vol. 59, no. 4, pp. 803–806, 1999.

[52] C. Abate-Shen and M. M. Shen, "Molecular genetics of prostate cancer," *Genes and Development*, vol. 14, no. 19, pp. 2410–2434, 2000.

[53] H. Y. Yeh, S. W. Cheng, Y. C. Lin, C. Y. Yeh, S. F. Lin, and V. W. Soo, "Identifying significant genetic regulatory networks in the prostate cancer from microarray data based on transcription factor analysis and conditional independency," *BMC Medical Genomics*, vol. 2, article 70, 2009.

[54] E. T. Keller, Z. Fu, and M. Brennan, "The biology of a prostate cancer metastasis suppressor protein: raf kinase inhibitor protein," *Journal of Cellular Biochemistry*, vol. 94, no. 2, pp. 273–278, 2005.

[55] R. G. Bedolla, Y. Wang, A. Asuncion et al., "Nuclear versus cytoplasmic localization of filamin a in prostate cancer: immunohistochemical correlation with metastases," *Clinical Cancer Research*, vol. 15, no. 3, pp. 788–796, 2009.

[56] C.-Y. A. Wong, H. Wuriyanghan, Y. Xie et al., "Epigenetic regulation of phosphatidylinositol 3,4,5-triphosphate-dependent rac exchanger 1 gene expression in prostate cancer cells," *Journal of Biological Chemistry*, vol. 286, pp. 25813–25822, 2011.

[57] A. Bratland, P. J. Boender, H. K. Hoifodt et al., "Osteoblast-induced EGFR/ERBB2 signaling in androgen-sensitiveprostate carcinoma cells characterized by multiplex kinase activity profiling," *Clinical & Experimental Metastasis*, vol. 26, pp. 485–496, 2009.

[58] Z. Ding, C. J. Wu, G. C. Chu et al., "SMAD4-dependent barrier constrains prostate cancer growth and metastatic progression," *Nature*, vol. 470, no. 7333, pp. 269–276, 2011.

[59] I. P. Serk, J. Zhang, K. A. Phillips et al., "Targeting Src family kinases inhibits growth and lymph node metastases of prostate cancer in an orthotopic nude mouse model," *Cancer Research*, vol. 68, no. 9, pp. 3323–3333, 2008.

# 19

# Distribution of *BoLA-DRB3* Allelic Frequencies and Identification of Two New Alleles in Iranian Buffalo Breed

**J. Mosafer,[1] M. Heydarpour,[2] E. Manshad,[1] G. Russell,[3] and G. E. Sulimova[4]**

[1] *Department of Animal Science, Ferdowsi University of Mashhad, P.O. Box 91775-1163, 9177948974 Mashhad, Iran*
[2] *Population Health Research Institute (PHRI), Department of Medicine, McMaster University, Hamilton, ON, Canada L8L2X2*
[3] *Moredun Research Institute, Pentlands Science Park, Midlothian, Penicuik EH26 0PZ, UK*
[4] *Vavilov Institute of General Genetics, Russian Academy of Sciences, Moscow 119991, Russia*

Correspondence should be addressed to J. Mosafer, mosafer_58@yahoo.com

Academic Editors: G. Füst and M. Ota

The role of the major histocompatibility complex (MHC) in the immune response makes it an attractive candidate gene for associations with disease resistance and susceptibility. This study describes genetic variability in the *BoLA-DRB3* in Iranian buffaloes. Heminested PCR-RFLP method was used to identify the frequency of *BoLA-DRB3* alleles. The *BoLA-DRB3* locus is highly polymorphic in the study herd (12 alleles). Almost 63.50% of the alleles were accounted for by four alleles (*BoLA-DRB3.2 *48*, **20*, **21*, and *obe*) in Iranian buffalo. The *DRB3.2 *48* allele frequency (24.20%) was higher than the others. The frequencies of the *DRB3.2 *20 and DRB3.2 *21* are 14.52 and 14.00, respectively, and *obe* and *gbb* have a new pattern. Significant distinctions have been found between Iranian buffalo and other cattle breed studied. In the Iranian buffaloes studied alleles associated with resistance to various diseases are found.

## 1. Introduction

The major histocompatibility complex (MHC) is a large cluster of tightly linked genes that play an important role in the immune system [1, 2]. The products of these genes are involved in the induction and regulation of immune response. The MHC spans approximately 4 Mb of the human genome, 1.5 Mb in mice and approximately 2.5 Mb of the cattle genome [3, 4]. In humans, the MHC is located on chromosome 6 whereas, in cattle, it is located on chromosome 23. It has been estimated that the mammalian MHC contains over 200 genes. The structure and organization of the MHC genes of cattle, known as the bovine leukocyte antigen (*BoLA*) complex, are very similar to those of the human MHC. The genes are organized into three distinct classes (class I, II, and III). Each of these classes is divided into regions and subregions, containing pseudo genes. The major difference between the organizations of the *BoLA* complex that of the human MHC and the *BoLA* complex is that found in two separate regions of the chromosome rather than a single cluster of genes seen in most mammals. The larger gene cluster is located at BTA23 band 22 and apparently contains all of the bovine class I and class III sequences and genes encoding both subunits of the classical class II proteins *DQ* and *DR*. The remaining *BoLA* class II loci (*DIB*, *DNA*, *DOB*, *DYA*, *DYB*, *TCP1*, *LMP2*, *LMP7*, and *TAP2*) are located in a cluster near the centromere at BTA23 band 12-13 [4]. Comparative analysis suggests that the disruption was likely caused by a single chromosomal inversion.

Of the class II genes, cattle express one *DR* gene pair (*DRA* and *DRB3*) and one or two *DQ* gene pairs per haplotype. The coding sequence of *DRA* is monomorphic, while the *DRB3* gene has over 103 identified alleles [5]. As a result of this polymorphism as well as its functional importance, the *DRB3* locus and its gene products are among the best defined in cattle. Associations between *BoLA* alleles and disease have also been identified for class I and class II genes. Class I associations include tick resistance [6], nematode egg worm counts [7, 8], resistance to persistent lymphocytosis caused by bovine leukemia virus (BLV) [9], chronic posterior spinal paresis (PSP) [10], Ketosis [11], resistance to dermatophilosis [12] and mastitis [11, 13–15].

Associations have been demonstrated between genes in this region and diseases such as decreased risk of cystic ovarian disease and retained placenta [16], resistance to persistent lymphocytosis caused by BLV [17], and mastitis [15, 16, 18]. Raising buffalo has changed from a novelty enterprise to a profitable one. Buffalo may be marketed for their meat and byproducts, for recreational hunting, and breeding stock. Research on buffalo is also important because buffaloes have a higher tolerance to cold temperatures than domestic cattle and therefore exhibit greater winter hardiness. Buffalos superior digestion of low-quality feeds also makes it better suited for production on marginal rangelands. Buffaloe live longer than domestic cattle. Buffalo cows remain productive until 20 years of age. The typical replacement rate for buffalo cows is 10 percent. Cows can be bred to calve at three years of age. The demand for buffalo meat has increased primarily because consumers perceive that it has less intramuscular than beef and pork. Some also believe that this means the meat is lower in cholesterol, though this has not been proven. Some consumers prefer the taste of buffalo over the taste of beef or pork. Other valuable byproducts from buffalo include mounted heads, skulls, and hides. In this study we analyzed the polymorphism of *BOLA-DRB3* gene in Iranian Mazandarani Buffalo using PCR-RFLP, in order to identify alleles of the gene.

## 2. Materials and Methods

*2.1. DNA Isolation.* Iranian Mazandarani Buffalo ($n = 100$) from Miankaleh Island in north of Behshahr from Mazandaran city were used in this study. Approximately 5–10 mL of blood was collected from each animal via the jugular or mammary vein, and aliquots of the whole blood were stored at −20°C. The DNA was isolated from the whole blood by a modified InstaGene protocol (Bio-Rad, Melville, NY). Briefly, the whole blood (0.3 mL) was washed once with 1 mL of PBS buffer, pH 7.2. InstaGene purification matrix (0.2 mL; Bio-Rad) was added to the pellet. The mixture was vortexed and incubated at 56°C for 60 min, followed by 15 min incubation at 99°C. The cell lysate was than centrifuged for 5 min at 7000 ×g. From the supernatant, 0.175 mL was withdrawn and diluted two-fold with 10x magnesium-free thermophilic buffer (500 Mm KCL, 100 mM Tris-HCL (PH 9.0), and 1% Triton x-100), and 2 $\mu$L of this mixture was used for DNA amplification.

*2.2. BoLA-DRB3.2 Gene Amplification.* DNA amplification of the *BoLA-DRB3.2* gene was achieved by a two-step PCR, using primers and methods designed for the *BoLA-DRB3* locus. The oligonucleotide primers HL030 (5′-ATCCTCTCTCTGCAGCACATTTCC-3′), HL031 (5′-TTTAATTCGCGCTCACCTCGCCGCT-3′), and HL032 (5′-TCGCCGCTGCACAGTGAAACTCTC-3′) were used. Reaction 1 was carried out in a total volume of 25 $\mu$L containing 200 $\mu$L of each dNTP, 0.5 $\mu$m of each primer of HL030 and HL031, 2.5 $\mu$L of 10x magnesium-free thermophilic buffer (500 mm KCL, 100 mm Tris-HCL (PH 9.0), and 1% Triton X-100), 2.5 $\mu$L of 25 mm $MgCl_2$, 13.5 $\mu$L of sterile $H_2O$, 5U of Taq DNA polymerase (Promega, San Diego, Calif) and 2 $\mu$L of genomic DNA. Each reaction mixture was overlaid with 20 $\mu$L of mineral oil. Following an initial denaturation step at 94.5°C for 270 s, the parameters for the thermocycler (Ericomp, San Diego) were set at 94.5°C for 90 s, 66°C for 120 s, and 72°C for 60 s. Ten cycles of DNA amplification were performed followed by one step of 72°C for 5 min. From this DNA amplification reaction mixture, 4 $\mu$L was used for the second PCR reaction. The second reaction was carried out in a total volume of 100 $\mu$L containing 200 $\mu$m of each dNTP, 0.5 $\mu$m of each primer of HL030 and HL032, 10 $\mu$L of 10x magnesium-free thermophilic buffer, 10 $\mu$L of 25 mM MgCl2, 59 $\mu$L of sterile $H_2O$, and 15 U of Taq DNA polymerase (Promega). Each reaction mixture was overlaid with 20 $\mu$L of mineral oil. Parameters for the thermocycler were set at 94.5°C for 90 s, 66°C for 30 s, for a total of 30 cycles, followed by a step of 72°C for 5 min.

*2.3. Restriction Endonuclease Digestion.* The PCR-amplified DNA fragments from the second PCR restriction were digested with restriction endonuclease *Rsa*I, *BstYI*, and *Hae*III (New England Biolabs, Beverly, Mass). For the restriction endonuclease digestion restriction, 10 $\mu$L of the second PCR restriction product was used for each digestion. Samples were digested with *Rsa*I or *Hae*III (20 units, New England Biolabs) for 2 h at 37°C in a total volume of 20 $\mu$L. Samples were digested with *BstYI* (20 units, New England Biolabs) for 2 h at 50°C followed by 15 min at 85°C in a total volume of 20 $\mu$L.

*2.4. Agarose Gel Electrophoresis.* Restriction enzyme digested samples were electrophoresed in 4% agarose (Metaphor; FMC Bioproducts, Rockland, Me) with TBE buffer (0.9 m tris base, 0.09 m boric acid, and 2.5 mm EDTA; pH 8.3). Gels were run at 100 V for 4 h and stained with ethidium bromide (1.0 $\mu$g 1 mL; Sigma Chemical Co., St. Louis, Mo). The DNA was visualized by transillumination (Foto dyne Inc., New Berlin, Wis) and photographed with type 55 polaroid film (Polaroid Corp., Cambridge, Mass). The negative of the type 55 polaroid film was scanned and analyzed with gel manager for windows (Biosystematica, Devon, UK). The *BoLA-DRB3.2* PCR-RFLP nomenclature was used for *BoLA-DRB3* exon 2 alleles as described by [19] and maintained in the BoLA immunopolymorphism database (http://www.ebi.ac.uk/ipd/mhc/bola/). New or unpublished allele types were identified using the restriction enzyme pattern as described by [19].

*2.5. Sequence Analysis.* Sequencing of new alleles was done by chain termination method on ABI 377 sequencing machine (GNTC Co., Germany). Sequencing results were analyzed using FASTA program, and the sequencing was compared against the other alleles by using BLAST at NCBI site (http://www.ncbi.nlm.nih.gov/).

## 3. Results

We used heminested PCR-RFLP method for identification of the frequency of *BoLA-DRB3* alleles in Iranian Mazandarani Buffaloes. PCR products were represented by 284 bp fragments as was expected on the basis of the nucleotide sequence

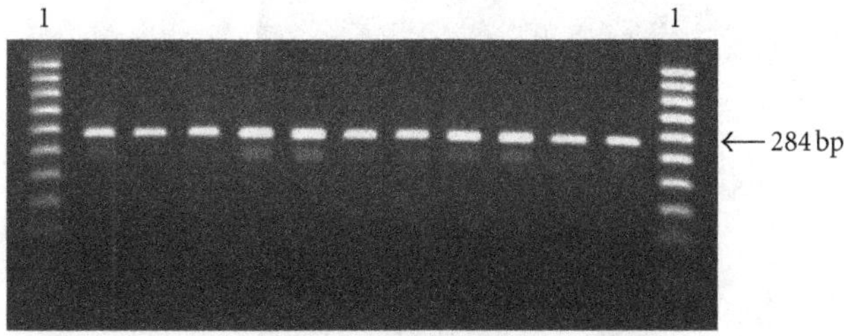


Figure 1: Heminested PCR products. Lane 1 is 50 bp molecular marker. The other lanes are PCRproducts of *BoLA-DRB3.2* with 284 bp size.

of the gene (Figure 1). The spectra of *Rsa*I, *Hae*III, and *BstYI* restriction sites are shown by [19]. Typical patterns of restriction of the amplified products with the endonucleases *Rsa*I, *Hae*III, and *BstYI* are shown in Figure 2. This is the first study of the DNA polymorphism of the *BoLA-DRB3* gene in Iranian buffaloes. We can demonstrate that the *BoLA-DRB3* locus is highly polymorphic in the studied herd. The *BoLA-DRB3* in Iranian Mazandarani Buffalo has 12 alleles (Table 1). Analyzing *BoLA-DRB3* of periparturient buffalo reported finding 12 allele types. More significant distinctions have been found between Iranian buffaloes and cattle breeds studied. Almost 63.50% of the alleles were accounted for by four alleles (*BoLA-DRB3.2 *48*, **20*, **21*, and *obe*) in Iranian Mazandarani Buffaloes. The frequencies of the *DRB3.2 *43*, **50* and *gbb* alleles were lower than the other alleles in this breed. In addition to these alleles we found two new alleles. The nucleotide sequences of the amplified exon 2 of these alleles were determined and submitted in the GenBank database under accession no. DQ187336 for *obe* allele and DQ187337 for *gbb* allele. Afterwards, they were compared to the *DRB3* alleles published by the *BoLA* Nomenclature Committee using the program BLAST from National Center for Biotechnology Information (http://www.ncbi.nih.gov/).

Table 1: Frequencies of *BoLA-DRB3.2* alleles in Iranian Mazandarani Buffalo ($n = 100$) as identified by polymerase chain reaction and restriction fragment length polymorphism analysis. Corresponding DNA patterns are shown at the bottom of the figures. Nomenclature of DNA patters is given corresponding to http://www.projects.roslin.ac.uk/bola/drb3pcr.html. Standard errors of the allele frequencies do not exceed 5%.

| Alleles | DRB3 PCR-RFLP | *Rsa*I/*BstYI*/ *Hae*III patterns | Frequencies (%) |
|---|---|---|---|
| DRB3 *3901 | *48 | Wba | 24.20 |
| *DRB3 *2301, *2901, *3601* | *20 | Ibb | 14.52 |
| *DRB3 *0801* | *21 | Ibe | 14.00 |
| *New* | new | Obe | 10.75 |
| *DRB3 *2601* | *19 | Sbb | 10.22 |
| *DRB3 *0401* | *13 | Hba | 9.14 |
| *DRB3 *2802* | *42 | Hbf | 8.51 |
| *DRB3 *3701* | *49 | Wbe | 2.70 |
| *DRB 07* | *37 | Oba | 2.70 |
| *DRB3 *25012* | *43 | Kbf | 1.08 |
| *DRB3 *4001* | *50 | Xba | 1.08 |
| *New* | new | gbb | 1.08 |

## 4. Discussion

This is the first study of the DNA polymorphism of the *BoLA-DRB3.2* gene in Iranian Mazandarani Buffalo. Previous work in *Bubalus bubalis* defined 8 *DRB* alleles among 75 animals from different buffalo populations in different countries [20], while a study of 25 unrelated Indian river buffalo revealed 22 *DRB* alleles [21]. These data, together with the present study, demonstrate that allelic frequencies of *BoLA-DRB3* appear to depend on the breed and population, likely a result of founder population structure and selection pressure. Our present study demonstrated that the *BoLA-DRB3* exon 2 is highly polymorphic in Iranian Mazandarani Buffalo. We found 12 PCR variants in Mazandarani Buffalo. The most frequently isolated alleles in our buffalo were *BoLA-DRB3.2 *48*, **20*, **21*, *obe*, **19*, **13*, **42* and these accounted for about 81% of the alleles in this population. But the alleles that Sena reported were not found in this study on Mazandarani Buffalo. The polymorphism of *BoLA-DRB3* by DNA sequence analyses of 18 African cattle investigated by [22]. They found 18 alleles was in small random sample of animals from two breeds. Then they found 30 alleles in *B. P. taurus* cattle in Africa and Zebu cattle [23]. A research on 127 Brahman cattle (zebu) in Martinique said that an amino acid sequence coded by the exon 2 of *BoLA-DRB3* gene associated with a *BoLA* class I specificity constitutes a likely genetic marker of resistance to dermatophilosis [12]. 13 alleles within eight major allelic families in African *Bos indicus* and *Bos Taurus* cattle were found by [23]. They found 7 alleles from *DQA3* within three major allelic families. 18 alleles in a research on 568 zebu Brahman cattle (*Bos indicus*) from Martinique (French West Indies) in *BoLA-DRB3* exon 2 were found by [12]. They said that 5 official alleles *DRB3.2 *0301*, **0302*, **0901*, **0902*, and **1202* correlate with the susceptibility. They found another correlation between susceptibility and the *BoLA-DQB *1804* allele. A research on zebu Brahman cattle in Martinique (FWI) showed that MHC molecules can control diseases such as dermatophilosis [12].

There are some articles about *BoLA-DRB3.2* in other cattle. Polymorphism of the *BoLA-DRB3* gene has been reported in the studies of Jersey [24]. This breed has a different allele and allele frequency profile from buffaloes. For example, the six most frequently detected alleles in Jersey cows are *BoLA-DRB3.2 *8*, **10*, **15*, **21*, **36*, and *ibe*, accounting for approximately 74% of the alleles in the population of the herd (172 animals). But the six most frequently detected alleles (*BoA-DRB3.2 *8*, **9*, **21*, **27*, **7*, and **24*) accounted for 70% of the alleles in a population of Japanese shorthorn cows [5]. Between these alleles only we saw *BoLA-DRB3.2 *21* on the Mazandarani Buffalo; therefore we can say that alleles and their frequency were different between buffaloes and other breeds of cattle. In Argentine Creole cows

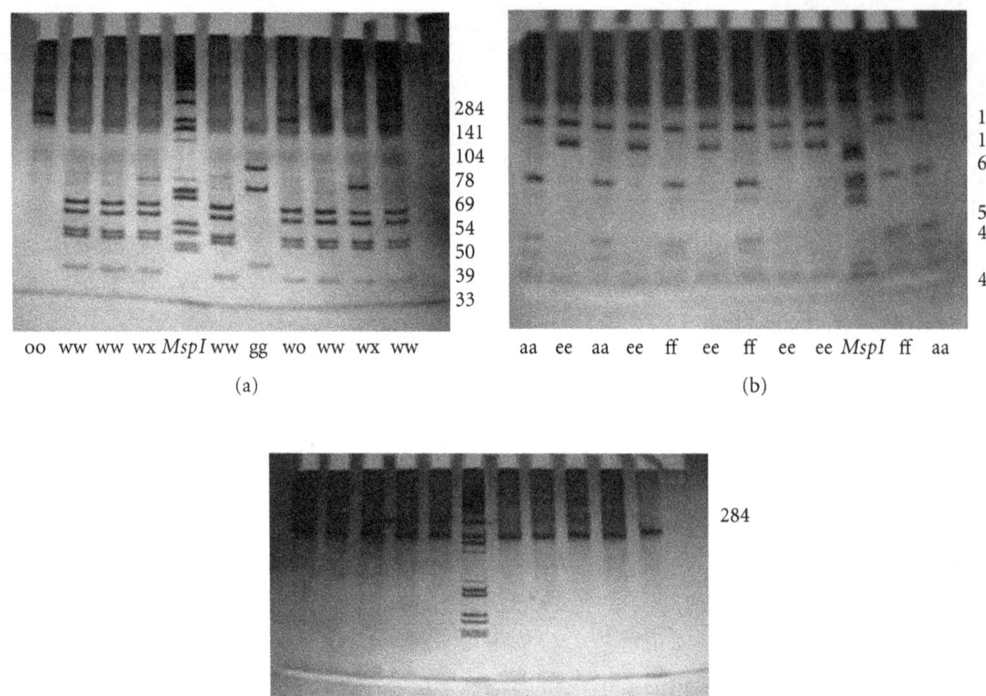


Figure 2: Electrophoresis in 8% polyacrylamide gel of exon 2 amplification products of gene *BoLA-DRB3* digested by endonucleases *Rsa*I (b), *Hae*III (a), and *BstYI* (c). *Msp*I fragments of plasmid pUC19 are used as a molecular marker. The length of fragments composing *Rsa*I, *Hae*III, or *Bst*YI DNA patterns is shown on pictures.

(194 animals) six most frequently detected alleles (*BoLA-DRB3.2* **15*, **18*, **24*, **20*, **27*, and **5*) that account for approximately 73% of the alleles in the herd and *BoLA-DRB3.2* **20* were found in Iranian Mazandarani Buffalo. In Russian Ayrshire cattle allele *DRB3.2* **7* is prevalent by 37.6%, and the combined frequency of alleles *DRB3.2* **7*, **28*, **10*, and **24* is 77% [19]. Alleles **22*, **24*, **11*, **16*, **18*, **23*, **8*, and **27* are the most frequent in Russian black pied cattle that were not found in Mazandarani Buffalo in this research [25]. In a study of ten breeds beef and dairy cattle were analyzed and *BoLA-DRB3.2* **5*, **29*, and **30* alleles were identified only in south Devon, Angus, Gelbvieh and the *BoLA-DRB3.2* **7* allele only in Angus, Gelbvieh, and Holstein Friesian cows. Significant associations have been made between *BoLA* genes and some infection diseases of cattle, particularly diseases that are prevalent during lactation. For example, [15] indicated that one *BoLA-DRB3* gene pattern in a study of Holstein cows ($n = 106$) is associated with resistance to *Staphylococcus aureus* mastitis. Three alleles (*DRB3.2* **11*, **23*, and **28*) determine resistance to leukemia and four (*DRB3.2* **8*, **16*, **22*, and **24*) are associated with susceptibility.

There was a lot research on *BoLA-DRB3.2* in Iranian cattle (Table 2). For example, [26] found 15 alleles in Iranian Sarabi cattle. Their frequency was between 2 and 23%, and 4 of these alleles (*BoLA-DRB3.2* **48*, **43*, **20*, and **19*) were found in Iranian Mazandarani Buffaloes in this study. They found 16 alleles in Iranian Najdi cattle that have 2–13% frequency and 1 of them (*BoLA-DRB3.2* **43*) was found in Iranian Mazandarani Buffalo. 19 alleles in Iranian Sistani cattle were found by [27] that have 1–22% frequency and 3 of them (*BoLA-DRB3.2* **37*, **21*, and **13*) were found in Iranian Mazandarani Buffalo. In addition, [28] found 19 alleles were found in Iranian Golpayegani cattle that have 2–14% frequency and 2 of them (*BoLA-DRB3.2* **20*, and **19*) were found in Iranian Mazandarani Buffalo. Also, [29] found 26 alleles in Iranian Holstein cattle that have 2–26.6% frequency and 5 of these alleles (*BoLA-DRB3.2* **49*, **37*, **21*, **20*, and **13*) were found in Iranian Mazandarani Buffalo. In this study we found that our data indicate that allelic frequencies of *BoLA-DRB3* may, at least to some extent, depend on the breed and geographical location. Similarities between the immune systems of cattle and buffalo and the overlapping range of pathogens that infect these two livestock species may support the view that disease associations discovered for the *BoLA* system can have practical application in buffalo. Continued analysis of *BoLA-DRB3* alleles in large agricultural populations may help reduce the spreading of alleles providing susceptibility to diseases such as mastitis or leukemia in buffalo herds. Thus, investigation of polymorphism of the *BoLA-DRB3* gene may have great practical as well as theoretical value.

TABLE 2: Frequencies of *BoLA-DRB3.2* alleles for the studied Iranian cattle breeds.

| *DRB3* alleles | Sarabi ($N = 52$) | Najdi ($N = 52$) | Sistani ($N = 49$) | Golpayegani ($N = 50$) |
|---|---|---|---|---|
| 2 | 10 | 2 | — | 4 |
| 3 | 2 | 2 | 1 | 2 |
| 4 | — | — | 2 | 3 |
| 7 | — | — | 4 | 11 |
| 8 | 2 | 8 | **22** | — |
| 10 | — | — | 4 | 6 |
| 11 | 18 | 11 | 5 | 7 |
| 12 | 10 | 5 | — | 6 |
| 13 | — | — | 3 | — |
| 14 | 2 | **13** | — | — |
| 15 | — | 5 | 8 | 2 |
| 16 | — | — | — | **14** |
| 17 | 2 | 2 | — | — |
| 19 | 2 | — | — | 10 |
| 20 | 2 | — | — | 2 |
| 21 | — | — | 2 | — |
| 22 | 2 | 2 | — | 2 |
| 23 | 15 | 11 | — | — |
| 24 | — | **13** | 2 | 2 |
| 25 | 3 | — | — | 2 |
| 28 | — | — | — | 8 |
| 29 | — | — | 1 | — |
| 31 | — | — | — | 4 |
| 34 | — | — | 21 | — |
| 35 | — | — | — | 3 |
| 36 | — | 11 | 2 | — |
| 37 | — | — | 1 | — |
| 43 | 6 | 3 | — | — |
| 44 | — | — | 6 | — |
| 45 | — | — | 2 | 6 |
| 47 | — | — | 3 | — |
| 48 | 5 | — | — | — |
| 51 | — | — | 1 | — |
| 52 | **23** | 6 | — | 6 |
| 53 | — | 5 | — | — |
| 54 | — | 5 | — | — |
| *X* | — | — | 8 | — |

## Acknowledgment

This work was supported by the Faculty of Agriculture of Mashhad University of Ferdowsi and Center of Excellence in Animal Science (Iran, Mashhad).

## References


[1] J. Klein, *Natural History of the Major Histocompatibility Complex*, John Wiley, New York, NY, USA, 1986.

[2] L. Andersson and C. J. Davies, "The major histocompatibility complex," in *Cell Mediated Immunity in Ruminants*, B. M. L. Goddeeris and W. I. Morrison, Eds., pp. 37–57, CRC Press, Boca Raton, Fla, USA, 1994.

[3] H. A. Lewin, "Genetic organization, polymorphism, and function of the bovine major histocompatibility complex," in *The Major Histocompatibility Complex Region of Domestic Animal Species*, L. B. Schook and S. J. Lamont, Eds., pp. 65–98, CRC Press, Boca Raton, Fla, USA, 1996.

[4] M. F. Rothschild, L. Skow, and S. J. Lamount, "The major histocompatibility complex and its role in disease resistance and immune responsiveness," in *Breeding for Disease Resistance in Farm Animals*, R. F. E. Axford, S.C. Bishop, F. W. Nicholas, and J. B. Owen, Eds., pp. 243–252, CAB International, Wallingord, Wash, USA, 2000.

[5] S. Takeshima, Y. Nakai, M. Ohta, and Y. Aida, "Short communication: characterization of *DRB3* alleles in the MHC of Japanese Shorthorn cattle by polymerase chain reaction-sequence-based typing," *Journal of Dairy Science*, vol. 85, no. 6, pp. 1630–1632, 2002.

[6] M. J. Stear, M. J. Newman, and F. W. Nicholas, "Tick resistance and the major histocompatibility system," *Australian Journal of Experimental Biology and Medical Science*, vol. 62, no. 1, pp. 47–52, 1984.

[7] M. J. Stear, T. J. Tierney, F. C. Baldock, S. C. Brown, F. W. Nicholas, and T. H. Rudder, "Class I antigens of the bovine major histocompatibility system are weakly associated with variation in faecal worm egg counts in naturally infected cattle," *Animal Genetics*, vol. 19, no. 2, pp. 115–121, 1988.

[8] M. J. Stear, D. J. S. Hetzel, S. C. Brown, L. J. Gershwin, M. J. Mackinnon, and F. W. Nicholas, "The relationships among ecto- and endoparasite levels, Class I antigens of the bovine major histocompatibility system, immunoglobulin E levels and weight gain," *Veterinary Parasitology*, vol. 34, no. 4, pp. 303–321, 1990.

[9] M. J. Stear, C. K. Dimmock, M. J. Newman, and F. W. Nicholas, "*BoLA* antigens are associated with increased frequency of persistent lymphocytosis in bovine leukaemia virus infected cattle and with increased incidence of antibodies to bovine leukaemia virus," *Animal Genetics*, vol. 19, no. 2, pp. 151–158, 1988.

[10] C. A. Park, H. C. Hines, D. R. Monke, and W. T. Threlfall, "Association between the bovine major histocompatibility complex and chronic posterior spinal paresis–a form of ankylosing spondylitis—in Holstein bulls," *Animal Genetics*, vol. 24, no. 1, pp. 53–58, 1993.

[11] C. M. Mejdell, O. Lie, H. Solbu, E. F. Arnet, and R. L. Spooner, "Association of major histocompatibility complex antigens (*BoLA-A*) with AI bull progeny test results for mastitis, ketosis and fertility in Norwegian Cattle," *Animal Genetics*, vol. 25, no. 2, pp. 99–104, 1994.

[12] J. C. Maillard, D. Martinez, and A. Bensaid, "An amino acid sequence coded by the exon 2 of the BoLA DRB3 gene associated with a BoLA class I specificity constitutes a likely genetic marker of resistance to dermatophilosis in Brahman Zebu cattle of Martinique (FWI)," *Annals of the New York Academy of Sciences*, vol. 791, pp. 185–197, 1996.

[13] H. Solbu, R. L. Spooner, and O. Lie, "A possible influence of the bovine major histocompatibility complex (*BoLA*) on mastitis," in *Proceedings of the 2nd World Congress of Genetics Applied to Livestock Production*, vol. 7, p. 368, Madrid, Spain, October 1982.

[14] O. Oddgeirsson, S. P. Simpson, A. L. Morgan, D. S. Ross, and R. L. Spooner, "Relationship between the bovine major histocompatibility complex (*BoLA*), erythrocyte markers and susceptibility to mastitis in Icelandic cattle," *Animal Genetics*, vol. 19, no. 1, pp. 11–16, 1988.

[15] Y. H. Schukken, D. J. Wilson, F. Welcome, L. Garrison-Tikofsky, and R. N. Gonzalez, "Monitoring udder health and milk quality using somatic cell counts," *Veterinary Research*, vol. 34, no. 5, pp. 579–596, 2003.

[16] S. Sharif, B. A. Mallard, B. N. Wilkie et al., "Associations of the bovine major histocompatibility complex *DRB3* alleles with occurrence of disease and milk somatic cell score in Canadian dairy cattle," *Animal Genetics*, vol. 29, no. 3, pp. 185–193, 1998.

[17] A. Xu, M. J. VanEijk, C. Parck, and H. A. Lewin, "Polymorphism in *BoLA-DRB3* exon 2 correletes with resistance to persistent lymphocytosis caused by bovine leukemia virus," *Journal of Immunology*, vol. 151, no. 12, pp. 6977–6985, 1993.

[18] A. Lunden, L. Andersson-Eklund, and L. Andersson, "Lack of association between bovine major histocompatibility complex class II polymorphism and production traits," *Journal of Dairy Science*, vol. 76, no. 3, pp. 843–852, 1993.

[19] M. J. Van Eijk, J. A. Stewart-Haynes, and H. A. Lewin, "Extensive polymorphism of the *BoLA-DRB3* gene distinguished by PCR-RFLP," *Animal Genetics*, vol. 23, no. 6, pp. 483–496, 1992.

[20] L. Sena, M. P. C. Schneider, B. Brenig, R. L. Honeycutt, J. E. Womack, and L. C. Skow, "Polymorphisms in MHC-DRA and -DRB alleles of water buffalo (Bubalus bubalis) reveal different features from cattle DR alleles," *Animal Genetics*, vol. 34, no. 1, pp. 1–10, 2003.

[21] S. De, R. K. Singh, and G. Butchaiah, "MHC-DRB exon 2 allele polymorphism in Indian river buffalo (Bubalus bubalis)," *Animal Genetics*, vol. 33, no. 3, pp. 215–219, 2002.

[22] S. Mikko and L. Andersson, "Extensive MHC class II DRB3 diversity in African and European cattle," *Immunogenetics*, vol. 42, no. 5, pp. 408–413, 1995.

[23] K. T. Ballingall, A. Luyai, and D. J. McKeever, "Analysis of genetic diversity at the *DQA* loci in African cattle: evidence for a *BoLA-DQA3* locus," *Immunogenetics*, vol. 46, no. 3, pp. 237–244, 1997.

[24] B. E. Gilliespie, B. M. Jayarao, H. H. Dowlen, and S. P. Oliver, "Analysis and frequency of bovine lymphocyte antigen *DRB3.2* alleles in Jersey cows," *Journal of Dairy Science*, vol. 82, no. 9, pp. 2049–2053, 1999.

[25] G. E. Sulimova, I. G. Udina, G. O. Shaikhaev, and I. A. Zakharov, "DNA polymorphism at the BOLA-DRB3 gene of cattle in relation to resistance and susceptibility to leukemia," *Journal of Genetics*, vol. 31, no. 9, pp. 1105–1109, 1995.

[26] F. Montazer Torbati, F. Eftekhar Shahroudi, M. R. Nassiry, A. Safanezhad, and M. B. Montazer Torbati, "Frequency of bovine lymphocyte antigen *DRB3.2* alleles in Sarabi cows," *Iranian Journal of Biotechnology*, vol. 2, no. 2, 2004.

[27] A. Mohammadi, M. R. Nassiry, J. Mosafer, M. R. Mohammadabadi, and G. E. Sulimova, "Distribution of BoLA-DRB3 allelic frequencies and identification of a new allele in the Iranian cattle breed Sistani (*Bos indicus*)," *Russian Journal of Genetics*, vol. 45, no. 2, pp. 224–229, 2009.

[28] J. Mosafer and M. R. Nassiry, "Identification of bovine lymphocyte antigen DRB3.2 alleles in Iranian golpayegani cattle by DNA test," *Asian-Australasian Journal of Animal Sciences*, vol. 18, no. 12, pp. 1691–1695, 2005.

[29] M. R. Nassiry, F. Eftekhar Shahroodi, J. Mosafer et al., "Analysis and frequency of bovine lymphocyte antigen (*BoLA-DRB3*) alleles in Iranian Holstein cattle," *Russian Journal of Genetics*, vol. 41, no. 6, pp. 664–668, 2005.

# Mutation at the Human D1S80 Minisatellite Locus

**Kuppareddi Balamurugan,[1] Martin L. Tracey,[2] Uwe Heine,[3] George C. Maha,[3] and George T. Duncan[4]**

[1] *School of Criminal Justice, University of Southern Mississippi, 118 College Drive # 5127, Hattiesburg, MS 39406, USA*
[2] *Department of Biology, Florida International University, University Park Campus, Miami, FL 33199, USA*
[3] *DNA Identification Testing Division, Laboratory Corporation of America, 1440 York Court Extension, Burlington, NC 27215, USA*
[4] *Broward County Sheriff's Office, Forensic Laboratory DNA Unit, Fort Lauderdale, FL 33301, USA*

Correspondence should be addressed to Kuppareddi Balamurugan, kuppareddi.balamurugan@usm.edu

Academic Editors: P. Momigliano Richiardi, P. Y. Woon, and N. Zhang

Little is known about the general biology of minisatellites. The purpose of this study is to examine repeat mutations from the D1S80 minisatellite locus by sequence analysis to elucidate the mutational process at this locus. This is a highly polymorphic minisatellite locus, located in the subtelomeric region of chromosome 1. We have analyzed 90,000 human germline transmission events and found seven (7) mutations at this locus. The D1S80 alleles of the parentage trio, the child, mother, and the alleged father were sequenced and the origin of the mutation was determined. Using American Association of Blood Banks (AABB) guidelines, we found a male mutation rate of $1.04 \times 10^{-4}$ and a female mutation rate of $5.18 \times 10^{-5}$ with an overall mutation rate of approximately $7.77 \times 10^{-5}$. Also, in this study, we found that the identified mutations are in close proximity to the center of the repeat array rather than at the ends of the repeat array. Several studies have examined the mutational mechanisms of the minisatellites according to infinite allele model (IAM) and the one-step stepwise mutation model (SMM). In this study, we found that this locus fits into the one-step mutation model (SMM) mechanism in six out of seven instances similar to STR loci.

## 1. Introduction

The human genome can be grossly partitioned into three categories: nonrepetitive (single copy sequences), moderately repetitive (families of retroposon-like sequences), and highly repetitive ("classical" satellite) DNA [1]. Minisatellites represent a class of tandem repeats with repeat units ranging from 6 to 100 bp and fall into the class of highly repetitive satellites. Total array sizes range from 0.5 to 30 kb. Minisatellite loci are found throughout the genome and are estimated to number in the thousands with a strong telomeric and strand bias [2]. They are generally clustered in subtelomeric and centromeric regions; however, it has been suggested that minisatellites show no obvious predisposition for accumulation in subtelomeric or centromeric regions but are associated with rates of high recombination and mutation in these regions [2–9].

Studies have shown that minisatellites may mutate by unexpectedly complex conversion-like events [10]. Higher mutational polar bias occurs at the termini at some, if not most minisatellites, as well as microsatellite loci [11]. Minisatellite loci generally are variable near one end of the array or in some cases both ends. It has been suggested that this implies a mutational mechanism driven by cis-acting elements and thus the existence of a localized mutation hotspot at one end of the array involving complex processes of gene conversion [12–14]. One caveat is that the control regions have not been identified and only associated flanking region markers have been found.

Little is known about the general biology of minisatellites [15]. The purposes of the study are (1) to examine the mutational rate and (2) to examine mutations from the D1S80 minisatellite locus by sequence analysis to elucidate the mutational process, as an example of a particular type of minisatellite array. The highly polymorphic minisatellite locus D1S80, (gene location: 1p36.32, GenBank sequence Accession # D28507) was first described by Nakamura in 1989. It is a minisatellite locus composed of short repeat units,

Table 1: Nucleotide sequences of observed repeat units. The consensus sequence represents the most common nucleotide observed in each position of the repeats. Twenty variations based on the consensus sixteen base repeat unit are tabulated. Each repeat unit is assigned a letter code. Dots (·) represent a match to the consensus sequence as represented by type H repeat unit. Letters represent nucleotide differences when compared to the consensus sequence and correspond to A, G, C, and T nucleotides. A (-) represents a missing nucleotide in the repeat unit A.

| Repeat type | Sequence | | | | | | | | | | | | | | | | |
|---|---|---|---|---|---|---|---|---|---|---|---|---|---|---|---|---|---|
| Type A | T | C | A | · | C | · | · | · | – | A | · | · | – | · | · | · | |
| Type B | A | C | A | · | · | · | · | · | · | A | · | · | · | · | · | · | |
| Type C | · | · | · | · | · | · | · | · | · | · | · | · | A | · | · | · | |
| Type D | · | · | A | · | · | · | · | · | · | · | · | · | A | · | · | · | |
| Type E | · | · | A | · | · | · | · | · | · | A | · | · | · | · | · | · | |
| Type F | · | · | · | · | · | · | · | · | · | A | · | · | · | · | · | · | |
| Type G | · | · | A | · | · | · | · | · | · | · | · | · | · | · | · | · | |
| Type H | G | A | G | G | A | C | C | A | C | C | G | G | C | A | A | G | (Consensus) |
| Type I | · | · | · | · | · | · | · | · | · | · | A | · | G | · | · | · | |
| Type J | · | · | · | A | · | · | · | · | · | · | A | · | G | · | · | · | |
| Type K | · | · | · | · | · | · | · | · | · | · | A | · | · | · | · | · | |
| Type L | · | · | · | · | · | · | · | · | · | T | · | · | · | · | · | · | |
| Type M | · | · | · | · | · | · | · | · | · | · | A | · | G | · | G | · | |
| Type N | · | · | · | · | · | · | · | · | · | G | · | · | · | · | · | · | |
| Type O | · | · | · | · | · | · | · | · | · | · | · | · | · | G | · | · | |
| Type P | · | · | A | · | · | · | · | · | · | · | · | · | A | · | G | · | |
| Type Q | · | · | A | · | · | · | · | · | G | · | · | · | A | · | · | · | |
| Type R | · | G | · | A | · | · | · | · | · | · | A | · | G | · | · | · | |
| Type S | · | · | A | · | · | · | · | · | · | · | A | · | G | · | · | · | |
| Type T | · | · | A | · | · | · | · | · | · | · | A | · | · | · | · | · | |
| Type U | · | · | A | · | · | · | · | · | · | · | A | · | A | · | · | · | |

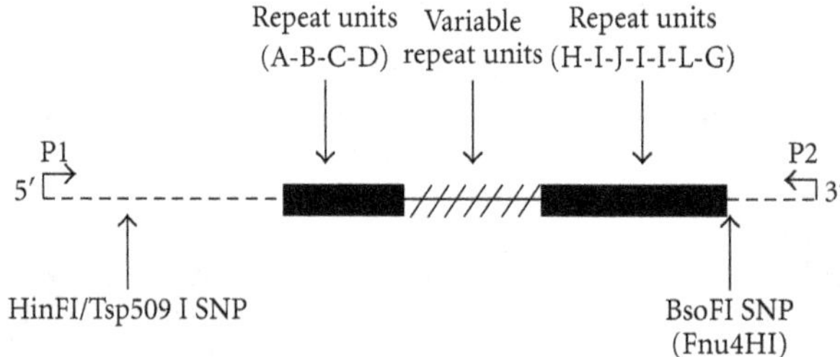


Figure 1: The general structure of the D1S80 locus is that of a monomeric type VNTR consisting of four 5′ repeat units (A-B-C-D) and seven 3′ repeat units (H-I-J-I-I-L-G) that are constant (motifs), with variable number of repeats in between. These two motifs are essentially identical in almost all alleles sequenced in all population groups so far (data not shown). There is a 132 base pair 5′ flanking region which includes the forward PCR primer (P1) and a 32-base pair 3′ flanking region which includes the reverse PCR primer (P2) sequence. The two-flanking restriction site polymorphism, HinfI (G↓ANTC), 58 base pairs from the 5′ end of the primer, and the Fnu4HI (GC↓NGC), the first base after the last repeat are shown by arrows.

typically less than or equal to 16 base pairs per unit [16, 17] (Table 1, Figure 1). It is located in the sub-telomeric region of chromosome 1 (chr1:2,390,716 - 2,391,193) roughly 2.4M bases from the p-telomere [18, 19]. The total range of repeat lengths (approx. 0.2 to 1kb) and the relatively small number of repeats (generally 14 to >40) within the unit make this locus suitable for PCR analysis [15, 20] (Figure 1). The observed heterozygosity has been reported as high as 90.5% and as low as 24% with greater than 27 alleles [21]. In most human populations, 18 and 24 repeat units are the most common and possibly the primordial alleles at this locus [22]. Because of the complexity of the repeat array structure, differences between the alleles become more evident when the arrays of repeat units are divided into common motifs. Six motifs ranging from three to nine repeat units have been identified. Allele frequencies and motifs characterized by the 18 and 24 structural alleles were described earlier [23, 24].

The D1S80 locus was previously used for forensic analysis due to its small size; however, this locus has recently been used as an associative tool in the elucidation of the chromosome 1p36 deletion syndrome which is one of the most common deletion syndromes located near a chromosomal terminus (one in 5000 births) [15, 25].

Recent and past work with respect to this locus has included the use of two single nucleotide polymorphisms (SNPs). These included rs16824398, which is a SNP that involves a HinfI restriction site at a nucleotide 58 bases downstream from the forward primer and another SNP in the 3′ flanking region that involves an Fnu4HI restriction site, at the base next to the last repeat. Interestingly, all 18-repeat alleles to date have been found to be associated with HinfI(+) and Fnu4HI(−) restriction site polymorphisms at the 5′ and 3′ ends, respectively (Figure 1). On the other hand, allele 24 is associated with HinfI(−) and Fnu4HI(+) polymorphism. Of the alleles tested, 98.5% exhibits linkage disequilibrium between two specific SNPs. If an allele is positive for HinfI, then it is negative for Fnu4HI, and if an allele is negative for HinfI, it is then positive for Fnu4HI. There is very strong linkage disequilibrium between the two SNPs [23, 26, 27]. In this study, we report the mutation rate and a nonpolarized mutational mechanism at the D1S80 minisatellite locus.

## 2. Materials and Methods

*2.1. Sample Collection and Parentage Analysis.* Buccal epithelial cell samples from 45,000 trios for paternity analysis were collected in quadruplicate from each individual using cotton-tipped swabs and air dried at ambient temperature. Each paternity short tandem repeat (STR) assay was done using duplicate independently prepared DNA extractions and assayed using proprietary inhouse STR multiplexes. Each multiplex included a shared locus (generally D1S80) with other multiplexes to confirm sample identity. D1S80 was assayed in combination with two to three other STR loci in multiplex reactions. The other STRs were selected to produce lower molecular weight fragments that would not interfere with analysis of the larger D1S80 amplified products. Amplified products were analyzed using nondenaturing high-resolution polyacrylamide gel electrophoresis with silver stain detection. Alleles were manually determined by comparison with allelic ladders containing all common D1S80 alleles. All STR-based parentage testing was performed with full knowledge and consent of the tested individuals or authorized parent/guardian. Appropriate institutional review board approval was obtained for sequence analysis of the samples.

*2.2. Parentage and Mutation Analysis.* To fully understand the implications and delineate the mutational mechanisms that the D1S80 locus undergoes, we examined the number of parent child allele transfers analyzed at this locus for the year 1996 from Laboratory Corporation of America, Burlington, NC. Observed mutations were confirmed as coming from concordant father or mother by analyzing short tandem repeat (STR) data from samples collected from alleged father, mother, and child using Amp*fl*STR Identifiler human identification kit (Applied Biosystems, Foster City, CA). This was essential to confirm that the mutation occurred in the D1S80 locus and was not the result of nonpaternity.

Several factors were used in the elucidation of the origin of the mutation between the alleged father and mother of the child. The first factor was the phase of the SNPs at the 5′- and 3′-flanking sequence of the repeat array. Another factor was a match of the repeat units of the alleles most likely to be associated between father and mother and finally we assumed that the minimal size change was the most probable one.

A mutation rate was then estimated by the number of samples sequenced that contain the mutation and compared to the number of trios sampled that year at Laboratory Corporation of America. Parentage calculations and mutation rate analysis were performed following guidelines put forth by the American Association of Blood Banks (AABB) [28].

*2.3. DNA Analysis and Sequencing.* Total genomic DNA was isolated from the trios that contained the mutation by standard organic extraction with phenol/chloroform/isoamyl alcohol followed by purification and concentration using Amicon Ultracel centrifugal filters (Millipore Corporation, Billerica, MA, USA). Quantitation of total genomic DNA was performed using Quantifiler human DNA quantitation kit (Applied Biosystems, Foster City, CA, USA). Paternity was confirmed by analyzing a set of fifteen short tandem repeat (STR) markers using the Identifiler human identification kit (Applied Biosystems, Foster City, CA, USA) as per manufacturer's recommendations. The samples were run on an automated 310 Genetic Analyzer and analyzed with GeneMapper ID software version 3.2 (Applied Biosystems, Foster City, CA, USA). The D1S80 locus was amplified using primers described by Kasai et al. [16]. Alleles were separated by 5% polyacrylamide gels and stained with ethidium bromide. Individual allelic bands were excised from the gel using disposable razor blades and suspended in 100 $\mu$L of Tris-EDTA (TE) buffer overnight. One to five $\mu$L of the gel extract was used for amplification of single alleles using the same primers described above. The alleles were sequenced using big dye terminator chemistry and ABI 310 genetic analyzer (Applied Biosystems, Foster City, CA, USA). Homozygous samples or samples that did not yield clean sequence were cloned into pBluescript plasmid vector. The DH5$\alpha$ competent cells were transformed using the recombinant vector and the cells were grown in LB/ampicillin agar plates. The resulting colonies were screened for the presence of the right size insert by directly sequencing the plasmid DNA isolated from minipreps of individual colonies. Sequencing of the cloned alleles was performed using big dye terminator chemistry and the sequence was analyzed using Applied Biosystems DNA sequencing analysis software version 5.2.

## 3. Results and Discussion

Paternity calculations were performed using AABB guidelines [28] and a residual likelihood ratio from a low of 180,848 to a high of 190,926,913,150 was observed. Mutations were identified as non-Mendelian inheritance of alleles differing in size in seven (7) out of a total of 45,000 trios (Table 2). Most of the mutations (6/7) involved a change of one repeat unit (Table 2). Although one mutation involved a two-repeat deletion, our observations support the strict step-wise replication slippage mutation (SSM) model according to which the majority of the mutational events at microsatellite loci involve gain or loss of one-repeat unit rather than the complex mutational events as illustrated by

TABLE 2: DNA sequence characteristics of seven trios illustrating the mutational event. The standard paternity test, called a trio, involves the child, mother, and alleged father. Allele represents the number of 16 base pairs repeat units making up the entire allele. The alphabets in each sample represent the arrangement of the repeat types making up each allele. The alphabet code for the repeat types and the corresponding bases is given in Table 1. HinfI and Fnu4HI are the two restriction site polymorphisms located on the 5′ and 3′ flanking regions of the repeats, respectively, and are concordant between the progenitor and the mutated allele. The star (*) represents the mutated allele and aligned with the progenitor wherever possible. The progenitors are identified based on the sequence characteristic, SNP polymorphism, and the smallest allele difference between the parent and child. For the four indeterminate mutations, the mutated allele is aligned with both the possible progenitors.

| | Sample | Allele | HinfI | Repeat types | Fnu4HI | Comments |
|---|---|---|---|---|---|---|
| 1 Trio | | | | | | |
| 1 | *Mother-18 allele 1* | *18* | *Positive* | *A B C D D E C H* ***H*¶** *I I H I J I I L G* | *Negative* | Possible progenitor |
| 1 | **Child-17 allele 1* | *17* | *Positive* | *A B C D D E C H I I H I J I I L G* | *Negative* | **Deletion of one repeat unit** |
| 1 | *Father-18 allele 1* | *18* | *Positive* | *A B C D D E C H* ***H*¶** *I I H I J I I L G* | *Negative* | Possible progenitor |
| 1 | Child-18 allele 2 | 18 | Positive | A B C D D E C H H I I H I J I I L G | *Negative* | |
| 1 | Mother-18 allele 2 | 18 | Positive | A B C D D E C H H I I H I J I I L G | *Negative* | |
| 1 | Father-22 allele 2 | 22 | Positive | A B C D D E C H H H I J I I I H I J I I L G | *Negative* | |
| 2 Trio | | | | | | |
| 3 | *Father-18 allele 1* | *18* | *Positive* | *A B C D D E C H H I I H I J I I L G* | *Negative* | Possible progenitor |
| 3 | **Child-19 allele 2* | *19* | *Positive* | *A B C D D* **D¶** *E C H H I I H I J I I L G* | *Negative* | **Duplication of one repeated unit** |
| 3 | *Mother-18 allele 1* | *18* | *Positive* | *A B C D D E C H H I I H I J I I L G* | *Negative* | Possible progenitor |
| 3 | Child-18 allele 1 | 18 | Positive | A B C D D E C H H I I H I J I I L G | *Negative* | |
| 3 | Mother-25 allele 2 | 25 | Negative | A B C D D D E F C G H I I I J H I J H I J I I L G | *Positive* | |
| 3 | Father-31 allele 2 | 31 | Positive | A B C D D E F C G H I I H I H I I K I K I K I I H I J I I L G | *Negative* | |
| 3 Trio | | | | | | |
| 7 | **Child-19 allele 1* | *19* | *Negative* | *A B C D E F C G H I I J H I J I I L G* | *Positive* | **Deletion of one repeated unit** |
| 7 | *Father-20 allele 1* | *20* | *Negative* | *A B C D E F C G H I I* ***I*¶** *J H I J I I L G* | *Positive* | Progenitor |
| 7 | Child-24 allele 2 | 24 | Negative | A B C D D E F C G H I I I J H I J H I J I I L G | *Positive* | |
| 7 | Mother-24 allele 1 | 24 | Negative | A B C D D E F C G H I I I J H I J H I J I I L G | *Positive* | Donor, child allele 2 |
| 7 | Father-24 allele 2 | 24 | Negative | A B C D D E F C G H I I I J H I J H I J I I L G | *Positive* | |
| 7 | Mother-31 allele 2 | 31 | Positive | A B C D D E F C G H I I H K I K H K I K I K I I H I J I I L G | *Negative* | |
| 4 Trio | | | | | | |
| 8 | **Child-21 allele 1* | *21* | *Negative* | *A B C D E F* **F¶** *C G H I I I J H I J I I L G* | *Positive* | **Duplication of one repeated unit** |
| 8 | *Mother-20 allele 1* | *20* | *Negative* | *A B C D E F C G H I I I J H I J I I L G* | *Positive* | Progenitor |
| 8 | Child-31 allele 2 | 31 | Positive | A B C D D E F C G H I I H I H I I K I K I K I I H I J I I L G | *Negative* | |
| 8 | Father-31 allele 2 | 31 | Positive | A B C D D E F C G H I I H I H I I K I K I K I I H I J I I L G | *Negative* | Donor, child allele 2 |
| 8 | Father-26 allele 1 | 26 | Positive | A B C D D E F C G H I I H I H I I I I L S J I I L G | *Negative* | |
| 8 | Mother-31 allele 2 | 31 | Positive | A B C D D E F C G H I I H K I K H K I K I K I I H I J I I L G | *Negative* | |

Table 2: Continued.

| | Sample | Allele | HinfI | Repeat types | Fnu4HI | Comments |
|---|---|---|---|---|---|---|
| 5Trio | | | | | | |
| 10 | *Mother-28 allele 2* | *28* | *Positive* | *A B C D D E C G H I K K I* **I¶ H¶** *I I H I J I H I J I I L G* | *Negative* | Possible progenitor |
| 10 | **Child-26 allele 1* | *26* | *Positive* | *A B C D D E C G H I K K I I I H I J I H I J I I L G* | *Negative* | **Deletion of two repeated units** |
| 10 | *Father-28 allele 1* | *28* | *Positive* | *A B C D D E C G H I K K I* **I¶ H¶** *I I H I J I H I J I I L G* | *Negative* | Possible progenitor |
| 10 | Child-28 allele 2 | 28 | Positive | A B C D D E C G H I K K I I H I I H I J I H I J I I L G | *Negative* | |
| 10 | Mother-24 allele 1 | 24 | Positive | A B C D D E C G H I K K I I K I I H I J I I L G | *Negative* | Mother's sequence does not match child 26 |
| 10 | Father-31 allele 2 | 31 | Positive | A B C D D E F C G H I I H K I K H K I K I K I I H I J I I L G | *Negative* | |
| 6Trio | | | | | | |
| 11 | *Mother-24 allele 2* | *24* | *Negative* | *A B C D D E F C G* **H¶** *I I I J H I J H I J I I L G* | *Positive* | Possible progenitor |
| 11 | **Child-23 allele 1* | *23* | *Negative* | *A B C D D E F C G I I I J H I J H I J I I L G* | *Positive* | **Deletion of one repeated unit** |
| 11 | *Father-24 allele 1* | *24* | *Negative* | *A B C D D E F C G* **H¶** *I I I J H I J H I J I I L G* | *Positive* | Possible progenitor |
| 11 | Child-24 allele 2 | 24 | Negative | A B C D D E F C G H I I I J H I J H I J I I L G | *Positive* | |
| 11 | Mother-18 allele 1 | 18 | Positive | A B C D D E C H H I I H I J I I L G | *Negative* | |
| 11 | Father-31 allele 2 | 31 | Positive | A B C D D E F C G H I I H K I K H K I K I K I I H I J I I L G | *Negative* | |
| 7Trio | | | | | | |
| 12 | **Child-19 allele 1* | *19* | *Positive* | *A B C D D E C H H I I H I J* **I¶** *I I L G* | *Negative* | **Duplication of one repeated unit** |
| 12 | *Father-18 allele 1* | *18* | *Positive* | *A B C D D E C H H I I H I J I I L G* | *Negative* | Progenitor |
| 12 | Child-20 allele 2 | 20 | Negative | A B C D E F C G H I I I J H I J I I L G | *Positive* | |
| 12 | Mother-20 allele 2 | 20 | Negative | A B C D E F C G H I I I J H I J I I L G | *Positive* | Donor, child allele 2 |
| 12 | Mother-18 allele 1 | 18 | Positive | A B C D D E C H H I I H I J I I L G | *Negative* | |
| 12 | Father-31 allele 2 | 31 | Positive | A B C D D E F C G H I I H I H I I K I K I K I I H I J I I L G | *Negative* | |

¶Represent the mutated region of the parental and child allele.

other minisatellites. We assumed that the smallest size change was most probable, thus the most parsimonious changes were assumed in our analysis of the mutational events [29, 30]. The phase of the HinfI and Fnu4HI restriction site polymorphisms of all mutated alleles of the children was concordant to the phase of the polymorphisms of the progenitor. Alec Jeffreys, who has examined these mutational events in detail, has stated that replication slippage and unequal crossover events which are intrinsic to the tandem repeat array are not primarily the mechanisms of mutation. Instead, he postulated that germline repeat instability which is regulated by cis-acting elements somewhere near the array may be responsible [31]. In our data, in four trio sets, it was not possible to ascertain which parent was the progenitor of the mutation. In the remaining three trios, we were able to ascertain two mutations to the father and one to the mother. Other studies have emphasized the origin of most mutations from paternal lineage due to more numbers of meiotic events [32–34], but we cannot make this prediction from our data since four of the seven mutations are indeterminate. The advantage of samples like this permits the analysis of true *in vivo* germline events rather than somatic mutational events.

Only two studies are reported that have dealt with the mutational events within this locus [35–37]. The first study [35] reported three mutations involving the loss or gain of three repeats to as much as a loss of six repeats. Of these events, two could be attributed to a maternal origin and the remaining event could not be discerned between mother and father. The data of our study does not support these findings where most mutational events involved one repeat loss or gain. Their conclusion was that paternal events were most prevalent in microsatellite loci, while in minisatellite loci maternal events were most prevalent. The second study dealt with computer simulations and a stepwise mutation model [37]. Jeffreys et al. found that minisatellites do not mutate by replication slippage and concluded that most mutations were exhibited in the male germline [14].

Mutation rates for minisatellites have been estimated to range from 0.5% to greater than 20% per generation in some studies and in another study the rate was reported as $0.53 \times 10^{-3}$ to $1.53 \times 10^{-3}$ [35]. Only one study up to this time has examined the D1S80 locus specifically and found a mutation rate of 3 out of 6153 meioses or $5 \times 10^{-4}$ [35]. We found 7 mutations out of 90,000 meioses (45,000 paternal and 45,000 maternal meioses). This then represented an effective overall mutation rate of approximately $7.77 \times 10^{-5}$. The mutation rate of $7.77 \times 10^{-5}$ represents the overall mutation rate rather than a rate specific to the male or female germline. Only one event could be narrowed down to the mother and two events attributable to the father. In the remaining four trios, we could not choose the progenitor because of the similarities of the maternal and paternal sequences. For the four indeterminate mutations, the AABB prorates them based on known mutation rates [28]. In this case, one adds 2.667 to the males and 1.333 to the female for estimated mutation rates of $1.04 \times 10^{-4}$ (4.667/45,000) and $5.18 \times 10^{-5}$ (2.333/45,000), respectively. Some mutations among the trios might not appear as Mendelian errors in the analysis since the scoring was based on a size difference between alleles (difference in the number of repeats between parent and child), and single base changes would not be detected, and the actual mutation rates could be slightly higher.

Several studies have examined mutational mechanisms according to infinite allele model (IAM) and the one-step stepwise mutation model (SMM). They have found that minisatellites sometimes fit both mechanisms; however, short tandem repeat loci (STR) were most similar to the simulation results under the SMM model of mutation. In this study, we found that this locus fit into the one-step stepwise mutation model (SMM) mechanism in six out of seven instances similar to STR loci [31, 37, 38]. Jeffreys, when using single-molecule polymerase chain reaction (SP-PCR) for elucidation of the minisatellite MS1, found an abundance of mutational events in sperm consisting of the deletion or addition of one-repeat unit [38].

Jeffreys et al. [31] reported polarized mutational events for the three minisatellite loci MS32, MS205, and MS31A. In this study, the authors reported that 93% of mutants lays within 20 repeats of the progenitor allele and 90% of the mutant alleles which had gained repeats showed extreme polarity, with repeat segments being added in the 5′ region of the progenitor allele. This finding is in contrast to our results in this study. All the mutated alleles have deletion or duplication in close proximity to the center of the repeats, while four repeats in the 5′ end and seven repeats in the 3′ end are relatively constant in both the progenitor and the mutated alleles.

In addition to the differences in the repeat regions of the progenitor and mutated alleles, Jeffreys et al. have also found that the polymorphisms flanking the minisatellite repeat array have failed to reveal exchange of flanking markers [31]. This is true with the results obtained from this study that the 5′ HinFI and the 3′ Fnu4HI restriction site polymorphisms are concordant with the progenitor and the mutated alleles. This phenomenon suggests that the 5′- and 3′-flanking regions and at least four repeats on either end of the repeat array are not involved in the mutational process. These differences in the region of the mutation in the repeat array indicate possible different mutational mechanisms in different minisatellite loci.

The data we have reported highlights the importance of a number of questions. (1) Do repeats have functions and do they have an effect on genome structure? (2) Are the allelic clades the result of constraints on mutation? (3) Is selection acting at the D1S80 locus?, and (4) Why is the D1S80 locus conserved among primates (unpublished data) including humans?

## Conflict of Interests

The authors declare no conflict of interests, financially or otherwise.

## Acknowledgments

The authors would like to thank Mrs Yasotha Balasubramaniam, Toronto, ON, Canada for the artwork and the four anonymous reviewers for their comments and suggestions.

This study was partly supported by funding provided by the National Institute of Justice, Department of Justice, Washington, DC, USA, under project number 2009-D1-BX-0293 to the University of Southern Mississippi and by a startup grant for one of the authors (K. Balamurugan) from the University of Southern Mississippi, Hattiesburg, MS, USA. Points of view in the document are those of the authors and do not necessarily represent the official view of the U.S. Department of Justice.

## References

[1] B. Lewin, *Genes IV.*, Oxford University Press, New York, NY, USA, 1990.

[2] G. F. Richard, A. Kerrest, and B. Dujon, "Comparative genomics and molecular dynamics of DNA repeats in eukaryotes," *Microbiology and Molecular Biology Reviews*, vol. 72, no. 4, pp. 686–727, 2008.

[3] D. L. Ellsworth, M. D. Shriver, and E. Boerwinkle, "Nucleotide sequence analysis of the apolipoprotein B 3' VNTR," *Human Molecular Genetics*, vol. 4, no. 5, pp. 937–944, 1995.

[4] R. Saferstein, *Forensic Science Handbook*, Regents/Prentice Hall, Englewood Cliffs, NJ, USA, 1993.

[5] J. A. L. Armour and A. J. Jeffreys, "Biology and applications of human minisatellite loci," *Current Opinion in Genetics and Development*, vol. 2, no. 6, pp. 850–856, 1992.

[6] J. A. L. Armour, P. C. Harris, and A. J. Jeffreys, "Allelic diversity at minisatellite MS205 (D16S309): evidence for polarized variability," *Human Molecular Genetics*, vol. 2, no. 8, pp. 1137–1145, 1993.

[7] A. J. Jeffreys, V. Wilson, and S. L. Thein, "Hypervariable 'minisatellite' regions in human DNA," *Nature*, vol. 314, no. 6006, pp. 67–73, 1985.

[8] W. Stephan, "Tandem-repetitive noncoding DNA: forms and forces," *Molecular Biology and Evolution*, vol. 6, no. 2, pp. 198–212, 1989.

[9] G. P. Smith, "Evolution of repeated DNA sequences by unequal crossover," *Science*, vol. 191, no. 4227, pp. 528–535, 1976.

[10] P. R. J. Bois, "Hypermutable minisatellites, a human affair?" *Genomics*, vol. 81, no. 4, pp. 349–355, 2003.

[11] C. B. Kunst and S. T. Warren, "Cryptic and polar variation of the fragile X repeat could result in predisposing normal alleles," *Cell*, vol. 77, no. 6, pp. 853–861, 1994.

[12] A. J. Jeffreys, A. MacLeod, K. Tamaki, D. L. Neil, and D. Monckton, "Minisatellite repeat coding as a digital approach to DNA typing," *Nature*, vol. 354, no. 6350, pp. 204–209, 1991.

[13] R. K. Wolff, R. Plaetke, A. J. Jeffreys, and R. White, "Unequal crossingover between homologous chromosomes is not the major mechanism involved in the generation of new alleles at VNTR loci," *Genomics*, vol. 5, no. 2, pp. 382–384, 1989.

[14] A. J. Jeffreys, M. J. Allen, J. A. L. Armour et al., "Mutation processes at human minisatellites," *Electrophoresis*, vol. 16, no. 9, pp. 1577–1585, 1995.

[15] I. C. Gray and A. J. Jeffreys, "Evolutionary transience of hypervariable minisatellites in man and the primates," *Proceedings of the Royal Society B*, vol. 243, no. 1308, pp. 241–253, 1991.

[16] K. Kasai, Y. Nakamura, and R. White, "Amplification of a variable number of tandem repeats (VNTR) locus (pMCT118) by the polymerase chain reaction (PCR) and its application to forensic science," *Journal of Forensic Sciences*, vol. 35, no. 5, pp. 1196–1200, 1990.

[17] Y. Nakamura, M. Carlson, K. Krapcho, and R. White, "Isolation and mapping of a polymorphic DNA sequence (pMCT118) on chromosome 1p [D1S80]," *Nucleic Acids Research*, vol. 16, no. 19, p. 9364, 1988.

[18] W. James Kent, C. W. Sugnet, T. S. Furey et al., "The human genome browser at UCSC," *Genome Research*, vol. 12, no. 6, pp. 996–1006, 2002.

[19] W. J. Kent, "BLAT—the BLAST-like alignment tool," *Genome Research*, vol. 12, no. 4, pp. 656–664, 2002.

[20] B. Budowle, R. Chakraborty, A. M. Giusti, A. J. Eisenberg, and R. C. Allen, "Analysis of the VNTR locus D1S80 by the PCR followed by high-resolution PAGE," *American Journal of Human Genetics*, vol. 48, no. 1, pp. 137–144, 1991.

[21] A. Sajantila, B. Budowle, M. Strom et al., "PCR amplification of alleles at the D1S80 locus: comparison of a Finnish and a North American Caucasian population sample, and forensic casework evaluation," *American Journal of Human Genetics*, vol. 50, no. 4, pp. 816–825, 1992.

[22] R. Deka, S. DeCroo, L. Jin et al., "Population genetic characteristics of the D1S80 locus in seven human populations," *Human Genetics*, vol. 94, no. 3, pp. 252–258, 1994.

[23] K. Balamurugan, R. Pomeroy, G. Duncan, and M. Tracey, "Investigating SNPs flanking the D1S80 locus in a Tamil population from India," *Human Biology*, vol. 82, no. 2, pp. 221–226, 2010.

[24] K. Balamurugan, N. Prabakaran, G. Duncan, B. Budowle, M. Tahir, and M. Tracey, "Allele frequencies of 13 STR loci and the D1S80 locus in a Tamil population from Madras, India," *Journal of Forensic Sciences*, vol. 46, no. 6, pp. 1515–1517, 2001.

[25] S. K. Shapira, C. McCaskill, H. Northrup et al., "Chromosome 1p36 deletions: the clinical phenotype and molecular characterization of a common newly delineated syndrome," *American Journal of Human Genetics*, vol. 61, no. 3, pp. 642–650, 1997.

[26] G. T. Duncan, K. Balamurugan, B. Budowle, and M. L. Tracey, "Hinf I/Tsp509 I and BsoF I polymorphisms in the flanking regions of the human VNTR locus D1S80," *Genetic Analysis*, vol. 13, no. 5, pp. 119–121, 1996.

[27] S. A. Limborska, A. V. Khrunin, O. V. Flegontova, V. A. Tasitz, and D. A. Verbenko, "Specificity of genetic diversity in D1S80 revealed by SNP-VNTR haplotyping," *Annals of Human Biology*, vol. 38, no. 5, pp. 564–569, 2011.

[28] Annual Report Summary for Testing in 2008, Prepared by the Relationship Testing Program Unit, http://www.aabb.org/sa/facilities/Documents/rtannrpt08.pdf.

[29] J. J. Wiens and M. R. Servedio, "Phylogenetic analysis and intraspecific variation: performance of parsimony, likelihood, and distance methods," *Systematic Biology*, vol. 47, no. 2, pp. 228–253, 1998.

[30] E. Sober, "The contest between parsimony and likelihood," *Systematic Biology*, vol. 53, no. 4, pp. 644–653, 2004.

[31] A. J. Jeffreys, K. Tamaki, A. MacLeod, D. G. Monckton, D. L. Neil, and J. A. L. Armour, "Complex gene conversion events in germline mutation at human minisatellites," *Nature Genetics*, vol. 6, no. 2, pp. 136–145, 1994.

[32] K. Tamaki, C. H. Brenner, and A. J. Jeffreys, "Distinguishing minisatellite mutation from non-paternity by MVR-PCR," *Forensic Science International*, vol. 113, no. 1–3, pp. 55–62, 2000.

[33] Y. E. Dubrova, G. Grant, A. A. Chumak, V. A. Stezhka, and A. N. Karakasian, "Elevated minisatellite mutation rate in the post-chernobyl families from Ukraine," *American Journal of Human Genetics*, vol. 71, no. 4, pp. 801–809, 2002.

[34] A. J. Jeffreys, P. Bois, J. Buard et al., "Spontaneous and induced minisatellite instability," *Electrophoresis*, vol. 18, no. 9, pp. 1501–1511, 1997.

[35] A. Sajantila, M. Lukka, and A. C. Syvänen, "Experimentally observed germline mutations at human micro- and minisatellite loci," *European Journal of Human Genetics*, vol. 7, no. 2, pp. 263–266, 1999.

[36] M. D. Shriver, *Origins and evolution of VNTR loci: the apolipoprotein B 3' VNTR*, Ph.D. thesis, The University of Texas Health Science Center at Houston, Houston, Tex, USA , 1993.

[37] M. D. Shriver, L. Jin, R. Chakraborty, and E. Boerwinkle, "VNTR allele frequency distributions under the stepwise mutation model: a computer simulation approach," *Genetics*, vol. 134, no. 3, pp. 983–993, 1993.

[38] I. Berg, R. Neumann, H. Cederberg, U. Rannug, and A. J. Jeffreys, "Two modes of germline instability at human minisatellite MS1 (locus D1S7): complex rearrangements and paradoxical hyperdeletion," *American Journal of Human Genetics*, vol. 72, no. 6, pp. 1436–1447, 2003.

# Effects of Ginseng and Echinacea on Cytokine mRNA Expression in Rats

**Deniz Uluışık and Ercan Keskin**

*Department of Physiology, Faculty of Veterinary Medicine, University of Selcuk, 42075 Selçuklu, Konya, Turkey*

Correspondence should be addressed to Deniz Uluışık, denizfedai@selcuk.edu.tr

Academic Editor: Yehuda Shoenfeld

The aim of the study was to determine the effect of ginseng and echinacea on the mRNA expression of IL-10, TNF-$\alpha$, and TGF-$\beta$1 in healthy rats. Six-week-old male Fischer 344 rats ($n = 48$) were used. The animals were divided into three equal groups, as follows: control (C); ginseng (G); echinacea (E). While the C group was fed a standard rat diet (Purina) *ad libitum* for a period of 40 days, the G and E groups animals received the same diet containing 0.5 g/kg of *Panax ginseng* root powder and 0.75 g/kg of *Echinacea purpurea* root powder, respectively. Blood samples were obtained from 8 rats in each group after 20 and 40 days of treatment, and the mRNA expression of IL-10, TNF-$\alpha$, and TGF-$\beta$1 was determined. After 20 days of treatment, the expression of IL-10 mRNA in the G group was different from the C group ($P < 0.05$); however, after 40 days of treatment, there was no difference between the groups. There was no difference after 20 and 40 days of treatment between the groups with respect to the expression of TGF-$\beta$1 mRNA. After 20 days of treatment, the expression of TNF-$\alpha$ mRNA in the E group was higher ($P < 0.05$) than the C group. After 40 days of treatment, the expression of TNF-$\alpha$ mRNA was similar in all of the groups. Based on the current study, the increase in expression of IL-10 mRNA in the G group and the increase in expression of TNF-$\alpha$ mRNA in the E group support the use of these plants for purposes of modulating the immune system. However, a more detailed study regarding the effects of ginseng and echinacea on these cytokines and other cytokines is needed.

## 1. Introduction

Ginseng refers to the root of several species in the plant genus *Panax* (C. A. Meyer Araliaceae), including *P. ginseng*, *P. japonicas*, *P. quinquefolium*, and *P. notoginseng* [1]. Recently, *P. ginseng* has been widely used worldwide as an ingredient in dietary health supplements and an additive in foods [2]. The glycosidal saponins (glycosylated steroids) known as ginsenosides are the main active components of ginseng [3]. Based on several studies, tonic, immunomodulatory, antimutagenic, and antiageing activities have been reported among the pharmacologic properties of ginseng [4, 5]. Clinical studies have also demonstrated that ginseng may improve immunostimulation, antitumor activity, cardiovascular function, antioxidant activity, hypoglycemic activity, and the pituitary-adrenocortical system [6–8].

Echinacea is another well-known herb with a worldwide reputation. There is widespread interest in the therapeutic and preventive potential of echinacea [7]. *Echinacea angustifolia*, *E. pallida*, and *E. purpurea* are the species most often used medicinally [9, 10]. It has been reported that echinacea has carbohydrate, glycoside, alkaloide, alkylamide, and polyacetylene structures as active ingredients. Echinacea is commonly used for the prevention and treatment of upper respiratory tract infections (URTIs), viral, bacterial, and fungal infections, complementary therapy for cancer chemotherapy to support the immune system, chronic fatigue syndrome, acquired immunodeficiency diseases (AIDs), and snake bites [11].

In the current study, we determined the expression of IL-10 and TGF-$\beta$1 mRNA as antiinflammatory cytokines and TNF-$\alpha$ as a proinflammatory cytokine in healthy rats fed ginseng and echinacea.

Table 1: Statistical analysis summary of results obtained from echinacea versus control after processing RT-PCR data through REST (Relative Expression Software Tool) in 20th day.

| Gene | Type | Reaction efficiency | Expression | Std. error | 95% CI | P (H1) | Result |
|---|---|---|---|---|---|---|---|
| GAPDH | REF | 0,91 | 1,000 | 0,501–2,050 | 0,313–2,909 | 1,000 | |
| IL-10 | TRG | 0,91 | 4,053 | 0,956–12,147 | 0,867–22,956 | 0,131 | |
| TGF-$\beta$1 | TRG | 1,0 | 2,569 | 0,715–8,246 | 0,334–19,656 | 0,472 | |
| TNF-$\alpha$ | TRG | 0,91 | 5,042 | 2,078–13,565 | 0,944–23,456 | 0,035 | UP |

Normalisation factor of the parameters used for analysis was 0.40.

Table 2: Statistical analysis summary of results obtained from ginseng versus control after processing RT-PCR data through REST (Relative Expression Software Tool) in 20th day.

| Gene | Type | Reaction efficiency | Expression | Std. error | 95% CI | P (H1) | Result |
|---|---|---|---|---|---|---|---|
| GAPDH | REF | 0,91 | 1,000 | 0,497–1,970 | 0,295–3,419 | 1,000 | |
| IL-10 | TRG | 0,91 | 12,304 | 4,545–33,116 | 2,140–61,238 | 0,041 | UP |
| TGF-$\beta$1 | TRG | 1,0 | 1,179 | 0,556–2,755 | 0,464–4,988 | 1,000 | |
| TNF-$\alpha$ | TRG | 0,91 | 8,468 | 4,847–19,217 | 2,357–24,843 | 0,065 | |

Normalisation factor of the parameters used for analysis was 0.18.

## 2. Materials and Methods

Ginseng and echinacea roots were commercial products (General Nutrition Products (GNC), Inc., 1050 Woodruff Road Greenville, SC, USA). In this study, 48 healthy Fischer 344 male rats were used. The rats were divided into three equal groups. The mean weights of the groups were similar. The animals were kept in individual cages for 40 days and fed *ad libitum*. The control group (C) was fed standard rat pellets, the ginseng group (G), and echinacea group (E) were fed pellets containing 0.5 g/kg of *P. ginseng* root powder and 0.75 g/kg of *E. purpurea* root powder, respectively. On the 20th and 40th days of the study, citrated blood samples were obtained from 8 animals in each group. The Ethical Committee of the Faculty of Veterinary Medicine (Report no. 2007/036) approved the study protocol.

*2.1. Total RNA Isolation and cDNA Synthesis.* Leucocyte isolation was performed using red blood cell lysis buffer (RBCLB; Roche Diagnostics, Mannheim, Germany). Total RNA was isolated using components and instructions from the High Pure RNA Isolation Kit (Roche Diagnostics, Mannheim, Germany) in the leucocyte pellet. The total elution volume of RNA was 50 mL in the final step. RNA was stored at −80°C until used in a cDNA preparation reaction. Total RNA was used as a template for reverse transcription (RT) in 20 mL volumes, as described in the 1st Strand cDNA Synthesis Kit for RT-PCR (Roche Diagnostics, Mannheim, Germany). In the RT reaction, we used random hexamer primers. The cDNA was stored at −20°C until used in real-time PCR in a LightCycler (LC; Roche Diagnostic Ltd., Lewes, UK).

*2.2. Real-Time PCR.* Real-time PCR was performed for the quantification of gene expression using a LC rapid thermal system (Roche) according to the manufacturer's instructions. A LightCycler FastStart DNA Master SYBR Green I kit (Roche Diagnostics, Mannheim, Germany) was used together with LightCycler-Primer Set Kits (Search-LC, Heidelberg, Germany) for IL-10, TNF-$\alpha$, and TGF-$\beta$1, and a housekeeping gene (glyceraldehyde phosphate dehydrogenase (GAPDH)). Reactions were performed in a 20 $\mu$L volume with 10 $\mu$L of master mix (2 $\mu$L of SybrGreen mix, 2 $\mu$L of specific primers for each cytokine, and 6 $\mu$L of sterile water) and 10 $\mu$L of cDNA template. The run was programmed according to the LightCycler-Primer Set Kits instructions. The first step was a 10 min denaturation at 95°C, followed by 35 cycles for GAPDH, 45 cycles for the remaining cytokines, a 95°C denaturation for 10 sec, 68°C–58°C annealing for 10 sec (decreasing 0.5°C/cycle), and finally a 72°C extension for 16 sec. The PCR products were subjected to melting curve-analyses starting with 58°C, and increasing to 95°C (0.1°C/sec) to confirm product specificity. The results were expressed as a ratio between the concentration of the housekeeping gene (GAPDH) and the relevant cytokine in each sample. RNA input was normalized by the average expression of the housekeeping genes encoding GAPDH.

The statistical differences among the groups were tested by REST software 2005 [12].

## 3. Results

In the current study, it was determined that the expression of IL-10 mRNA was not different between the C and E groups (Table 1 and Figure 1); however, the expression of IL-10 mRNA increased significantly in the G group when compared to the C group (Table 2 and Figure 2, $P < 0.05$) as a result of feeding 20 days with feeds including ginseng and echinacea. The expression of IL-10 mRNA on day 40 of treatment was not different between the groups (Tables 3 and 4 and Figures 3 and 4). It was also shown that the expression of IL-10 mRNA in each group on days 20 and 40 of treatment was not statistically different.

TABLE 3: Statistical analysis summary of results obtained from echinacea versus control after processing RT-PCR data through REST (Relative Expression Software Tool) in 40th day.

| Gene | Type | Reaction efficiency | Expression | Std. Error | 95% CI | P (H1) | Result |
|---|---|---|---|---|---|---|---|
| GAPDH | REF | 0,91 | 1,000 | 0,270–3,369 | 0,086–11,659 | 1,000 | |
| IL-10 | TRG | 0,91 | 1,707 | 0,369–5,595 | 0,233–11,811 | 0,263 | |
| TGF-$\beta$1 | TRG | 0,91 | 0,837 | 0,404–1,686 | 0,248–3,303 | 0,508 | |
| TNF-$\alpha$ | TRG | 0,91 | 1,587 | 0,583–4,967 | 0,307–8,978 | 0,241 | |

Normalisation factor of the parameters used for analysis was 1.00.

TABLE 4: Statistical analysis summary of results obtained from ginseng versus control after processing RT-PCR data through REST (Relative Expression Software Tool) in 40th day.

| Gene | Type | Reaction efficiency | Expression | Std. Error | 95% CI | P (H1) | Result |
|---|---|---|---|---|---|---|---|
| GAPDH | REF | 0,91 | 1,000 | 0,443–2,742 | 0,230–7,522 | 1,000 | |
| IL-10 | TRG | 0,91 | 1,533 | 0,394–4,632 | 0,200–6,599 | 0,502 | |
| TGF-$\beta$1 | TRG | 0,91 | 1,522 | 0,446–8,461 | 0,263–25,741 | 0,565 | |
| TNF-$\alpha$ | TRG | 0,91 | 0,897 | 0,463–1,689 | 0,246–2,596 | 0,876 | |

Normalisation factor of the parameters used for analysis was 1.46.

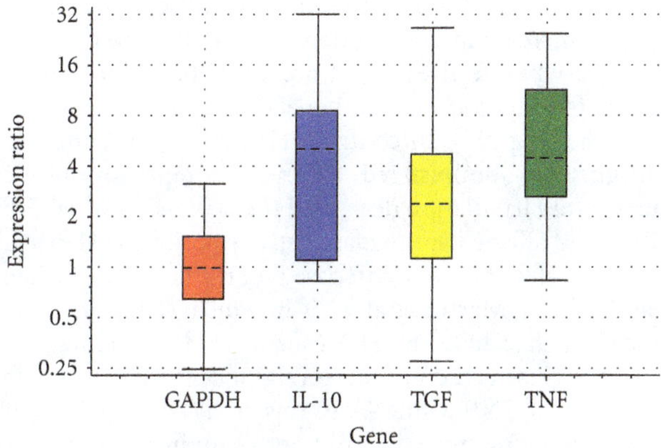


FIGURE 1: Relative expressions of IL-10, TGF-$\beta$1, and TNF-$\alpha$ obtained from echinacea versus control after processing RT-PCR with REST in 20th day.

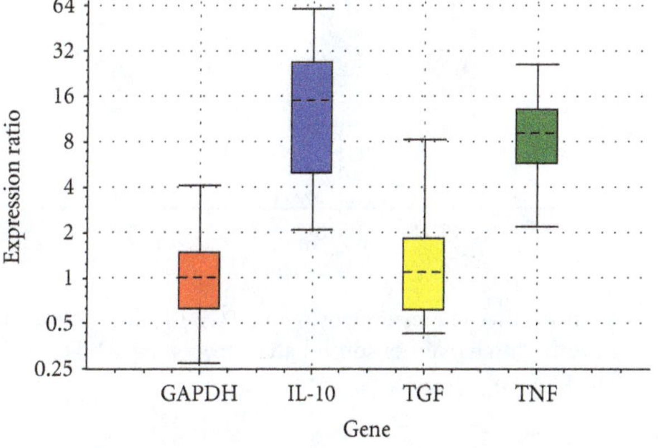


FIGURE 2: Relative expressions of IL-10, TGF-$\beta$1, and TNF-$\alpha$ obtained from ginseng versus control after processing RT-PCR with REST in 20th day.

There was no difference between the groups with respect to the expression of TGF-$\beta$1 mRNA on the 20th and 40th days of treatment (Tables 1, 2, 3, and 4 and Figures 1, 2, 3, and 4). It was also shown that the expression of TGF-$\beta$1 mRNA in each group on days 20 and 40 of treatment was not statistically different.

The expression of TNF-$\alpha$ mRNA in the E group on the 20th day of treatment was significantly higher than the C group (Table 1 and Figure 1, $P < 0.05$). In addition, there was no significant difference between groups with respect to the expression of TNF-$\alpha$ mRNA on the 40th day of treatment (Tables 3 and 4 and Figures 3 and 4). In contrast, the expression of TNF-$\alpha$ mRNA obtained on days 20 and 40 of treatment did not differ significantly for the 3 groups as a function of time.

## 4. Discussion

The administration of ginseng for 20 days increased the expression of IL-10 mRNA significantly ($P < 0.05$) when compared to the C group (Table 2 and Figure 2), and this increase was in agreement with the findings obtained by Liou et al. [13] and Wang et al. [14] using mouse cell cultures in which a ginseng extract was used.

In contrast, Larsen et al. [15] reported that ginseng extract did not change IL-10 production in polymorphonuclear leukocytes, while Lee et al. [16] showed that the addition of ginseng extract in human monocytic cell cultures decreased the expression of IL-10. The equilibrium that should exist between pro- and anti-inflammatory cytokines in various infections and inflammatory states is critical with respect to host defense. Indeed, an uncontrolled and

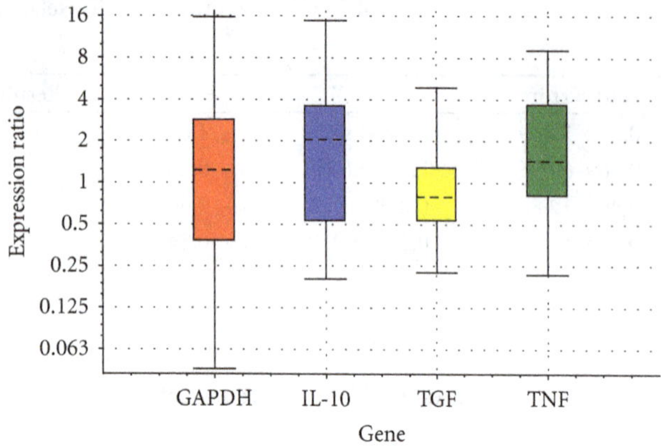


FIGURE 3: Relative expressions of IL-10, TGF-$\beta$1, and TNF-$\alpha$ obtained from echinacea versus control after processing RT-PCR with REST in 40th day.

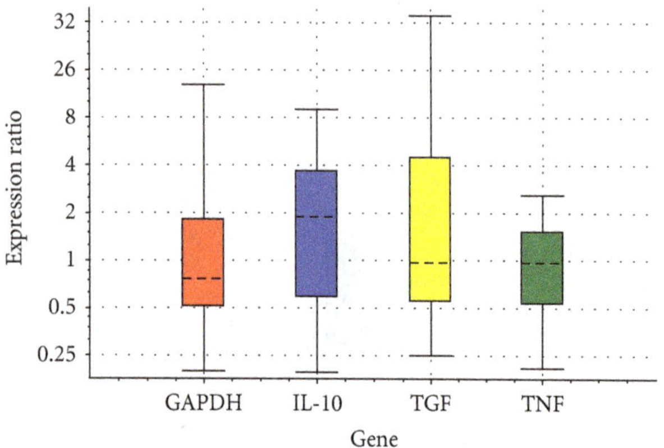


FIGURE 4: Relative expressions of IL-10, TGF-$\beta$1, and TNF-$\alpha$ obtained from ginseng versus control after processing RT-PCR with REST in 40th day.

excessive immune response may result in more harm than an infection or various noxious agents cause in humans [17]. The increase in the expression of IL-10 and suppression of IL-2 production in T cells following ginseng application and the corresponding limited proliferation in B and T cell lines and decrease in immunoglobulin production in B cells support this opinion [13].

In addition, it is difficult to explain the lack of difference in IL-10 mRNA expression in the G group compared to the C group on day 40 of treatment (Table 4 and Figure 4), because the effect of ginseng treatment on the expression of IL-10 mRNA as a function of time, if any, is unknown.

The expression of IL-10 mRNA of the E group on the 20th day of the study was not different from the C group (Table 1 and Figure 1). While Gertsch et al. [18] defined that the addition of echinacea extract in human peripheral mononuclear cell cultures increased IL-10 expression, Hwang et al. [19] reported that the addition of echinacea to mouse splenocyte cell cultures increased IL-10 expression, and Raduner et al. [20] concluded that the addition of echinacea alkylamide to human peripheral whole blood cell cultures increased the expression of IL-10. In contrast, Guiotto et al. [21] showed that oral echinacea in various doses in humans decreased the expression of IL-10 and Zhai et al. [22] reported that there was no difference in the expression of IL-10 by LPS stimulation in mononuclear leucocyte cell cultures obtained from mice that had received an oral echinacea extract when compared to the control group. Echinacea or alkylamides first stimulate a proinflammatory response, then an antiinflammatory response; this time-dependent bimodal effect is important in immunomodulation [23].

While variations in TGF-$\beta$ mRNA following echinacea treatment were not observed in the current study, Randolph et al. [23] reported that the expression of TGF-$\beta$ mRNA decreased in cell cultures which were treated with different echinacea extracts (*E. purpurea* and *E. angustifolia*). While a significant increase or decrease was not observed in the TGF-$\beta$ level following ginseng treatment in the current study, different statements suggested that the TGF-$\beta$1 level increased or decreased in the research conducted with single or combined application of various saponins that are active and contained in this plant or in the root of the plant.

Kanzaki et al. [24] stated that ginseng saponin application in human fibroblast cultures increased the TGF-$\beta$1 level. An increase in TGF-$\beta$1 depends on saponin; therefore, the materials effective in extracellular matrix formation, such as fibronectin, may be beneficial in tissue repair and wound healing [24]. In contrast, Han et al. [25] concluded that ginseng administered to mice intraperitoneally 24 hours before irradiation decreased the expression of TGF-$\beta$1 mRNA and affected antioxidant activity positively; Han et al. [25] supported their opinion based on the findings of Chang et al. [26] and Martin et al. [27] in which TGF-$\beta$ increased radicals, such as ROS and HO, causing oxidative damage.

While ginseng did not cause a significant alteration in the expression of TNF-$\alpha$ mRNA in this study, Wang et al. [28] stated that a ginseng extract did not result in any alterations in the expression of TNF-$\alpha$ mRNA in mouse macrophages. In contrast, Rivera et al. [29] reported that ginsenoside Rb1 increased TNF-$\alpha$ titration in lymphocyte cultures obtained from mice after parvovirus inoculation, while Song et al. [30] concluded that there was an increase in the expression of TNF-$\alpha$ mRNA by ginseng extract application in murine peritoneal macrophage cell cultures. In contrast, Huang et al. [31] stated that the addition of ginseng extract in human lymphocyte cell cultures decreased the expression of TNF-$\alpha$, and Kim et al. [32] showed that the addition of ginsenoside Rb1 to human peripheral mononuclear cell cultures decreased the expression of TNF-$\alpha$ and there was also a decrease in the expression of TNF-$\alpha$ when given orally to mice with arthritis. In these studies with ginseng, different results and decisions should be evaluated with various ginseng content, different systems, and varying amounts.

Randolph et al. [23] and Gertsch et al. [18] stated that the addition of echinacea extract to human monocytic cell cultures and echinacea alkylamide to human macrophage cell cultures increased the expression of TNF-$\alpha$ mRNA, which supported the increase in the expression of TNF-$\alpha$ mRNA in the E group in the current study. In parallel with these findings, it was determined that echinacea root powder or extracts increased the TNF-$\alpha$ level in cell cultures treated with LPS [22] and in infected cell cultures [33]. In contrast, oral echinacea in humans in various doses [21] and the addition of echinacea extract to human peripheral whole blood cultures [20] decreased the TNF-$\alpha$ level.

In the studies which showed that echinacea extracts or preparations decreased or increased cytokines, such as TNF-$\alpha$, it was suggested that this material modulated the immune system both as a stimulator and as an inhibitor via CB2 receptors [18]. It was determined that the different results obtained with the echinacea studies were correlated with the content from different kinds and parts of this plant [20].

When the increase in the expression of IL-10 mRNA by ginseng was taken into consideration in this study, it can be concluded that this plant is effective in immunomodulation. In the E group of this study, the increase in the expression of TNF-$\alpha$ mRNA was considered with some cytokines (IL-1, IL-6, NO, and IFN-$\gamma$) not measured in our study, supporting the opinion that echinacea modulates and stimulates the immune system. Because no variations occurred in the expression of TGF-$\beta$1 mRNA by ginseng or echinacea in the amounts used in the current study, it cannot be concluded that these plants have no effect on TGF-$\beta$1.

Given the complexity of the immune system, the interaction with other systems and the multifunctionality of measured cytokines were taken into consideration, more detailed and different studies should be conducted to confirm the results.

## Conflict of Interests

None of the authors of this paper has a financial or personal relationship with other people or commercial identity mentioned in their paper that might lead to a conflict of interests for any of the authors.

## Acknowledgment

This study was supported by The Scientific Research Projects Coordination Unit of Selcuk University (Project no. 06102036).

## References

[1] T. K. Yun, "Brief introduction of Panax ginseng C.A. Meyer.," *Journal of Korean Medical Science*, vol. 16, pp. S3–S5, 2001.

[2] S. Sakamoto, B. Pongkitwitoon, S. Nakamura, K. Maenaka, H. Tanaka, and S. Morimoto, "Efficient silkworm expression of single-chain variable fragment antibody against ginsenoside Re using Bombyx mori nucleopolyhedrovirus bacmid DNA system and its application in enzyme-linked immunosorbent assay for quality control of total ginsenosides," *Journal of Biochemistry*, vol. 148, no. 3, pp. 335–340, 2010.

[3] A. Wilkie and C. Cordess, "Ginseng—a root just like a carrot?" *Journal of the Royal Society of Medicine*, vol. 87, no. 10, pp. 594–595, 1994.

[4] D. Kiefer and T. Pantuso, "Panax ginseng," *American Family Physician*, vol. 68, no. 8, pp. 1539–1542, 2003.

[5] T. K. Lee, R. M. Johnke, R. R. Allison, K. F. O'Brien, and L. J. Dobbs, "Radioprotective potential of ginseng," *Mutagenesis*, vol. 20, no. 4, pp. 237–243, 2005.

[6] S. A. Susin, H. K. Lorenzo, N. Zamzami et al., "Mitochondrial release of caspase-2 and -9 during the apoptotic process," *Journal of Experimental Medicine*, vol. 189, no. 2, pp. 381–393, 1999.

[7] K. I. Block and M. N. Mead, "Immune system effects of echinacea, ginseng, and astragalus: a review," *Integrative Cancer Therapies*, vol. 2, no. 3, pp. 247–267, 2003.

[8] J. Z. Luo and L. Luo, "Ginseng on hyperglycemia: effects and mechanisms," *Evidence-Based Complementary and Alternative Medicine*, vol. 6, pp. 423–427, 2009.

[9] R. Bauer, P. Remiger, and E. Alstat, "Alkamides and caffeic acid derivatives from the roots of Echinacea tennesseensis," *Planta Medica*, vol. 56, no. 6, pp. 533–534, 1990.

[10] I. Mistrikova and S. Vaverkova, "Morphology and anatomy of *Echinacea purpurea, E. angustifolia, E. pallida* and *Parthenium integrifolium*," *Biologia*, vol. 62, no. 1, pp. 2–5, 2007.

[11] S. Goldhaber-Fiebert and K. J. Kemper, "Echinacea (E. angustifolia, E. pallida, and E. Purpurea)," The Longwood Herbal Task Force, pp. 1-24, 1999, http://www.longwoodherbal.org/echinacea/echinacea.pdf.

[12] R. Atencia, F. J. Bustamante, A. Valdivieso et al., "Differential expression of viral PAMP receptors mRNA in peripheral blood of patients with chronic hepatitis C infection," *BMC Infectious Diseases*, vol. 7, article 136, 2007.

[13] C. J. Liou, W. C. Huang, and J. Tseng, "Short-term oral administration of ginseng extract induces type-1 cytokine production," *Immunopharmacology and Immunotoxicology*, vol. 28, no. 2, pp. 227–240, 2006.

[14] Y. Wang, D. Peng, W. Huang, X. Zhou, J. Liu, and Y. Fang, "Mechanism of altered TNF-$\alpha$ expression by macrophage and the modulatory effect of Panax notoginseng saponins in scald mice," *Burns*, vol. 32, no. 7, pp. 846–852, 2006.

[15] M. W. Larsen, C. Moser, N. Høiby, Z. Song, and A. Kharazmi, "Ginseng modulates the immune response by induction of interleukin-12 production," *Acta Pathologica, Microbiologica et Immunologica*, vol. 112, no. 6, pp. 369–373, 2004.

[16] H. C. Lee, R. Vinodhkumar, J. W. Yoon, S. K. Park, C. W. Lee, and H. Y. Kim, "Enhanced inhibitory effect of ultra-fine granules of red ginseng on LPS-induced cytokine expression in the monocyte-derived macrophage THP-1 cells," *International Journal of Molecular Sciences*, vol. 9, no. 8, pp. 1379–1392, 2008.

[17] S. Mocellin, F. Marincola, C. R. Rossi, D. Nitti, and M. Lise, "The multifaceted relationship between IL-10 and adaptive immunity: putting together the pieces of a puzzle," *Cytokine and Growth Factor Reviews*, vol. 15, no. 1, pp. 61–76, 2004.

[18] J. Gertsch, R. Schoop, U. Kuenzle, and A. Suter, "Echinacea alkylamides modulate TNF-$\alpha$ gene expression via cannabinoid receptor CB2 and multiple signal transduction pathways," *FEBS Letters*, vol. 577, no. 3, pp. 563–569, 2004.

[19] S. A. Hwang, A. Dasgupta, and J. K. Actor, "Cytokine production by non-adherent mouse splenocyte cultures to Echinacea extracts," *Clinica Chimica Acta*, vol. 343, no. 1-2, pp. 161–166, 2004.

[20] S. Raduner, A. Majewska, J. Z. Chen et al., "Alkylamides from Echinacea are a new class of cannabinomimetics: cannabinoid type 2 receptor-dependent and -independent immunomodulatory effects," *Journal of Biological Chemistry*, vol. 281, no. 20, pp. 14192–14206, 2006.

[21] P. Guiotto, K. Woelkart, I. Grabnar et al., "Pharmacokinetics and immunomodulatory effects of phytotherapeutic lozenges

(bonbons) with *Echinacea purpurea* extract," *Phytomedicine*, vol. 15, no. 8, pp. 547–554, 2008.

[22] Z. Zhai, D. Haney, L. Wu et al., "Alcohol extracts of Echinacea inhibit production of nitric oxide and tumor necrosis factor-alpha by macrophages in vitro," *Food and Agricultural Immunology*, vol. 18, no. 3-4, pp. 221–236, 2007.

[23] R. K. Randolph, K. Gellenbeck, K. Stonebrook et al., "Regulation of human immune gene expression as influenced by a commercial blended Echinacea product: preliminary studies," *Experimental Biology and Medicine*, vol. 228, no. 9, pp. 1051–1056, 2003.

[24] T. Kanzaki, N. Morisaki, R. Shiina, and Y. Saito, "Role of transforming growth factor-$\beta$ pathway in the mechanism of wound healing by saponin from Ginseng Radix rubra," *British Journal of Pharmacology*, vol. 125, no. 2, pp. 255–262, 1998.

[25] Y. Han, S. J. Son, M. Akhalaia et al., "Modulation of radiation-induced disturbances of antioxidant defense systems by ginsan," *Evidence-based Complementary and Alternative Medicine*, vol. 2, no. 4, pp. 529–536, 2005.

[26] C. M. Chang, A. Limanni, W. H. Baker et al., "Bone marrow and splenic granulocyte-macrophage colony-stimulating factor and transforming growth factor-$\beta$ mRNA levels in irradiated mice," *Blood*, vol. 86, no. 6, pp. 2130–2136, 1995.

[27] M. Martin, J. L. Lefaix, and S. Delanian, "TGF-$\beta$1 and radiation fibrosis: a master switch and a specific therapeutic target?" *International Journal of Radiation Oncology Biology Physics*, vol. 47, no. 2, pp. 277–290, 2000.

[28] H. Wang, J. K. Actor, J. Indrigo, M. Olsen, and A. Dasgupta, "Asian and Siberian ginseng as a potential modulator of immune function: an in vitro cytokine study using mouse macrophages," *Clinica Chimica Acta*, vol. 327, no. 1-2, pp. 123–128, 2003.

[29] E. Rivera, F. E. Pettersson, M. Inganäs, S. Paulie, and K. O. Grönvik, "The Rb1 fraction of ginseng elicits a balanced Th1 and Th2 immune response," *Vaccine*, vol. 23, no. 46-47, pp. 5411–5419, 2005.

[30] J. Y. Song, S. K. Han, E. H. Son, S. N. Pyo, Y. S. Yun, and S. Y. Yi, "Induction of secretory and tumoricidal activities in peritoneal macrophages by ginsan," *International Immunopharmacology*, vol. 2, no. 7, pp. 857–865, 2002.

[31] X. X. Huang, M. Yamashiki, K. Nakatani, T. Nobori, and A. Mase, "Semi-quantitative analysis of cytokine mRNA expression induced by the herbal medicine Sho-saiko-to (TJ-9) using a gel doc system," *Journal of Clinical Laboratory Analysis*, vol. 15, no. 4, pp. 199–209, 2001.

[32] S. J. Kim, H. J. Jeong, B. J. Yi et al., "Transgenic Panax ginseng inhibits the production of TNF-$\alpha$, IL-6, and IL-8 as well as COX-2 expression in human mast cells," *American Journal of Chinese Medicine*, vol. 35, no. 2, pp. 329–339, 2007.

[33] J. A. Rininger, S. Kickner, P. Chigurupati, A. McLean, and Z. Franck, "Immunopharmacological activity of Echinacea preparations following simulated digestion on murine macrophages and human peripheral blood mononuclear cells," *Journal of Leukocyte Biology*, vol. 68, no. 4, pp. 503–510, 2000.

# A Genetic Approach to Spanish Populations of the Threatened *Austropotamobius italicus* Located at Three Different Scenarios

**Beatriz Matallanas, Carmen Callejas, and M. Dolores Ochando**

*Departamento de Genética, Facultad de Ciencias Biológicas, Universidad Complutense de Madrid, C/José Antonio Novais 2, 28040 Madrid, Spain*

Correspondence should be addressed to M. Dolores Ochando, dochando@bio.ucm.es

Academic Editors: N. Kouprina, S. Mastana, and M. Ota

Spanish freshwater ecosystems are suffering great modification and some macroinvertebrates like *Austropotamobius italicus*, the white-clawed crayfish, are threatened. This species was once widely distributed in Spain, but its populations have shown a very strong decline over the last thirty years, due to different factors. Three Spanish populations of this crayfish—from different scenarios—were analysed with nuclear (microsatellites) and mitochondrial markers (*COI* and *16S* rDNA). Data analyses reveal the existence of four haplotypes at mitochondrial level and polymorphism for four microsatellite loci. Despite this genetic variability, bottlenecks were detected in the two natural Spanish populations tested. In addition, the distribution of the mitochondrial haplotypes and SSR alleles show a similar geographic pattern and the genetic differentiation between these samples is mainly due to genetic drift. Given the current risk status of the species across its range, this diversity offers some hope for the species from a management point of view.

## 1. Introduction

Spain is the country with the greatest biodiversity of Europe, with around 80,000 catalogued taxa. The maritime barrier of the Mediterranean, the land barrier of the Pyrenees in the North, and the country's orographic and climate peculiarities, invest it with unique biogeographic characteristics. Therefore, the country's large number of endemic—specially freshwater—species makes it a biodiversity hot spot [1].

At present, Spanish freshwater ecosystems are suffering great modification at the hands of climate change, environmental degradation, habitat fragmentation, the rise in human demand for water, and a range of human activities. Together, these factors have contributed to a notable increase in the size of Spain's arid and semiarid regions, and to changes in its biodiversity [2].

In 2008 more than 80% of Spanish endemisms were reported to suffer some level of threat in the IUCN Red List. At present, the assessments for some of these species have got worse. Among the macroinvertebrates, the crayfish *Austropotamobius italicus* was listed as vulnerable in 2008, but in 2010 it has been categorized as endangered [3].

*Austropotamobius italicus* was once a cornerstone of Iberian freshwater ecosystems with large populations widely distributed throughout most of the country's limestone basins. Indeed, it was absent in the more western areas, the highest mountain ranges, and the subdesert areas of the southeast and River Ebro valley. The dramatic decline in its numbers all over its Spanish range is the result of a combination of the factors mentioned above, as well as of the introduction of exotic crayfish species and the related spread of crayfish plague (caused by the fungus *Aphanomyces astaci*). As a consequence, only around 1000 small populations now remain in Spain (Alonso, pers. com.) occupying marginal areas or short stretches of watercourses usually isolated from the main river systems [4].

At present, restoration programs are based mainly on translocation of individuals from other natural or farmed populations and are limited by the low number and abundance of existing populations. Besides, it should be taken

TABLE 1: Collection sites and genetic variability of *A. italicus* populations studied in the present work. Columns show, respectively, Code, Population, Watershed, Drainage Direction, and Collection sites they come from, H: hatchery. S: number of segregating sites; h: number of mtDNA haplotypes found; Hd and $\pi$: haplotype and nucleotide diversity, respectively. Last four columns refer to relative frequencies of haplotypes found in each population.

| | Code | Population | Watershed/drainage direction | Collection sites | S | h | Hd | $\pi$ | Relative frequencies | | | |
|---|---|---|---|---|---|---|---|---|---|---|---|---|
| | | | | | | | | | Hap_1 | Hap_2 | Hap_3 | Hap_4 |
| 1 | GRA | Brook Ermitas | Guadalquivir/Atlantic (West) | Albuñuelas/Granada | 0 | 1 | 0 | 0 | 1 | 0 | 0 | 0 |
| 2 | NAV | River Ega | Ebro/Mediterranean (East) | Estella/Navarra | 1 | 2 | 0.2 | 0.00008 | 0 | 0.9 | 0 | 0.1 |
| 3 | RIL | River Gallo | Crayfish hatchery, (H) | Guadalajara/ Castilla—La Mancha | 1 | 2 | 0.556 | 0.0022 | 0.5 | 0 | 0.5 | 0 |

into account that availability of individuals for restocking purposes needs to be substantially increased by either traditional hatcheries or extensive ponds [5]. These action plans consider several factors such as the risk of transmission of crayfish plague, the risk of survival when establishing new populations, the characteristics of water bodies to be restored, or the distribution of exotic species in those areas [5, 6].

Notwithstanding, a major goal of such programs should also preserve genetic variability—the basis for viability and future evolution of populations [7, 8]. Indeed, knowledge of the levels and patterns of distribution of the genetic diversity is critical when making conservation management decisions. In other words, effective long-term conservation planning must incorporate genetic information [9] because the loss of genetic variation and inbreeding depression put wildlife populations at an increased risk [10].

In this context, our group is conducting a comprehensive study on the genetic variation and its distribution in *Austropotamobius italicus* populations from Spain. Our previous survey, by random amplified polymorphic DNA (RAPD) fingerprinting, detected a certain degree of polymorphism in some of the populations tested [11].

Three molecular markers were used in the present study, two of them mitochondrial and the other one, nuclear. This approach combines the advantages of both methods. It is clear that the mitochondrial genome of animals is an excellent target for genetic analysis because of its lack of introns, its limited exposure to recombination, and its haploid mode of inheritance [12]. Mitochondrial DNA (mtDNA) has proved to be powerful for genealogical and evolutionary studies of animal populations. Otherwise, microsatellite loci are highly polymorphic markers, distributed throughout the nuclear genome and generally not linked to loci under strong selection [13]. These codominant markers have revealed substantial variation in species with low variability in other nuclear markers [14] and have been used to study genetic differentiation among closely related populations [15].

Taking into account all above, our aim was to study the genetic variability of three Spanish populations of white-clawed crayfish belonging to three different scenarios—protected area, crashed population, and hatchery.

## 2. Material and Methods

*2.1. Samples.* A total of 45 individuals of *Austropotamobius italicus* were collected from three different populations (Table 1). One of them, located in a protected area: NAV, a native population. A second population, RIL, from a crayfish hatchery, maintained with a high effective number. Thirdly, GRA also a native population whose number crashed during the 1990s due to several pathologies.

For each population ten individuals were studied employing two mitochondrial markers: cytochrome oxidase subunit I (*COI*) and *16S* rDNA gene. For SSR analysis fifteen individuals from each of the three populations sampled were used.

*2.2. DNA Isolation, Amplification, and Sequencing.* Genomic DNA was extracted from 20–50 mg of claw muscle or peri-opod tissues (without killing the animal) using the DNeasy Blood and Tissue Kit from Qiagen (Valencia, CA, USA) and resuspended in Tris-EDTA (10 mM; 1 mM; pH 8.0).

DNA concentration and purity were estimated by absorbance at 260–280 nm in a NanoDrop ND-100 (NanoDrop Technologies, Wilmintong, USA) spectrophotometer. Its integrity was verified by 0.8% agarose gels in Tris-EDTA buffer (10 mM; 1 mM; pH8). Gels were stained with ethidium bromide (1 $\mu$g/mL) and visualized with UV light trans-illuminator. All the polymerase chain reactions (PCR, [16]) were carried out using a Lab Cycler (SensoQuest, Göttingen, Germany).

A fragment from the mtDNA *COI* gene was amplified in a final volume of 50 $\mu$L with 25 ng of total DNA, 1x reaction buffer, 2 mM $MgSO_4$, 200 $\mu$M of each dNTP, 15 $\mu$g of BSA, 1 $\mu$M of each primer, and 1 U of Vent DNA polymerase (New England Biolabs, Ipswich, MA, USA). The primers used were *COI* Scylla [17] and LCO [18]. The optimal PCR programme include an initial denaturation step of 94°C for 5 min followed by 44 cycles of 94°C for 45 s, 53°C for 1 min, and 72°C for 1 min 30 s, and a final extension step of 72°C for 10 min.

The selective amplification of a segment of the mtDNA *16S* rDNA was performed with the primers 1472 and Tor12sc [19], applying the following PCR conditions: an initial denaturation step of 95°C for 2 min followed by 9 cycles of 95°C for 25 s, 57°C for 30 s, and 72°C for 150 s, then 29 cycles of 95°C for 35 s, 54°C for 30 s, and 72°C for 150 s, finally an extension step of 72°C for 10 min. Each reaction, with a final volume of 50 $\mu$L, contained 25 ng of total DNA, 1x reaction buffer, 2.2 mM $MgCl_2$, 200 $\mu$M of each dNTP, 330 $\mu$g of BSA, 0.44 $\mu$M of each primer, and 1 U of AmpliTaq DNA polymerase (Applied Biosystems, USA).

Double-stranded amplified products for both mitochondrial markers were purified with the High Pure PCR Product Purification Kit (Boehringer-Manheim) and used as templates for sequencing reactions. These reactions were carried out with the "BIG Dye Terminator Cycle Sequencing Ready Reaction Kit" (Applied Biosystems, Inc., USA) on a 3730 DNA Analyzer (Applied Biosystems, Inc., USA), using the primers employed for the amplification step, at the Genomic Unit of The Complutense University of Madrid.

The SSR study included five loci. All primers used were developed by Gouin et al.: Ap1, Ap2, Ap3, Ap5 reverse, Ap6 [20], and Ap5 forward [21]. SSR loci were amplified using a forward labelled primer with one of the Applied Biosystems fluorochromes 6-FAM, PET or VIC.

QIAGEN Multiplex PCR kit (Qiagen, Hilden, Germany) was used to amplify the Ap1, Ap2, Ap3, and Ap6 SSR loci. Reaction with a final volume of 6.5 $\mu$L contained 5 ng of total DNA, 3.25 $\mu$L of 2x QIAGEN Multiplex PCR Master Mix, 0.1 $\mu$M of Ap1 and Ap6 primers, and 1 $\mu$M of Ap2 and Ap3 primers. Amplification conditions were 95°C for 10 min followed by 36 cycles of 94°C for 30 s, 62°C for 60 s, and 72°C for 60 s, with an extension final step of 72°C for 10 min. Ap5 locus was amplified in a final volume of 6.5 $\mu$L with 35 ng of total DNA, 1x reaction buffer, 1.9 mM $MgCl_2$, 8 $\mu$M of each dNTP, 0.2 $\mu$M of forward labelled primer, 0.4 $\mu$M of reverse primer, and 0.5 U of AmpliTaq DNA polymerase (Applied Biosystems, USA). The optimal PCR programme for this locus include an initial denaturation step of 94°C for 10 min followed by 15 cycles of 94°C for 30 s, 55°C for 35 s, and 72°C for 50 s, then 25 cycles of 94°C for 30 s, 60°C for 35 s and 72°C for 50 s, finally an extension step of 72°C for 10 min.

PCR products were run with the internal size standard GeneScan 500 LIZ (Applied Biosystems, USA) on a 3730 DNA Analyser (Applied Biosystems, USA). Allele size was determined through Peak Scanner Software v1.0 (Applied Biosystems, USA).

*2.3. mtDNA Alignment and Sequence Analysis.* The nucleotide sequences of the mitochondrial DNA were aligned using CLUSTAL W software [22] and edited with BioEdit v 7.0.9.0 [23]. After alignment and edition, amplified fragments, *16S* (1317 bp) and *COI* (1184 bp), were also used together to obtain a single sequence of 2501 bp length. The genetic diversity estimates (haplotype diversity, H; nucleotide diversity, $\pi$; number of segregating sites, S) were calculated using DnaSP v 5.10.01 programme [24].

$F_{ST}$ pairwise genetic distances [25], which quantify how genetic diversity is partitioned within and between populations, and gene flow (Nm), were estimated through DnaSP v 5.10.01 software package [24]. Principal component analysis (PCA) [26] was performing using NTSYSpc v2.10q software package [27] to visualize the grouping populations. Finally, haplotype frequencies for each mitochondrial gene were geographically depicted for each population using PhyloGeoViz v 2.4.4 [28].

*2.4. Microsatellite Analysis.* Genetic diversity was quantified as the mean number of observed alleles per locus ($A$), effective number of alleles per locus ($n_e$), the observed heterozygosity ($H_o$) and the Hardy-Weinberg expected heterozygosity ($H_e$) [29] through Popgene software [30]. Genepop v4 [31] was employed to estimate deviations from Hardy-Weinberg equilibrium across populations and across loci using the Markov chain method (10000 iterations). The software also tested the linkage disequilibrium across all populations and estimated the inbreeding coefficient $F_{IS}$ within each population by the Weir and Cockerham [32] method.

To assess the effects of genetic drift and mutation in the structure of these populations, two statistics of genetic differentiation, $F_{ST}$ [32] and $Rho_{ST}$ [33], were calculated (Genepop software). $R$-statistics were expected to be larger than $F$-statistics when stepwise-like mutations have contributed to population differentiation [34]. Otherwise, if both statistics are similar, genetic drift is considered the main force for genetic differentiation. Wilcoxon's signed ranked test was performed to assess differences between $F_{ST}$ and $Rho_{ST}$ estimates.

Genetic structure of populations was inferred using the model-based clustering algorithms implemented in STRUCTURE v2.0 [35]. For parameter estimations, the admixture model with correlated allele frequencies was used (with 100000 MCMC iterations of burn-in length and 100000 after-burning repetitions).

In order to analyse the homogeneity of samples, a correspondence analysis (CA)—on the matrix of allele counts per sample, both at the population and the individual levels—was conducted using Genetix v4.05.2 software [36]. This technique is especially useful when the number of available loci is limited.

## 3. Results

*3.1. mtDNA Analysis.* A 2501 bp (1184-nucleotide sequence from *COI* gene and 1317 bp from *16S* rDNA gene) fragment was obtained from 30 individuals. Sequence analysis revealed four single nucleotide polymorphisms (SNPs), three of them informative under parsimony (Table 2). Four haplotypes were identified, three at high or intermediate frequencies (Haplotypes 1–3) and the remaining one (Hap_4), at low frequency. Likewise, Haplotypes 1 and 2, both differing in a transition in position 1536, account for 80% of individuals (Table 2).

Regarding the populations, NAV and RIL show genetic diversity at mtDNA gene level since two different haplotypes were detected in each sample (Hap_1 and Hap_3 in RIL population and Hap_2 and Hap_4 in NAV sample). The

Table 2: Haplotypes found in *A. italicus* Spanish populations. Columns 2–5 refer to position of the SNP in the 2501 nt sequence. Bold numbers—first line—: SNP informative under parsimony. Freq: frequency in percentage of each haplotype in the total of individuals analysed. Num. Indiv.: number of individuals represented by each haplotype. Num. pop: number of populations in which they were detected.

| | **531** | **1536** | 2071 | **2319** | Freq% | Num. Indiv | Num. Pop |
|---|---|---|---|---|---|---|---|
| Hap_1 | C | T | A | A | 50 | 15 | 2 |
| Hap_2 | C | C | A | A | 30 | 9 | 1 |
| Hap_3 | T | C | C | G | 16.67 | 5 | 1 |
| Hap_4 | T | C | A | G | 3.33 | 1 | 1 |

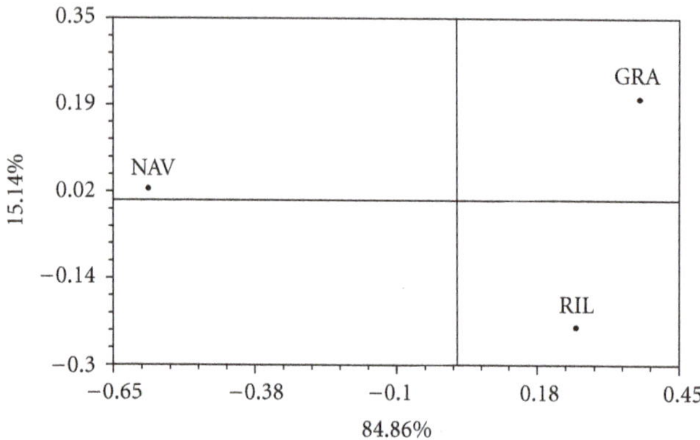


Figure 1: Results of the PCA analysis based on mtDNA sequences from three samples of *A. italicus*. Eigenvalues for each principal component are listed besides each axis.

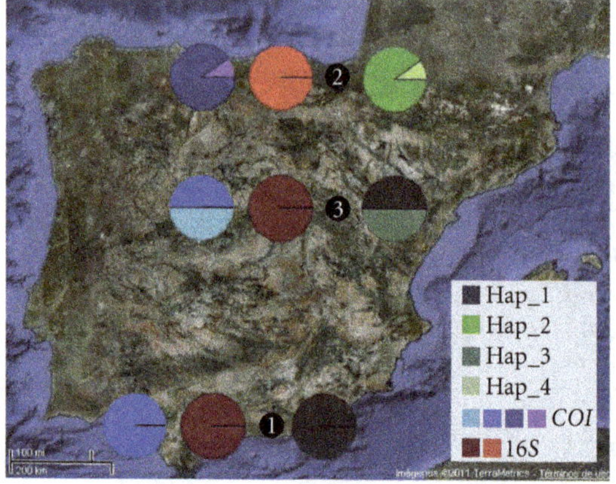


Figure 2: Geographic localization and genetic composition of 30 individuals from 3 different populations. Each population is shown as 3 pie charts representing membership proportions. In blue series: *COI* mtDNA gene, red series: *16S* mtDNA gene and green series as total mtDNA.

highest haplotype and nucleotide diversity were found in the farmed population (RIL) (Table 1).

As a whole, $F_{ST}$ value revealed significant population differentiation at the mtDNA level ($F_{ST} = 0.8547$). The inferred Nm value was 0.09 that reduces to 0.03 when only the two natural populations are taking into account. The highest $F_{ST}$ genetic distances were found between comparisons of NAV population and the other two samples ($F_{ST\ NAV\text{-}GRA} = 0.952$, $F_{ST\ NAV\text{-}RIL} = 0.855$).

Relationships among these three populations were visualized by the principal component analysis (Figure 1). The first PCA axis explains 84.86% of variance and reveals two well-separated groups: NAV (mainly Hap_2) and, GRA and RIL (mostly Hap_1). The second PCA axis explains 15.14% and disjoined populations with Hap_1 into two different groups: GRA (Hap_1) and RIL (Haplotypes 1 and 3).

As shown in Figure 2, haplotypes found were not evenly distributed across samples. At *COI* level, GRA and RIL samples shared one of the haplotypes found, although GRA was monomorphic whereas RIL also presented a private haplotype. In addition, NAV population held two different and exclusive haplotypes. Nonetheless, only two different groups were observed at rDNA *16S* gene, since a single mutation separated the two haplotypes found. Thus, GRA and RIL shared the same haplotype while NAV sample held the other one.

*3.2. Microsatellite Analysis.* A total of 45 individuals were analysed through five SSR loci. Four of which were polymorphic—Ap1, Ap2, Ap3, and Ap6—whereas locus Ap5 was monomorphic.

The total number of alleles detected for the three Spanish populations was 33. All samples had private alleles—GRA: Ap2: 190, 196, and 200; Ap3: 126, and 192; Ap6: 352, 354, 356 and 368. NAV: Ap3: 178, and 188. RIL: Ap2: 108; Ap3: 150; Ap6: 362 and 380—usually at low frequencies. The number of alleles per locus ranged from 1 (locus Ap5) to 8 (locus Ap6) and their frequencies are shown in Figure 3.

Parameters of genetic diversity are displayed in Table 3. GRA population had the highest values for the mean observed allelic diversity per locus ($A$) and the effective number of alleles ($n_e$) while NAV sample showed the lowest ones. In all populations, the average observed heterozygosity ($H_o$) was lower than average expected heterozygosity ($H_e$), mainly in NAV sample where a large homozygote excess was observed. The $F_{IS}$ values, as expected, confirm the results ($F_{IS}$ values, Table 3).

Significant deviations for Hardy-Weinberg expectations were found for Ap2 and Ap3 in all populations, as well as for Ap6 locus, excepting RIL sample. No significant linkage disequilibrium was detected between pairs of loci in these populations.

Clustering analysis by STRUCTURE (Figure 4) revealed that the three geographic groups (GRA, NAV, and RIL) indeed represented four genetically distinct populations.

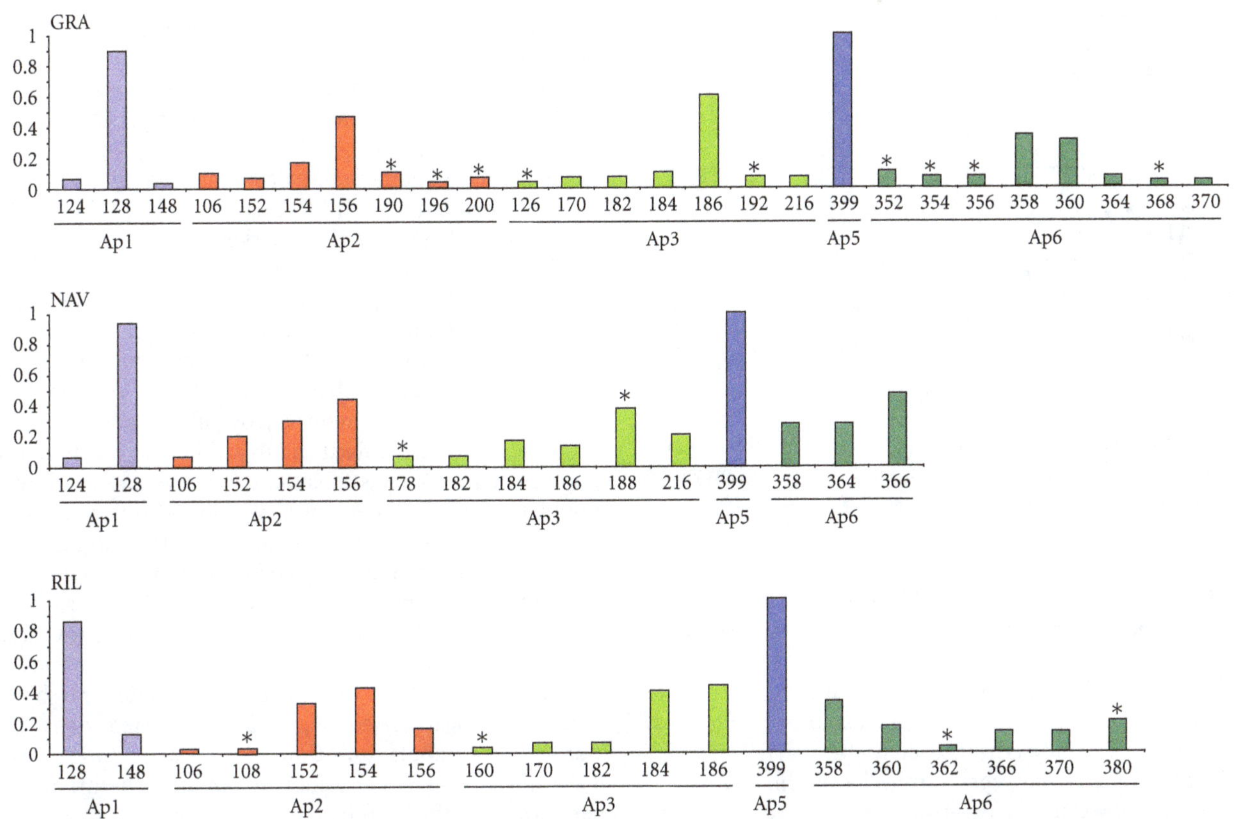


FIGURE 3: Allele frequencies in the populations of *A. italicus* for each microsatellite locus. Asterisk indicates private alleles.

TABLE 3: Genetic variation for five microsatellite loci in three Spanish populations of *A. italicus*. % HW: % loci in Hardy-Weinberg equilibrium; *A*: mean number of observed alleles; $n_e$: effective number of alleles; $H_o$: observed mean heterozygosity; $H_e$: expected mean heterozygosity; $F_{IS}$: inbreeding coefficient.

| Population | Code | Parameter | | | | | |
|---|---|---|---|---|---|---|---|
| | | % HW | $A$ | $n_e$ | $H_o$ | $H_e$ | $F_{IS}$ |
| GRA | 1 | 20 | 5.2000 | 2.5677 | 0.2133 | 0.4745 | 0.5591 |
| NAV | 2 | 20 | 3.2000 | 2.4785 | 0.0800 | 0.4579 | 0.8303 |
| RIL | 3 | 40 | 3.8000 | 2.5551 | 0.2267 | 0.4818 | 0.5383 |

The correspondence analysis (CA, Figure 5) agreed with STRUCTURE analysis. The first axe (70% of the inertia) clearly separated NAV sample from the remaining two studied populations, as the mtDNA markers did. The second axis (around 30% of the inertia) disjoined farmed population (RIL) from GRA, although a narrow area exists where individuals belonging to these two populations overlap.

According to Wilcoxon's signed ranks test, Rho$_{ST}$ and $F_{ST}$ values were similar ($P > 0.05$). The highest $F_{ST}$ genetic distance was found between GRA and NAV samples ($F_{ST} = 0.1030$) and the lowest between GRA and RIL populations ($F_{ST} = 0.0518$).

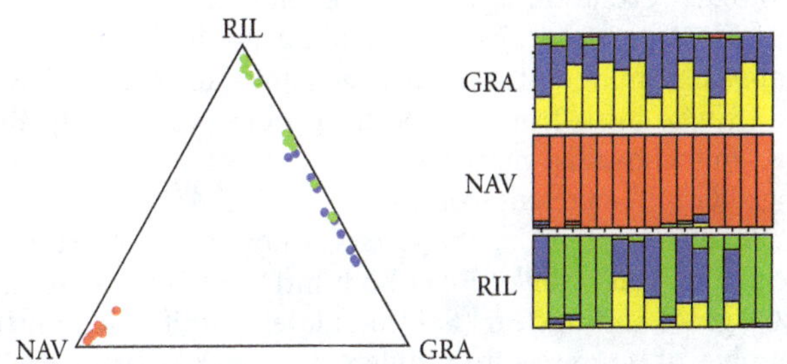


FIGURE 4: Summary of the clustering results for the *A. italicus* data assuming three populations. Each point shows the mean estimated ancestry for an individual in the sample. For a given individual, the values of the three coefficients in the ancestry vector $q_{(i)}$ were given by the distances to each of the three sides of the equilateral triangle.

## 4. Discussion

The main goal of the present work was the study of the genetic variability of three Spanish samples (Table 1)—located at distinct scenarios—of white-clawed crayfish, a threatened freshwater species in the Iberian Peninsula. Since the genetic diversity of a species reflects the influence

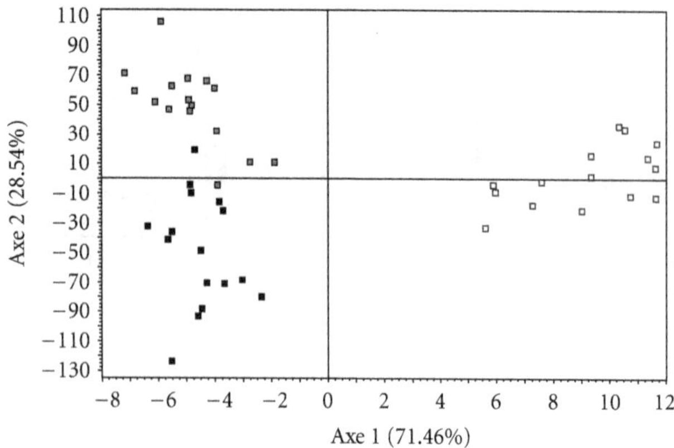


FIGURE 5: Correspondence analysis: projection of the individuals on the plane defined by the first two axes. Black, white, and grey dots codifying for GRA, NAV, and RIL populations, respectively.

of both historical and recent evolutionary events, a double approach was used to perform this task. On one hand, most phylogeographic studies of animals have relied on the analysis of mtDNA sequence variation due to its unique attributes and the different mutation rates compared to most nuclear genes. Its analysis has proven useful in defining major phylogeographic assemblages within species, including the European freshwater species—complex of *Austropotamobius* [37–39]. On the other hand, SSR nuclear loci have high mutation rates and tend to recover genetic variability quickly after the action of processes that affect it negatively. Thus, the molecular footprints on these SSR loci should be less long standing than in mitochondrial genome [40]. In this way, the SSR markers have been useful for addressing questions relating to current population structure of many freshwater species [41], including crayfish [42].

With respect to both analysed mtDNA sequences, *COI* gene resulted more sensitive for detecting genetic variability than *16S* rDNA. *COI* is a powerful marker for the study of the genetic variation at the intraspecific level in crayfish [43, 44] and other crustaceans [45–47] because its rate of molecular evolution is about threefold greater than that of *16S* rDNA gene [48]. Notwithstanding, given that the entire nonrecombining mitochondrial genome can be considered as a single locus from a genealogical perspective [49], the two mitochondrial markers—*COI* and *16S*—are discussed together in the present work.

Results indicate that the species exhibits in Spain a certain degree of genetic diversity at both mtDNA (*16S* rDNA and *COI* gene) and nuclear (SSR loci) level, despite the limited number of samples—from different scenarios—analysed in this study.

Four different mitochondrial haplotypes (Table 2) have been found, while for years the lack of genetic variability at mitochondrial level in Spanish populations of *A. italicus* was an accepted hypothesis [38, 50]. The degree of genetic diversity (Hd, $\pi$, Table 1) is higher than the previously reported for Iberian crayfish [37, 39] and similar, or even higher, to values obtained for others European populations with mitochondrial markers [40, 51]. Concerning microsatellite loci, four out of five markers tested were polymorphic, though most alleles showed low frequencies. The heterozygosity was relatively low albeit the mean number of alleles per locus detected was higher compared with other European populations [42, 52, 53] (Table 3, Figure 3). The existence of genetic variability in Spanish populations of this crayfish corroborates our former surveys, also at nuclear level, trough RAPD and ISSR markers [2, 11].

As shown in Figure 1, evidence for genetic differentiation among these three populations at mitochondrial level occurs. The distribution of mitochondrial haplotypes shows a clear geographic pattern (Figure 2) where Northern Spanish population do not share the haplotypes present in the Southern one, according to the reported by other authors [37, 39, 54]. The existence of fixed mtDNA differences between the populations analysed is consistent with severe restrictions on population size and geographic isolation. Given that mtDNA contains about one-fourth of the genetic variation included in the nuclear genome, large portions of the haplotypes can be wiped out during bottleneck events [55, 56].

Data from SSR analysis seem to bear out the above pattern. Although a certain degree of genetic variability was found, the four polymorphic SSR markers show a single allele nearly fixed (Figure 3) as expected for recently bottlenecked populations. Comparisons between $Rho_{ST}$ and $F_{ST}$ values indicate that genetic differentiation among these samples can be attributed to genetic drift. As shown in the STRUCTURE and correspondence analyses (Figures 4 and 5), the geographic labels matched very closely to the genetic clustering. The Northern population (NAV) is clearly separated from the other two. These analyses also highlight that some specimens from different populations are genetically similar at nuclear level, since some individuals from RIL sample are included in GRA cluster (Figure 4). The close relationship between GRA and RIL (hatchery) specimens could be explained by human translocations—a common practice in Spain since the 19th century—from GRA area to the source populations that gave rise to the current farmed sample or by the existence of a common ancestor population with a wider distribution in the far past.

Focusing on the Southern population, GRA shows a unique mitochondrial haplotype and a single allele per SSR locus at high or very high frequency, whereas the other ones remain at low frequencies (Table 3). The effective number of SSR alleles was also lower than expected, indicating a homozygote excess in the sample. The single haplotype found at mtDNA level suggest a strong bottleneck, proposed by some authors for this species [49, 50] around the last glaciation. In addition, SSR outcomes reveal a decline in population size recently. As a matter of fact, in the last two decades this population had suffered a drastic regression in number due to high mortalities caused by *Saprolegnia* spp., severe climatic droughts during the 1990s and a big flood at 1997 [57]. The genetic drift caused by bottlenecks was intensified, thus some haplotypes/alleles have been eventually lost while others became fixed according to its frequency [58]. Notwithstanding the presence of private SSR alleles is indicative of a recent increasing in GRA population size. The Andalusian Regional Government started at 2002 a conservation and management program of the white-clawed crayfish [59] that contemplates the development of emergency plans for drought and disease among other measures. Though the GRA sample is not directly situated in a protected area, possibly this population is benefiting from the strategies adopted in surrounding areas, including control/eradication of alien species.

The Northern population (NAV), although located in a protected area since 1996, showed the lowest variability at number of alleles per locus and exclusive alleles (Figure 3). These results indicate a recent and strong bottleneck since allelic diversity seems to be one of the most sensitive methods for detecting relatively recent demographic bottlenecks [13]. It is because rare alleles, which contribute little to the average heterozygosity, are easily lost during population size constrictions [58, 60]. In addition, the rate of inbreeding, unavoidable in small populations, is not negligible ($F_{IS}$ = 0.8303, Table 3). Inbreeding reduces fitness [61] and usually enhances susceptibility to infectious diseases [62]. It is known that late in 20th century, high mortalities occurred in this region mainly due to the crayfish plague caused by *Aphanomyces astaci* [63]. Although two haplotypes were found, the mitochondrial analysis also supports the existence of a more ancient bottleneck because the 90% of crayfish shared the same mtDNA haplotype (Tables 1 and 2). NAV sample comes from a waterbody where *Pacifastacus leniusculus* also inhabits [64, 65] so, competition for space and diseases such as aphanomycosis could be some of the causes of the low genetic variability found in this population. Thus, genetic studies are necessary to ensure this species' future even in protected areas and to guarantee the existence of suitable levels of genetic variability in crayfish populations.

The hatchery population (RIL) exhibits the highest genetic diversity at mitochondrial level and moderate values for all the SSR variability parameters (Tables 1 and 3). The variability found may be explained by the fact that it was established in the 1980s with crayfishes from distinct basins and since then, it was kept under favourable conditions which have allowed to maintain a high population density [11, 66].

At present, Spanish populations of white-clawed crayfish are in regression due to environmental changes among other factors [2, 5, 67]. The current levels of genetic variability of *A. italicus* in Spain are affected by successive and drastic bottlenecks and consequently, by the action of the genetic drift, enhanced in these small and fragmented populations. However, ancient historical events such as population fragmentation, recolonizations from refugia during the ice ages [37, 38, 50, 68], or the formation of fluvial basins, must also have influenced their present structure, as well as it has been demonstrated in other species [69–71].

Our results underscore the usefulness of employing both mitochondrial and nuclear markers to assess current levels of genetic variability of the populations analysed as well as their genetic structure. Genetic information of the present study should be taken into account for future conservation plans. Though Spanish populations are in decline, a certain degree of genetic diversity has been detected. Given the pattern of the genetic variability found, it would be advisable an increase of within-population heterozygosity without eroding the differentiations that characterize the genetic structure of these Spanish samples. In this way, future works with more samples are needed in order to confirm these results and to provide guidelines about restocking purposes in each area. Hence, the crayfish hatchery analysed in this study (RIL) could be suitable for restocking the Southern population (GRA) but it is not fitted for the Northern one (NAV) that has to be considered as a different management unit [72], given its particular genetic characteristics.

## Acknowledgments

Thanks are due to the following people for help in sampling the populations: Mari Cruz Cano, Javier Diéguez-Uribeondo, José María Gil, and José Luis Múzquiz. This work was funded by the "Convenio de Colaboración" (no. 4152634) between the Ministry of Medio Ambiente y Medio Rural y Marino and The Complutense University.

## References

[1] R. A. Mittermeier, P. Robles, M. Hoffman et al., *Hotspots Revisited: Earth's Biologically Richest and Most Endangered Terrestrial Ecoregions*, Conservation International, Washington, DC, USA, 2005.

[2] C. Callejas, B. Beroiz, F. Alonso, A. Vivero, B. Matallanas, and M. D. Ochando, "Preserving the biodiversity of freshwater ecosystems in a scenario of increasing desertification: lesson from genetics," in *Handbook of Environmental Research*, A. Edelstein, Ed., pp. 261–291, Nova Science Publishers, 2009.

[3] L. Füreder, F. Gherardi, D. Holdich, J. Reynolds, P. Sibley, and C. Souty-Grosset, "*Austropotamobius pallipes*," in *IUCN 2010: IUCN Red List of Threatened Species. Version 2010.4*, 2010, http://www.iucnredlist.org/.

[4] R. Martinez, E. Rico, and F. Alonso, "Characterisation of *Austropotamobius italicus* (Faxon 1914) populations in a Central Spain area," *Buletin Francais de la Pêche et la Pisciculture*, vol. 370-371, pp. 43–56, 2003.

[5] F. Alonso, C. Temiño, and J. Diéguez-Uribeondo, "Status of the white-clawed crayfish, *Austropotamobius pallipes* (Lereboullet, 1858), in Spain: distribution and legislation," *BFPP: Bulletin Francais de la Peche et de la Protection des Milieux Aquatiques*, vol. 73, no. 356, pp. 31–54, 2000.

[6] J. Diéguez-Uribeondo, A. Rueda, E. Castien, and J. C. Bascones, "A plan of restoration in Navarra for the native freshwater crayfish species of Spain, *Austropotamobius pallipes*," *BFPP: Bulletin Francais de la Peche et de la Protection des Milieux Aquatiques*, vol. 70, no. 347, pp. 625–637, 1997.

[7] J. Avise, *Phylogeography: The History and Formation of Species*, Harvard University Press, Cambridge, Mass, USA, 2000.

[8] C. Moritz, K. McGuigan, and L. Bernatchez, "Conservation of freshwater fishes: integrating evolution and genetics with ecology," in *Conservation of Freshwater Fishes: Options for the Future*, M. J. Collares-Pereira, I. G. Cowx, and M. M. Coelho, Eds., pp. 293–310, Blackwell Science, Oxford, UK, 2002.

[9] C. Souty-Grosset, F. Grandjean, R. Raimond, M. Frelon, C. Debenest, and M. Bramard, "Conservation genetics of the white-clawed crayfish *Austropotamobius pallipes*: the usefulness of the mitochondrial DNA marker," *BFPP: Bulletin Francais de la Peche et de la Protection des Milieux Aquatiques*, vol. 70, no. 347, pp. 677–692, 1997.

[10] D. H. Reed and R. Frankham, "Correlation between fitness and genetic diversity," *Conservation Biology*, vol. 17, no. 1, pp. 230–237, 2003.

[11] B. Beroiz, C. Callejas, F. Alonso, and M. D. Ochando, "Genetic structure of Spanish white-clawed crayfish (*Austropotamobius pallipes*) populations as determined by RAPD analysis: reasons for optimism," *Aquatic Conservation*, vol. 18, no. 2, pp. 190–201, 2008.

[12] C. Saccone, C. De Giorgi, C. Gissi, G. Pesole, and A. Reyes, "Evolutionary genomics in Metazoa: the mitochondrial DNA as a model system," *Gene*, vol. 238, no. 1, pp. 195–209, 1999.

[13] M. Pascual, C. F. Aquadro, V. Soto, and L. Serra, "Microsatellite variation in colonizing and palearctic populations of *Drosophila subobscura*," *Molecular Biology and Evolution*, vol. 18, no. 5, pp. 731–740, 2001.

[14] S. J. Goodman, "Patterns of extensive genetic differentiation and variation among European harbor seals (*Phoca vitulina vitulina*) revealed using microsatellite DNA polymorphisms," *Molecular Biology and Evolution*, vol. 15, no. 2, pp. 104–118, 1998.

[15] M. T. Koskinen, J. Nilsson, A. J. Veselov, A. G. Potutkin, E. Ranta, and C. R. Primmer, "Microsatellite data resolve phylogeographic patterns in European grayling, *Thymallus thymallus*, Salmonidae," *Heredity*, vol. 88, no. 5, pp. 391–401, 2002.

[16] K. Mullis, F. Faloona, S. Scharf, R. K. Saiki, G. Horn, and H. Erlich, "Specific enzymatic amplification of DNA in vitro: the polymerase chain reaction," *Cold Spring Harbor Symposia on Quantitative Biology*, vol. 51, no. 1, pp. 263–273, 1986.

[17] D. Gopurenko, J. M. Hughes, and C. P. Keenan, "Mitochondrial DNA evidence for rapid colonisation of the Indo-West Pacific by the mudcrab *Scylla serrata*," *Marine Biology*, vol. 134, no. 2, pp. 227–233, 1999.

[18] O. Folmer, M. Black, W. Hoeh, R. Lutz, and R. Vrijenhoek, "DNA primers for amplification of mitochondrial cytochrome c oxidase subunit I from diverse metazoan invertebrates," *Molecular Marine Biology and Biotechnology*, vol. 3, no. 5, pp. 294–299, 1994.

[19] C. R. Largiadèr, F. Herger, M. Lörtscher, and A. Scholl, "Assessment of natural and artificial propagation of the white-clawed crayfish (*Austropotamobius pallipes* species complex) in the Alpine region with nuclear and mitochondrial markers," *Molecular Ecology*, vol. 9, no. 1, pp. 25–37, 2000.

[20] N. Gouin, F. Grandjean, and C. Souty-Grosset, "Characterization of microsatellite loci in the endangered freshwater crayfish *Austropotamobius pallipes* (Astacidae) and their potential use in other decapods," *Molecular Ecology*, vol. 9, no. 5, pp. 636–637, 2000.

[21] N. Gouin, C. Souty-Grosset, A. Ropiquet, and F. Grandjean, "High dispersal ability of *Austropotamobius pallipes* revealed by microsatellite markers in a French brook," *BFPP: Bulletin Francais de la Peche et de la Protection des Milieux Aquatiques*, no. 367, pp. 681–689, 2002.

[22] J. D. Thompson, D. G. Higgins, and T. J. Gibson, "CLUSTAL W: improving the sensitivity of progressive multiple sequence alignment through sequence weighting, position-specific gap penalties and weight matrix choice," *Nucleic Acids Research*, vol. 22, no. 22, pp. 4673–4680, 1994.

[23] T. A. Hall, "BioEdit: a user-friendly biological sequence alignment editor and analysis program for Windows 95/98/NT," *Nucleic Acids Symposium Series*, vol. 41, pp. 95–98, 1999.

[24] P. Librado and J. Rozas, "DnaSP v5: a software for comprehensive analysis of DNA polymorphism data," *Bioinformatics*, vol. 25, no. 11, pp. 1451–1452, 2009.

[25] R. R. Hudson, M. Slatkin, and W. P. Maddison, "Estimation of levels of gene flow from DNA sequence data," *Genetics*, vol. 132, no. 2, pp. 583–589, 1992.

[26] P. H. A. Sneath and R. R. Sokal, *Numerical Taxonomy: The Principles and Practice of Numerical Classification*, Freeman, San Francisco, Calif, USA, 1973.

[27] F. J. Rohlf, *NTSYSpc: Numerical Taxonomy System, Ver. 2.10q*, Exeter, Setauket, NY, USA, 2000.

[28] Y. H. E. Tsai, "PhyloGeoViz: a web-based program that visualizes genetic data on maps," *Molecular Ecology Resources*, vol. 11, no. 3, pp. 557–561, 2011.

[29] M. Nei, "Estimation of average heterozygosity and genetic distance from a small number of individuals," *Genetics*, vol. 89, no. 3, pp. 583–590, 1978.

[30] F. C. Yeh, R. C. Yang, T. B. J. Boyle, Z. H. Ye, and J. X. Mao, *POPGENE: The User-Friendly Shareware for Population Genetic Analysis*, Molecular Biology and Biotechnology Centre, University of Alberta, Alberta, Canada, 1997.

[31] M. Raymond and F. Rousset, "GENEPOP (version 1.2): population genetics software for exact tests and ecumenicism," *Journal of Heredity*, vol. 86, pp. 248–249, 1995.

[32] B. S. Weir and C. C. Cockerham, "Estimating F-statistics for the analysis of population structure," *Evolution*, vol. 38, no. 6, pp. 1358–1370, 1984.

[33] Y. Michalakis and L. Excoffier, "A generic estimation of population subdivision using distances between alleles with special reference for microsatellite loci," *Genetics*, vol. 142, no. 3, pp. 1061–1064, 1996.

[34] M. Slatkin, "A measure of population subdivision based on microsatellite allele frequencies," *Genetics*, vol. 139, no. 1, pp. 457–462, 1995.

[35] D. Falush, M. Stephens, and J. K. Pritchard, "Inference of population structure using multilocus genotype data: linked loci and correlated allele frequencies," *Genetics*, vol. 164, no. 4, pp. 1567–1587, 2003.

[36] K. Belkhir, P. Borsa, L. Chikhi, N. Raufaste, and F. Bonhomme, "GENETIX, logiciel sous WindowsTM pour la génétique des populations," Laboratoire Génome, Populations, Interactions CNRS UMR 5000, Université de Montpellier II, Montpellier, France, 2000.

[37] J. Diéguez-Uribeondo, F. Royo, C. Souty-Grosset, A. Ropiquet, and F. Grandjean, "Low genetic variability of the white-clawed crayfish in the Iberian Peninsula: its origin and management implications," *Aquatic Conservation*, vol. 18, no. 1, pp. 19–31, 2008.

[38] P. Trontelj, Y. MacHino, and B. Sket, "Phylogenetic and phylogeographic relationships in the crayfish genus *Austropotamobius* inferred from mitochondrial *COI* gene sequences," *Molecular Phylogenetics and Evolution*, vol. 34, no. 1, pp. 212–226, 2005.

[39] F. Grandjean, M. Frelon-Raimond, and C. Souty-Grosset, "Compilation of molecular data for the phylogeny of the genus Austropotamobius: one species or several?" *BFPP: Bulletin Francais de la Peche et de la Protection des Milieux Aquatiques*, no. 367, pp. 671–680, 2002.

[40] J. C. Avise, *Molecular Markers, Natural History and Evolution*, Sinauer Associates, Inc., Sunderland, Mass, USA, 2nd edition, 2004.

[41] O. Aung, T. T. T. Nguyen, S. Poompuang, and W. Kamonrat, "Microsatellite DNA markers revealed genetic population structure among captive stocks and wild populations of mrigal, Cirrhinus cirrhosus in Myanmar," *Aquaculture*, vol. 299, no. 1–4, pp. 37–43, 2010.

[42] N. Gouin, F. Grandjean, and C. Souty-Grosset, "Population genetic structure of the endangered crayfish *Austropotamobius pallipes* in France based on microsatellite variation: biogeographical inferences and conservation implications," *Freshwater Biology*, vol. 51, no. 7, pp. 1369–1387, 2006.

[43] H. C. Shull, M. Pérez-Losada, D. Blair et al., "Phylogeny and biogeography of the freshwater crayfish Euastacus (Decapoda: Parastacidae) based on nuclear and mitochondrial DNA," *Molecular Phylogenetics and Evolution*, vol. 37, no. 1, pp. 249–263, 2005.

[44] M. Versteegen and S. Lawler, "Population genetics of the Murray River crayfish Euastacus armatus," *Freshwater Crayfish*, vol. 11, pp. 146–157, 1997.

[45] P. A. Haye, I. Kornfield, and L. Watling, "Molecular insights into Cumacean family relationships (Crustacea, Cumacea)," *Molecular Phylogenetics and Evolution*, vol. 30, no. 3, pp. 798–809, 2004.

[46] J. C. Meyran, M. Monnerot, and P. Taberlet, "Taxonomic status and phylogenetic relationships of some species of the genus *Gammarus* (Crustacea, Amphipoda) deduced from mitochondrial DNA sequences," *Molecular Phylogenetics and Evolution*, vol. 8, no. 1, pp. 1–10, 1997.

[47] J. C. Meyran and P. Taberlet, "Mitochondrial DNA polymorphism among alpine populations of *Gammarus lacustris* (Crustacea, Amphipoda)," *Freshwater Biology*, vol. 39, no. 2, pp. 259–265, 1998.

[48] N. Knowlton and L. A. Weigt, "New dates and new rates for divergence across the Isthmus of Panama," *Proceedings of the Royal Society B*, vol. 265, no. 1412, pp. 2257–2263, 1998.

[49] J. C. Avise, "Twenty-five key evolutionary insights from the phylogeographic revolution in population genetics," in *Phylogeography of Southern European Refugia*, S. Weiss and N. Ferrand, Eds., pp. 7–21, Springer, Dordrecht, The Netherlands, 2007.

[50] F. Grandjean, N. Gouin, C. Souty-Grosset, and J. Diéguez-Uribeondo, "Drastic bottlenecks in the endangered crayfish species *Austropotamobius pallipes* in Spain and implications for its colonization history," *Heredity*, vol. 86, no. 4, pp. 431–438, 2001.

[51] C. Pedraza-Lara, F. Alda, S. Carranza, and I. Doadrio, "Mitochondrial DNA structure of the Iberian populations of the white-clawed crayfish, *Austropotamobius italicus italicus* (Faxon, 1914)," *Molecular Phylogenetics and Evolution*, vol. 57, no. 1, pp. 327–342, 2010.

[52] S. Baric, A. Höllrigl, L. Füreder, J. Petutschnig, and J. Dalla Via, "First analysis of genetic variability in Carinthian populations of the white-clawed crayfish *Austropotamobius pallipes*," *BFPP: Bulletin Francais de la Peche et de la Protection des Milieux Aquatiques*, no. 380-381, pp. 977–990, 2006.

[53] S. Zaccara, F. Stefani, and G. Crosa, "Diversity of mitochondrial DNA of the endangered white-clawed crayfish (*Austropotamobius italicus*) in the Po River catchment," *Freshwater Biology*, vol. 50, no. 7, pp. 1262–1272, 2005.

[54] S. Baric, A. Höllrigl, L. Füreder, and J. Dalla Via, "Mitochondrial and microsatellite DNA analyses of *Austropotamobius pallipes* populations in South Tyrol (Italy) and Tyrol (Austria)," *BFPP: Bulletin Francais de la Peche et de la Protection des Milieux Aquatiques*, no. 376-377, pp. 599–612, 2005.

[55] S. Bertocchi, S. Brusconi, F. Gherardi, F. Grandjean, and C. Souty-Grosset, "Genetic variability of the threatened crayfish *Austropotamobius italicus* in Tuscany (Italy): implications for its management," *Fundamental and Applied Limnology*, vol. 173, no. 2, pp. 153–164, 2008.

[56] B. Matallanas, M. D. Ochando, A. Vivero, B. Beroiz, F. Alonso, and C. Callejas, "Mitochondrial DNA variability in Spanish populations of A. italicus inferred from the analysis of a *COI* region," *Knowledge and Management of Aquatic Ecosystems*, vol. 401, article 30, 2011.

[57] J. M. Gil, *Situación, biología y conservación del cangrejo de río autóctono (Austropotamobius pallipes) en la provincia de Granada*, Ph.D. thesis, University of Granada, Granada, Spain, 1999.

[58] P. R. England, G. H. R. Osler, L. M. Woodworth, M. E. Montgomery, D. A. Briscoe, and R. Frankham, "Effects of intense versus diffuse population bottlenecks on microsatellite genetic diversity and evolutionary potential," *Conservation Genetics*, vol. 4, no. 5, pp. 595–604, 2003.

[59] Andalusian Regional Government, http://www.juntadeandalucia.es/medioambiente/site/web/menuitem.a5664a214f3c3df81d8899661525ea0/?vgnextoid =2566722ff100a110VgnVCM1000000624e50aRCRD&vgnextchannel=bc21d8c67de3e010VgnVCM1000000624e50aRCRD&lr=lang_es.

[60] C. C. Spencer, J. E. Neigel, and P. L. Leberg, "Experimental evaluation of the usefulness of microsatellite DNA for detecting demographic bottlenecks," *Molecular Ecology*, vol. 9, no. 10, pp. 1517–1528, 2000.

[61] R. Frankham, J. D. Ballou, and D. A. Briscoe, *Introduction to Conservation Genetics*, Cambridge University Press, Cambridge, Mass, USA, 2002.

[62] S. J. O'Brien and J. F. Evermann, "Interactive influence of infectious disease and genetic diversity in natural populations," *Trends in Ecology and Evolution*, vol. 3, no. 10, pp. 254–259, 1988.

[63] J. Diéguez-Uribeondo and A. Rueda, "Nuevas esperanzas para el cangrejo de río autóctono," *Quercus*, vol. 97, pp. 8–12, 1994.

[64] R. Asensio, "Breve historia reciente de la pesca de cangrejos en Álava," El periódico de Álava, July 17th 2003.

[65] Government of Navarra, http://www.navarra.es/NR/rdonlyres/ACF581F9-C847-4537-9679-4023E21EF5F8/179072/2011_Normasdepesca1.pdf.

[66] B. Beroiz, *Caracterización morfométrica y genética de las poblaciones españolas de cangrejo de río* Austropotamobius pallipes *(Lereboullet, 1858) mediante el uso de marcadores moleculares*, Ph.D. thesis, University Complutense of Madrid, Madrid, Spain, 2004.

[67] J. Galindo, B. Nebot, J. C. Delgado, and M. Chirosa, "Alarma tras la primera radiografía del cangrejo de río en Andalucía," *Quercus*, vol. 206, pp. 50–51, 2003.

[68] F. Grandjean and C. Souty-Grosset, "Mitochondrial DNA variation and population genetic structure of the white-clawed crayfish, *Austropotamobius pallipes*," *Conservation Genetics*, vol. 1, no. 4, pp. 309–319, 2000.

[69] G. M. Hewitt, "Some genetic consequences of ice ages, and their role in divergence and speciation," *Biological Journal of the Linnean Society*, vol. 58, no. 3, pp. 247–276, 1996.

[70] G. M. Hewitt, "Speciation, hybrid zones and phylogeography—or seeing genes in space and time," *Molecular Ecology*, vol. 10, no. 3, pp. 537–549, 2001.

[71] C. Callejas and M. D. Ochando, "Phylogenetic relationships among Spanish Barbus species (Pisces, Cyprinidae) shown by RAPD markers," *Heredity*, vol. 89, no. 1, pp. 36–43, 2002.

[72] C. Moritz, "Applications of mitochondrial DNA analysis in conservation: a critical review," *Molecular Ecology*, vol. 3, no. 4, pp. 401–411, 1994.

# Influence of Codon Bias on Heterologous Production of Human Papillomavirus Type 16 Major Structural Protein L1 in Yeast

**Milda Norkiene and Alma Gedvilaite**

*Institute of Biotechnology, Vilnius University, Graiciuno 8, Vilnius, Lithuania*

Correspondence should be addressed to Alma Gedvilaite, agedv@ibt.lt

Academic Editors: V. Dvornyk, S. Y. Morozov, and R. D. Possee

Heterologous gene expression is dependent on multistep processes involving regulation at the level of transcription, mRNA turnover, protein translation, and posttranslational modifications. Codon bias has a significant influence on protein yields. However, sometimes it is not clear which parameter causes observed differences in heterologous gene expression as codon adaptation typically optimizes many sequence properties at once. In the current study, we evaluated the influence of codon bias on heterologous production of human papillomavirus type 16 (HPV-16) major structural protein L1 in yeast by expressing five variants of codon-modified open reading frames (OFRs) encoding HPV-16 L1 protein. Our results showed that despite the high toleration of various codons used throughout the length of the sequence of heterologously expressed genes in transformed yeast, there was a significant positive correlation between the gene's expression level and the degree of its codon bias towards the favorable codon usage. The HPV-16 L1 protein expression in yeast can be optimized by adjusting codon composition towards the most preferred codon adaptation, and this effect most probably is dependent on the improved translational elongation.

## 1. Introduction

The production of functional proteins in heterologous hosts is an important issue of modern biotechnology. However, often it is difficult to generate recombinant proteins outside their original context. Protein expression is dependent of multistep processes involving regulation at the level of transcription, mRNA turnover, protein translation, and posttranslational modifications leading to the formation of a stable product [1]. Currently, it is accepted that codon bias has a crucial role in heterologous gene expression and that nonoptimal codon content can limit gene expression due to the shortage of available tRNAs in the heterologous host resulting in slowed elongation of the nascent peptide or premature termination of translation [2–5]. The observation that highly expressed genes have strong codon bias towards "preferred" codons is used to substitute the codons throughout the length of the target sequence into preferred high-frequency codons from the expression host. These changes improved the synthesis and yield of some heterologous proteins in different organisms [1, 5, 6].

Nonetheless, the codons that are adapted to the efficient elongation of endogenous genes may not always correspond to the efficient codons for heterologous genes, because overexpression often causes amino acid starvation what leads to changes and disbalance of charged tRNA pools [5]. It was shown that endogenous genes encoding amino acid biosynthetic enzymes that are essential during amino acid starvation preferentially use codons that are poorly adapted to the typical pool of charged tRNAs but are well adapted to starvation-induced tRNA pools [7, 8]. For this reason, the approach of codon bias optimization by adjusting codon usage to match cellular tRNA abundances in standard conditions now is changing to "codon harmonization" [9]. This new approach puts some nonpreferred codons in positions that correspond to predicted protein domain boundaries, and "codon sampling" adjusts the codon usage to reflect the overall usage in the target genome [10]. In the absence of tRNA abundance estimates, codon frequencies in the target genome are sometimes used.

Despite accumulating information about the impact of codon bias on the heterologous gene's expression, it is still unclear what approach for codon adaptation should be used designing putative transgene sequences as these approaches have not been systematically compared against each other. The codon adaptation typically optimizes many sequence properties at once, and in most cases, it is difficult to determine which parameter causes the observed differences in heterologous gene expression.

It was shown that papillomavirus late mRNAs may not be efficiently translated in undifferentiated cells due to a mismatch of codon usage and tRNA availability [11]. Optimization of human papillomavirus (HPV) L1 genes by introducing favorable human and plant codons improved production of L1 protein in human and plant cells [12, 13].

In the current study, we aimed to evaluate the influence of codon bias on heterologous generation of HPV type 16 (HPV-16) major structural protein L1 in yeast by expressing five different codon-modified open reading frames (OFRs) encoding HPV-16 L1 protein.

## 2. Materials and Methods

*2.1. Generation of HPV-16 L1 Protein Expression Plasmids.* All DNA manipulations were performed according to standard procedures [14]. Enzymes and kits for DNA manipulations were purchased from Termo Scientific Fermentas (Vilnius, Lithuania). The codons in the sequence of native HPV-16 isolate 114/K L1 gene [15] L1-Pv were either optimized according to the frequency analysis of codons determined (http://www.kazusa.or.jp/codon/) for overall yeast *S. cerevisiae* proteins (L1-Sc), plant proteins (i.e., *Solanum tuberosum* L1-Pl, EMBL accession no. AJ313181), and mammalian proteins (*Homo sapiens* L1-Hm, accession no. AJ313179) or adapted for expression in *E. coli* (L1-Ec) allowing some deviations from the strictly optimized codon usage. The genes encoding L1-Sc and L1-Ec of the HPV-16 were synthesized in GenScript (Piscataway, NJ, USA). The genes encoding L1-Pv, L1-Pl, and L1-Hm [12] were kindly provided by Martin Müller (German Cancer Research Center).

The L1-encoding ORFs were cloned into the yeast vector pFX7 [16], allowing the selection of yeast transformants by permitting resistance to formaldehyde [17]. The sequences of the inserted L1-encoding genes were verified by DNA sequencing. The generated plasmids were transformed into yeast *Saccharomyces cerevisiae* strain AH22-214 (*a, leu2 his4*).

*2.2. Expression of HPV-16-Derived L1 Proteins in Yeast.* Yeast transformants harboring plasmids with genes encoding HPV-16 L1-Sc, L1-Pl, L1-Ec, L1-Hm, and L1-Pv proteins were grown in 15 mL YEPD medium (yeast extract 1%, peptone 2%, and glucose 2%) supplemented with 5 mM formaldehyde overnight at 30°C. The synthesis of recombinant proteins was induced after transferring yeast cells into induction medium 20 mL YEPG (yeast extract 1%, peptone 2%, and galactose 3%) supplemented with 5 mM formaldehyde and culturing for additional 18 h. Yeast biomass harboring recombinant proteins was harvested by centrifugation and stored at −20°C before use. Ten transformants from every group were analyzed.

*2.3. RNA Extraction and Northern Blot Analysis.* Total yeast RNA was isolated 4 hours after induction by the method described earlier [18]. The Northern blot analysis was performed by the separation of 15 $\mu$g of total RNA on a 1% agarose gel containing 2.2 M formaldehyde, followed by transfer to a Hybond-N+ filter (Amersham Biosciences, Little Halfont, England) and hybridization. The DNA templates of all five L1-encoding ORFs and PGK1 ORF used for probes were generated by PCR amplification and labeled with [$\alpha$-33P]-dATP (Hartman Analytic, Braunschrveig, Germany) by random priming of the DNA probe using a DecaLabel DNA Labeling Kit (Thermo Scientific Fermentas) according to the manufacturer's instructions. The blots were washed extensively at 65°C and L1 and PGK1 transcripts were identified by phosphorimaging.

*2.4. Preparation of Yeast Lysates SDS-PAGE and Western Blot Analysis.* 10–20 mg of yeast cell pellets was resuspended in 10 volumes (vol/wt) of DB150 buffer (150 mM NaCl, 1 mM $CaCl_2$, 0.001% Trition X-100 in 10 mM Tris/HCl-buffer, pH 7.2) and 1 mM PMSF. An equal volume of glass beads was added and the cells were lysed by vortexing at high speed, 8 times for 30 sec, with cooling on ice for 30 sec between each vortexing. Then an equal volume of 2 × SDS-PAGE sample buffer (125 mM Tris-HCl, pH6.8, 20% glycerol, 8% SDS, 150 mM DTT, 0.01% bromophenol blue) was added directly to the same tube, mixed and boiled immediately at 100°C for 10 minutes. 4–10 $\mu$L of the prepared whole cell lysate was loaded onto SDS-polyacrylamide gel (up to 20 $\mu$g protein in each lane) and sodium dodecylsulfate polyacrylamide gel electrophoresis (SDS-PAGE) was run in SDS-Tris-glycine buffer. Western blot analyses were performed according to methods described previously [19]. As a primary antibody for the immunodetection of HPV-16 L1 protein, mouse polyclonal antibody generated in-house (dilution 1 : 1000) was used. As a secondary antibody, goat antimouse IgG antibody conjugated to horseradish peroxidase diluted 1 : 3000 (Bio-Rad, Hercules, CA, USA) was used. The gels were scanned and 1 or 2 proteins bands (~50 kDa and 26 kDa) were used to determine the ratio of these proteins in the lines for evaluation of loaded yeast lysates quantitative differences. The quantitative evaluation of protein band was performed using the ImageQuant TL 1D gel analysis software (GE Healthcare).

## 3. Results and Discussion

*3.1. Description of Codon-Optimized HPV-16 L1 Genes.* The impact of synonymous codon bias on heterologous production of HPV-16 L1 protein in yeast cells was studied by expressing five L1-encoding ORFs composed of different codons. All five ORFs encoding L1 protein of HPV-16 isolate 114/K [15] were placed under the control of galactose inducible promoter using the same insertion site in the vector

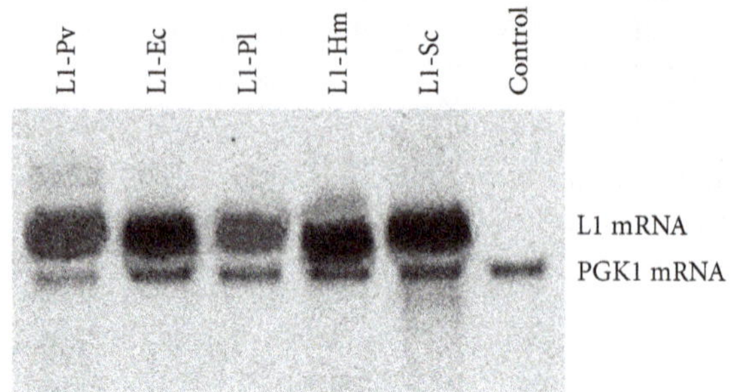


FIGURE 1: Analysis of HPV-16 L1 mRNA expression in transformed yeast cells by Northern blot. Total yeast RNA was isolated 4 h after induction from yeast transformants expressing five different HPV-16 L1 ORFs and control yeast cells transformed with the empty pFX7 vector. Fifteen micrograms of total RNA was loaded per lane and hybridized with L1-Pv, L1-Ec, L1-Pl, L1-Hm, and PGK1-control cDNA probes labeled with [$\alpha$-33P]-dATP by random priming. The data from one representative experiment are shown. The 4 independent experiments with other randomly picked transformants in every group were performed with similar results.

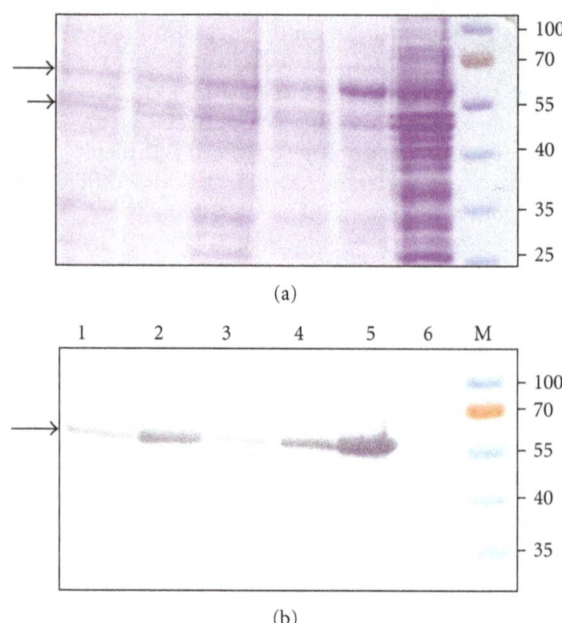


FIGURE 2: Analysis of HPV-16 L1 expression in yeast by SDS-PAGE (a) and Western blot with mouse polyclonal antibody against HPV-16 L1 protein (b). The same samples were run on each gel. In lanes: (1) crude lysate of yeast expressing ORF L1-Pv; (2) crude lysate of yeast expressing ORF L1-Ec; (3) crude lysate of yeast expressing ORF L1-Pl; (4) crude lysate of yeast expressing ORF L1-Hm; (5) crude lysate of yeast expressing ORF L1-Sc; (6) negative control sample from crude lysate of *S. cerevisiae* cells transformed with the empty vector pFX7; and M: prestained protein weight marker (Thermo Scientific Fermentas). Long arrow points to the band with HPV16 L1 protein. The protein band (~50 kDa) pointed with short arrow was used to determine the ratio of this protein in the lines for evaluation of loaded yeast lysates quantitative differences using the ImageQuant TL 1D gel analysis software (GE Healthcare). The ratio of ~50 kDa protein band in lines was 1.77, 1.00, 1.69, 1.74, 1.30, and 4.44 accordingly. The data from one representative experiment are shown. The expression level of L1 proteins in 10 randomly picked transformants in every group was alike.

pFX7 [16]. The pFX7 vectors with inserted different L1-encoding ORFs were transformed into yeast strain AH22-214 and the expressions of L1 proteins were analyzed by both Northern blot and Western blot.

The codons in the authentic L1 ORF of HPV-16 isolate 114/K (L1-Pv) were optimized for either *S. cerevisiae* (L1-Sc), plant (*Solanum tuberosum;* L1-Pl), and *Homo sapiens* cells (L1-Hm) or adapted for expression in *E.coli* (L1-Ec) allowing some deviations from the strict usage of high-frequency codons to reflect the overall codon usage in the target genome. Few deviations from the strict usage of optimized codons were made also to allow the insertion or removal of recognition sites for restriction endonucleases. The majority of codons in used OFRs were modified as compared to the native HPV-16 L1 gene (L1-Pv): in L1-Sc ORF, 46.3%; in L1-Pl ORF, 51.1%; in L1-Ec ORF, 60.2%; and in L1-Hm ORF, 78.6% codons were modified while the encoded protein sequence remained unchanged (Table 1). All upstream and downstream noncoding sequences were removed in all the constructs analyzed in this study. In addition to the adjustment of codon composition, the introduced changes were likely to affect all known and unknown negative regulatory elements present in the authentic HPV-16 L1-Pv ORF.

*3.2. Analysis of HPV-16 L1 Expression in Yeast.* To determine whether the alterations of the primary sequence of the L1 mRNAs introduced by the codon changes also affected the state and the level of transcripts, total yeast RNA was isolated from yeast cells 4 h after induction and the L1 mRNA transcription was analyzed by Northern blot. The transcripts of all five HPV-16 L1 ORFs were detected in the respective transformed yeast cells (Figure 1). Despite slightly weaker signal of L1-Pl and L1-Pv transcripts detected in Northern blot analysis, the overall mRNA produced from all five used ORFs was highly abundant and did not show apparent degradation products (Figure 1). This confirmed that codon modifications have not affected significantly the levels of transcription and the stability of HPV-16 L1 mRNA in the used yeast expression system. In contrast, the expression of L1-Pv, L1-Pl and L1-Hm in plants [13] and mammalian cells [12] was encountered with the instability of mRNA: the L1-Pv transcript was not detectable in Northern blot and most of the respective L1-Pl and L1-Hm mRNA was found to be degraded. The stability of L1-Pl and L1-Hm transcripts in plant was improved only by adding the translational enhancer 5′-leader sequence of tobacco mosaic virus [13].

In the next step, the production of heterologous HPV-16 L1 protein in transformed yeast was analyzed by both SDS-PAGE and Western blot (Figure 2). The HPV-16 L1 protein was expressed by all constructs encoding five different ORFs as demonstrated by the immunoreactivity of the respective protein bands in Western blot using HPV-16 L1-protein specific antibodies (Figure 2(b)). Western blot analysis of yeast cell lysates revealed different production levels of HPV-16 L1 protein: the highest production of L1 protein was

Table 1: Codon usage in HPV-16 L1-Sc, L1-Pl, L1-Ec, Li, L1-Hm, and L1-Pv ORFs.

| Amino acids | Codons | Number of the indicated codons in the ORF | | | | |
|---|---|---|---|---|---|---|
| | | L1-Pv | L1-Ec | L1-Pl | L1-Hm | L1-Sc |
| Ala | GCU | 10 | 20 | 29 | | 30 |
| | GCC | 6 | 4 | | 30 | |
| | GCA | 14 | | 1 | | |
| | GCG | | 6 | | | |
| Arg | AGA | 4 | 1 | | | 19 |
| | AGG | 4 | | 19 | 19 | |
| | CGU | 2 | 13 | | | |
| | CGC | 4 | 2 | | | |
| | CGA | 4 | 1 | | | |
| | CGG | 1 | 2 | | | |
| Asn | AAU | 21 | | 28 | | 27 |
| | AAC | 7 | 28 | | 28 | 1 |
| Asp | GAU | 18 | | 27 | | 27 |
| | GAC | 9 | 27 | | 27 | |
| Cys | UGU | 9 | 1 | 12 | | 12 |
| | UGC | 3 | 11 | | 12 | |
| Gln | CAA | 11 | 19 | 19 | | 19 |
| | CAG | 8 | | | 19 | |
| Glu | GAA | 14 | 20 | 19 | | 20 |
| | GAG | 6 | | 1 | 20 | |
| Gly | GGU | 15 | 32 | | | 35 |
| | GGC | 9 | 2 | | 35 | |
| | GGA | 8 | 1 | 35 | | |
| | GGG | 3 | | | | |
| His | CAU | 8 | | 10 | | 10 |
| | CAC | 2 | 10 | | 10 | |
| Ile | AUU | 12 | | 21 | | 22 |
| | AUC | | 21 | 1 | 22 | |
| | AUA | 10 | 1 | | | |
| Lys | AAA | 27 | 32 | 34 | | 34 |
| | AAG | 7 | 2 | | 34 | |
| Leu | UUA | 23 | 1 | | | |
| | UUG | 5 | 1 | | | 43 |
| | CUU | 3 | | 43 | | |
| | CUC | | | | | |
| | CUA | 7 | 8 | | | |
| | CUG | 5 | 33 | | 43 | |
| Met | AUG | 10 | 10 | 10 | 10 | 10 |
| Phe | UUU | 23 | 1 | 24 | | 24 |
| | UUC | 1 | 23 | | 24 | |
| Pro | CCU | 17 | | | | |
| | CCC | 5 | | | 37 | |
| | CCA | 15 | 3 | 37 | | 37 |
| | CCG | | 34 | | | |

Table 1: Continued.

| Amino acids | Codons | Number of the indicated codons in the ORF | | | | |
|---|---|---|---|---|---|---|
| | | L1-Pv | L1-Ec | L1-Pl | L1-Hm | L1-Sc |
| Ser | AGU | 7 | 1 | | | |
| | AGC | 2 | 2 | | 33 | |
| | UCU | 13 | 1 | | | 31 |
| | UCC | 4 | 23 | | | |
| | UCA | 7 | 3 | 33 | | 2 |
| | UCG | | 3 | | | |
| Thr | ACU | 14 | 2 | 41 | | 41 |
| | ACC | 7 | 34 | | 41 | |
| | ACA | 19 | 3 | | | |
| | ACG | 1 | 2 | | | |
| Trp | UGG | 7 | 7 | 7 | 7 | 7 |
| Tyr | UAT | 15 | | | | 22 |
| | UAC | 7 | 22 | 22 | 22 | |
| Val | GUU | 17 | 30 | 32 | | 32 |
| | GUC | 2 | 1 | | | |
| | GUA | 10 | | | | |
| | GUG | 3 | 1 | | 32 | |

observed in the lysate of transformed yeast expressing ORF L1-Sc (Figure 2(b), lane 5) and the moderate L1 production was detected in yeast expressing ORF L1-Ec and ORF L1-Hm (Figure 2(b), lanes 2 and 4, resp.). The diversity of HPV-16 L1 expression levels detected in the Western blot analysis but not in Northern blot suggested that the main differences in L1 protein expression levels may be addressed to the translation efficiency, rather than to gene transcription changes.

The optimization of codons for the yeast cells in the L1-Sc construct clearly had the favorable effect on L1 production. The expression of L1 protein encoded by the construct L1-Sc, carrying *S. cerevisiae* optimized codons, was the most successful with clearly detectable protein band in both SDS-PAGE and Western blot (Figures 2(a) and 2(b), lane 5). In contrast, the expression of L1-Pl construct has proven to be the most inefficient as only a faint signal of L1 protein was detected in the Western blot (Figure 2(b), lane 3). Surprisingly, the sequence of L1-Pl was the most homologous to the L1-Sc sequence because only 5 amino acids were coded by different codons in this construct. Thus, despite these small differences between the sequences encoding ORFs L1-PI and L1-Sc, there were significant differences in the production levels of L1 protein. The levels of generated HPV-16 L1 protein using other three constructs were higher than that of L1-Pl but lower than L1-Sc and could be lined from L1-Pv to L1-Hm and L1-Ec by the increasing intensity of the signal in the Western blot (Figure 2(b), lines 1, 4, and 2, resp.).

The plant and *S.cerevisiae*-optimized L1 ORFs (L1-PI and L1-Sc) sequences were comprised of similar low GC content (GC content of 35,9% and 33.0%, resp.) even lower than that found in authentic HPV-16 L1-Pv ORF (GC content of 38.7%). Meanwhile, both human- and *E. coli*-adapted L1 ORFs (L1-Hm and L1-Ec) had significantly higher GC content (65.1% and 54.5% GC, resp.). Thus, it was assumed that the efficiency of translation was not influenced by the differences in the GC content of mRNA. Most likely, the effective translation of studied HPV-16 L1 mRNAs was related to the codon usage and the availability of tRNA pool recognizing synonymous codons rather than the mRNA primary structure. Moreover, the usage of GGA (Gly), CUU (Leu) and TCA (Ser) codons as a single opportunity distinguished the L1-Pl construct from the most efficient L1-Sc and other constructs (Table 1). In the L1-Pv construct 8/35 GGA (Gly), 3/43 CUU (Leu) and 7/33 TCA (Ser) codons were also used but only in parallel with other synonymous codons and so it may loosen the pressure that was obvious for the expression of L1-Pl construct (Table 1). Although the expression level of L1-Pv protein was low, it was higher than expression of the L1-Pl despite the usage of numerous other nonfavorable codons. It is not clear which one or all three codons had this translation limiting effect because according to the frequency analysis of codons determined for overall yeast proteins (http://www.kazusa.or.jp/codon/) these codons GGA (Gly), CUU (Leu), and TCA (Ser) are moderately used. However, in highly expressed yeast genes CUU (Leu) codons are not used but few GGA (Gly) and TCA (Ser) could be found [20, 21]. Previous studies have shown that the pools of available charged tRNR may change under starvation conditions in *E. coli* [7, 8]. The results obtained in our study did not exclude the possibility that the availability of charged tRNR in yeast was also changed because of amino acid starvation caused by the expression of heterologous protein. On the other hand, the results of our study suggested that the pools of the most favorable codons in yeast were not affected and did not limit the expression level of the L1-Sc construct. Our results also supposed that despite the high toleration for adaptation of various codons used throughout the length of the sequence of heterologously expressed genes in transformed yeast, there was a significant positive correlation between the gene's expression level and the degree of its codon bias towards the favorable codon usage.

In conclusion, the HPV-16 L1 protein expression in yeast can be optimized by adjusting codon composition towards an efficient codon usage and this effect most probably is dependent on the improved translational elongation.

## Conflict of Interests

The authors declare that they have no conflict of interests.

## Acknowledgments

This work was supported by the Lithuanian Science Council (Grant no. AUT-16/2010) and the Agency for Science, Innovations and Technology (Grant no. 31V-116).

## References

[1] E. Angov, "Codon usage: nature's roadmap to expression and folding of proteins," *Biotechnology Journal*, vol. 6, no. 6, pp. 650–659, 2011.

[2] J. F. Kane, "Effects if rare codon clusters on high-level expression of heterologous proteins in *Echerichia coli*," *Current Opinion in Biotechnology*, vol. 6, no. 5, pp. 494–500, 1995.

[3] E. Goldman, A. H. Rosenberg, G. Zubay, and F. W. Studier, "Consecutive low-usage leucine codons block translation only when near the 5' end of a message in *Echerichia coli*," *Journal of Molecular Biology*, vol. 245, no. 5, pp. 467–473, 1995.

[4] S. Kanaya, Y. Yamada, Y. Kudo, and T. Ikemura, "*Bacillus subtilis* tRNAs: gene expression level and species-specific diversity of codon usage based on multivariate analysis," *Gene*, vol. 238, no. 1, pp. 143–155, 1999.

[5] C. Gustafsson, S. Govindarajan, and J. Minshull, "Codon bias and heterologous protein expression," *Trends in Biotechnology*, vol. 22, no. 7, pp. 346–353, 2004.

[6] Z. Zhou, P. Schnake, L. Xiao, and A. A. Lal, "Enhanced expression of a recombinant malaria candidate vaccine in Escherichia coli by codon optimization," *Protein Expression and Purification*, vol. 34, no. 1, pp. 87–94, 2004.

[7] J. Elf, D. Nilsson, T. Tenson, and M. Ehrenberg, "Selective charging of tRNA isoacceptors explains patterns of codon usage," *Science*, vol. 300, no. 5626, pp. 1718–1722, 2003.

[8] K. A. Dittmar, M. A. Sørensen, J. Elf, M. Ehrenberg, and T. Pan, "Selective charging of tRNA isoacceptors induced by amino-acid starvation," *EMBO Reports*, vol. 6, no. 2, pp. 151–157, 2005.

[9] E. Angov, C. J. Hillier, R. L. Kincaid, and J. A. Lyon, "Heterologous protein expression is enhanced by harmonizing the codon usage frequencies of the target gene with those of the expression host," *PLoS ONE*, vol. 3, no. 5, Article ID e2189, 2008.

[10] J. B. Plotkin and G. Kudla, "Synonymous but not the same: the causes and consequences of codon bias," *Nature Reviews Genetics*, vol. 12, no. 1, pp. 32–42, 2011.

[11] J. Zhou, W. J. Liu, S. W. Peng, X. Y. Sun, and I. Frazer, "Papillomavirus capsid protein expression level depends on the match between codon usage and tRNA availability," *Journal of Virology*, vol. 73, no. 6, pp. 4972–4982, 1999.

[12] C. Leder, J. A. Kleinschmidt, C. Wiethe, and M. Müller, "Enhancement of capsid gene expression: preparing the human papillomavirus type 16 major structural gene l1 for DNA vaccination purposes," *Journal of Virology*, vol. 75, no. 19, pp. 9201–9209, 2001.

[13] S. Biemelt, U. Sonnewald, P. Galmbacher, L. Willmitzer, and M. Müller, "Production of human papillomavirus type 16 virus-like particles in transgenic plants," *Journal of Virology*, vol. 77, no. 17, pp. 9211–9220, 2003.

[14] J. Sambrook and D. Russell, *Molecular Cloning: A Laboratory Manual*, Cold Spring Harbor Laboratory Press, New York, NY, USA, 2001.

[15] R. Kirnbauer, J. Taub, H. Greenstone et al., "Efficient self-assembly of human papillomavirus type 16 L1 and L1-L2 into virus-like particles," *Journal of Virology*, vol. 67, no. 12, pp. 6929–6936, 1993.

[16] K. Sasnauskas, O. Buzaite, F. Vogel et al., "Yeast cells allow high-level expression and formation of polyomavirus-like particles," *Biological Chemistry*, vol. 380, no. 3, pp. 381–386, 1999.

[17] K. Sasnauskas, R. Jomantiene, A. Januska, E. Lebediene, J. Lebedys, and A. Janulaitis, "Cloning and analysis of a Candida maltosa gene which confers resistance to formaldehyde in *Saccharomyces cerevisiae*," *Gene*, vol. 122, no. 1, pp. 207–211, 1992.

[18] M. E. Schmitt, T. A. Brown, and B. L. Trumpower, "A rapid and simple method for preparation of RNA from *Saccharomyces cerevisiae*," *Nucleic Acids Research*, vol. 18, no. 10, pp. 3091–3092, 1990.

[19] A. Gedvilaite, A. Zvirbliene, J. Staniulis, K. Sasnauskas, D. H. Krüger, and R. Ulrich, "Segments of puumala hantavirus nucleocapsid protein inserted into chimeric polyomavirus-derived virus-like particles induce a strong immune response in mice," *Viral Immunology*, vol. 17, no. 1, pp. 51–68, 2004.

[20] A. Hoekema, R. A. Kastelein, M. Vasser, and H. A. de Boer, "Codon replacement in the PGK1 gene of Saccharomyces cerevisiae: experimental approach to study the role of biased codon usage in gene expression," *Molecular and Cellular Biology*, vol. 7, no. 8, pp. 2914–2924, 1987.

[21] J. L. Bennetzen and B. D. Hall, "Codon selection in yeast," *Journal of Biological Chemistry*, vol. 257, no. 6, pp. 3026–3031, 1982.

# Permissions

The contributors of this book come from diverse backgrounds, making this book a truly international effort. This book will bring forth new frontiers with its revolutionizing research information and detailed analysis of the nascent developments around the world.

We would like to thank all the contributing authors for lending their expertise to make the book truly unique. They have played a crucial role in the development of this book. Without their invaluable contributions this book wouldn't have been possible. They have made vital efforts to compile up to date information on the varied aspects of this subject to make this book a valuable addition to the collection of many professionals and students.

This book was conceptualized with the vision of imparting up-to-date information and advanced data in this field. To ensure the same, a matchless editorial board was set up. Every individual on the board went through rigorous rounds of assessment to prove their worth. After which they invested a large part of their time researching and compiling the most relevant data for our readers. Conferences and sessions were held from time to time between the editorial board and the contributing authors to present the data in the most comprehensible form. The editorial team has worked tirelessly to provide valuable and valid information to help people across the globe.

Every chapter published in this book has been scrutinized by our experts. Their significance has been extensively debated. The topics covered herein carry significant findings which will fuel the growth of the discipline. They may even be implemented as practical applications or may be referred to as a beginning point for another development. Chapters in this book were first published by Hindawi Publishing Corporation; hereby published with permission under the Creative Commons Attribution License or equivalent.

The editorial board has been involved in producing this book since its inception. They have spent rigorous hours researching and exploring the diverse topics which have resulted in the successful publishing of this book. They have passed on their knowledge of decades through this book. To expedite this challenging task, the publisher supported the team at every step. A small team of assistant editors was also appointed to further simplify the editing procedure and attain best results for the readers.

Our editorial team has been hand-picked from every corner of the world. Their multi-ethnicity adds dynamic inputs to the discussions which result in innovative outcomes. These outcomes are then further discussed with the researchers and contributors who give their valuable feedback and opinion regarding the same. The feedback is then collaborated with the researches and they are edited in a comprehensive manner to aid the understanding of the subject.

Apart from the editorial board, the designing team has also invested a significant amount of their time in understanding the subject and creating the most relevant covers. They scrutinized every image to scout for the most suitable representation of the subject and create an appropriate cover for the book.

The publishing team has been involved in this book since its early stages. They were actively engaged in every process, be it collecting the data, connecting with the contributors or procuring relevant information. The team has been an ardent support to the editorial, designing and production team. Their endless efforts to recruit the best for this project, has resulted in the accomplishment of this book. They are a veteran in the field of academics and their pool of knowledge is as vast as their experience in printing. Their expertise and guidance has proved useful at every step. Their uncompromising quality standards have made this book an exceptional effort. Their encouragement from time to time has been an inspiration for everyone.

The publisher and the editorial board hope that this book will prove to be a valuable piece of knowledge for researchers, students, practitioners and scholars across the globe.

# List of Contributors

**R. Vodouhe**
Bioversity International, Office of West and Central Africa, 08 BP 0931 Cotonou, Benin

**A. Dansi**
Laboratory of Agricultural Biodiversity and Tropical Plant Breeding, Department of Genetics, Faculty of Sciences and Technology (FAST), University of Abomey-Calavi (UAC), 071BP28 Cotonou, Benin
Department of Crop Science (DCS), Crop, Aromatic and Medicinal Plant Biodiversity Research and Development Institute (IRDCAM), 071BP28 Cotonou, Benin

**R. Słopien and A. Warenik-Szymankiewicz**
Department of Gynecological Endocrinology, Poznan University of Medical Sciences, Ul. Polna 33, 60-535 Poznan, Poland

**A. Słopien**
Department of Child and Adolescent Psychiatry, Poznan University of Medical Sciences, Ul. Szpitalna 27/33, 60-572 Poznan, Poland

**A. Rozycka, M. Lianeri and P. P. Jagodzinski**
Department of Biochemistry and Molecular Biology, Poznan University of Medical Sciences, Ul. Swiecickiego 6, 60-781 Poznan, Poland

**Olivia Sheppard, Frances K. Wiseman, Aarti Ruparelia and Elizabeth M. C. Fisher**
Department of Neurodegenerative Disease, UCL Institute of Neurology, Queen Square, London WC1N 3BG, UK

**Victor L. J. Tybulewicz**
Division of Immune Cell Biology, MRC National Institute for Medical Research, The Ridgeway, Mill Hill, London NW7 1AA, UK

**Corrado Romano**
Unit of Pediatrics and Medical Genetics, I.R.C.C.S. Associazione Oasi Maria Santissima, 94018 Troina, Italy

**Carmelo Schepis**
Unit of Dermatology, I.R.C.C.S. Associazione Oasi Maria Santissima, 94018 Troina, Italy

**Mulatu Geleta and Tomas Bryngelsson**
Department of Plant Breeding and Biotechnology, Swedish University of Agricultural Sciences, P.O. Box 101, 230 53 Alnarp, Sweden

**Amy Adamson and Dennis La Jeunesse**
Department of Biology, University of North Carolina at Greensboro, Greensboro, NC 27402, USA

**Dibyendu Talukdar**
Department of Botany, R.P.M. College, University of Calcutta, Uttarpara, West Bengal, Hooghly 712 258, India

**Irma Sanchez-Ramos, Ismael Cross and Laureana Rebordinos**
Laboratorio de Genetica, Universidad de Cadiz, Poligono Rıo San Pedro s/n, 11510 Puerto Real, Spain

**Jaroslav Macha and Vladimir Krylov**
Department of Cell Biology, Faculty of Science, Charles University in Prague, Prague 2, Vinicna 7, 12843 Prague, Czech Republic

**Gonzalo Martınez-Rodrıguez**
Instituto de Ciencias Marinas de Andalucıa, Consejo Superior de Investigaciones Cientıficas, Republica Saharaui, no. 2, 11510 Puerto Real, Spain

**Linnea Asplund**
Department of Ecology and Evolution, Uppsala University, SE-752 36 Uppsala, Sweden
Department of Crop Production Ecology, Swedish University of Agricultural Sciences, SE-750 07 Uppsala, Sweden

**Matti W. Leino**
Swedish Museum of Cultural History, SE-643 98 Julita, Sweden
IFM Molecular Genetics, Linkoping University, SE-581 83 Linkoping, Sweden

**Jenny Hagenblad**
Department of Ecology and Evolution, Uppsala University, SE-752 36 Uppsala, Sweden
IFM Molecular Genetics, Linkoping University, SE-581 83 Linkoping, Sweden
Department of Biology, Norwegian University of Science and Technology, NO-7491 Trondheim, Norway

**M. Y. Rafii**
Institute of Tropical Agriculture, Universiti Putra Malaysia, 43400 Serdang, Selangor, Malaysia
Department of Crop Science, Faculty of Agriculture, Universiti Putra Malaysia, 43400 Serdang, Selangor, Malaysia

**M. Sohrabi, M. M. Hanafi and A. Siti Nor Akmar**
Institute of Tropical Agriculture, Universiti Putra Malaysia, 43400 Serdang, Selangor, Malaysia

**M. A. Latif**
Department of Crop Science, Faculty of Agriculture, Universiti Putra Malaysia, 43400 Serdang, Selangor, Malaysia
Bangladesh Rice Research Institute (BRRI), Gazipur 1701, Bangladesh

**Zbigniew Rusinowski and Olga Domeradzka**
Department of Ornamental Plants, Warsaw School of Life Sciences, Faculty of Horticulture and Landscape Architecture, Warsaw University of Life Sciences, Nowoursynowska 159, 02776 Warsaw, Poland

**Fabio Coppede**
Faculty of Medicine, University of Pisa, 56126 Pisa, Italy
Genetics and Epigenetics of Complex Disease Program, Department of Neuroscience (DAI Neuroscience), Pisa University Hospital, Via S. Giuseppe 22, 56126 Pisa, Italy

**F. Ahmed**
Department of Crop Science, Faculty of Agriculture, Universiti Putra Malaysia, 43400, Serdang, Selangor, Malaysia

**M. S. Mazid**
Institute of Tropical Agriculture, Universiti Putra Malaysia, 43400 UPM Serdang, Selangor, Malaysia

**M. E. Ali**
Institute of Nano Electronic Engineering (INNE), Universiti Malaysia Perlis, 01000 Kangar, Malaysia

**M. Y. Omar**
Department of Biology, Faculty of Science, Universiti Putra Malaysia, 43400, Serdang, Selangor, Malaysia

**S. G. Tan**
Department of Cell and Molecular Biology, Faculty of Biotechnology and Molecular Science, Universiti Putra Malaysia, 43400, Serdang, Selangor, Bangladesh

**Ghorban Elyasi Zarringhabaie**
Department of Animal Science, East Azarbaijan Research Center for Agriculture and Natural Resources, Tabriz, Iran

**Arash Javanmard**
Department of Genomics, Agricultural Biotechnology Research Institute for Northwest and West of Iran, Tabriz, Iran

**Ommolbanin Pirahary**
Animal Science Section, East Azerbaijan Jahad-e-Keshavarzi Organization, Tabriz, Iran

**Guang-Li Yang**
Department of Life Sciences, Shangqiu Normal University, Shangqiu 476000, China

**Dong-Li Fu, Yu-Tao Wang, Shu-Ru Cheng, Su-Li Fang and Yu-Zhu Luo**
Gansu Province Key Laboratory of Herbivorous Animal Biotechnology, Gansu Agricultural University, Lanzhou 730070, China

**Xia Lang**
Lanzhou Institute of Animal and Veterinary Pharmaceutics Sciences, Chinese Academy of Agricultural Sciences, Lanzhou 730050, China

**Victor Olusegun Oyetayo**
Department of Microbiology, Federal University of Technology, P.M.B. 704, Akure, Nigeria

**Ederson Akio Kido, Jose Ribamar Costa Ferreira Neto, Roberta Lane de Oliveira Silva, Valesca Pandolfi, Sergio Crovella and Ana Maria Benko-Iseppon**
Department of Genetics, Federal University of Pernambuco (UFPE), 50670-901 Recife, PE, Brazil

**Ana Carolina Ribeiro Guimaraes, Daniela Truffi Veiga and Sabrina Moutinho Chabregas**
Biotechnology Division, Sugarcane Technology Center (CTC), 13400-970 Piracicaba, SP, Brazil

**Carlos Roberto Arias**
Institute of Information Systems and Applications, National Tsing Hua University, Hsinchu 30013, Taiwan
Facultad de Ingenieria, Universidad Tecnologica Centroamericana, Tegucigalpa 11101, Honduras

**Hsiang-Yuan Yeh**
Computer Science Department, National Tsing Hua University, Hsinchu 30013, Taiwan

**Von-Wun Soo**
Institute of Information Systems and Applications, National Tsing Hua University, Hsinchu 30013, Taiwan
Computer Science Department, National Tsing Hua University, Hsinchu 30013, Taiwan

**J. Mosafer and E. Manshad**
Department of Animal Science, Ferdowsi University of Mashhad, P.O. Box 91775-1163, 9177948974 Mashhad, Iran

**M. Heydarpour**
Population Health Research Institute (PHRI), Department of Medicine, McMaster University, Hamilton, ON, Canada

**G. Russell**
Moredun Research Institute, Pentlands Science Park, Midlothian, Penicuik EH26 0PZ, UK

**G. E. Sulimova**
Vavilov Institute of General Genetics, Russian Academy of Sciences, Moscow 119991, Russia

**Kuppareddi Balamurugan**
School of Criminal Justice, University of Southern Mississippi, 118 College Drive # 5127, Hattiesburg, MS 39406, USA

**Martin L. Tracey**
Department of Biology, Florida International University, University Park Campus, Miami, FL 33199, USA

**Uwe Heine and George C. Maha**
DNA Identification Testing Division, Laboratory Corporation of America, 1440 York Court Extension, Burlington, NC 27215, USA

**George T. Duncan**
Broward County Sheriff's Office, Forensic Laboratory DNA Unit, Fort Lauderdale, FL 33301, USA

**Milda Norkiene and Alma Gedvilaite**
Institute of Biotechnology, Vilnius University, Graiciuno 8, Vilnius, Lithuania

**Deniz Uluısık and Ercan Keskin**
Department of Physiology, Faculty of Veterinary Medicine, University of Selcuk, 42075 Selcuklu, Konya, Turkey

**Beatriz Matallanas, Carmen Callejas and M. Dolores Ochando**
Departamento de Genetica, Facultad de Ciencias Biologicas, Universidad Complutense de Madrid, C/Jose Antonio Novais 2, 28040 Madrid, Spain

www.ingramcontent.com/pod-product-compliance
Lightning Source LLC
Chambersburg PA
CBHW080657200326
41458CB00013B/4893
*9781632390073*